HANDBOOK OF ROTORDYNAMICS

Other Books of Interest

Baumeister and Marks • MARKS' STANDARD HANDBOOK FOR MECHANICAL ENGINEERS
Brady and Clauser • MATERIALS HANDBOOK
Bralla • HANDBOOK OF PRODUCT DESIGN FOR MANUFACTURING
Brunner • HANDBOOK OF INCINERATION SYSTEMS
Corbitt • STANDARD HANDBOOK OF ENVIRONMENTAL ENGINEERING
Elliot • STANDARD HANDBOOK OF POWERPLANT ENGINEERING
Freeman • STANDARD HANDBOOK OF HAZARDOUS WASTE TREATMENT AND DISPOSAL
Ganic and Hicks • THE McGRAW-HILL HANDBOOK OF ESSENTIAL ENGINEERING INFORMATION AND DATA
Gieck • ENGINEERING FORMULAS
Harris • HANDBOOK OF NOISE CONTROL
Harris and Crede • SHOCK AND VIBRATION HANDBOOK
Hicks • STANDARD HANDBOOK OF ENGINEERING CALCULATIONS
Jones • DIESEL PLANT OPERATIONS HANDBOOK
Juran and Gryna • JURAN'S QUALITY CONTROL HANDBOOK
Kurtz • HANDBOOK OF APPLIED MATHEMATICS FOR ENGINEERS AND SCIENTISTS
Karassik, et al. • PUMP HANDBOOK
Parmley • STANDARD HANDBOOK OF FASTENING AND JOINING
Rohsenow, Hartnett, and Ganic • HANDBOOK OF HEAT TRANSFER FUNDAMENTALS
Rohsenow, Hartnett, and Ganic • HANDBOOK OF HEAT TRANSFER APPLICATIONS
Rosaler and Rice • STANDARD HANDBOOK OF PLANT ENGINEERING
Rothbart • MECHANICAL DESIGN AND SYSTEMS HANDBOOK
Shigley and Mischke • STANDARD HANDBOOK OF MACHINE DESIGN
Tuma • HANDBOOK OF NUMERICAL CALCULATIONS IN ENGINEERING
Young • ROARK'S FORMULAS FOR STRESS AND STRAIN
Grimm and Rosaler • HANDBOOK OF HVAC DESIGN
Wadsworth • HANDBOOK OF STATISTICAL METHODS FOR ENGINEERS AND SCIENTISTS
Maynard • INDUSTRIAL ENGINEERING HANDBOOK

HANDBOOK OF ROTORDYNAMICS

Fredric F. Ehrich Editor-in-Chief

McGRAW-HILL, INC.

New York St. Louis San Francisco Auckland Bogota
Caracas Lisbon London Madrid Mexico Milan
Montreal New Delhi Paris San Juan São Paulo
Singapore Sydney Tokyo Toronto

Library of Congress Cataloging-in-Publication Data

Rotordynamics / Fredric F. Ehrich, editor.
p. cm.
Includes bibliographical references.
ISBN 0-07-019330-4
1. Rotors—Dynamics. I. Ehrich, Fredric F.
TJ1058.R673 1992
621.8—dc20

91-20752
CIP

1 2 3 4 5 6 7 8 9 0 DOC/DOC 9 6 5 4 3 2 1

ISBN 0-07-019330-4

The sponsoring editor for this book was Robert W. Hauserman, the editing supervisor was Kimberly A. Goff, and the production supervisor was Suzanne W. Babeuf.

This book was set in Times Roman. It was composed by Universities Press.

Printed and bound by R. R. Donnelley and Sons.

CONTENTS

CONTRIBUTORS

Dr. Dara W. Childs, *Director, Turbomachinery Laboratory, Texas A & M University, College Station, Tex.* (CHAP. 1)

Dr. Stephen H. Crandall, *Ford Professor of Engineering, Department of Mechanical Engineering, Massachusetts Institute of Technology, Cambridge, Mass.* (CHAP. 2)

Dr. Fredric F. Ehrich, *Staff Engineer, GE Aircraft Engines, Lynn, Mass.* (CHAP. 1)

Dr. Ronald L. Eshelman, *President & Director, Vibration Institute, Willowbrook, Ill.* (CHAP. 4)

Dr. Edgar J. Gunter, *Professor, Department of Mechanical & Aerospace Engineering, University of Virginia, Charlottesville, Va.* (CHAP. 3)

Charles Jackson, P.E., *Turbomachinery Consultant, Texas City, Tex.* (CHAPS. 3 & 4)

Dr. Harold D. Nelson, *Professor, Department of Engineering, Texas Christian University, Tempe, Ariz.* (CHAP. 2)

Melvin A. Prohl, *Engineering Consultant, Boxford, Mass.* (CHAP. 1)

PREFACE

Although there are a great many excellent academic text books in the field of vibrations (not the least of which is the venerable, but still fresh, *Mechanical Vibrations* by my own mentor, the late J. P. Den Hartog) and its important branch of rotordynamics, we have perceived a need for a comprehensive book on the subject in a handbook format. We then brought together the contributions of the most distinguished engineers, scientists, consultants, and analysts from industry, academia, and research institutions in the four major aspects of the field: design, analysis, manufacturing (balance and alignment), and test of high-speed rotors. Although all four fields are treated comprehensively, we believe that the treatment of the area of design is quite unique since it deals not only with the conceptual framework of rotordynamics, but also with strategies that might be used by the designer of high-speed rotors in a variety of apparatus, operating regimes and environments.

While the primary focus of the handbook has been to supply the designers, the analysts, the manufacturers, and the developers of high-speed rotating machinery with the methodology and data they might require, we have kept in mind that in recent years a great many of the tasks have been packaged in comprehensive computer programs. Although our purpose was not to make the handbook an operating manual for such computer programs, we hope that it will be useful in providing the understanding of the underlying principles, concepts, and approaches and will make the user an informed and more skillful user of those computer programs.

Because of its special emphasis on providing understanding as well as methodology and data, we believe that this book will also prove to be useful as a textbook in educating and training in academia and the industries where rotating machines are designed, developed, manufactured, used, and maintained.

I would like to personally express my appreciation to the General Electric Company which accommodated my special effort in editing and contributing to this handbook in addition to my regular duties as a practicing engineer and made possible the rich career experience that provided the foundation for that contribution. It pleases me to think that I follow in a long and honorable tradition of distinguished contributors to the science and understanding of vibration in rotating machinery such as W. Campbell, A. L. Kimball, B. L. Newkirk, M. Prohl (also a contributor to this handbook), and J. Alford who made their careers at the General Electric Company.

Fredric Ehrich

HANDBOOK OF ROTORDYNAMICS

CHAPTER 1

VIBRATION CONSIDERATIONS IN THE DESIGN OF ROTATING MACHINERY

Melvin A. Prohl
Engineering Consultant

Dr. Fredric F. Ehrich
Staff Engineer
GE Aircraft Engines

Dr. Dara W. Childs
Director
Turbomachinery Laboratory
Texas A & M University

1.1 INTRODUCTION

The end-use functional requirements of a device must inevitably be the primary concern of the designer. But, as is the case for most power generation, power conditioning, and power absorption equipment where the design involves rotating elements, the designer must anticipate that the durability, reliability, performance, environmental acceptability, and overall user satisfaction are inextricably linked to the vibration characteristics implicit in the design.

In an earlier era, when product requirements had not yet pressed designs to the outer bounds of light weight, high speed, small clearance, long life, high reliability, and exacting environmental acceptability, it may have been sufficient to give only cursory attention to vibration design and to depend on field experience and case-by-case problem solving to evolve a vibration-trouble-free design. But the inevitable growth in user requirements and expectations and the advances in vibration technology now permit and, indeed, require exacting and complete consideration of vibration characteristics of the design in its basic conception and definition.

This chapter deals with perspectives that will enable the designer to anticipate most of the vibration phenomena that can manifest themselves in rotating machinery and describes approaches and procedures in the design process and alternative configurations to ensure that the vibration phenomena are manifest at acceptable levels in the final product, as is ultimately experienced in the course of its service.

Since we must deal with the generality of an enormous variety of devices and range of design approaches for any individual device, the design process is not a direct one—there is no single formula or universal design scheme that can be specified. Instead we approach the problem by considering three principal types of vibration phenomena—transverse or lateral or radial or flexural; torsional; and longitudinal—and two major classes of stimuli—forced vibration and self-excited vibration.

For flexural vibrations, we consider the most commonly encountered sources of excitation and several strategies for dealing with these phenomena in the design, including various categories of design configuration, components, and arrangements to carry out these strategies, as summarized in Table 1.1. As a topic subsidiary to forced vibrations, in Sec. 1.7 we treat the subject of nonlinear forced vibration in rotating machinery. Although the basic excitation mechanisms and some strategies for their elimination are the same as for simple linear forced vibration response, the nonlinearities act to alter the peak response amplitudes and frequencies of vibration and the rotational speeds at which those peaks may occur (often in unfavorable directions), so that special precautionary design measures are often advisable.

TABLE 1.1 Flexural Vibration of the Rotor System

Stimulus	
Forced vibration	Self-excited vibration instabilities
Rotor unbalance, rotor bow, etc. (Sec. 1.2)	Whirling or whipping (Sec. 1.6.2), parametric instability (Sec. 1.6.3), stick-slip or chatter (Sec. 1.6.4), instabilities in forced vibrations (Sec. 1.6.5)
Design strategies for dealing with vibration	
1. Place the system's critical speeds out of the steady-state operating range (Sec. 1.3).	1. Raise the system's natural frequencies to minimize the degree of supercriticality (for supercritical whirling or whipping phenomena).
2. Introduce damping to reduce critical response peaks (Sec. 1.4.4).	2. Introduce damping to cancel the destabilizing forces in the operating range.
3. Minimize the excitation, i.e., balancing and aligning rotors (Secs. 1.4.7 and 1.4.8; Chap. 3).	3. Diminish the destabilizing forces or eliminate the destabilizing mechanism.
4. Isolate the apparatus from the environment (Sec. 1.5).	

Torsional and longitudinal vibrations are dealt with in Sec. 1.8.

The ramifications of system components such as bearings, dampers, seals, and couplings are treated in Sec. 1.4. Note that we do *not* deal with vibration problems unique to individual rotor or stator components (such as turbomachinery blades or disks) although we do, directly or by inference, deal with problems of adjacent or peripheral components whose vibration may be derivative of (i.e., is driven by) the rotor system vibration.

A versatile, accurate, and speedy computational tool for prediction of the vibration characteristics of any hypothesized design is an indispensable tool in the design process. Although the design process is too complex to conceive of automating in a direct sequence from specification to finished design definition, the direct sequence can be approximated by an efficient and speedy iteration process which starts with posing design alternatives, then analytically evaluating them, and subsequently selecting and quantitatively refining the design that comes closest to meeting the design requirements. Such analytic tools are described in Chap. 2, and their availability and use are presumed in this chapter on design.

The quality of the design of any apparatus is no better than the quality with which the apparatus is manufactured. The converse is also true, since the quality of the manufacturing process is very much predicated on the design. This interdependence is particularly true of the vibration performance of rotating machinery. The key manufacturing processes which affect vibration behavior are the balancing of the rotor and the alignment of rotor components. These manufacturing sequences are, in turn, greatly dependent on the design provisions made for balancing and for the assembly and alignment sequence and for the stability of that alignment over time. The technology of balance and alignment and some implicit design requirements and approaches are dealt with in Chap. 3.

Even as far as the art and science of vibration technology have progressed, we cannot yet claim a state of absolute perfection in the design process, and the development of complex and sophisticated high-performance equipment should generally include sequences of experimental development, prototype proof testing, and continuous or periodic monitoring of vibration performance in service. Chapter 4 describes the technology required for these testing activities and provisions in the design to anticipate and facilitate such testing.

1.2 BASIC CONCEPTS—CRITICAL SPEEDS AND RESPONSE TO UNBALANCE

1.2.1 The Jeffcott Rotor

To establish some important concepts and definitions and to provide information of general utility, it is useful to consider the *Jeffcott rotor,* a much simplified model of a high-speed rotor that retains many of the essential characteristics of more complex systems in its response to unbalance. In the Jeffcott rotor (Fig. 1.1), the central mass is carried on a flexible massless shaft supported in two bearings. Two conditions of bearing support are considered: rigid bearing supports, as in Fig. 1.1*a*, and flexible bearing supports (represented by an axisymmetric spring and dashpot array), as in Fig. 1.1*b*.

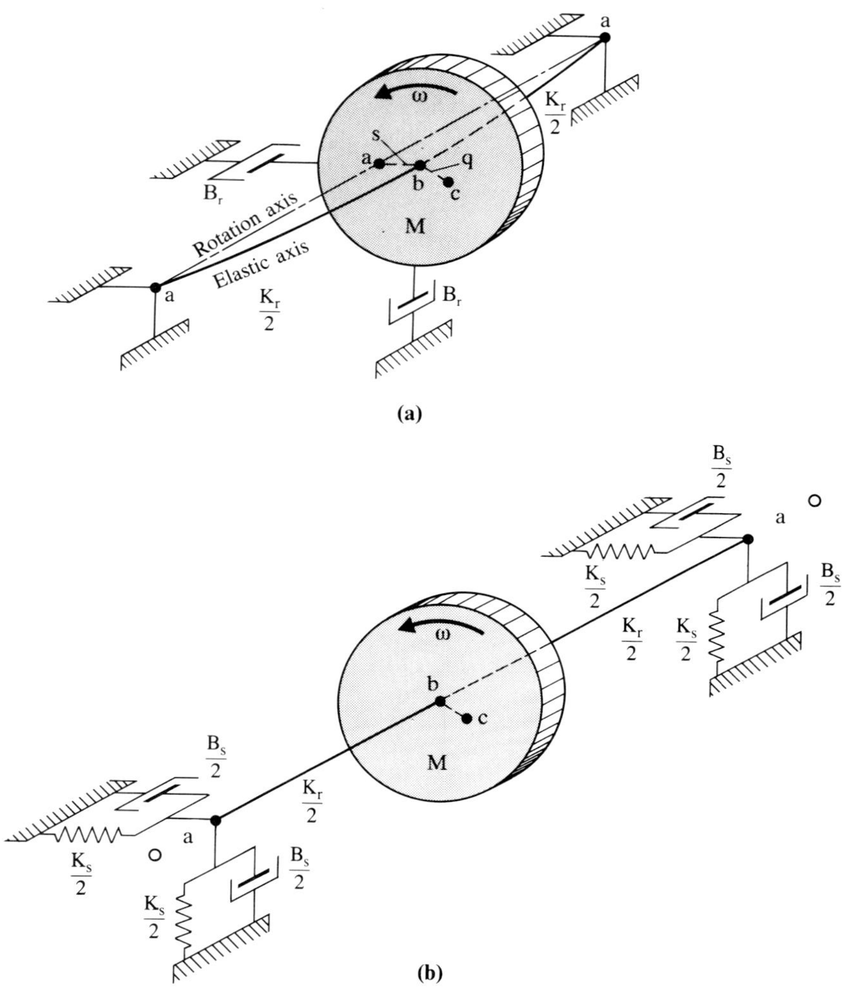

FIGURE 1.1 (*a*) Jeffcott rotor on rigid bearing supports; (*b*) Jeffcott rotor on flexible bearing supports.

The Jeffcott Rotor on Rigid Bearing Supports

In this model (Fig. 1.1*a*) an infinitely rigid simple support is assumed for the rotor bearings. The only damping B_r available is that derived from the fluid medium surrounding the rotor. Two types of unbalance are considered separately, as shown in Fig. 1.1*a*: unbalance due to mass eccentricity q and unbalance due to shaft bow s.

Case 1: Mass Eccentricity. The center of mass c of the rotor disk is offset from the elastic axis of the shaft by a radial distance q. The elastic axis intercepts the disk at point b, which is called the *elastic center* of the rotor disk. At standstill point b coincides with the axis of rotation a-a (shaft bow $s = 0$). As rotational speed increases, the centrifugal force of the eccentric mass M causes point b to move outward and whirl about the axis of rotation with a whirl radius r_b. A steady-state solution of the differential equations of motion for the system yields for the response to unbalance (Myklestad, 1956).

$$w_{b1} = \frac{r_b}{q} = \frac{\tau^2}{[(1-\tau^2)^2 + (2\zeta\tau)^2]^{1/2}} \tag{1.1}$$

$$\lambda_b = \arctan \frac{-2\zeta\tau}{1-\tau^2} \tag{1.2}$$

The whirl amplification factor w_{b1} (the whirl radius r_b normalized by the mass eccentricity q) and the phase angle λ_b (the angle by which the displacement vector lags the unbalance vector) are functions of two dimensionless variables: the *speed ratio* τ, which is equal to the rotational speed ω divided by the undamped critical speed ν

$$\tau = \frac{\omega}{\nu} \tag{1.3a}$$

where

$$\nu = \left(\frac{K_r}{M}\right)^{1/2} \tag{1.3b}$$

where K_r is the rotor stiffness and M is the rotor mass, and the *damping ratio* ζ, which is equal to the system damping B_r divided by the critical damping $2M\nu$. Critical damping is the maximum value of damping above which the response of the system to an impulsive load is no longer oscillatory.

$$\zeta = \frac{B_r}{2M\nu} \tag{1.4}$$

The whirl amplification factor w_{b1} and the phase angle λ_b are plotted versus the speed ratio τ in Fig. 1.2*a* and *b*. For low values of damping ζ and operation close to the critical speed, the whirl amplification factor w_{b1} becomes very large. With zero damping it becomes infinite at the critical speed ($\tau = 1$). The phase angle is equal to 90° at the critical speed—a fact that is useful in rotor balancing. At very high speeds, the whirl amplification factor approaches unity and the phase angle approaches 180°, which means that the center of mass, point c, has approached coincidence with the center of rotation. Thus supercritical operation for the Jeffcott rotor on rigid bearings poses no problem other than accommodating the peak amplitudes encountered on passage through the critical-speed region.

The peak values of the whirl amplification factor in Fig. 1.2*a* occur at values of the speed ratio τ_p larger than unity when damping is present.

$$\tau_p = \frac{1}{(1-2\zeta)^{1/2}} \qquad \text{when} \qquad \zeta < \left(\frac{1}{2}\right)^{1/2} = 0.707 \tag{1.5}$$

When $\zeta \geq 0.707$, there are no peaks and the whirl amplification approaches unity as the speed ratio approaches infinity.

A better understanding of Fig. 1.2*a* and *b* may be gained by constructing a displacement and force diagram for the system. In Fig. 1.3*a* force and

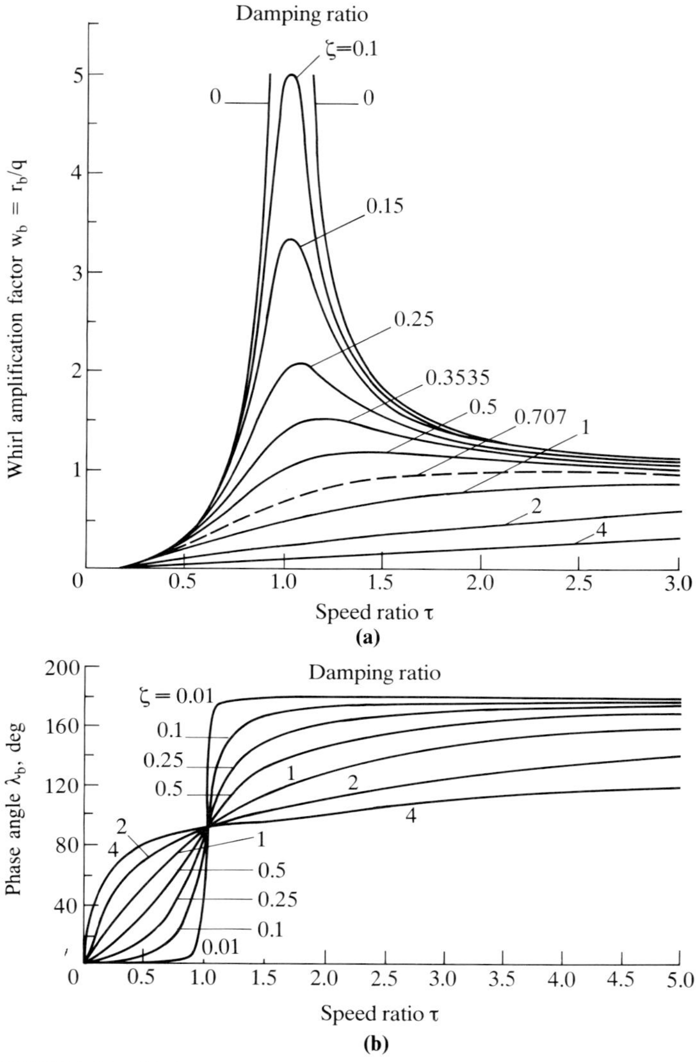

FIGURE 1.2 (*a*) Response to unbalance (mass eccentricity q) versus rotational speed—Jeffcott rotor on rigid bearing supports; (*b*) phase angle versus rotational speed—Jeffcott rotor on rigid bearing supports; (*c*) response to unbalance (shaft bow s) versus rotational speed—Jeffcott rotor on rigid bearing supports.

displacement vectors are shown for one typical point in the plot of Fig. 1.2*a* ($\tau = 1.1$ and $\zeta = 0.15$). The motion of the rotor is a stable whirling about the centerline of rotation (*a-a*), with point *b* (the elastic center) and point *c* (the mass center) describing circular orbits. The resultant centrifugal force is the vector sum of the inertia force F_i and the unbalance force F_u. It is equal in magnitude and

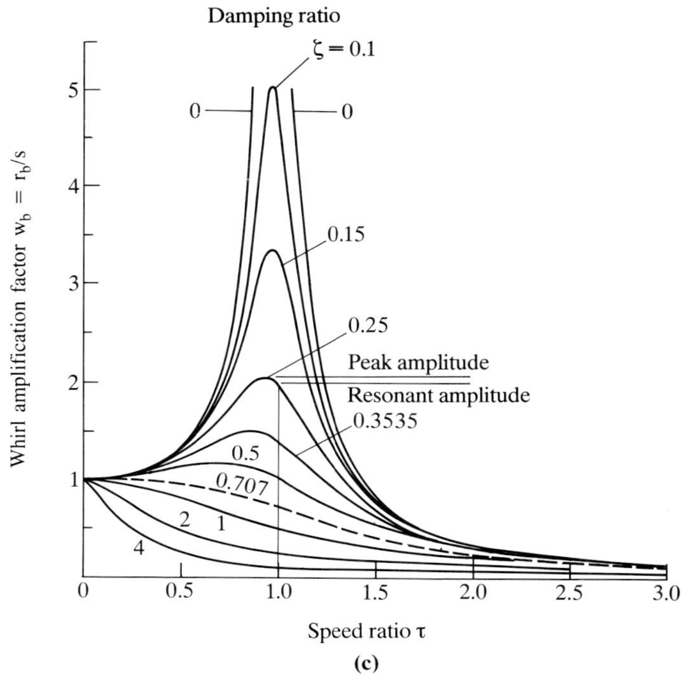

FIGURE 1.2 (*Continued.*)

opposite in direction to the resultant restraining force, which is the vector sum of the radial spring force F_s and the tangential damping force F_d.

$$F_u = M\omega^2 q \tag{1.6}$$

$$F_i = M\omega^2 r_b \tag{1.7}$$

$$F_s = K_r r_b \tag{1.8}$$

$$F_d = \omega B_r r_b \tag{1.9}$$

The two resultant forces are offset one to the other, and this negative torque is balanced by the driving torque on the rotor.

The damping term $2\zeta\tau$ in Eqs. 1.1 and 1.2 has a physical significance in that it is equal to the ratio of the damping force F_d to the spring force F_s:

$$2\zeta\tau = \frac{\omega B_r}{K_r} = \frac{F_d}{F_s} \tag{1.10a}$$

The reaction of the bearings on the rotor is the restraining force R_0, as shown in Fig. 1.3*a*:

$$R_0 = (F_s^2 + F_d^2)^{1/2} \tag{1.10b}$$

A dimensionless bearing reaction factor is derived from Eq. 1.10*a* after

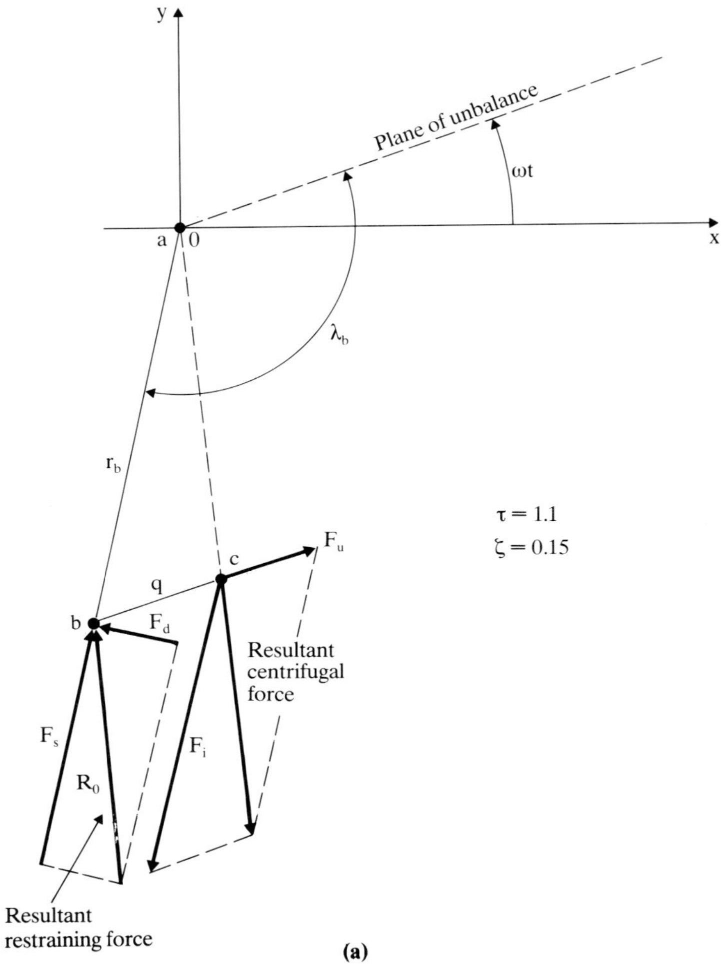

(a)

FIGURE 1.3 Vector diagram, displacements and forces: (*a*) Jeffcott rotor on rigid bearing supports (mass eccentricity q); (*b*) Jeffcott rotor on rigid bearing supports (shaft bow s).

substitutions are made for F_s and F_d (Eqs 1.8 and 1.9):

$$\frac{R}{K_r q} = w_{b1}[1 + (2\zeta\tau)^2]^{1/2} \tag{1.10c}$$

For the undamped system ($\zeta = 0$), the bearing reaction factor is equal to the whirl amplification factor w_{b1}. At very high speeds ($\tau \gg 1$) where the mass center c coincides with the rotation axis a-a and the elastic center b whirls with a radius q, the reaction force R_0 becomes equal to $K_r q$. This is the spring force required to bend the elastic axis by the amount q.

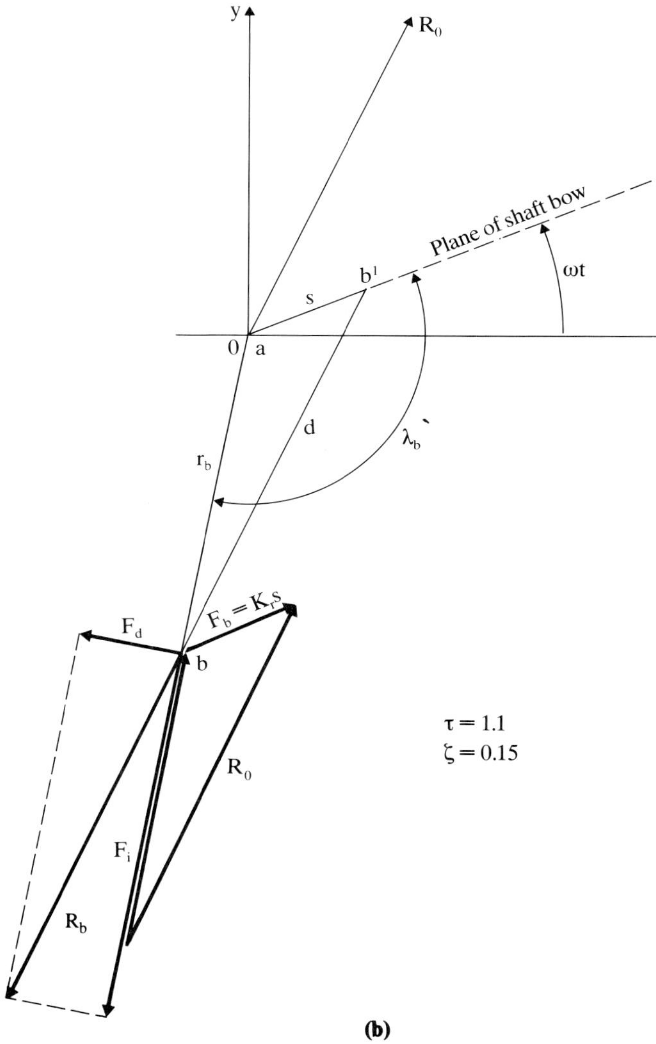

FIGURE 1.3 (*Continued.*)

The effect of the damping term in Eq. 1.10*b* is to increase the bearing reaction factor, the percentage change increasing with increasing speed. At very high speeds ($\tau \gg 1$), the bearing reaction factor approaches an infinite value.

It is of interest to examine the nature of the critical speed more closely. For the hypothetical case where the rotor has no unbalance ($q = 0$) and no damping ($\zeta = 0$), the balance between the inertia force and the spring force is given by

$$M\omega^2 r_b = K_r r_b \tag{1.11}$$

At the critical speed, where $\omega = (K_r/M)^{1/2}$, this force balance is satisfied for all values of the whirl radius r_b. In other words, there exists a state of indefinite equilibrium—which, of course, is upset in any real situation with even the slightest amount of unbalance and damping. The *critical speed* of a rotor can therefore be defined as the speed at which the inertia forces and elastic restoring forces are in perfect balance for an undamped system.

If a radial sinusoidal force with amplitude proportional to the frequency squared is applied to the stationary rotor mass of the Jeffcott rotor, the resulting planar beam vibration is governed by the same equations that apply to the whirling motion of the unbalanced rotor, and its response will be quite similar—a fact that can be quite useful in understanding the behavior and in experimental modeling. Remember that this analogy has certain important limitations:

1. It does not properly accommodate the gyroscopic stiffening of the system associated with the mass moment of inertia of the rotor.

2. It does not properly accommodate the internal damping of the rotor. For the planar vibration of the nonrotating rotor, internal damping serves to limit vibration amplitude. For a spinning rotor which is whirling in a circular orbit, the rotor acts as a frozen, deflected beam with no flexing of the shaft fibers, so internal damping has no energy dissipation effect. In fact, internal damping can lead to instability and large vibration amplitudes, as is described in Sec. 1.6.2.

3. Some of the key system parameters, such as the stator stiffness and damping, are derived from the rotation of the rotor (as in fluid film bearings) so that the nonrotating planar vibration case does not really represent the rotating system.

Case 2: Shaft Bow. In this case the center of mass c is assumed to be coincident with the elastic center b (mass eccentricity $q = 0$). The rotor would be in perfect balance except for a permanent bow in the elastic axis of the shaft which could be due to physical damage or thermal distortion during operation or a manufacturing deviation. Thus, at standstill, the elastic center b is assumed to be displaced from the rotation axis a-a by a radial distance s, the amount of bow in the massless shaft, as shown in Fig. 1.1*a*. As speed increases from standstill, point b moves away from its initial position b' under the influence of centrifugal force and whirls about the axis of rotation with a whirl radius r_b. A steady-state solution of the differential equations of motion for the system yields the whirl amplification factor w_{b2}.

$$w_{b2} = \frac{r_b}{s} = \frac{1}{[(1-\tau^2)^2 + (2\zeta\tau)^2]^{1/2}} \tag{1.12a}$$

The phase angle λ_b is given by Eq. 1.2 as in case 1. The whirl amplification factor w_{b2} (the whirl radius r_b normalized by the shaft bow s) is a function of the speed ratio τ and the damping ratio ζ (as in case 1) and is plotted in Fig. 1.2*c*. A comparison of Fig. 1.2*c* for case 2 with Fig. 1.2*a* for case 1 suggests that:

1. As the speed ratio τ approaches zero, the whirl amplification factor w_b tends to zero for case 1 and to unity for case 2. For very high rotational speeds ($\tau \gg 1$) the reverse is true; i.e., the whirl amplification factor approaches unity for case 1 and zero for case 2.
2. At the undamped critical speed ($\tau = 1$), identical values of the whirl amplification factor are obtained for the two cases.

In Fig. 1.3*b* a displacement and force diagram has been plotted for one typical point in the plot of Fig. 1.2*c* ($\tau = 1.1$ and $\zeta = 0.15$). Some additional insight into the nature of the solution for the bowed rotor may be gained from the diagram. A trigonometric solution for the elastic displacement d of the elastic center (from b' at standstill to b at speed ratio τ) which makes use of Eq. 1.12*a* for the whirl radius r_b and Eq. 1.2 for the phase angle λ_b yields

$$d = r_b[\tau^4 + (2\zeta\tau)^2]^{1/2} \tag{1.12b}$$

The resultant force R_b at the elastic center b is the vector sum of the inertia force F_i given by Eq. 1.7 and the damping force F_d given by Eq. 1.9:

$$R_b = (F_i^2 + F_d^2)^{1/2} = K_r r_b[\tau^4 + (2\zeta\tau)^2]^{1/2} \tag{1.12c}$$

A comparison of Eq. 1.12*b* and *c* shows that the resultant force R_b is proportional to the resultant elastic displacement d, with a proportionality factor equal to the rotor stiffness K_r. The resultant force vector and the resultant displacement vector are collinear, as shown in Fig. 1.3*b*.

The bearing reaction force R_0 acting on the rotor is equal and opposite in direction to the resultant force R_b acting at the elastic center. The moment due to the offset of these forces is balanced by the driving torque. The dimensionless bearing reaction factor from Eqs. 1.12*c*, 1.12*a*, and 1.1 is then

$$\frac{R_0}{K_r s} = w_{b2}\tau^2\left[1 + \left(\frac{2\zeta}{\tau}\right)^2\right]^{1/2} = w_{b1}\left[1 + \left(\frac{2\zeta}{\tau}\right)^2\right]^{1/2} \tag{1.12d}$$

The bearing reaction factor of case 2 for the undamped system ($\zeta = 0$) is equal to the whirl amplification factor w_{b1} as for case 1.

The effect of the damping term in Eq. 1.12*d* is to increase the bearing reaction factor, but, unlike in case 1, the percentage change due to damping decreases as the speed increases. With damping, the bearing reaction factors for cases 1 and 2 are identical at only two speed ratios, $\tau = 0$ and $\tau = 1$ (the undamped critical speed). At very large values of the speed ratio ($\tau \gg 1$), the elastic center b for case 2 moves to the rotation axis *a-a,* and the damping cases to act. Thus at this limit of the speed ratio the reaction force R_0 is simply equal to the spring force $K_r s$ that is required to completely straighten the bowed rotor.

For case 2 there is also an analogy with planar beam vibration. If a radial sinusoidal force of constant amplitude is applied to the stationary mass of the Jeffcott rotor, the resulting planar beam vibration is governed by the same equations that apply to the whirling motion of the bowed rotor. All the limitations of this analogy cited under case 1 apply to case 2.

The analysis of the Jeffcott rotor in rigid bearing supports has two deficiencies from the designer's point of view: (1) The amount of damping available from the fluid surrounding the rotor in any practical high-speed machinery arrangement is usually negligible. (2) The support for the rotor provided by the bearings and by the stationary structure is never infinitely rigid. Bearing flexibility and the damping associated with this flexibility are always present to some degree, and this bearing damping or support damping is essential for proper operation of the rotor near a critical speed, as we show in the next section, which considers flexible bearing supports and damping associated with the flexible stator system.

The Jeffcott Rotor on Flexible Bearing Supports

In this model both the stator and the rotor are treated as flexible elements. The only damping in the system is that associated with the bearing support. Damping from the fluid surrounding the rotor is considered negligible, as is generally the case for any practical high-speed system.

First, we derive certain basic calculation parameters from a special case of the system shown in Fig. 1.1*b*, where the rotor is assumed to be infinitely rigid ($K_r = \infty$). This special case is mathematically identical to the system of Fig. 1.1*a* (the flexible rotor on rigid bearing supports). The undamped critical speed μ is now given by

$$\mu = \left(\frac{K_s}{M}\right)^{1/2} \tag{1.13}$$

where K_s is the stiffness of the support and M is the mass of the rotor.

The new speed ratio σ is equal to the rotational speed ω divided by the undamped critical speed μ

$$\sigma = \frac{\omega}{\mu} \tag{1.14}$$

and the new damping ratio η is equal to the support damping B_s divided by the critical damping $2M\mu$:

$$\eta = \frac{B_s}{2M\mu} \tag{1.15}$$

If σ is substituted for τ and if η is substituted for ζ, then Fig. 1.2*a* and *b* and Eqs. 1.1, 1.2, and 1.5 (applying to the flexible rotor on rigid bearing supports) can be used to represent the response to unbalance of the rigid rotor on flexible supports.

A steady-state solution of the differential equations of motion for the complete system of Fig. 1.1*b* (flexible rotor and flexible supports) yields for the response to unbalance

$$w_a = \frac{r_a}{q} = \frac{\sigma^2}{[c_1^2 + (c_3c_2)^2]^{1/2}} \tag{1.16}$$

$$\lambda_a = \arctan\left(\frac{-c_3c_2}{c_1}\right) \tag{1.17}$$

$$w_b = \frac{r_b}{q} = w_a[(1+\kappa)^2 + (c_3\kappa)^2]^{1/2} \tag{1.18}$$

$$\lambda_b = \arctan\frac{-c_3}{(1+\kappa)c_1 + c_3^2\kappa c_2} \tag{1.19}$$

where $c_1 = 1 - (1+\kappa)\sigma^2$
$c_2 = 1 - \sigma^2\kappa$
$c_3 = 2\sigma\eta$

The whirl amplification factor w_a is equal to the whirl radius r_a at the bearing location, normalized by the mass eccentricity q. The phase angle λ_a is the relative angle between the displacement vector r_a and the unbalance vector q. The whirl amplification factor w_b is equal to the whirl radius r_b normalized by the mass eccentricity q. The phase angle λ_b is the relative angle between the displacement vector r_b and the unbalance vector q. They are all functions of three dimensionless variables: the speed ratio σ, the damping ratio η, and the stiffness ratio κ. The *stiffness ratio* is equal to the support stiffness K_s divided by the rotor stiffness K_r:

$$\kappa = \frac{K_s}{K_r} \tag{1.20}$$

The ratio κ may also be thought of as a *flexibility ratio*, defined by the ratio of the static deflection δ_r in the rotor to the static deflection δ_s in the stator, as deflected by the rotor's weight:

$$\kappa = \frac{\delta_r}{\delta_s} \tag{1.21a}$$

where $\delta_r = \dfrac{W}{K_r}$ (1.21*b*)

$\delta_s = \dfrac{W}{K_s}$ (1.21*c*)

W = rotor weight

It is not possible to represent the response to unbalance as given by Eqs. 1.16 through 1.19 by a single set of plots, as in Fig. 1.2*a* and *b*. Therefore a representative plot of the whirl amplification factors (w_a and w_b) and the phase angles (λ_a and λ_b) as functions of the speed ratio is presented in Fig. 1.4 for a single combination of the stiffness ratio and damping ratio ($\kappa = 2$ and $\eta = 1$). Of particular interest in Fig. 1.4 are the peak values of the whirl amplification factors for the bearings and the rotor mass ($w_a = 1.26$ and $w_b = 5.01$), the corresponding phase angles ($\lambda_a = -143°$ and $\lambda_b = -102°$), and the speed ratio at which these values occur ($\sigma_p = 0.65$). The speed at which the whirl amplitude peaks is commonly referred to as the *damped*, or *actual, critical speed*.

Designers often use the undamped rigid bearing critical speed ν as their point of reference. Thus in Fig. 1.4 a scale for the speed ratio τ is included in addition to the scale for the speed ratio σ. The relationship between these two speed ratios is given by

$$\tau = \kappa^{1/2}\sigma \tag{1.22}$$

In Fig. 1.5 a vector diagram is displayed for the displacements and forces in the system of Fig. 1.4 at the damped critical speed ($\sigma_p = 0.65$). As in the simpler system of Fig. 1.3 (the rotor on rigid bearings), the motion of this rotor at any given speed is a stable whirling about the centerline of rotation (o–o). Point a (the journal center), point b (the elastic center), and point c (the mass center) all whirl in circular orbits with a fixed spatial relationship between the points. The bearing spring force F_s and the bearing damping force F_d act at the journal center while the centrifugal inertia force F_i and the centrifugal unbalance force F_u act at the center of mass. The resultant bearing force and the resultant centrifugal force are equal in magnitude and opposite in direction, but offset one to the other. The torque due to this offset plus the torque due to the bearing damping force is balanced by the driving torque on the rotor.

The centrifugal forces F_u and F_i are defined by Eqs. 1.6 and 1.7, whereas the

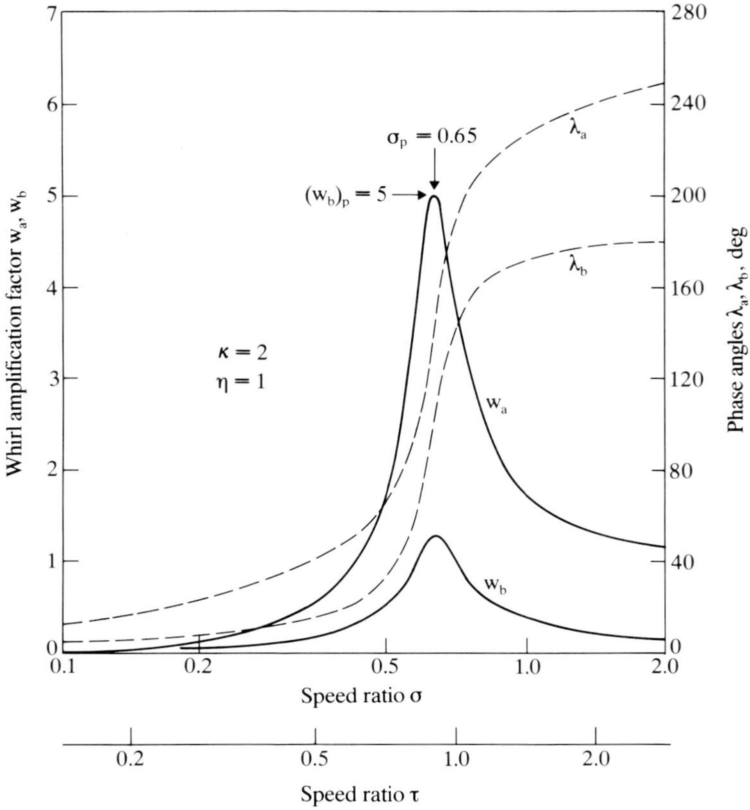

FIGURE 1.4 Response to unbalance versus rotational speed—Jeffcott rotor on flexible bearing supports.

bearing forces are now given by

$$F_s = K_s r_a \tag{1.23}$$

$$F_d = \omega B_s r_a \tag{1.24}$$

Also

$$\frac{F_d}{F_s} = \frac{\omega B_s}{K_s} = 2\sigma\eta \tag{1.25}$$

Of particular importance to the designer is the peak response to unbalance of the rotor over the complete speed range and the value of the speed at which peaking occurs. In Fig. 1.6 the peak value of the whirl amplification factor $(w_b)_p$ has been plotted as a function of the damping ratio η for a series of constant values of the stiffness ratio κ.

Two important concepts are derived from this plot: (1) For any constant value of the damping ratio η there is a dramatic increase in the whirl amplification

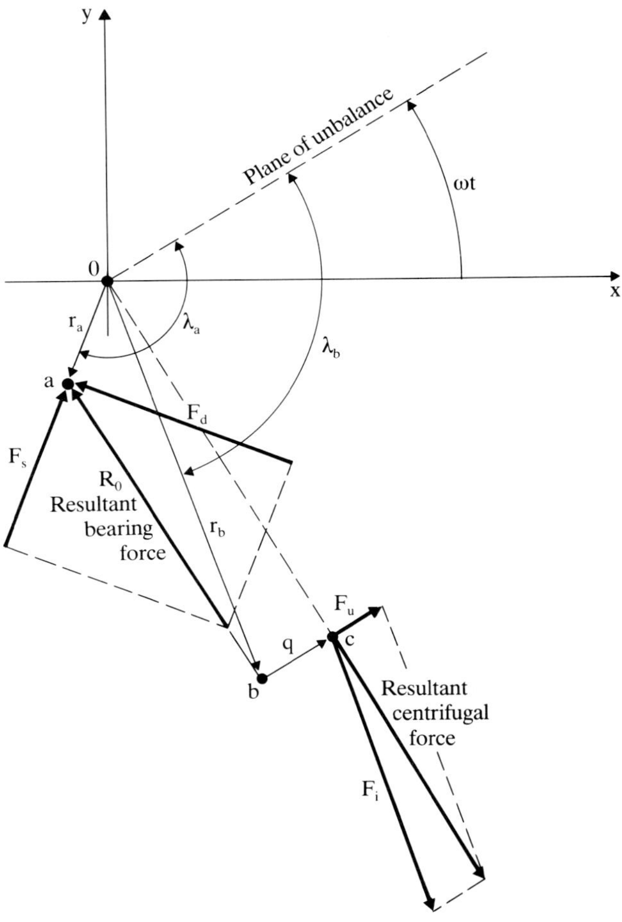

FIGURE 1.5 Vector diagram, displacements and forces—Jeffcott rotor on flexible bearing supports.

factor $(w_b)_p$ with an increase of the stiffness ratio κ. In other words, the greater the stiffness of the support or the greater the flexibility of the rotor, the more responsive the rotor becomes to unbalance. (2) For any assigned value of the stiffness ratio, a minimum value of the whirl amplification factor occurs at some intermediate level of the damping ratio. As the damping ratio tends toward zero, the whirl amplification factor for all values of the stiffness ratio becomes infinite because of the absence of any system damping. Also as the damping ratio tends toward infinity, the whirl amplification factor for all values of the stiffness ratio again becomes infinite because of the rigidity of the bearing support and the consequent loss of damping action. In Fig. 1.6 the optimum value of the damping factor for minimum response to unbalance varies from 0.707 for $\kappa = 0$ to 2.14 for $\kappa = 16$.

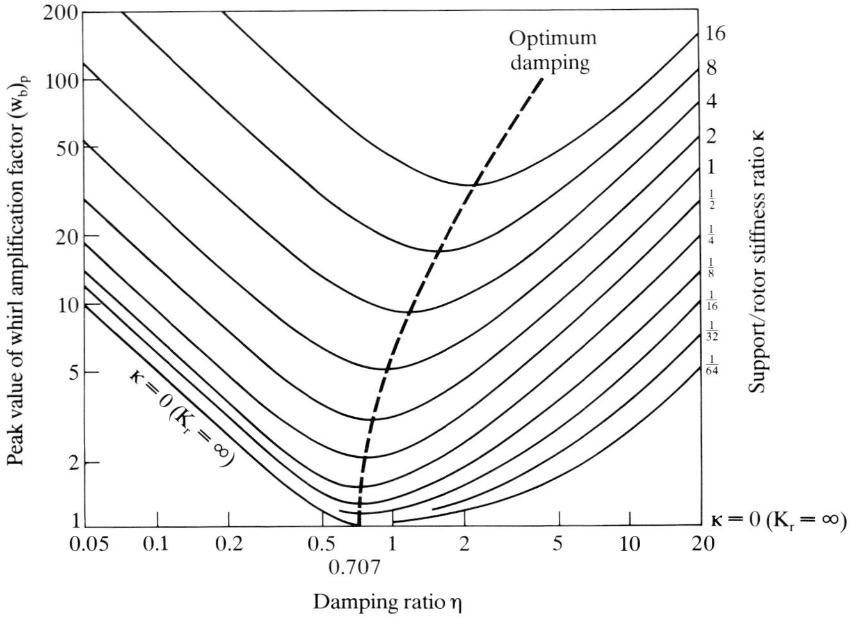

FIGURE 1.6 Peak response to unbalance versus damping ratio—Jeffcott rotor on flexible bearing supports.

In conjunction with Fig. 1.6 the value of the speed ratio σ_p at which the peak value of the whirl amplification factor occurs has been plotted in Fig. 1.7 against the damping ratio η for constant values of the stiffness ratio κ. As the damping ratio approaches zero, the speed ratio σ_p has a limit of

$$\sigma_p = \left(\frac{1}{1+\kappa}\right)^{1/2} \tag{1.26}$$

which is the undamped critical speed. As the damping ratio approaches infinity, the limit value for the speed ratio σ_p is

$$\sigma_p = \left(\frac{1}{\kappa}\right)^{1/2} \tag{1.27}$$

If reference to the undamped rigid bearing critical speed ν is desired, then (by virtue of Eq. 1.22) the limit values for the speed ratio τ_p become

$$\tau_p = \left(\frac{\kappa}{1+\kappa}\right)^{1/2} \qquad \text{as } \eta \to 0 \tag{1.28}$$

$$\tau_p = 1 \qquad \text{as } \eta \to \infty \tag{1.29}$$

The curves for the zero value of the stiffness ratio κ in Figs. 1.6 and 1.7 are derived from the special case of the rigid rotor on flexible supports. With an

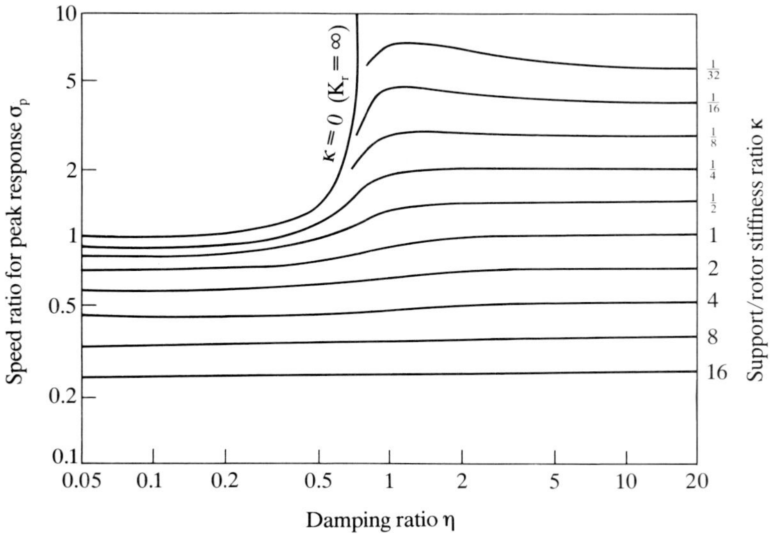

FIGURE 1.7 Speed ratio at peak response versus damping ratio—Jeffcott rotor on flexible bearing supports.

increase of the damping ratio the peak value of the whirl amplification factor decreases to unity at $\eta = 0.707$ and holds at unity for η greater than 0.707, while the speed ratio σ_p increases from unity to an infinite value at $\eta = 0.707$ and holds at infinity for η greater than 0.707.

The zero value for the stiffness ratio results from setting K_r equal to infinity (the rigid rotor). A zero value for K_s, which would also make the stiffness ratio zero, is not admissible for the simplified mathematical formulations with which we are dealing, since the undamped rigid rotor critical speed μ would become equal to zero and the design parameters σ and η would become infinite.

The critical damping value $2M\mu$, which is the normalizing constant in the expression for the damping ratio η, was derived from the rigid rotor solution of the system of Fig. 1.1*b* where the rotor spring constant K_r is assumed infinite. Although this critical damping value provides a meaningful and a proper point of reference, the concept of critical damping no longer applies when a finite value is assigned to K_r. For this case the response of the rotor mass to an impulsive load will always be oscillatory, no matter how large the support damping B_s might be.

1.2.2 Rotor of Uniform Cross Section

The effect of mass distributed along the length of the rotor span is missing from the characteristics of the Jeffcott rotor as described in Sec. 1.2.1. With distributed mass there will be an infinite number of higher critical speeds in addition to the fundamental or first critical speed. Each critical speed has a characteristic mode shape. The simple model selected to display the effect of distributed mass is that of a uniform bar of circular cross section that is carried in bearings at each end, as

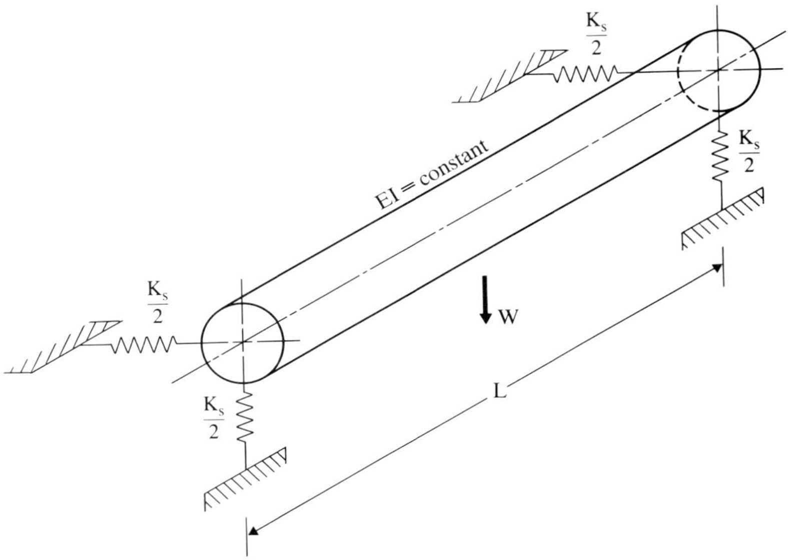

FIGURE 1.8 Rotor of uniform cross-section on flexible bearing supports.

shown in Fig. 1.8. The flexibility at the bearings is represented by an axisymmetric array of springs without damping.

Effect of Support Flexibility on Critical Speeds

Linn and Prohl (1951) presented information on the effect of bearing flexibility for the first four critical speeds of the undamped rotor system of Fig. 1.8. based on a solution of the differential equation for the elastic curve of the whirling rotor. In Fig. 1.9 the ratio of the critical speed with bearing flexibility to the first critical speed ω/ω_1 on rigid bearings ($K_s = \infty$) is plotted against the dimensionless function $[\delta_s/(g/\omega_1^2)]^{1/2}$, where δ_s represents the flexibility of the bearing supports (Eq. 1.21*c*) and g/ω_1^2 represents the flexibility of the rotor.

The first critical speed ω_1 on rigid bearings is given by

$$\omega_1 = \pi^2 \left(\frac{gEI}{WL^3}\right)^{1/2} \tag{1.30}$$

where g = acceleration due to gravity
E = modulus of elasticity
I = moment of inertia of cross section
W = total weight of span
L = span length

For any assigned value of the rigid bearing first critical speed ω_1, the first and second critical speed ratios show a continuing decrease with increasing bearing flexibility. At the same time, the associated mode shapes gradually transform, and for large values of bearing flexibility the mode shapes for these critical speeds become the first and second rigid-body mode shapes of the system, as shown in

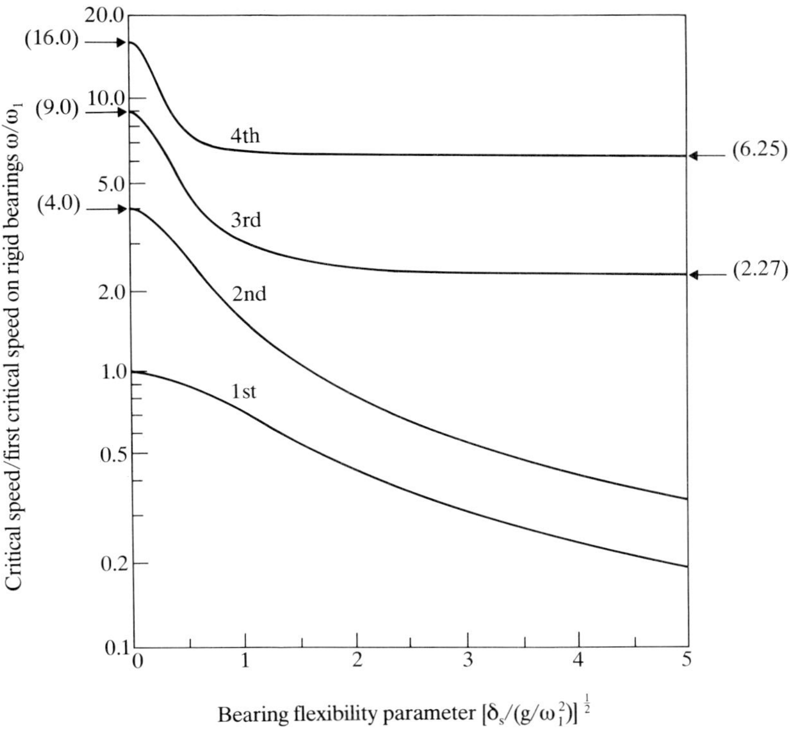

FIGURE 1.9 Critical speeds of a uniform rotor with bearing flexibility.

Fig. 1.9. The numerical ratio between the two critical speeds in Fig. 1.8 becomes equal to $3^{1/2}$ for large values of bearing flexibility. This is the theoretical ratio for the rigid-body modes. As the bearing flexibility approaches infinity, these rigid-body critical speeds both approach zero.

The third and fourth critical speed ratios first exhibit a rapid decline in value with increasing bearing flexibility and then level out and asymptotically approach values of 2.27 and 6.25. These two modes are sometimes referred to as the first and second "free-free" modes of the system, to reflect the fact that there is no restraint at either support point.

Comparison with the Jeffcott Rotor: The First Critical Speed

The rigid bearing critical speed ν of the Jeffcott rotor may be expressed in terms of the static deflection δ_r of the rotor on rigid bearings

$$\nu = \left(\frac{g}{\delta_r}\right)^{1/2} \tag{1.31}$$

Solving for δ_r in Eq. 1.31 and substituting this result in Eq. 1.28 yield the critical speed ratio

$$\frac{\omega}{\nu} = \frac{1}{(1 + \delta_s/\delta_r)^{1/2}} = \frac{1}{(1 + \delta_s \nu^2/g)^{1/2}} \tag{1.32}$$

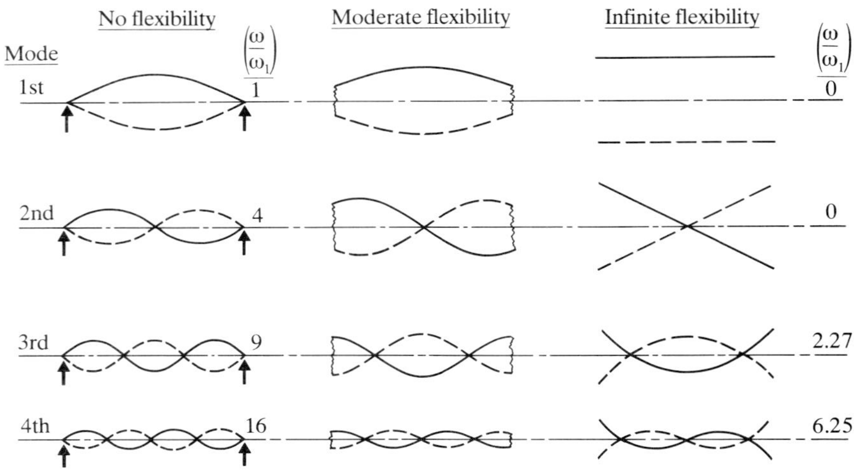

FIGURE 1.10 Mode shapes of a uniform rotor with bearing flexibility.

If Eq. 1.32 for the Jeffcott rotor is compared with the plot in Fig. 1.9 for the first critical speed of the rotor with uniform mass distribution, it is found that there is a maximum deviation of only 2.5 percent (Linn and Prohl, 1951). This close correspondence of results for the two rotor models having quite different mass distributions suggests that the rigid bearing critical speed of a rotor span provides a reliable assessment of span stiffness or flexibility. This fact has long been recognized by designers. In the preliminary design of a given class or type of machinery, limits placed on single-span values of rigid bearing critical speeds can be useful with regard to the proper placement of critical speeds.

Response to Unbalance

If damping is added to the spring supports of the uniform rotor of Fig. 1.8, then the response of this rotor to unbalance while passing through the first critical speed will, in a qualitative way, relate to the Jeffcott rotor and be influenced by the stiffness ratio and damping ratio in a manner similar to that indicated by Fig. 1.6. For all quantitative evaluations and for studying the higher critical speeds, recourse should be made to more precise numerical simulations, such as are described in Chap. 2 to assess rotor response.

Bearing location can have an important effect on the response to unbalance. If, e.g., the bearings of the uniform rotor were moved inboard to the nodal points of the first mode of the free-free rotor, as shown in Fig. 1.10, then damping in the supporting structure would be ineffective and excessive response to unbalance would result for operation close to the free-free critical speed.

1.3 PLACEMENT OF CRITICAL SPEEDS

The proper placement of critical speeds is an important element in the procedure for designing rotors to have good vibrational characteristics. For modest design changes in an established line of machinery, a relatively simple check of the undamped critical speeds, using proven assumptions for the support stiffness provided by the bearings, may be sufficient to ensure satisfactory operation. For new machinery and new applications, the use of more thorough modeling together with appropriate computational techniques and computer software as described in Chap. 2 is advisable. The evaluation of the damped response of the rotor and the associated stationary structure to rotor unbalance is of particular importance. The possibility of rotor instability as a source of excitation, as described in Sec. 1.6, should also be considered.

The focus of this section is on design considerations for ensuring satisfactory operation in the presence of forced synchronous vibrations resulting from residual unbalance in the rotor. Such residual unbalance may be due to a lack of coincidence of the center-of-mass axis with the elastic axis or to a bow in the elastic axis of the rotor or to a combination of the two asymmetries. Rotor bow may result from a temporary lack of symmetry in the temperature distribution in the rotor, or it can be caused, e.g., by improperly fabricated couplings or splines. Although there is a difference in the response curves away from critical speeds for the two types of residual unbalance, as described in Sec. 1.2.1, the ramifications of placement of critical speeds are identical.

1.3.1 Single-Shaft Rotors

A generic representation of a single-shaft single-span two-bearing machine is shown in Fig. 1.11*a*. In this system, the stator and its mounts as well as the rotor are assumed to be axisymmetric. All the elements are present that must be considered for proper placement of critical speeds relative to operating speeds—not only the mass and stiffness characteristics of the rotor but also the dynamic characteristics of the structure supporting the rotor—the bearings, the stator, and the stator mounts. Since the stator is most often considerably stiffer than the rotor because of its larger diameter, the stator is treated as a rigid body in the model used in this section.

The modes of vibration of the system of Fig. 1.11*a* may be divided into two categories: modes where the rotor behaves essentially as a rigid body and rotor flexural modes where the rotor whirls with a characteristic deflection curve.

There are two rigid-body modes for the machine of Fig. 1.11*a*. In the first or fundamental mode, the support points of the rotor move in phase with one another, and the locus of the rotor's whirling motion is that of a cylinder. In the second rigid body mode at a somewhat higher frequency, the support points of the rotor move out of phase with one another, and there is a node near or at midspan. The locus of the rotor's whirling motion is that of two cones, point to point. Usually these modes occur at frequency levels significantly below the frequency of the first flexural mode of the rotor. If the stator mass has the same magnitude as the rotor mass or less, then stator mounts must be relatively soft to give a low natural frequency. If the stator mass (which may include some foundation mass) is very large relative to the rotor mass, then low rigid-body mode frequencies will naturally prevail, even with stiff mounts. When the rotor is operated at the speeds at which these rigid-body modes are excited, the rotor and stator vibrate in phase with very little relative motion between them, provided that the bearings have adequate stiffness. There will be no significant closing up of clearances between rotor and stator, and the internal integrity of the rotating machinery will not be impaired. However, transmission of undesirable vibration to the environment may occur as described in Sec. 1.5.1.

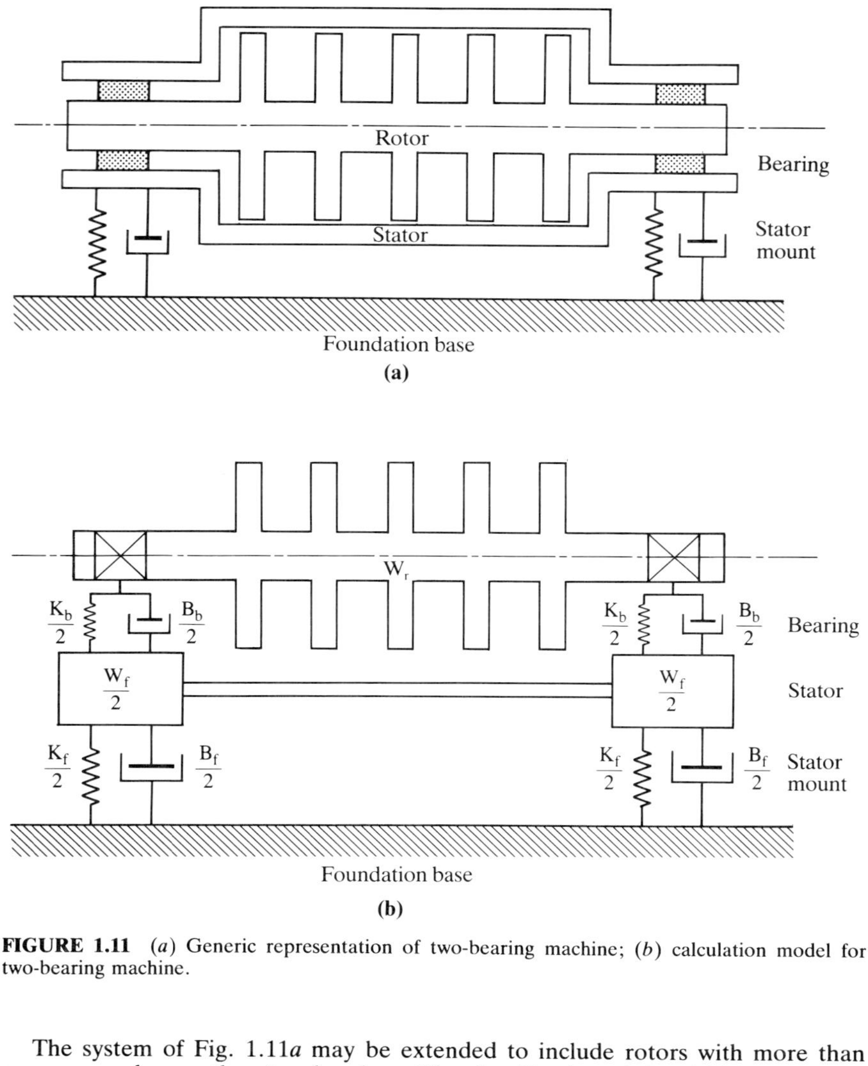

FIGURE 1.11 (*a*) Generic representation of two-bearing machine; (*b*) calculation model for two-bearing machine.

The system of Fig. 1.11*a* may be extended to include rotors with more than one span and more than two bearings. The classification of the rigid-body modes becomes more complex, but their significance for design remains the same.

It is the rotor flexural modes which require more careful design consideration. At each of these flexural modes the rotor whirls with a characteristic deflection shape, and the rotor and stator can vibrate out of phase with one another. There is the potential for large relative motion between the rotor and the stator. When operation is at or through a flexural critical speed with a high peak level of rotor whirl, rubbing of seals and other close-clearance elements may take place with the resultant loss of functionality and efficiency. In extreme cases, there can be major damage to the internal elements of the machine and in some cases to the supporting structure as well.

Three case studies are presented, each of which employs a different design option for the placement of rotor flexural critical speeds. Each of these design options is typical of and applicable to a broad class of rotating machinery.

Case 1. Single-shaft, single-span, two-bearing rotor with *transcritical* operation. The operating speed range includes the first flexural critical speed of the rotor as well as the lower-frequency rigid-body modes.

Case 2. Single-shaft, two-span, three-bearing rotor with *intercritical* operation. The operating speed range lies between the second and third flexural critical speeds of the rotor.

Case 3. Single-shaft, single-span, two-bearing rotor with *subcritical* operation. The operating speed range lies above the rotor rigid-body modes and below the first flexural critical speed of the rotor.

The terms *transcritical, intercritical*, and *subcritical* as used here apply to the rotor flexural critical speeds. The dynamic models for these three case studies have purposely been kept simple but still reasonably representative of the actual construction. A more exact and more detailed treatment is essential for most practical rotor systems. The refined analytical techniques described in Chap. 2 should be employed to accurately assess the location of the critical speeds and the peak levels of response.

Case 1: Transcritical Operation

The machine chosen to represent this case is a high-pressure steam turbine of a cross-compound marine propulsion unit. The rotor configuration is depicted in Fig. 1.12. The rotor is quite stiff as measured against customary impulse turbine

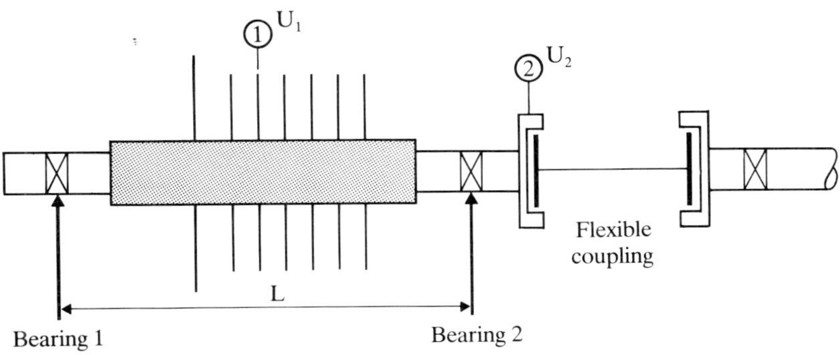

FIGURE 1.12 High-pressure steam-turbine rotor.

construction, and the span between bearings when evaluated in rigid bearings has its first or lowest critical speed N_1 at 5291 rpm. Speed N_1 is an important design parameter and has long been used by steam turbine designers as a measure of rotor stiffness. Note that high values of rotor stiffness relative to the effective support stiffness result in lower peak values of the whirl amplification factor, as shown in Fig. 1.6 for the Jeffcott rotor. The rotor is supported on journal bearings which are solidly attached to a stiff and relatively massive stator (the ratio of stator mass to rotor mass is 8.17). The stator is supported on "soft" structural steel and has a natural frequency N_f of 1500 cycles/min. A peak amplification factor Q_f of 10 is assumed for the effective damping of the stator system, where

$$Q_f = \frac{1}{2\zeta_f} \tag{1.33}$$

implying a value of 5 percent for the critical damping ratio ζ_f. The rotor is connected to a speed reduction gear by a gear-tooth type of flexible coupling with a floating distance piece. The rotor is assumed to be isolated from the driven member.

To simplify the analysis, the bearings, stator, and stator mounts are replaced by an idealized rotor support structure under each bearing, as depicted in Fig. 1.11*b*. One half of the stator mass (some foundation mass may be included) is lumped under each bearing. Each bearing connection between the rotor and stator masses is represented by a spring with a constant $K_b/2$ and a dashpot with a constant $B_b/2$. Similarly, each mount between the stator mass and the foundation base is represented by a spring-and-dashpot combination with spring constant $K_f/2$ and a damping constant $B_f/2$. This damped mass-elastic structure at each bearing may be replaced by an equivalent spring constant $K_s/2$ and an equivalent damping constant $B_s/2$, as defined in Sec. 1.4.1 by Eqs. 1.44 and 1.45. These equivalent spring and damping constants vary with the rotational speed. The pertinent numerical factors which define the system of case 1 are listed in Table 1.2.

A first step in the design procedure for the placement of critical speeds is the calculation of the critical speeds of the undamped system. The mode shapes for the first and second rotor flexural critical speeds are given in Fig. 1.13. The unit is required to operate at any speed up to a maximum of 6661 rpm. The first critical speed of the undamped system of 3861 rpm occurs at 58 percent of maximum speed, and the second critical speed of 7561 rpm occurs at 114 percent of maximum speed.

Of special significance to the designer is the response to unbalance. This is particularly true when transcritical operation is involved. It is apparent from the mode shapes at the critical speeds that unbalance in the midspan region of the rotor will be of concern at the first critical speed. At the second critical speed, it is

TABLE 1.2 System Characteristics
Case 1: High-pressure steam turbine

	Rotor	Foundation
Weight	$W_r = M_r g = 2148$ lb	$W_f = M_f g = 17{,}560$ lb
Critical speed	$N_1 = 5291$ rpm	$N_f = 1500$ cycles/min
Stiffness	$K_b = 1.461 \times 10^6$ lb/in	$K_f = 1.122 \times 10^6$ lb/in
Damping	$\omega B_b/K_b = 1.7$	$Q_f = 10$
Length	$L = 66$ in	
Unbalance location (type)	**Mass eccentricity**	**Unbalance amount**
Midspan (static)	$q' = 1$ mil	$U_1 = 34.4$ oz · in
Coupling (static)	$q' = 1$ mil	$U_2 = 34.4$ oz · in
Reference: Figs. 1.11*b* and 1.12		

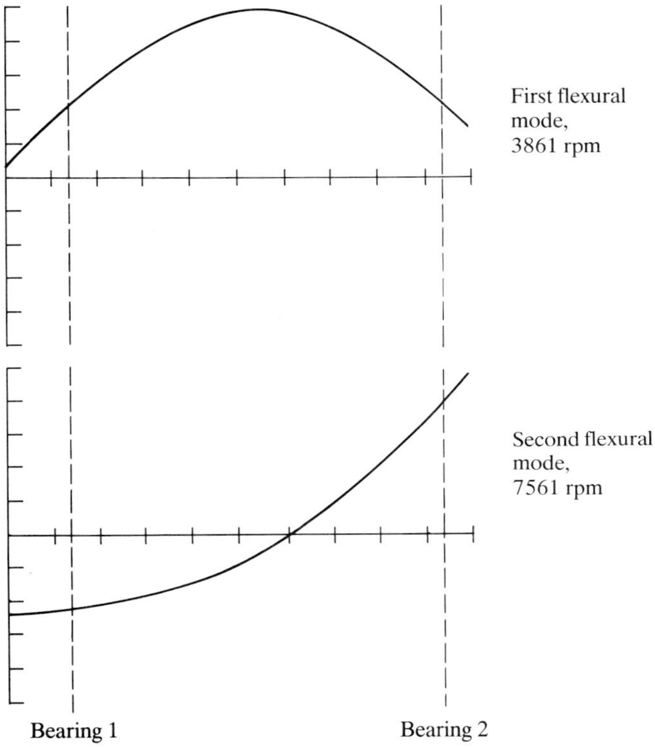

FIGURE 1.13 Undamped mode shapes—high-pressure steam turbine.

the unbalance at the rotor ends and, in particular, at the flexible coupling end that is important. The results of response calculations have been plotted in Fig. 1.14. The whirl radius r_1 in mils at the midspan location (1) has been plotted against speed over a range encompassing the rigid body modes, the first flexural mode, and the start of the second flexural mode for unbalance U_1 at the same midspan location. The whirl radius r_2 in mils for the coupling location (2) has also been plotted against speed for unbalance U_2 at the coupling location. The amounts of unbalance U_1 and U_2 used in the calculation have been selected assuming an eccentricity q' of 1 mil of the total rotor mass:

$$U = \frac{16W_r}{1000} q' \tag{1.34}$$

where U = unbalance, oz · in
q' = eccentricity, mils
W_r = rotor weight, lb

Because of the unit value assigned to q', the whirl radii r_1 and r_2 are equivalent in an approximate sense to the whirl amplification factor derived in Sec. 1.2.1 for the Jeffcott rotor. The rigid-body modes occur close to the stator natural

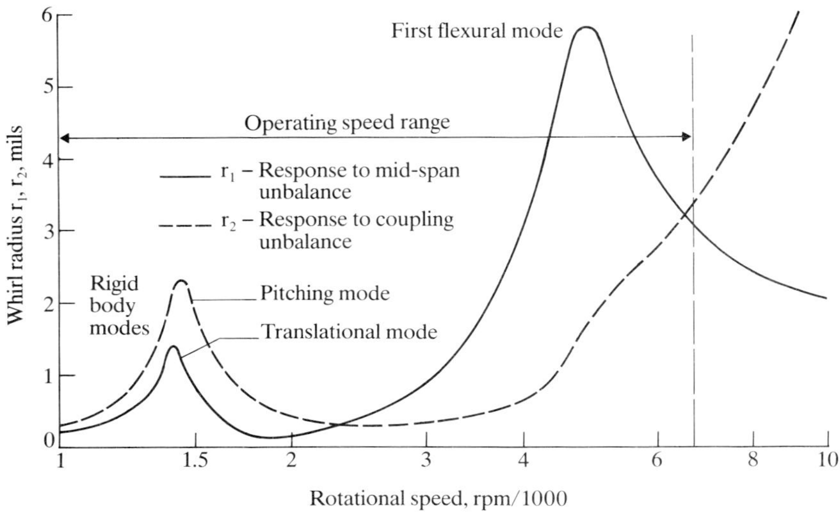

FIGURE 1.14 Response to unbalance—high-pressure steam turbine.

frequency of 1500 cycles/min with the pitching mode frequency only slightly greater than the translational frequency. This anomaly is due to the fact that the distributed stator mass has been lumped at the bearing locations. Because of the large ratio of stator mass to rotor mass, the effect of the distributed rotor mass is largely overshadowed. If all the mass in the system were uniformly distributed, the frequency ratio of the pitching mode to the translational mode would equal $3^{1/2}$. The peak response at the first flexural mode occurs at 4800 rpm, which is 24 percent greater than the undamped critical speed. This disparity is due to the high level of bearing damping. With precision high-speed balancing, the residual unbalance in the rotor body may be reduced to 0.10 to 0.01 of the amount derived from the 1-mil eccentricity assumption (Eq. 1.34). With this quality of balance, the maximum radius of whirl at the first flexural in Fig. 1.14 is reduced to a value between 0.58 and 0.058 mil. Operation should be entirely satisfactory at the first flexural critical speed and of no concern whatsoever at the rigid-body modes. This has been borne out by the successful operating record of several hundred machines of this design. Only in a few isolated instances where poor operating practice prevailed and attempts were made to run the rotor up to full operating speed with an excessive thermal bow did operating damage occur. The calculations on which the response curves of Fig. 1.14 are based show that a thermal bow equivalent to an eccentricity q' in the vicinity of 1 mil is sufficient to produce a damaging level of dynamic bearing loading.

Fortunately, coupling unbalance has relatively little effect on the response at the first flexural critical speed. However, if transcritical operation through the second critical speed were to be considered for this type of machine, then special attention would have to be paid to the coupling balance. For the gear-tooth coupling with a floating distance portrayed in Fig. 1.12, random variations in unbalance due to small but unavoidable errors in the centering of the distance piece will be present. This unbalance could cause a problem with excessive whirl amplitude at the coupling overhang and very high dynamic bearing reactions at higher speeds of operation.

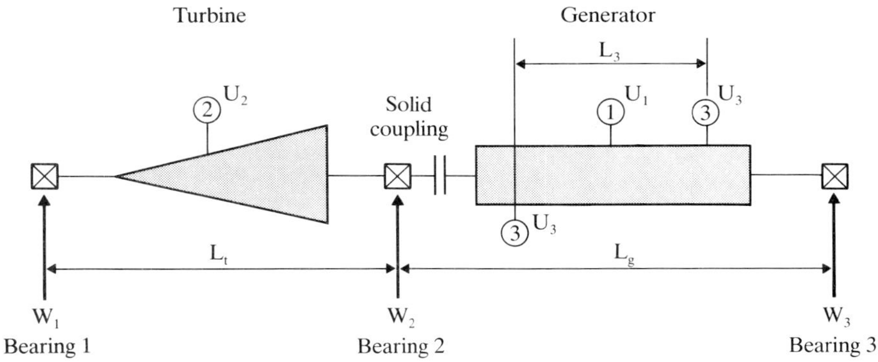

FIGURE 1.15 Turbine-generator set rotor.

Case 2: Intercritical Operation
Large steam-turbine generator sets may have as many as five or six solidly coupled spans with bearings between adjacent spans and at each end of the rotor. Individual spans will generally have different mass-elastic properties, and this gives rise to a large number of critical speeds, each successive critical speed being characterized, in general, by a dominant amplitude in the span of the next highest stiffness. For simplicity, a three-bearing, two-span arrangement has been chosen for study (see Fig. 1.15). The journal bearings are supported in sturdy bearing pedestals anchored to a massive reinforced-concrete foundation. Because of the massive foundation the effective spring and damping constants at the rotor journals may simply be taken as equal to the dynamic constants for the bearings, as explained in Sec. 1.4.1 under "Foundation Considerations." For the range of Sommerfeld number applying to this case, a value for $(C/W)K_{av}$ equal to 3 was selected (see Fig. 1.24) and for B_{av}/K_{av} a value of 1.7 (see Fig. 1.30). The bearing

TABLE 1.3 System Characteristics

Case 2: Turbine generator set

Bearing load	Bearing spring constant	Bearing damping factor
$W_1 = 4800$ lb	$K_1 = 2.32 \times 10^6$ lb/in	$\omega B_1/K_1 = 1.7$
$W_2 = 19{,}300$ lb	$K_2 = 5.78 \times 10^6$ lb/in	$\omega B_2/K_2 = 1.7$
$W_3 = 9800$ lb	$K_3 = 3.22 \times 10^6$ lb/in	$\omega B_3/K_3 = 1.7$

Span	Weight	Length	Rigid bearing critical speed
Turbine	$W_t = 14{,}000$ lb	$L_t = 145.3$ in	$N_{1t} = 1934$ rpm
Generator	$W_g = 19{,}500$ lb	$L_g = 184.8$ in	$N_{1g} = 1512$ rpm

Unbalance location (type)	Mass eccentricity	Unbalance amount	Moment arm
Generator (static)	$q' = 1$ mil	$U_1 = 312$ oz · in	
Turbine (static)	$q' = 1$ mil	$U_2 = 224$ oz · in	
Generator (dynamic)	$q' = 0.5$ mil	$U_3 = 156$ oz · in	$L_3 = 80$ in
Reference: Fig. 1.15			

constants derived from these dimensionless parameters are assumed to be axisymmetric and are listed in Table 1.3 along with other pertinent rotor information.

The first three undamped critical speeds calculated with the bearing stiffnesses given in Table 1.3 are 1556, 2064, and 4587 rpm, and the associated mode shapes are shown in Fig. 1.16. There is a large window between the second and third

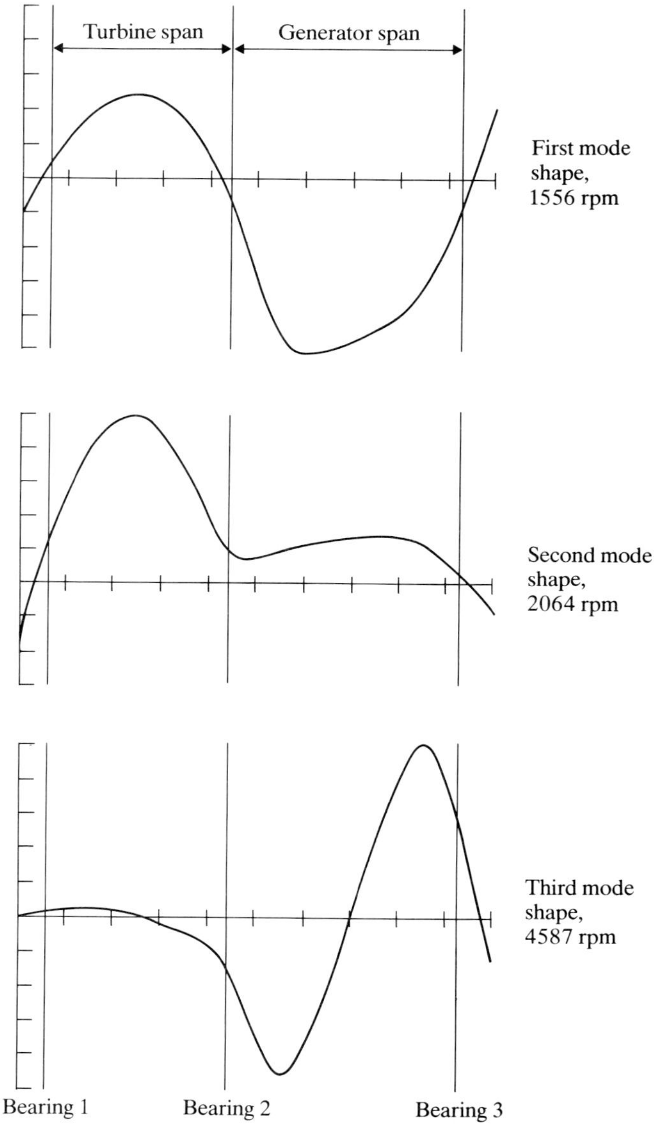

FIGURE 1.16 Undamped mode shapes of turbine-generator set.

critical speeds which safely accommodates the constant running speed of 3600 rpm with a margin for emergency trip action at 10 percent overspeed.

Calculated values of whirl radii due to unbalance, r_g in the generator and r_t in the turbine, are plotted as a function of rotational speed in Fig. 1.17. Three

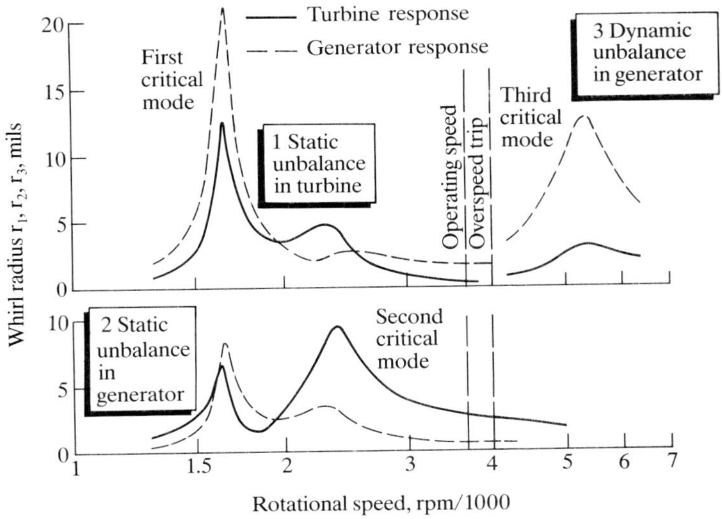

FIGURE 1.17 Response to unbalance in turbine-generator set.

different unbalances—U_1, U_2, and U_3—as shown in Fig. 1.15 are considered separately. Unbalances U_1 and U_2 are each single-point (or *static*) unbalances in the generator and turbine spans, respectively, the values of each of these unbalances corresponding to a mass eccentricity of the span of 1 mil in accordance with Eq. 1.34. In the third instance, there is a pair of single-point unbalances, each with a magnitude U_3, in the generator which are 180° out of phase with each other and produce a moment (or *dynamic*) unbalance equal to U_3L_3. The whirl radii plotted in Fig. 1.17 apply to the numbered locations of the unbalance as indicated by the numerical subscripts. These whirl radii are related in an approximate way to the whirl amplification factor of the Jeffcott rotor, as explained in case 1. The greatest value of whirl radius in the system ($r_{g1} = 21$) occurs in the generator span at the first critical speed due to the static unbalance U_1 in the generator span. This sensitivity to unbalance is related to the fact that the generator span with a rigid bearing critical speed of only 1512 rpm has a very low stiffness relative to the support stiffness provided by the journal bearings and the associated supporting structure. (Note the effect of the large value of the stiffness ratio κ on the whirl amplification factor of the Jeffcott rotor in Fig. 1.6.) At the second critical speed, the maximum response is in the turbine span with static unbalance U_2. To significantly excite the third critical speed requires a moment unbalance in the generator, typified here by U_3L_3.

Note that the peak values of response for the damped system occur at speeds somewhat higher than the undamped critical speeds—the increases being 5, 11, and 13 percent for the first, second, and third critical speeds, respectively.

Because of the responsiveness of the rotor, particularly at the first critical speed, great care must be exercised in the no-load run-up to operating speed to avoid damaging vibration caused by temporary thermal bowing of the rotor. Vibration monitoring instrumentation such as described in Chap. 4 is very important.

Bearing alignment is of great importance in multiple-span, solidly coupled rotors. Bearings should be properly spaced so that small changes in elevation will not produce excessive changes in bearing loading. Lightly loaded journal bearings may become unstable and completely alter the vibrational behavior of the system or produce substantial changes in the values of certain critical speeds.

Case 3: Subcritical Operation

A lightweight, high-speed, single-span, two-bearing unit has been selected for this case study. While this rotor is not a copy of any particular machinery element, it is similar to a rotor which might be found in a small aircraft gas-turbine engine. The rotor configuration is shown in Fig. 1.18. The hollow construction yields light weight with high lateral stiffness and a very high first flexural critical speed, the first flexural critical speed N_1 on rigid bearings being 26,610 rpm. The rotor is assumed to be supported on rolling-element bearings without damping, and the bearings are attached to a stator of weight equal to the rotor. The stator is supported on damped flexible mounts and has a natural frequency N_f of 8654 cycles/min. As a measure of the damping of the stator system, a peak amplification factor Q_f of 10 equivalent to 5 percent of critical damping is assumed.

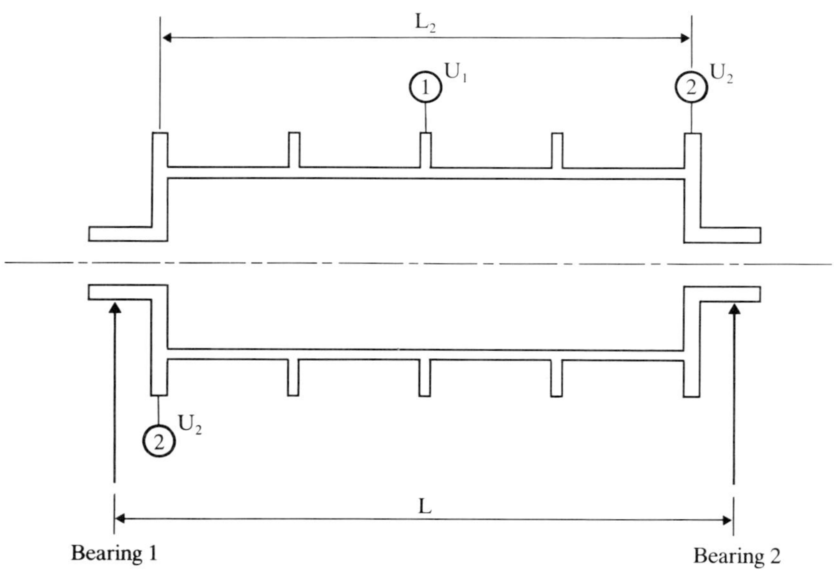

FIGURE 1.18 Small aircraft engine rotor.

As in case 1, the bearings, stator, and stator mounts are replaced by an idealized support structure under each bearing, as depicted in Fig. 1.11*b*. The equivalent spring constants $K_s/2$ and the equivalent damping constants $B_s/2$ are

TABLE 1.4 System Characteristics
Case 3: Small aircraft engine

	Rotor	Foundation
Weight	$W_r = M_r g = 188$ lbs	$W_f = M_f g = 188$ lb
Critical speed	$N_1 = 26{,}610$ rpm	$N_f = 8654$ cycles/min
Stiffness	$K_b = 1.20 \times 10^6$ lb/in	$K_f = 0.40 \times 10^6$ lb/in
Damping	$\omega B_b / K_b = 0$	$Q_f = 10$
Length	$L = 37.4$ in	

Unbalance location (type)	Mass eccentricity	Unbalance amount	Moment arm
Midspan (static)	$q' = 1$ mil	$U_1 = 3.0$ oz · in	
Rotor ends (dynamic)	$q' = 0.5$ mil	$U_2 = 1.5$ oz · in	$L_2 = 37.4$ in
Reference: Figs. 1.11*b* and 1.18			

calculated from Eqs. 1.44 and 1.45 with the bearing damping constants $B_b/2$ taken to be zero, as is appropriate for rolling-element bearings in the absence of special bearing dampers. This idealized support structure is assumed to be axisymmetric. The pertinent numerical constants which define the system of case 3 are given in Table 1.4.

The mode shapes for the first three undamped critical speeds are shown in Fig. 1.19. The translational rigid-body critical speed occurs at 5802 rpm, the pitching rigid-body critical speed at 7161 rpm, and the first flexural critical speed of the rotor at 19,910 rpm. Shown also in Fig. 1.19 are the relative magnitudes of the whirl radii of the rotor r_a and the stator r_f at the bearing locations. For the rigid-body modes, the difference between the whirl radii is small which makes possible a safe transition through these critical speeds. At the first flexural critical speed of the rotor, the rotor and stator whirl radii are out of phase with a large differential between them. This fact, combined with the deflected shape of the rotor, makes rubbing of internal seals a real possibility with static unbalance in the rotor. Calculated results for the damped response to unbalance are plotted in Fig. 1.20. The whirl radius r_1 in mils at the midspan location (1) is plotted over a speed range which includes the rigid-body mode critical speeds and the first flexural critical speed for a static unbalance U_1 at the midspan location. The whirl radius r_2 at the rotor ends (2) is also plotted for a dynamic unbalance $U_2 L_2$. The amounts of unbalance given in Table 1.4 have been adjusted as in the other case studies so that the whirl radii provide a rough assessment of the whirl amplification factor. The large value of the whirl radius for the first flexural critical speed, approximately 28, is a consequence of the absence of bearing damping. However, for subcritical operation in the speed range between 10,000 and 15,000 rpm where whirl amplitudes are very low, this lack of bearing damping is of no concern.

If operation were desired at the first flexural critical speed of the rotor, the introduction of squeeze-film dampers of the construction shown in Fig. 1.37 and described in Sec. 1.4.4 would greatly reduce the whirl amplification. This a direct consequence of the large differential between the whirl radii of the rotor and the stator at the bearing locations, as shown in Fig. 1.19. It is also apparent from Fig. 1.19 that the introduction of bearing damping will have relatively little impact on the rigid-body modes.

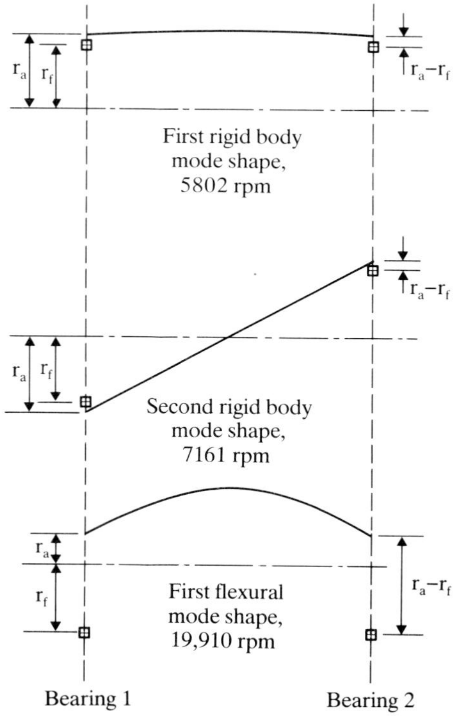

FIGURE 1.19 Undamped mode shapes for small aircraft engine.

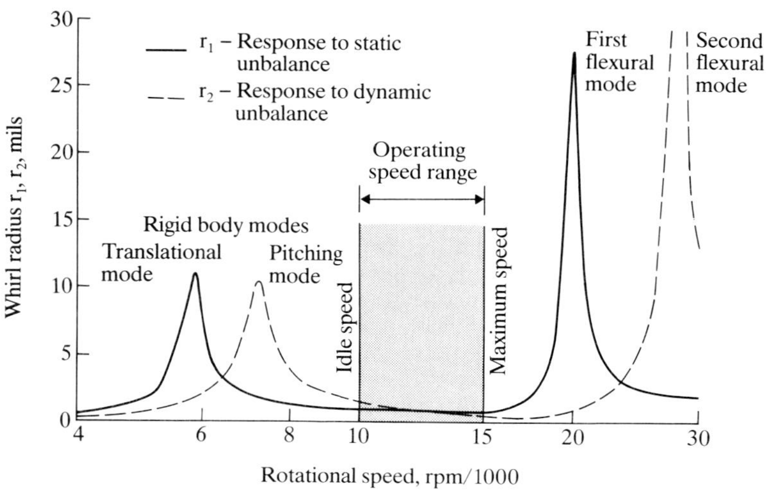

FIGURE 1.20 Response to unbalance for small aircraft engine.

Since all the damping in the system as defined in Table 1.4 is derived from the stator and its mounts, the peak amplitudes are approximately proportional to the peak amplification factor Q_f of the stator system. Halving the value of Q_f, that is, doubling the actual damping, would essentially cut all the peak amplitudes in half but at the same time would broaden all the resonant peaks. Damping in the stationary structure must be sufficient to provide acceptable levels of vibration when the rotor is accelerated through the rigid-body modes to the operating speed range. The placement of the operating speed range between the rigid-body modes of the rotor and the first flexural mode has three very real advantages:

1. The arrangement provides a broad operating speed range free of vibration resonances.

2. The approach affords the possibility of *avoiding* high-speed multiplane balancing since the system generally requires only a low-speed rigid-body balance. This can be an important boon in the manufacture and maintenance of a mass-produced machine with widely dispersed use.

3. Where operation at or near the first flexural critical speed is not required, the arrangement permits the use of rolling-element bearings without the complication of adding bearing dampers, since bearing damping is not needed while traversing the rotor rigid-body low-speed critical speeds.

This critical speed placement strategy has had wide application. A common example is the ordinary electric motor. The stiff and compact rotor has a high value of the first flexural critical speed; and if the motor stator is mounted on soft, damped springs to provide vibration isolation, then the generic response to unbalance characteristics shown in Fig. 1.20 will apply.

1.3.2 Multiple Rotors

In a machine with more than one rotor, each of which operates at a different speed and each of which is relatively isolated from the others (i.e., unbalance in one shaft tends *not* to excite critical response in any other shaft), the designer may deal with the placement of critical speeds of each rotor as a separate and distinct problem. But in situations in which the shafts are intimately coupled (e.g., coaxial shafts with intershaft bearings), the possibility of interaction must be considered. That is, excitation at frequencies of rotation within the operating range of each individual shaft must be considered as possible excitation for the natural frequencies of the other shaft(s). When warranted by the available precision of computation, it may be appropriate to modify the natural frequency of shaft A for the effects of rotation at speed Ω_A (such as gyroscopic stiffening) when being excited at a frequency Ω_B associated with unbalance on shaft B.

1.4 BEARINGS, ROTOR DAMPERS, SEALS, COUPLINGS, AND ROTOR DESIGN

1.4.1 Journal Bearings

Journal bearings have had a long history of application in high-speed rotating machinery, such as steam turbines. Long before the dynamic characteristics of the bearing were understood, the high level of inherent damping of the oil film made possible the successful operation of high-speed flexible rotors. Only in the last two or three decades have the dynamic characteristics of the journal bearing been properly understood and systematically evaluated. This section focuses on two characteristics of particular importance to rotor vibration response—the dynamic stiffness and damping coefficients of some representative bearing types as reported in recent literature. For the detailed design of journal bearings and the calculation of steady running characteristics, the reader is referred to existing handbooks, e.g., Wilcock and Booser (1957), Pincus and Sternlicht (1961), Booser (1983).

Bearing Types

A great variety of bearing geometries and types have been used over the years Schematic diagrams are given for six representative types in Fig. 1.21*a* through *f*.

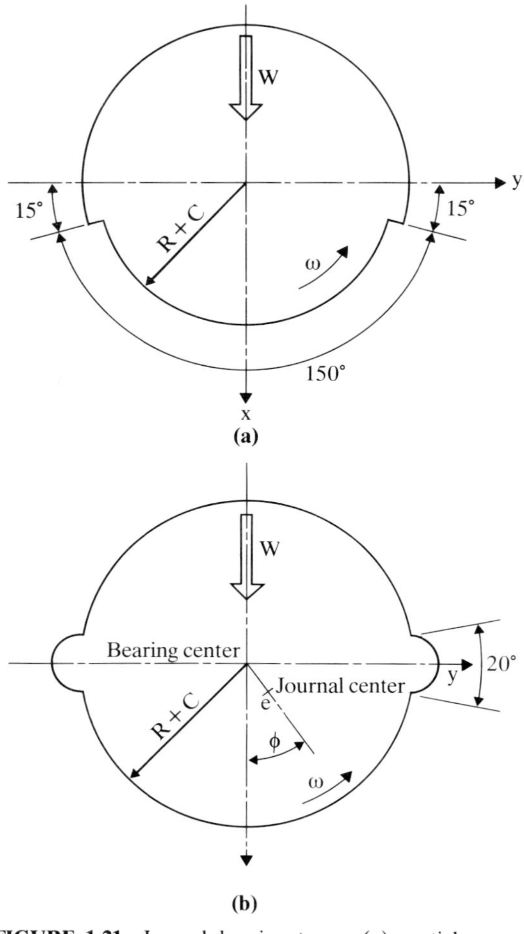

FIGURE 1.21 Journal bearing types: (*a*) partial arc, (*b*) two-axial-groove, (*c*) elliptical, (*d*) three-lobe, (*e*) offset cylindrical, (*f*) tilting-pad.

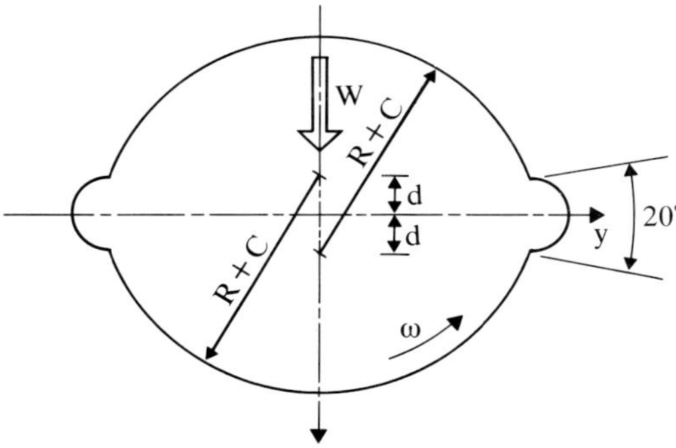

(c)

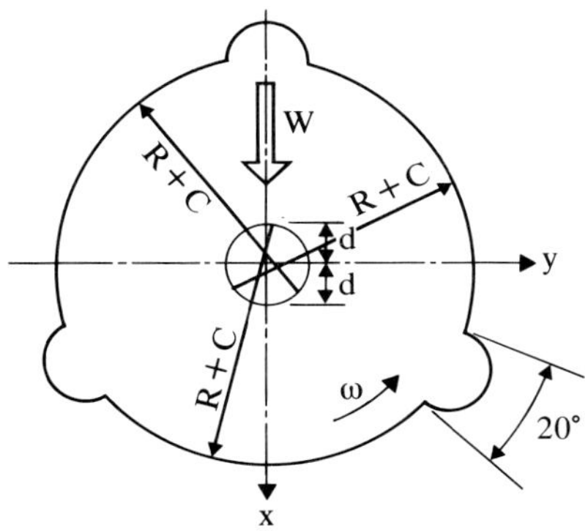

(d)

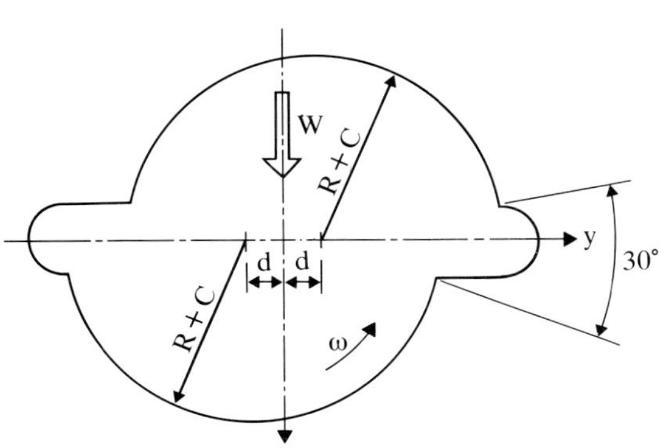

(e)

FIGURE 1.21 (*Continued.*)

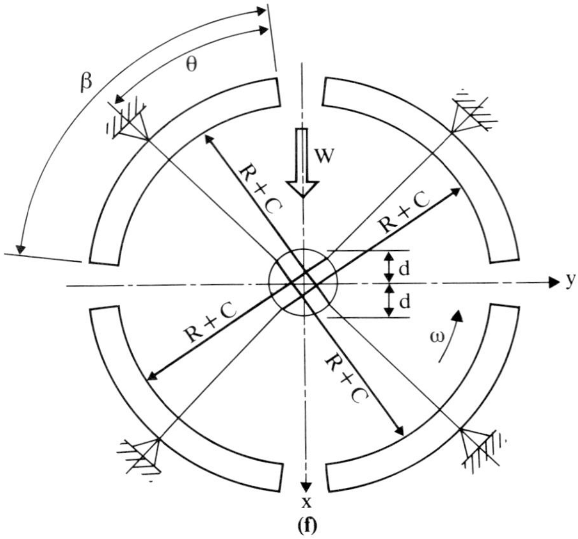

FIGURE 1.21 (*Continued.*)

The first two types, the partial-arc and the two-axial-groove bearings, are both versions of the plain cylindrical bearing. The next three types—elliptical, three-lobe, and offset cylindrical—all incorporate preload. The preload is specified by the ratio of the dimension d to the clearance C. A customary value of the preload is 0.5. Bearings with preload tend to run with a greater minimum film thickness for a given bearing size and operating parameters. When the direction of the steady load on the bearing varies with respect to the coordinate axes (x and y), the three-lobe bearing is preferred over the elliptical or offset cylindrical types.

The bearings mentioned above, with fixed geometry, may exhibit instability under certain operating conditions. The tilting-pad bearing of Fig. 1.21*f* is highly stable and should be employed when unstable operation is a possibility. Four-, five-, and six-pad arrangements are commonly used. The load direction may be between two pads or central to a pad. This bearing may be designed with or without preload.

Dynamic Stiffness and Damping Coefficients

Hagg and Sankey (1956) provided an early determination of dynamic bearing characteristics for a number of bearing types, using a special test rig. Selected values of rotating load and steady load were applied. From measurements made of the elliptical orbit of the journal, the stiffness constants K_1 and K_2 and the damping constants B_1 and B_2 were calculated relative to the minor and major axes of the ellipse (labeled by subscripts 1 and 2, respectively). In Fig. 1.22 these constants are plotted in dimensionless form as a function of Sommerfeld's number S for the centrally loaded partial arc bearing of Fig. 1.21*a* where $L/D = 1$ and

$$S = \left(\frac{\mu N}{P}\right)\left(\frac{R}{C}\right)^2 \tag{1.35}$$

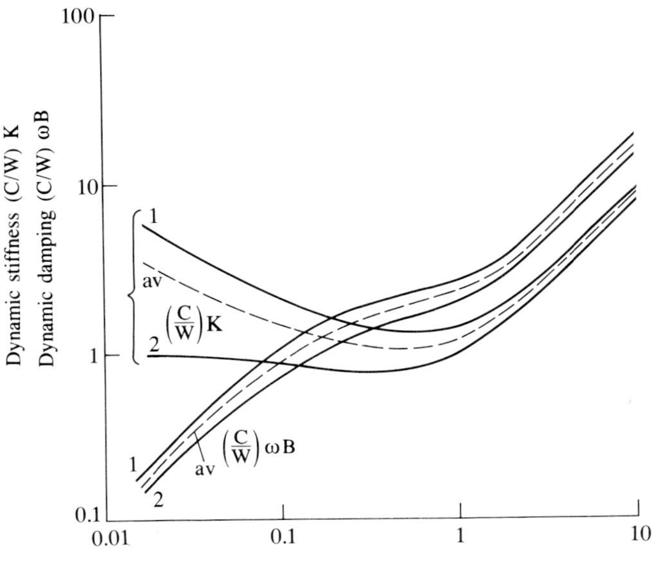

FIGURE 1.22 Dynamic stiffness and damping versus Sommerfeld's number for centrally loaded partial-arc bearing for $L/D = 1$ (Hagg and Sankey, 1956).

$$P = \frac{W}{LD} \tag{1.36}$$

where C = radial clearance
R = journal radius
$R + C$ = radius of curvature of bearing surface
d = preload dimension
m = preload = d/C
W = steady load
μ = absolute viscosity
N = rotational speed = $\omega/2\pi$
P = bearing pressure
L = axial length
D = journal diameter

Note that the axes (1 and 2) of the elliptical orbit bear no fixed relationship to the vertical and horizontal axes (x and y) of Fig. 1.21*a*.

The Hagg and Sankey information has been presented for its historical significance. Unfortunately, the experimental approach does not permit the evaluation of the cross-coupling terms which are an important aspect of the dynamic stiffness and damping characteristics of the bearing.

All the recent evaluations of stiffness and damping coefficients are based on a numerical solution of the well-known Reynolds equation. The rotor amplitude is assumed to be sufficiently small that fluid forces may be replaced by their gradients around the steady-state operating eccentricity. Thus the forces become proportional to the vibratory displacement and velocity, with the coefficients of proportionality being the stiffness and damping coefficients, respectively. Relative to the x and y coordinate system as shown in Fig. 1.21, the reactive fluid-film forces F_x and F_y become

$$F_x = -K_{xx}X - B_{xx}\frac{dX}{dt} - K_{xy}Y - B_{xy}\frac{dY}{dt} \tag{1.37}$$

$$F_y = -K_{yx}X - B_{yx}\frac{dX}{dt} - K_{yy}Y - B_{yy}\frac{dY}{dt} \tag{1.38}$$

Lund and Thomsen (1978) provide tabulated values of the eight coefficients in Eqs. 1.37 and 1.38 as a function of Sommerfeld's number for the four bearing geometries depicted in Fig. 1.21*b* through *e* where $L/D = 0.5$ and 1.0. Figures 1.23*a* and *b* and 1.24*a* and *b* present these coefficients as dimensionless numbers versus Sommerfeld's number for two of the geometries where $L/D = 1$. The stiffness coefficients for the two-axial-groove bearing are given in Fig. 1.23*a*, and the damping coefficients are given in Fig. 1.23*b*. The stiffness coefficients for the elliptical bearing are given in Fig. 1.24*a* and the damping coefficients in Fig. 1.24*b*. Note that some of the coefficients become negative at higher values of the Sommerfeld number.

Use of the eight-coefficient data makes possible the determination of the influence of journal bearing characteristics on the stability of the rotor system as well as the steady-state response to unbalance. It is the cross-coupling terms which appear in Eqs. 1.37 and 1.38 for bearings of fixed geometry that produce the destabilizing effect.

However, if the operating parameters of the journal bearings are such that the bearings are known to be operating in a stable regime, the response to unbalance can be calculated more simply by assuming an axisymmetric support condition—an assumption which is often sufficient for design work. A single average value of the stiffness coefficient and a single average value of the damping coefficient required for this simplified calculation may be determined from the eight-coefficient data by postulating that the journal is whirling in a circular orbit of radius r_a. Then the reactive forces of the oil film F_r (radial) and F_t (tangential) for any angle ω_t are given by

$$F_r = F_x \cos \omega t + F_y \sin \omega t \tag{1.39}$$

$$F_t = -F_x \sin \omega t + F_y \cos \omega t \tag{1.40}$$

Substituting the expressions for F_x and F_y in the equations for F_r and F_t, with $x = r_a \cos \omega t$ and $y = r_a \sin \omega t$, and integrating to obtain the average values for F_r and F_t over 1 yield

$$\frac{(F_r)_{\text{av}}}{r_a} = K_{\text{av}} = \frac{K_{xx} + K_{yy}}{2} + \frac{\omega(B_{xy} - B_{yx})}{2} \tag{1.41}$$

$$\frac{(F_t)_{\text{av}}}{r_a} = \omega B_{\text{av}} = \frac{\omega(B_{xx} + B_{yy})}{2} + \frac{-K_{xy} + K_{yx}}{2} \tag{1.42}$$

Lund (1964) presents data for tilting-pad journal bearings with 4, 5, 6, and 12 pads, all centrally pivoted ($\theta/\beta = 0.50$), and for various values of the ratio L/D. The effects of preload and pad inertia are also treated. The four-pad bearing data for $L/D = 1$ and for zero preload are presented in Fig. 1.25. The cross-coupling

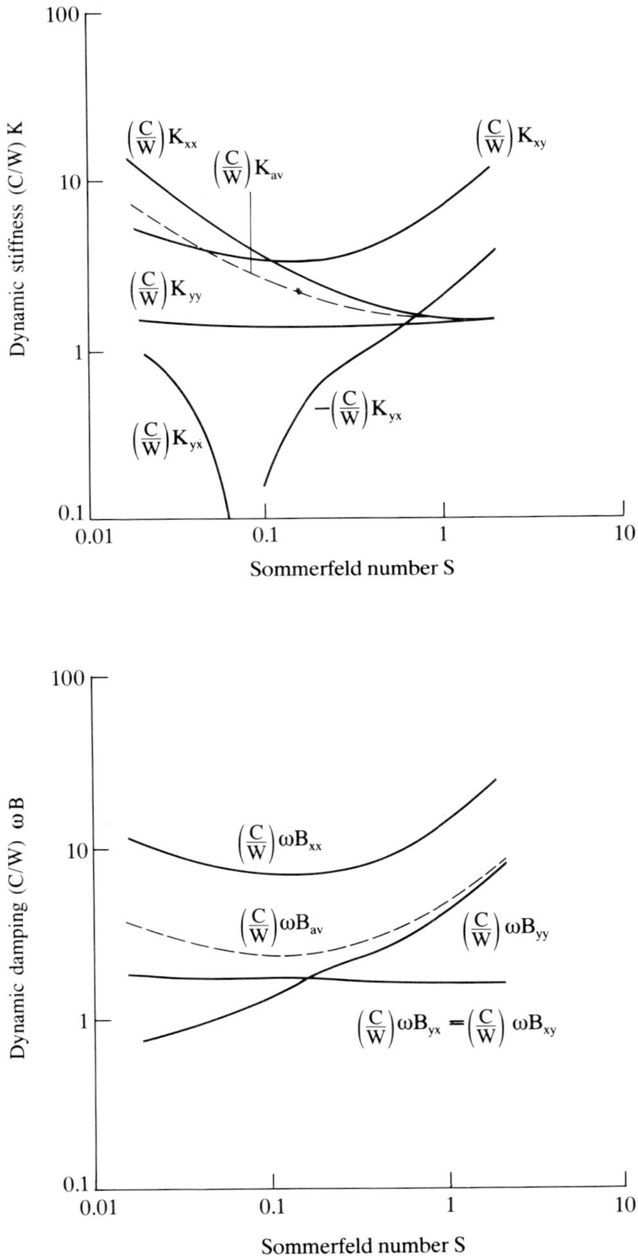

FIGURE 1.23 (*a*) Dynamic stiffness and (*b*) dynamic damping versus Sommerfeld's number for two-axial-groove bearings for $L/D = 1$ (Lund and Thomsen, 1978).

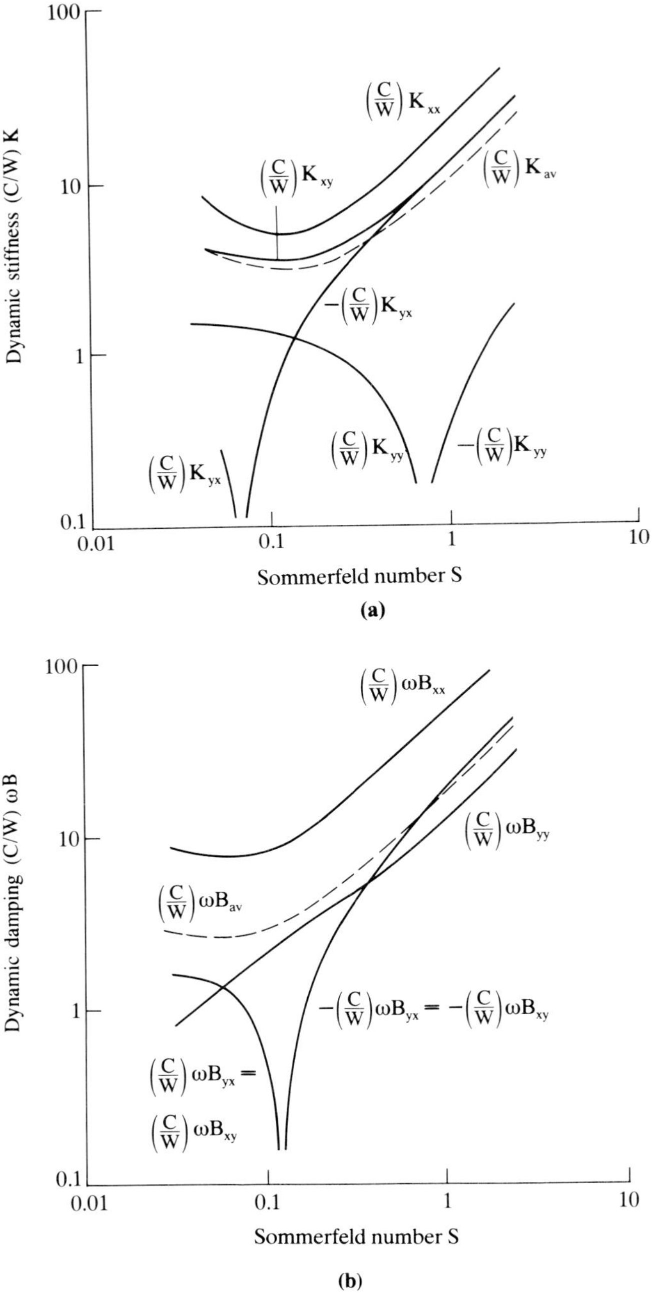

FIGURE 1.24 (*a*) Dynamic stiffness and (*b*) dynamic damping versus Sommerfeld's number for elliptical bearings for $L/D = 1$ (Lund and Thomsen, 1978).

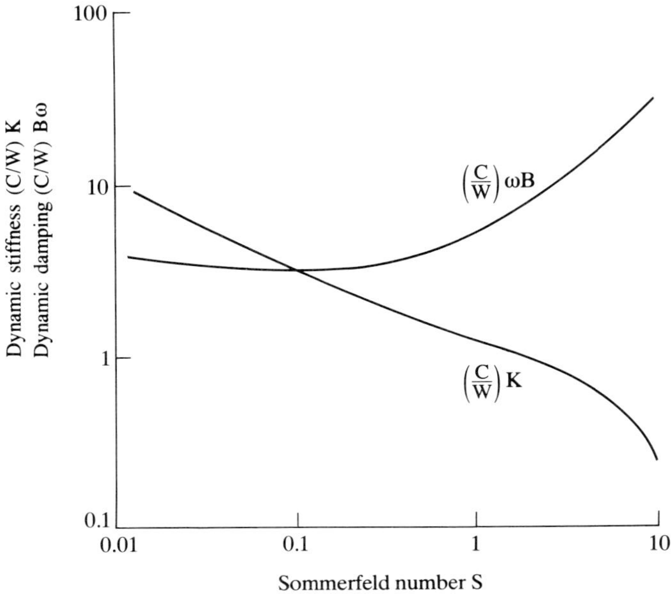

FIGURE 1.25 Dynamic stiffness and damping versus Sommerfeld's number for the four-tilting-pad bearing with central pivot for $L/D = 1$ (Lund, 1964).

terms are absent—evidence of the inherent stability of the tilting-pad bearing (when the small effects of pad inertia and pivot friction are neglected). Generally there will be four coefficients: two direct-coupled stiffness coefficients and two direct-coupled damping coefficients. Because the four-pad bearing is axisymmetric there is only one stiffness and one damping coefficient.

Orcutt (1967) provides information for a four-pad bearing with offset pivot ($\theta/\beta = 0.55$), with $L/D = 1$, and with preloads of 0 and 0.5. The offset-pivot design has superior hydrodynamic characteristics, but its use is restricted to machines with one direction of rotation. Dynamic coefficients are plotted in Fig. 1.26 for zero preload and for laminar flow. Due to the offset pivot, there is a marked difference in the coefficients compared to the data for the centrally pivoted design (Fig. 1.25) at the higher Sommerfeld numbers, particularly with regard to the stiffness coefficient.

Orcutt (1967) points out that the laminar-flow assumption is appropriate for conventional bearings which have small clearance and which are lubricated with hydrocarbon oils whose kinematic viscosity is high. If bearings are lubricated with liquid metals, water, or other low-kinematic-viscosity lubricants and if the rotational speed is high, turbulence occurs in the oil film when the Reynold number (Re) reaches 1500. The effect of Reynold numbers up to 16,000 on both steady-state and dynamic characteristics is given in the Orcutt (1987) paper:

$$\mathrm{Re} = \frac{\pi \mathrm{DNC}}{\mu/\rho} \tag{1.43}$$

where ρ = density of lubricant and μ/ρ = kinematic viscosity of lubricant.

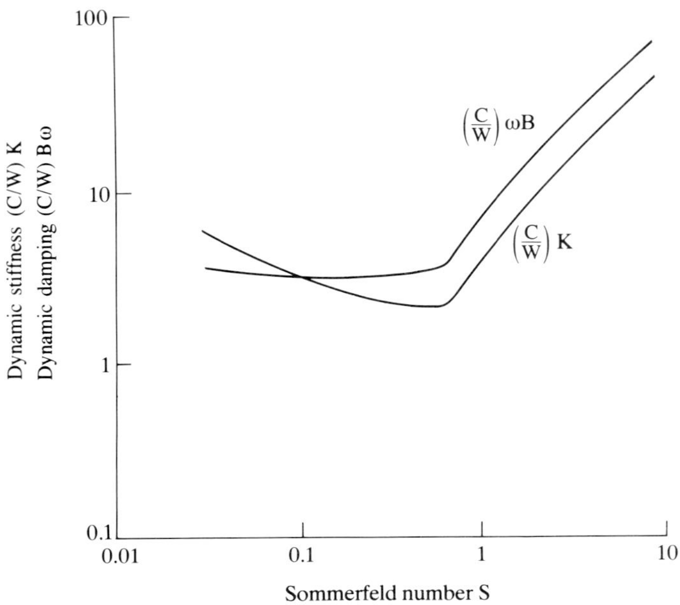

FIGURE 1.26 Dynamic stiffness and damping versus Sommerfeld's number for the four-tilting-pad bearing with offset pivot for $L/D = 1$ (Orcutt, 1967).

Foundation Considerations

For machinery such as steam turbines where the journal bearings are firmly attached to relatively massive stator and/or foundation elements, the dynamic support at the rotor journals usually may be represented solely by the stiffness and damping of the bearing oil film. The conditions under which this generalization is applicable can best be understood by considering the idealized support structure shown in Fig. 1.27. Assuming that this structure replaces each of the individual spring and dashpot arrays supporting the Jeffcott rotor as depicted in Fig. 1.1*b* and noting that the system as modified is still axisymmetric, we see that the effective spring constant K_s and the effective damping constant B_s for the new support arrangement are given by

$$\frac{K_s}{K_b} = \frac{C_1 C_3 + C_2 C_4}{C_3^2 + C_4^2} \tag{1.44}$$

$$\frac{\omega B_s}{K_b} = \frac{C_2 C_3 - C_1 C_4}{C_3^2 + C_4^2} \tag{1.45a}$$

$$\frac{B_s}{B_b} = \frac{1}{G} \frac{\omega B_s}{K_b} \tag{1.45b}$$

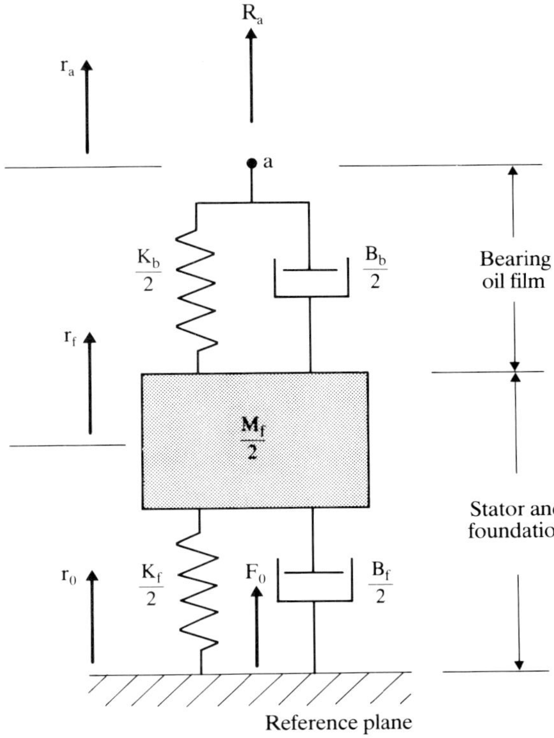

FIGURE 1.27 Idealized support structure.

where $G = \dfrac{\omega B_b}{K_b} = \dfrac{\omega B_{\text{av}}}{K_{\text{av}}}$

$$C_1 = 1 - 2GD\left(\frac{\omega}{\omega_f}\right) - \left(\frac{\omega}{\omega_f}\right)^2$$

$$C_2 = G + 2D\left(\frac{\omega}{\omega_f}\right) - G\left(\frac{\omega}{\omega_f}\right)^2$$

$$C_3 = 1 + H - \left(\frac{\omega}{\omega_f}\right)^2$$

$$C_4 = GH + 2D\left(\frac{\omega}{\omega_f}\right)$$

$\omega_f = \left(\dfrac{K_f}{M_f}\right)^{1/2}$ = undamped natural frequency of foundation mass

$H = \dfrac{K_b}{K_f}$ = ratio of bearing stiffness to foundation stiffness

$D = \dfrac{B_f}{2M_f\omega_f}$ = ratio of actual damping to critical damping of foundation

$Q = \dfrac{1}{2D}$ = amplification factor of foundation

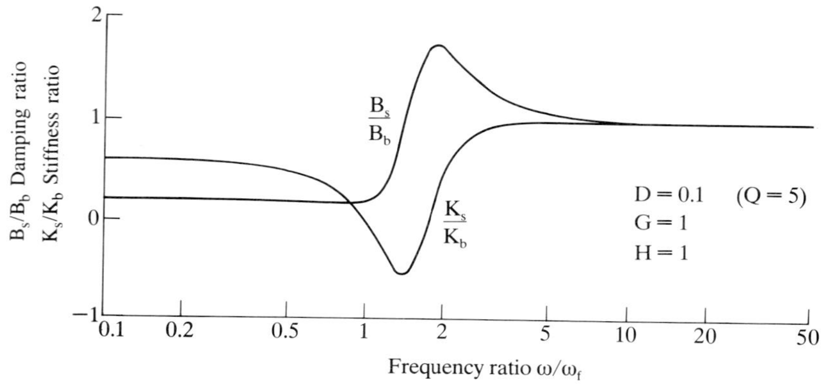

FIGURE 1.28 Foundation stiffness ratio and damping ratio for $G = 1$, $H = 1$, $D = 0.1$ ($Q = 5$).

The ratios K_s/K_b and B_s/B_b have been plotted in Fig. 1.28 as a function of the frequency ratio ω/ω_f for the following values of the dimensionless parameters in the governing equations:

$$G = 1 \qquad H = 1 \qquad D = 0.1 \qquad \text{or} \qquad Q = 5$$

A second plot in Fig. 1.29 shows the effect of changing the stiffness ratio to $H = 0.1$.

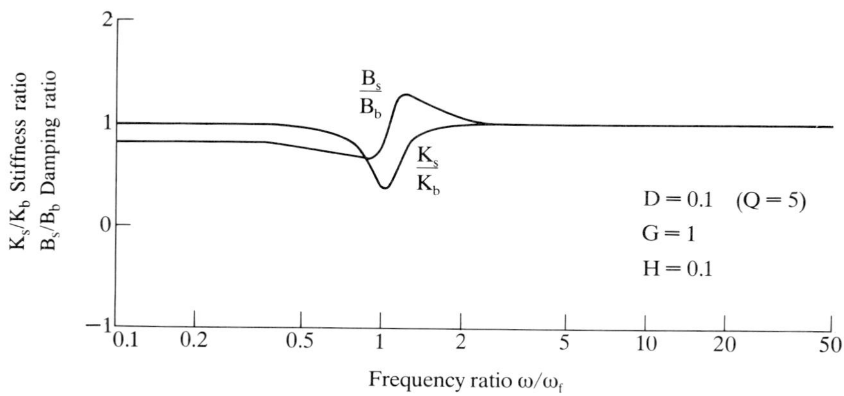

FIGURE 1.29 Foundation stiffness ratio and damping ratio for $G = 1$, $H = 0.1$, $D = 0.1$ ($Q = 5$).

The selection of a unit value for G is based on the plots in Fig. 1.30, where G is plotted versus Sommerfeld's number for the various bearing types. Values of G generally fall in the range of 0.7 to 2.0 except for two of the bearings—the two-axial-groove bearing and the tilting-pad bearing with central pivot—which exhibit a continuously rising characteristic with Sommerfeld's number.

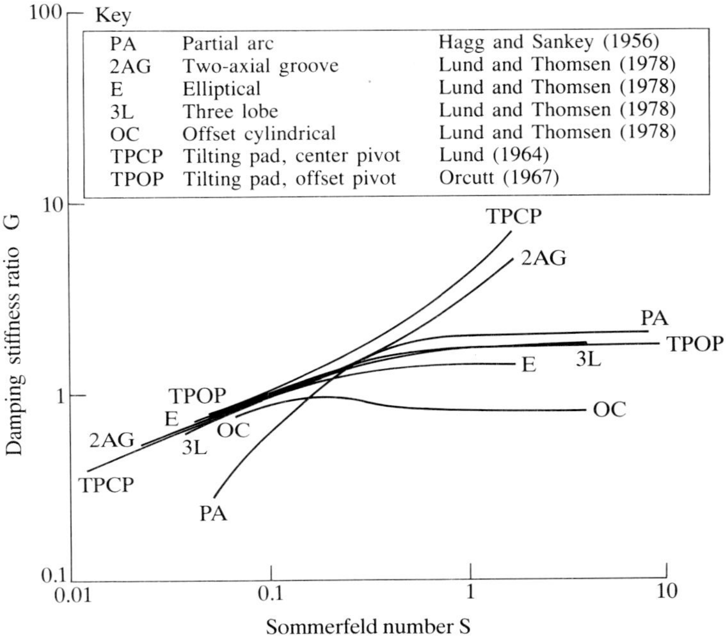

FIGURE 1.30 Damping/stiffness ratio parameter G versus Sommerfeld's number for various bearing types.

In both Figs. 1.28 and 1.29 the ratios K_s/K_b and B_s/B_b approach unity for values of the frequency ratio greater than unity and remain at the unit value for all higher values of the frequency ratio. This means that the foundation mass M_f can be treated as a rigid platform for the bearings. K_s becomes equal to the bearing stiffness K_b, and B_s equal to the bearing damping B_b. Regardless of the values of the basic parameters G, H, and D, this result will always apply at some sufficiently large value of the frequency ratio.

The key parameter for establishing the critical frequency ratio at which $K_s = K_b$ and $B_s = B_b$ is the stiffness ratio H:

$$H = \frac{K_b}{K_f} = \frac{K_b}{M_f \omega_f^2} \tag{1.46}$$

With given values for K_b and ω_f^2 in Eq. 1.46 a large value for the mass M_f ensures a small value of H and a critical frequency ratio close to unity. (Compare results in Fig. 1.29 with those in Fig. 1.28.)

When the running speed is close to the foundation frequency, there is a pronounced effect on K_s and B_s due to the resonance involving the foundation mass. K_s can become negative as in Fig. 1.28, which simply means that the support behaves as a mass rather than a spring. B_s always remains positive since the assumed system is basically stable.

For values of the frequency ratio less than about 0.5, the ratios K_s/K_b and B_s/B_b approach asymptotic values which are functions of the foundation system parameters and are defined by

$$\frac{K_s}{K_b} = \frac{1 + H + G^2H}{(1 + H)^2 + G^2H^2} \tag{1.47}$$

$$\frac{B_s}{B_b} = \frac{1}{(1 + H)^2 + G^2H^2} \tag{1.48}$$

Importance of Alignment (Multiple Spans)
For all multiple-span rotors which are solidly coupled, a necessary element of the design procedure is the calculation of rotor deflection and bearing reactions due to gravity as a function of the elevations of the various bearings relative to the in-line alignment of the bearings. Bearing elevations are then adjusted based on the calculations to provide a proper distribution of rotor weight between the various bearings. The aim is to have a sufficiently high level of bearing pressure on each bearing that stable operation will be ensured. For elliptical bearings on 3600 rpm equipment, a minimum level of 125 lb/in^2 is suggested. The upper limit of bearing pressure is usually set by a calculation of the minimum desired film thickness. If the bearing pressure falls below 100 lb/in^2, a tilting-pad bearing may be needed to ensure stable operation.

The rate of change of the reaction at a given bearing with the change in elevation at that bearing or an adjacent bearing is important. Too steep a gradient may require impractable tolerances on bearing lineup. This condition may occur, for example, if bearings are placed close together between two large in-line machinery elements. If differential thermal changes take place in the support structure of the bearings, then partial or complete unloading of one of the bearings may result. Undesirable vibration can result from the instability and "whipping" action of the unloaded bearing. Such vibration may be very difficult to correct if the thermal changes in the bearing support structure are tied to the normal operating cycle of the machinery. If bearing unloading is complete, changes in critical speed and vibration mode shape may also occur. The possibility of substituting one larger bearing in place of two smaller bearings should be investigated in the design phase.

1.4.2 Rolling-Element Bearings

Rolling-element bearings are used in many classes of rotating machinery where compact, high-load-capacity rotor support systems are required, where heat rejection requirements must be minimal, and where the bearings' capability of sustaining full-load capacity down to zero speed is a desired or required feature. The bearings may be grease-packed, but more usually they require a supply of fluid lubricant to lubricate the bearings and to serve as a medium for heat rejection. These bearings are usually found in rotating machinery aboard transportation vehicles such as aircraft where the movement of the vehicle entails shock and maneuver loads which must be sustained by the rotor's bearing system. For the detailed design of rolling-element bearings, the reader is referred to existing handbooks, e.g., Shaw and Macks (1949) and publications by the bearing manufacturers. The two major disadvantages in using rolling-element bearings are as follows:

1. In rolling-element bearings there is always rolling contact, and there are some aspects of rubbing or sliding contact between the rollers and the stator and

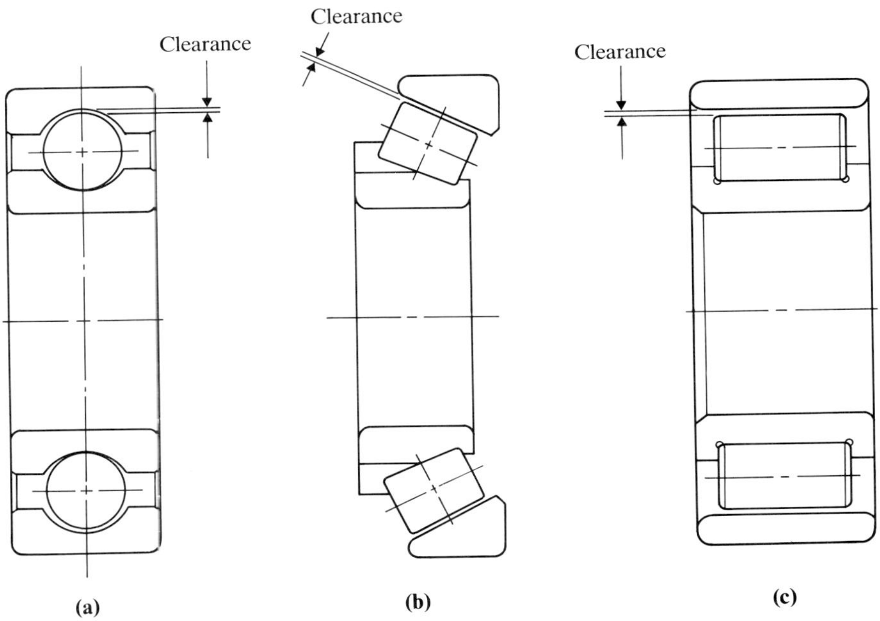

FIGURE 1.31 Classes of rolling-element bearings: (*a*) ball bearing, (*b*) tapered roller bearing, (*c*) cylindrical roller bearing.

rotor races and the cages, with life-limiting consequences. In fluid film bearings, there is no metal-to-metal contact between the rotor and the stator during normal operation, and there is no life-limiting wear.

2. Rolling-element bearings have essentially no intrinsic external damping, and where damping is necessary to good rotordynamic operation, some damping element(s) must be designed into the apparatus. Fluid film bearings have, intrinsic to their operation, significant external damping that serves in good stead for limiting critical speed peak amplitude and for delaying the onset of any unstable whirling.

There are two major classes of rolling-element bearings: those with spherical balls and those with tapered or cylindrical rollers, shown schematically in Fig. 1.31. Tapered roller and spherical ball (and other specialized types such as angular-contact ball) bearings sustain both radial and axial loads, while cylindrical roller bearings sustain only radial loads. Ideally, a rotor should have only one support point at which axial load is supported, to avoid the indeterminate forces that a redundant axial load support system would entail. One design approach is to systematically assemble a specified preload into a pair of thrust bearings on a given shaft. But a more usual approach is to limit any shaft to a single axial-load-sustaining bearing and to make other bearing(s) on the shaft a type that will sustain only radial loads (e.g., one with cylindrical rollers).

Rotordynamic Aspects of Rolling-Element Bearings

As a consequence of the features of rolling-element bearings and the circumstances of their application, there are certain unique aspects to the rotordynamic behavior of the rotors which they support. These five issues must be anticipated and accommodated in any design.

First, as noted above, rolling-element bearings have essentially no intrinsic damping. In supercritical shaft designs, any requirement for external shaft damping to attenuate critical speed peak amplitude or to forestall unstable operation must be accommodated by incorporating some additional damping element, typically either squeeze-film fluid dampers or solid-state dampers.

Second, a unique advantage of rolling-element bearings is the absence of any destabilizing forces in the bearings such as those which induce oil whip in rotors supported by hydrodynamic bearings. That is not to say that rotors supported by rolling-element bearings are immune from instability, for several other destabilizing mechanisms can cause unstable operation independent of bearing type, such as hysteretic whirl and whirl due to fluids trapped inside cylindrical rotor cavities, as is reviewed in Sec. 1.6.

Third, to avoid compression of the rolling elements and premature fatigue of the rolling elements and/or races associated with differential thermal expansion of the races during operation over a range of environmental temperatures, rolling-element bearings must be designed with radial clearance between the rolling elements and the races. Such bearing clearance may compromise bearing life if unloaded operation of the bearing occurs and the balls of rollers skid. It may also result in an effective softening of the bearing support stiffness and a very significant reduction in natural frequencies and critical speed. Moreover, the nonlinear nature of the effect of clearance on bearing support stiffness may cause many anomalies in vibration response to unbalance, such as hysteresis and jumps in vibration amplitude (see, e.g., Ehrich and O'Connor, 1967), pseudo-critical-speed peaks at whole-number multiples of the critical speed, and generation of a multiplicity of unexpected (sideband and harmonic multiple) frequencies which may excite undesirable resonances in the system's static parts, as summarized in Sec. 1.7. These effects of clearance may be effectively eliminated by designing some deviation from circularity into one of the races (e.g., two, three, or more lobes) so that at least two limited sectors of the bearing circumference provide positive zero-clearance roller contact. But the race in these close-clearance contact areas is designed so that it may be elastically deformed to allow for transit of the rolling elements without excessive compressive load being applied.

Fourth, as in hydrodynamic film bearings, it is customary to include in rotordynamic calculations the inherent stiffness of rolling-element bearings. This level of stiffness is significant, particularly when one is dealing with stiff rotors. Since it is not convenient to generalize the stiffness information because of the number of variables involved, a typical calculation is given for a sample roller bearing whose dimensional data are summarized in Table 1.5.

TABLE 1.5 Dimensional Data for a Sample Roller Bearing

Races		Rollers	
Bore diameter	110 mm	Number of rollers	28
Outer diameter	155 mm	Roller diameter	11 mm
Width of races	20 mm	Overall length	11 mm
		Effective length	8.27 mm
		Diametral clearance	0.01493 mm

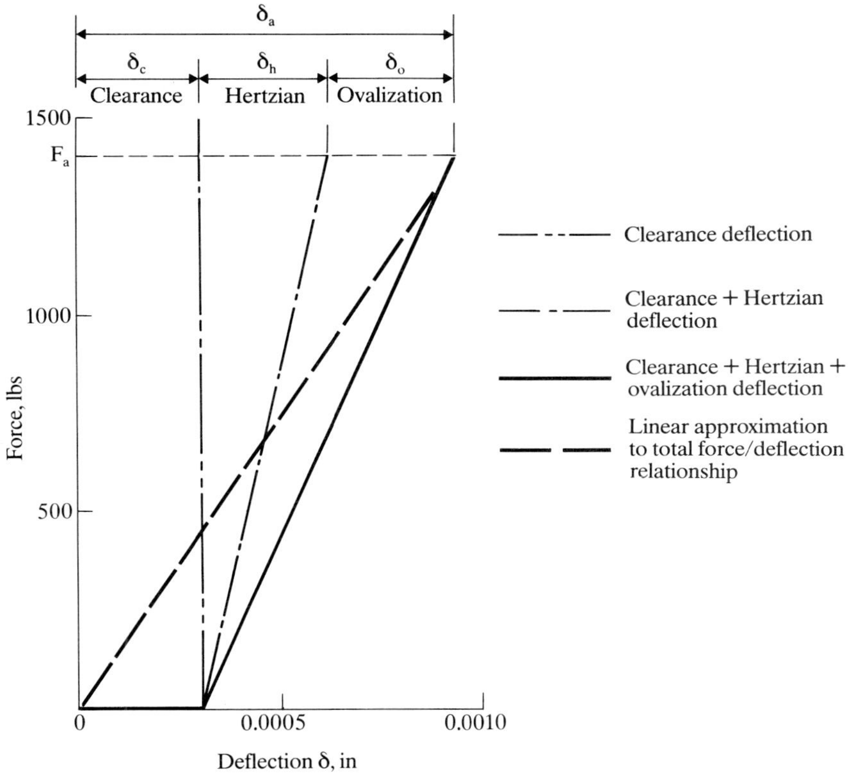

FIGURE 1.32 Stiffness characteristics of the sample roller bearing (see Table 1.5 for dimensional data).

Figure 1.32 shows a plot of force versus deflection for the sample bearing. There are three principal contributions to the deflection. The first is deflection through the radial clearance δ_c (where $\delta_c = 0.000294$ in). The second is deflection δ_h in line with the vertical force F_a due to the elastic compressions in the rollers and the inner and outer races at the points of contact, as depicted in Fig. 1.33. It is assumed that the inner surface of the inner race and the outer surface of the outer race remain circular for the purposes of this particular calculation. On this basis, the elastic compression associated with any given roller is proportional to the cosine of the angle between the roller and the vertical axis and varies from a maximum at the bottom position on the vertical centerline to zero at the horizontal centerline. The radial forces required to produce these elastic compressions in the rollers are determined by the well-known equations due to Hertz (see, e.g., Shaw and Macks, 1949, or Roark and Young, 1975). The force F_a is then equal to the vector sum of these radial forces. The hertzian equations are nonlinear, but for the sample calculation where $F_a = 1400$ lb the deviation from linearity is small. The average force-deflection characteristic is represented in Fig.

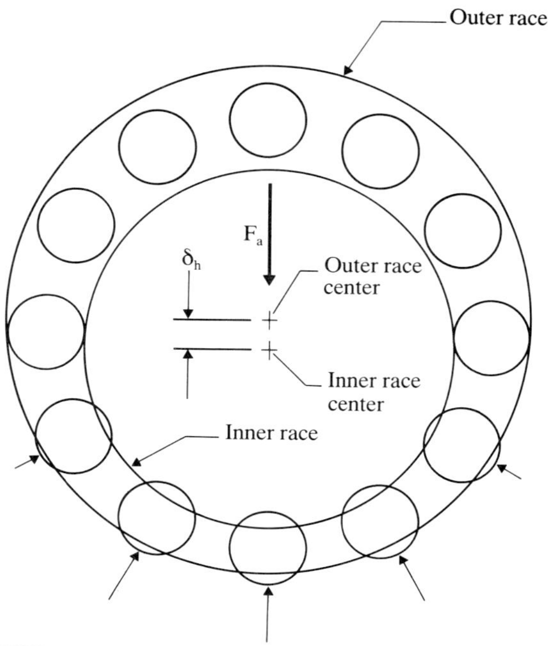

FIGURE 1.33 Bearing rolling-element (hertzian) deflection under load.

1.32 by a gradient of

$$\frac{F_a}{\delta_h} = \frac{1400}{0.000319} = 4.39 \times 10^6 \text{ lb/in}$$

The third contribution to deflection is the deflection δ_o due to ovalization of the bearing races associated with the nonuniform load distribution of the rollers. Included in this calculation are the elastic characteristics of the races, the rotor (which may be hollow) carrying the inner race, and the structure surrounding the outer race. The ovalization produces a change in the load distribution employed to evaluate δ_h, and an iterative solution is required to obtain the additional deflection δ_o to be added to δ_h. The combined effect of hertzian compression and ovalization is shown in Fig. 1.32 with a gradient equal to

$$\frac{F_a}{\delta_h + \delta_o} = \frac{1400}{0.000319 + 0.000320} = 2.19 \times 10^6 \text{ lb/in}$$

The total deflection δ_a due to the force F_a is therefore composed of three components: the radial clearance δ_c, the hertzian compression δ_h, and the ovalization δ_o. Because of the presence of the bearing clearance, the overall stiffness characteristic as defined in Fig. 1.32 is nonlinear. Figure 1.32 includes a linear approximation of the overall bearing stiffness characteristic with a gradient of

$$\frac{F_a}{\delta_a} = \frac{F_a}{\delta_c + \delta_h + \delta_o}$$

$$= \frac{1400}{0.000294 + 0.000319 + 0.000320} = 1.50 \times 10^6 \text{ lb/in}$$

If, in the course of the design process of placement of critical speeds, it is found that softer support stiffnesses are required than those provided by the bearings, it may be necessary to include in the design a flexible member at each bearing support whose stiffness can be accurately tailored to the values needed to achieve the required system natural frequencies. A typical flexible bearing support is shown in Fig. 1.34.

Fifth, in the case of hydrodynamic bearings, the rotating elements of the bearing are fixed to the rotor and can be balanced as part of the rotor system. In rolling-element bearings, the rollers and the cage or spacer rotate as a separate assembly at a speed approximately one-half that of the rotor. The rotor race may be balanced with the rotor, but the cage and roller assembly present a separate and additional balancing requirement. This need not be an imposing problem—rolling-element bearings are generally manufactured extremely precisely and have small diameters compared with the structure of the rotor which they support and therefore operate at a lower linear speed. Nevertheless, there have been reports of unusual asynchronous response (e.g., Yamamoto, 1959) attributed to slight variations in ball diameters and to cage imbalance.

1.4.3 Magnetic Bearings

Magnetic bearings can be used in high-speed rotors to improve stability by eliminating hydrodynamic bearings (Foster et al., 1986; Cataford and Lancee, 1986), without introducing life-limited rolling-element bearings. The magnetic bearing levitates the rotor by attraction created by opposed electromagnets, as illustrated in Fig. 1.35. As a passive device, the attraction mode of levitation is inherently unstable and would pull the rotor into contact with the magnet. However, magnetic bearings use measurements of the displacement between the rotor and the magnet and *actively* control the forces acting on the rotor, as illustrated in Fig. 1.36. Forces that are proportional to relative displacement yield effective stiffnesses; forces that vary in proportion to the relative velocity yield effective damping so that both stiffness *and* damping may be manipulated. Like tilting-pad bearings, magnetic bearings develop no destabilizing cross-coupled stiffness coefficients. In the normal sense of bearing analysis, they are inherently stable. They can, however, become unstable in the sense of a classical linear, sampled-data feedback control system. Magnetic bearings develop lower stiffnesses than hydrodynamic bearings. Incorporating magnetic bearings on a rotor markedly increases the rotor diameter and length at the bearing location. These changes in the rotor, in combination with a reduced bearing stiffness, typically yield a slight increase in the first critical speed, however, the second, third, and fourth critical speeds may be reduced considerably.

Because of the soft bearings and the controllable system damping, near-optimum bearing damping can be developed in the sense of Black (1975) and Barrett et al. (1978). Although not yet widely applied, magnetic bearings offer

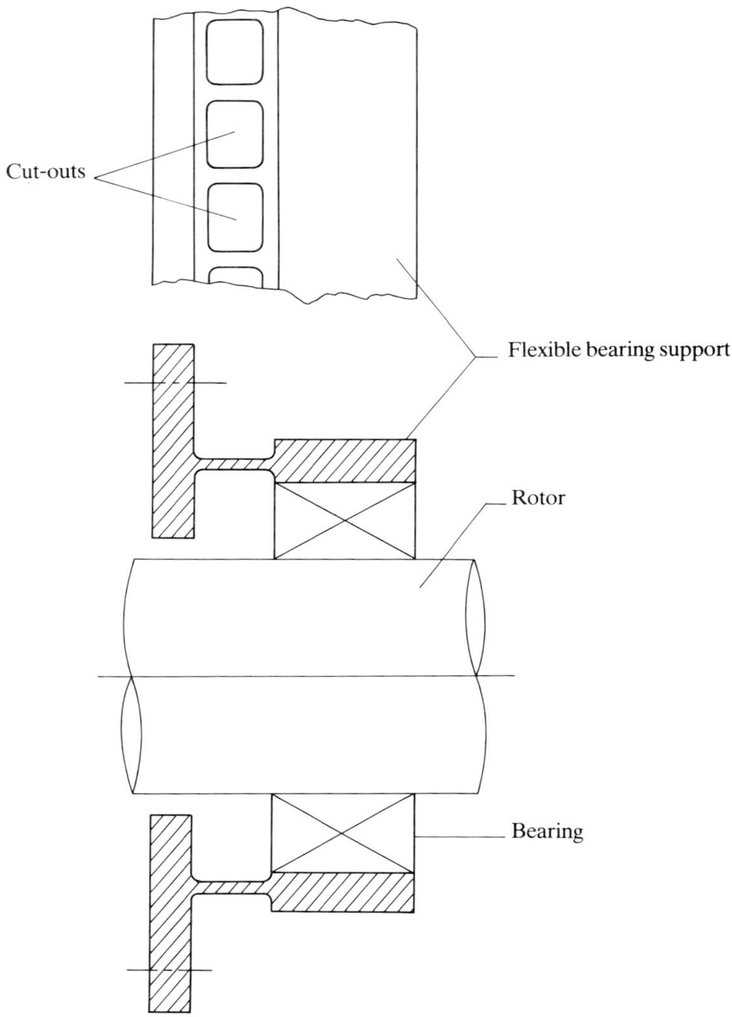

FIGURE 1.34 Typical flexible bearing support structure.

the prospects of designing stable rotors with sufficient damping to overcome destabilizing forces and even retrofit modifications of existing rotors to stabilize otherwise unstable systems.

1.4.4 Rotor Dampers

Squeeze-Film Dampers
The introduction of additional external damping is one of the general procedures for decreasing critical speed peak response and for eliminating rotordynamic

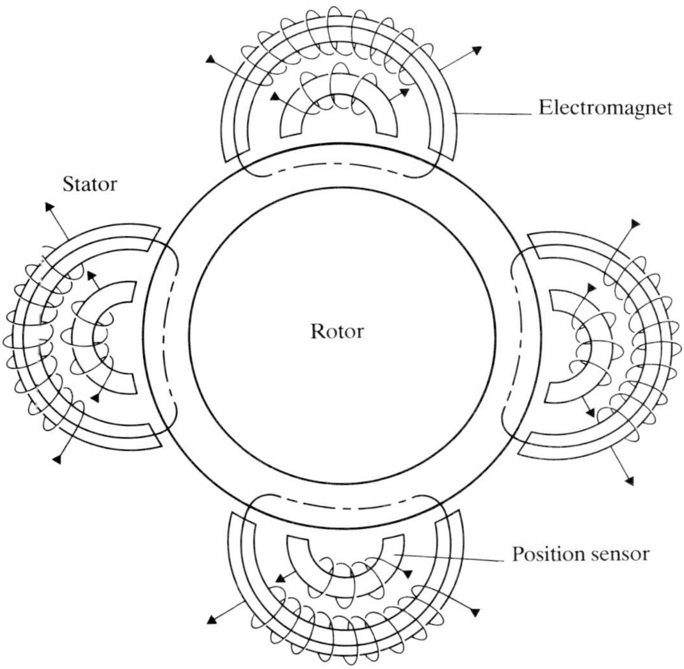

FIGURE 1.35 Magnetic bearing schematic (Cataford and Lancee, 1986).

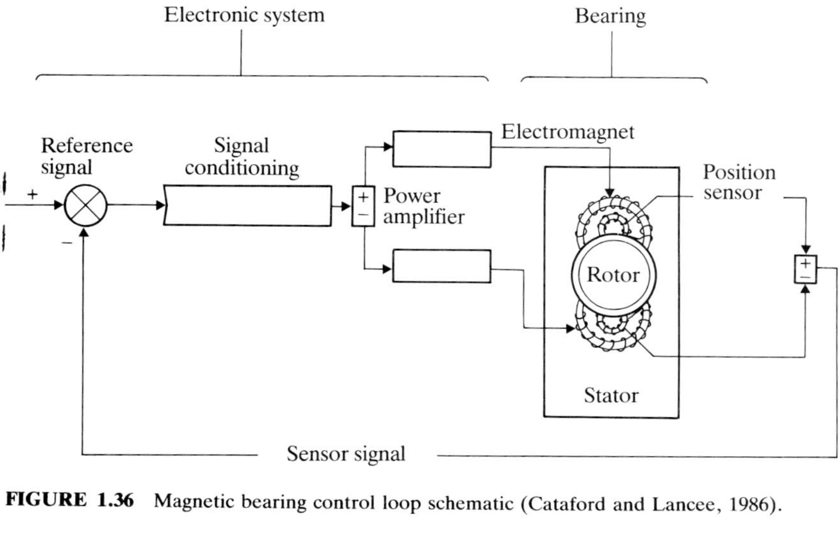

FIGURE 1.36 Magnetic bearing control loop schematic (Cataford and Lancee, 1986).

instability problems. The installation of "squeeze-film dampers" at the bearing supports represents the most common approach for adding damping to an unstable turbomachinery unit. Figure 1.37 illustrates such a damper. The outer

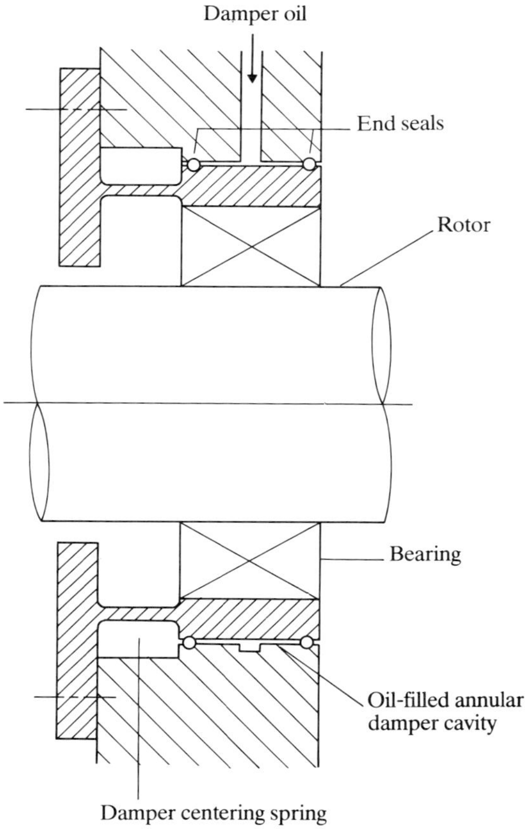

FIGURE 1.37 Squeeze-film damper.

race of a rolling-element bearing or the bearing segment of a hydrodynamic bearing is mounted in the inner element of the damper. A fluid film separates the inner and outer damper elements. Unlike a journal bearing, the inner damper element does not rotate, and there is no net fluid rotation to develop destabilizing cross-coupled stiffness coefficients. Dampers develop radial stiffness as a consequence of orbital motion of their inner element. For the purpose of stability improvement, most dampers use centering springs as illustrated; however, dampers without centering springs may be used for general vibration control if the damper is not located in a primary bearing support location. Dampers are generally supplied with oil at the center either through holes or through a central supply groove. Exit leakage at the ends is controlled by end seals which can either be O-ring or piston-ring designs.

The spring constant K_{sf} and the damping constant B_{sf} of the squeeze-film damper may be derived from a solution of the Reynold equation for a nonrotating journal bearing assuming laminar flow, full cavitation, a circular whirl orbit, and an inertialess oil film. The stiffness and damping of a squeeze film with no end flow (i.e., perfect and seals)—in what is referred to as *long journal bearing theory*—are found to be

$$K_{sf} = \frac{24R^3L\mu\omega\epsilon}{C^3(2+\epsilon^2)(1-\epsilon^2)} \tag{1.49}$$

$$B_{sf} = \frac{12\pi R^3L\mu}{C^3(2+\epsilon^2)(1-\epsilon^2)^{1/2}} \tag{1.50}$$

where R = squeeze-film radius
L = axial length of damper
C = radial clearance
ω = whirl speed
μ = oil viscosity
ϵ = eccentricity ratio: orbit radius/radial clearance

Similarly, for unrestricted end flow—in what is referred to as *short journal bearing theory*—the stiffness and damping of the squeeze film are found to be

$$K_{sf} = \frac{2RL^3\mu\omega\epsilon}{C^3(1-\epsilon^2)^2} \tag{1.51}$$

$$B_{sf} = \frac{\pi RL^3\mu}{2C^3(1-\epsilon^2)^{3/2}} \tag{1.52}$$

Figure 1.38 covers the application of Eqs. 1.49 through 1.52 for four different combinations of damper seals and oil feeds and shows the axial-pressure profiles in the active arc of the damper. With the center-groove oil feed (cases 3 and 4), the pressure at the center is equalized circumferentially and maintains the pressure level in the cavitated region. With the center hole feed, the connection with the cavitated region is broken, and squeeze-film pressure at a different level can develop at the center of the damper.

The stiffness and damping constants expressed in dimensionless form are plotted against eccentricity in Fig. 1.39 for the long journal bearing and in Fig. 1.40 for the short journal bearing. Both constants increase sharply at high values of eccentricity and tend to infinity at a unit value of eccentricity, where the rotor almost contacts the outer wall of the oil-film chamber.

Care must be exercised in using the simplified and approximate results expressed by Eqs. 1.49 through 1.52. Marmol and Vance (1978) have provided a comprehensive damper analysis based on the Reynold equation, and they show that more sophisticated fluid dynamic models are required for dampers operating at high speeds and high values of the squeeze-film Reynold number

$$\text{Re} = \frac{\rho\omega C^2}{\mu} \tag{1.53}$$

where ρ = oil density. For example, fluid inertia forces become significant at higher values of the squeeze-film Reynold number (San Andres and Vance, 1986*a, b*).

Case No.	Sealing and feeding configuration	Pressure profile	Spring and damping constants
1	Closed ends center hole feed	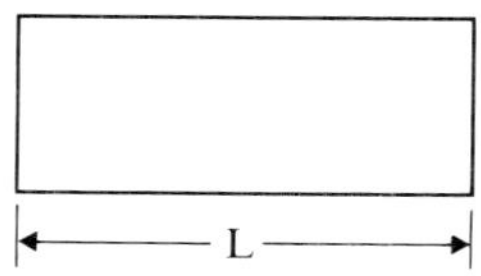	Long journal bearing theory
2	Open ends center hole feed	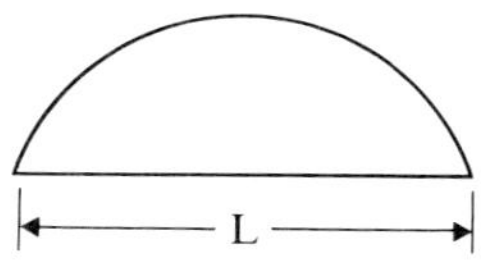	Short journal bearing theory
3	Open ends center groove feed	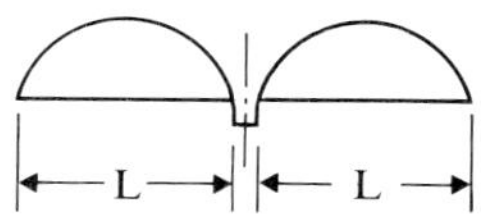	Short journal bearing theory
4	Closed ends center groove feed	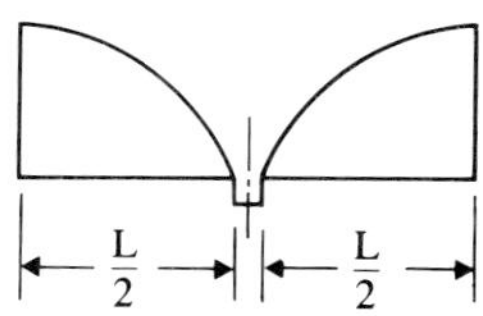	Short journal bearing theory Equivalent to Case 2

FIGURE 1.38 Squeeze-film damper seal and feed configurations applying long journal bearing theory (Eqs. 1.49 and 1.50, and short journal bearing theory (Eqs. 1.51 and 1.52).

Dampers must be designed with considerable care. They have the potential for significantly either improving or degrading response and stability characteristics of a rotor.

The damper must be located at a position along the rotor span which has a significant amount of relative dynamic deflection between the rotor and stator, to provide for damping action in any of the modes to be damped. That is, a damper must *not* be located at a nodal point in any of the modes that must be suppressed. The support stiffness of the centering spring and the clearance of the damper must be chosen carefully. If the clearance is too large or too small, stability and response can be markedly degraded. Gunter et al. (1987) provide an excellent discussion of this problem. Additional references can be found in Barrett et al. (1979).

Solid-State Dampers

Squeeze-film dampers, although they constitute an excellent approach to systematic and controlled vibration suppression, may be inconvenient or too costly to use in particular applications. A simple alternative is the application of a solid

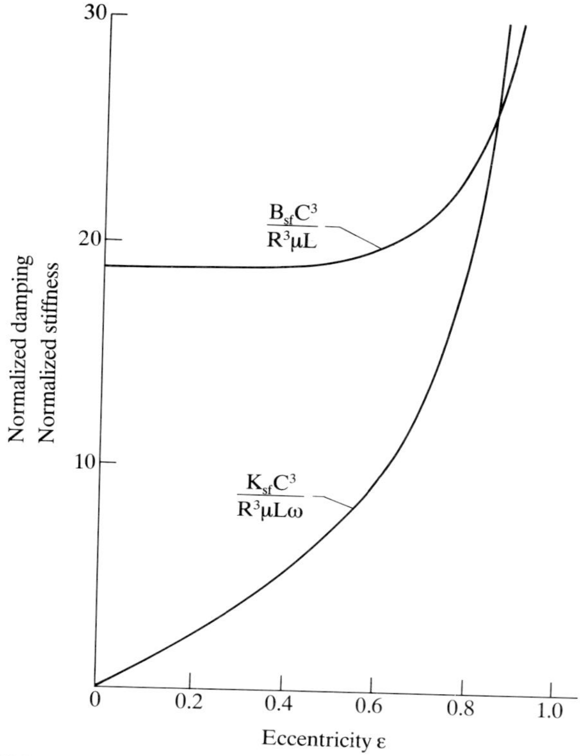

FIGURE 1.39 Squeeze-film damper constants versus eccentricity for long journal bearing.

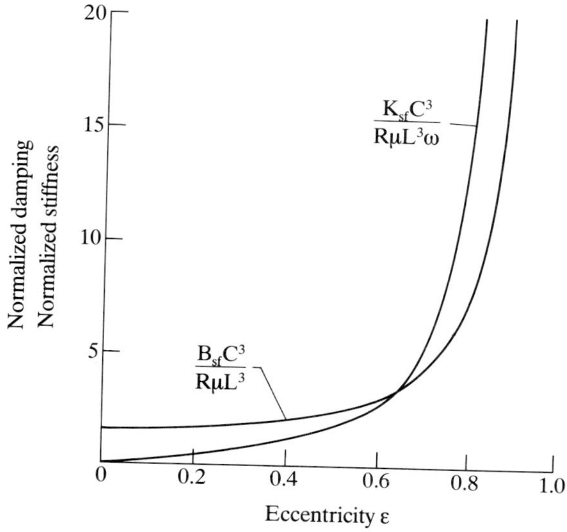

FIGURE 1.40 Squeeze-film damper constants versus eccentricity for short journal bearing.

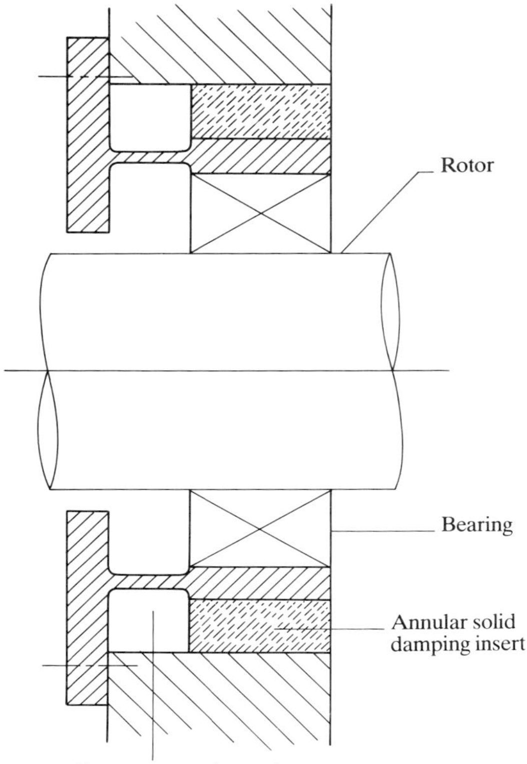

FIGURE 1.41 Solid-state damper.

damping compound such as elastomeric in an annular space in parallel with a centering spring, as shown in Fig. 1.41. The analogy to the squeeze-film bearing of Fig. 1.37 is clear, with the solid compound now substituted for the oil-filled cavity. Offsetting its obvious advantage of simplicity are its shortcomings—there is no fluid medium to carry away any generated heat, and the damping material may have a finite life and require periodic replacement, particularly if it operates in a hostile environment that is often typical of a rotor's shafting.

Active Dampers
Active dampers are devices in which forces, energized and controlled externally, can be applied to the rotor. If the amplitude, frequency, and phase of those applied forces can be manipulated on the basis of feedback signals of the rotor's deflection, then the equivalent of a damping force may be achieved. Magnetic bearings, although primarily conceived to provide basic rotor support, contain all the necessary ingredients to also supply artificial or active damping, as summarized in Sec. 1.4.3. Muszynska and Bentley (1987) have experimented successfully with a prototype of an active clearance control system which uses fluid jets in a concept attributed to Brown and Hart (1986). The fluid jets are

injected tangentially into a sealed cavity in a direction opposite to the shaft's rotation. If the jet velocities are controlled by a feedback servomechanism which senses rotor deflection, then stabilizing or damping forces can be generated by a mechanism which is the mirror image of the destabilizing forces generated by naturally generated corotational swirl.

Although not yet widely used, early prototype and experimental applications suggest that it has great promise.

1.4.5 Seal Construction and Operating Considerations

Floating Contact Seals

Floating contact seals are provided in fluid-handling machinery to both contain the process gas or fluid and separate it from the bearing lubricant. Figure 1.42 illustrates a typical floating seal configuration which seals on a cylindrical surface. The process gas is at the suction pressure P_s while oil is supplied at pressure $P_s + \Delta P$. The seal segments consist of outer and inner seals which are spring-loaded against each other. The preload of the spring causes the lapped external faces of the seal segments to be in contact with the seal-cartridge housing. An antirotation pin is provided to prevent rotation of the seal segments. The oil enters the seal cartridge between the two seal segments and then leaks axially along the shaft. Most of the oil is recovered; however, a portion is generally lost to the process gas stream. Oil seals are of interest in rotordynamic analysis of compressors because the fluid annuli between the inner and outer seals act as plain journal bearings. The ratios of radial clearance to journal radius for

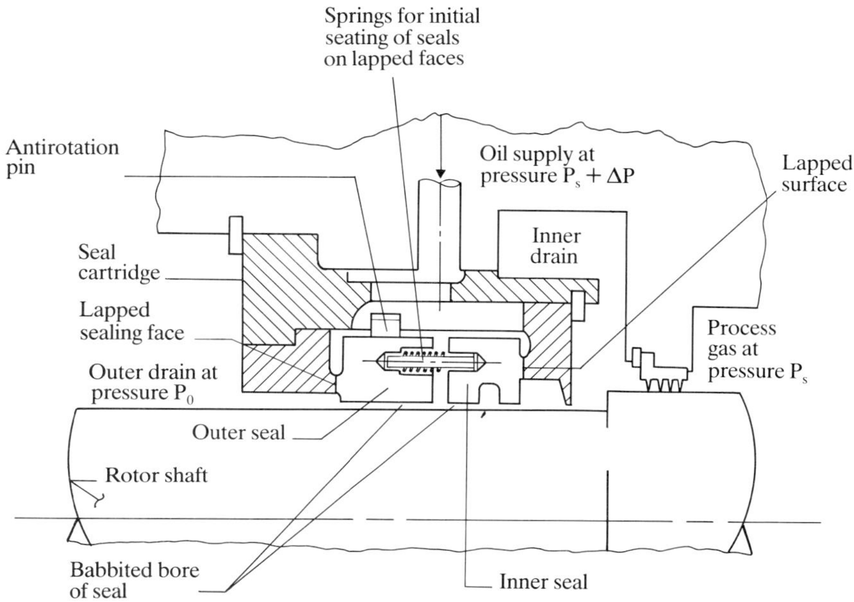

FIGURE 1.42 Oil-bushing breakdown seal (Kirk, 1986b).

these seals are comparable to those for a plain journal bearing. Experience indicates that the seals are not cavitated; hence they have negligible load capacity. Difficulties arise in predicting the static location of the seal rings relative to the rotor, as can be seen by looking at the force balance of Fig. 1.43. The

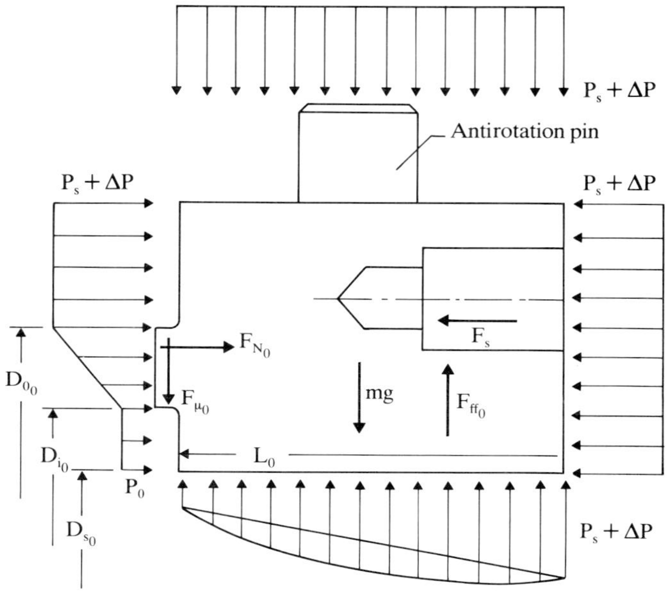

FIGURE 1.43 Force balance for a typical outer seal ring (Kirk, 1986b).

fluid-film force F_{ff0} developed by the relative position and motion of the bearing and seal ring is reacted by the friction force $F_{\mu 0}$ at the contact face. The friction force is defined in terms of the friction factor μ and normal force F_{N0} by

$$F_{\mu 0} = \mu F_{N0} \tag{1.54}$$

The normal force F_{N0} depends on the axial pressure balance and the preload spring force. A "balance seal" has a low normal force. For Fig. 1.43, a minimum value for thrust would be achieved by reducing the radius of the contact face until $D_{i0} = D_{s0}$. This would still leave an unbalanced pressure force on the seal ring.

Spiral-grooved face seals which seal on an annular face provide an alternative to the radial seal systems for primary gas sealing in high-pressure applications (Sedy, 1979, 1980). Figure 1.44 illustrates in (*a*) an assembly view of the two mating seal faces which are axially preloaded against each other by spring loading and in (*b*) the spiral-grooved sealing face on one of the mating rings. The spiral grooves create a separation force between the two mating faces that causes the seals to operate at extremely tight clearances between the rotating and stationary faces (0.0025 to 0.008 mm). These tight clearances result in very low leakage rates.

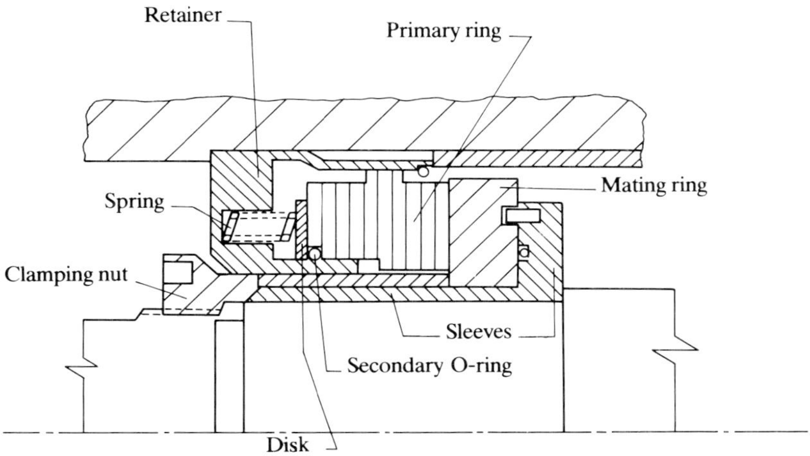

(a)

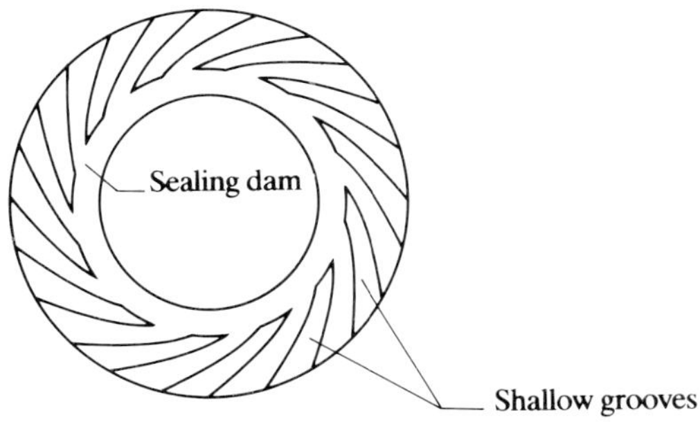

(b)

FIGURE 1.44 (*a*) Basic components of a spiral-grooved gas seal (Sedy, 1980); (*b*) spiral-grooved sealing face (Sedy, 1980).

The forces in this class of seals generally do not materially affect the natural frequency or the external damping of the system, so they are not usually involved in the rotordynamic design for response to forced vibration. But, under certain circumstances (e.g., when the friction force in the contacting support face locks the seal in an eccentric position), significant destabilizing forces can be generated. These circumstances are discussed in Sec. 1.6.2.

Labyrinth Seals

Figure 1.45 illustrates a typical labyrinth seal (in this case in a centrifugal compressor). The eye-packing seal restricts leakage along the front face of the impeller; the shaft-seal labyrinth restricts leakage along the shaft to the back side of the preceding impeller; and the balance piston drops the compressor discharge pressure back to the inlet pressure on a through-flow or in-line machine. On a

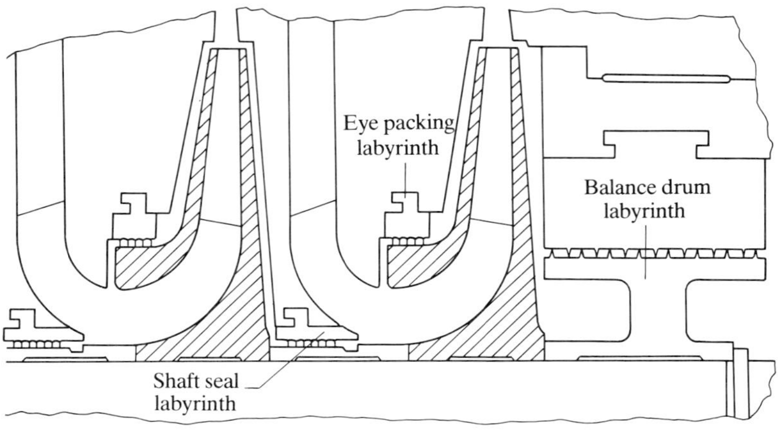

FIGURE 1.45 Compressor labyrinth seals.

back-to-back compressor rotor, about one-half of the machine's developed pressure is dropped across the center labyrinth. Note that the leakage flow on the back side of most impellers proceeds radially outward; however, on the last-stage impeller, leakage flow proceeds radially inward before entering the balance-drum labyrinth. As with floating seals, the fluid forces generally do not materially affect the natural frequency or the external damping of the system, so they are not usually involved in the rotordynamic design for response to forced vibration. But, under certain circumstances (e.g., high fluid swirl levels in the seal cavity), significant destabilizing forces can be generated. These circumstances are discussed in Sec. 1.6.2.

1.4.6 Couplings

Couplings are used to join the rotors of individual subassemblies, components, or subsystems of rotating machinery. If properly designed, couplings can decrease the sensitivity to eccentricity and/or angular misalignment of the centerlines of the components being coupled as well as permit relative axial motion between the rotors. There are at least three generic types of couplings.

The first is the single fixed joint or solid coupling, very often a spline connection that is locked up tight on a combination of pilot diameters and spline tooth contact surfaces to give the effect of a solid shaft. Such a joint must be assembled from parts with a high degree of precision and/or have provision for shimming or clocking adjustment in the course of assembly, to ensure sufficiently small angular misalignment and radial eccentricity between the centerlines of the two components being joined to give vibration below the allowable threshold. Where three or more bearings support the combined components (i.e., in a statically redundant rotor support system), it is advisable to provide for relatively flexible shafting on either side or both sides of the joint that will relieve the reaction forces of such a statically indeterminate mounting system, since such support reaction forces rotate in operation and excite vibration in the stator system. Care must also be taken to lock the joint securely, so that relative motion between contact surfaces is inhibited. Any relative motion between the contact surfaces of the rotor parts can cause rotordynamic instability (see "Hysteretic Whirl" in Sec. 1.6.2).

The second type of coupling is the single working joint. A splined joint with crowned teeth and a crowned tip diameter can allow large relative axial motion and modest amounts of angular misalignment between the centerlines of the components being coupled. It may be necessary to lubricate the spline to inhibit hysteretic whirl (Marmol et al., 1979). A joint composed of two orthogonal diametral hinges (referred to as a *Hookes joint*) will accommodate only angular misalignment. If the hinges are composed of flex plates, then lubrication will generally be unnecessary. Such double-hinged joints do not transmit rotational motion smoothly and can excite torsional vibration at twice the rotation frequency.

The third type of coupling is a distance piece with a joint at either end. Such an arrangement is capable of accommodating the entire range of possible misalignments—relative axial motion, angular misalignment, and radial eccentricity between the centerlines of the coupled components. The same types of elements (sliding spline or flex plate, etc.) as are used in single joints can be used with the same cautions and limitations. The distance piece constitutes an additional rotordynamic element with its own characteristic natural frequencies that must be accommodated and with its own requirements for balancing.

1.4.7 Rotor Design

Rotor Geometric Stability

Assembly Stability. If a rotor is assembled from many individual parts, it is subject to several potential vibrational hazards.

First, the relative concentricity of the individual rotor parts is controlled by a rabbet (often with an interfering shrink fit at assembly) or an axial toothed coupling. In either case, the assembly must be locked up with a single center tie bolt or multiple individual bolts (which may be body-bound) at an intermediate diameter. Either arrangement provides a normal force at the contact surfaces which prevents relative motion by way of surface friction. There are invariably internal forces present in the rotor, generated by internal unbalances and/or thermal effects, which act to unseat the contacts. If the normal forces produced by the bolt(s) at the contact surfaces are insufficient to counteract the internal forces that tend to unseat the joint, a permanent shift in their relative position may take place with a consequent shift in unbalance.

Second, any relative motion at the contact interface involves a destabilizing energy dissipation which can give rise to an unstable vibration of the rotor, termed *hysteretic whirl* (see Sec. 1.6.2).

Third, if keyways are used to transmit torque through assembled joints, some care must be taken to ensure that no major axial asymmetry (i.e., having dissimilar bending stiffnesses about two orthogonal diametral axes) is introduced to the rotor shafting which might give rise to parametric instability (see Sec. 1.6.3).

Operational Stability. Any rotor, whether monolithic or assembled, whose operational temperature environment varies significantly from ambient conditions is subject to consequent unbalance conditions. Two primary circumstances can give rise to shifts in unbalance.

First, if key elements of the rotor are not homogeneous and/or isotropic, changes in temperature may induce thermal distortions which change the rotor balance. Alternatively or additionally, internal stresses in the rotor from its in-process heat treatment may be relieved in the course of operation, also engendering a shift in unbalance.

Second, when a hot rotor is shut down, gravitionally forced convective air currents will tend to cool the bottom side of the rotor more rapidly than the upper side, inducing in the rotor a thermal bow. If the machine is started during this cooling period (when the rotor is neither uniformly hot nor uniformly cool), it will suffer the consequences of the thermally induced unbalance. If the condition cannot be accommodated in design by incorporating short unsupported spans of the rotor, then restarts may have to be restricted to very short intervals or very long intervals to avoid the intermediate period when a bowed rotor can be expected. As an ultimate extreme (which may actually be standard practice in some steam-turbine uses), provision may have to be made to continue to motor the rotor slowly during cooldown so that it cools uniformly.

Production and Field Balance Provisions

Provision must be made in the design of any high-speed rotor to balance the rotor after its final assembly, as a component and/or as a fully assembled system in situ in the operational system, either as the final step in its manufacturing cycle or in the field as a part of its routine maintenance or in correction of a problem in the course of service. When the appropriate balancing scheme and the number and location of balancing planes have been selected (see Chap. 3) and the range of prospective balance correction has been estimated, then the design must include provision for making such balance corrections to the rotor. This involves provision for addition and/or subtraction of small quantities of weight to or from the rotor at as large a radius as is possible, with latitude for adding or subtracting that weight at any required circumferential angle. The location and nature of the balance correction provision must also accommodate the circumstances of the prospective balancing action. Balancing in the field must be made feasible with a minimum of machine disassembly.

A typical approach to balance correction is the provision of an otherwise nonfunctional rib or annular protrusion in the balance planes from which material can be ground or machined away at the appropriate circumferential locations. This approach has the virtue of not requiring any disassembly/reassembly operation. It also ensures that there will be no shifting of the unbalance in the course of high-speed operation. This approach has been adapted to automated balance correction where material is removed from the rotor by laser melting while rotating in the balance machine. Material removal schemes do have the limitation that balance correction cannot be undone where rebalance of a rotor becomes necessary. Care must be exercised to avoid introduction of stress concentration in highly stressed areas of the rotor in the course of material removal.

Another approach to balance correction is that provision is made for addition of balance weights to the rotor in the balance planes at the appropriate circumferential locations. This can be done by providing for assembly of washers of variable number of variable weight under the head or nuts of rotor assembly bolts, or by selection of intentionally graded weights of rotor parts such as turbomachinery blades. Where it is inconvenient or inappropriate to provide multiple weights in fine enough gradation to accomplish the required balancing precision, the effect may be accomplished by using two standard weights, each having a weight equal to one-half the maximum correction weight required. The effect of a weight correction less than the maximum can then be obtained by offsetting the two weights at circumferential angles $\pm\theta$ symmetrically about the radial line at which the net correction is to be made. The effective correction is then the full weight times cos θ—giving the full correction with $\theta = 0°$ and zero correction with $\theta = 90°$.

1.4.8 Sensors for in Situ Balancing, Condition Monitoring, and Diagnostics

The balance procedure also requires one or more sensors of vibration activity of the rotor. These may be transducers that sense the instantaneous relative distance between the rotor and the stator at some sensitive axial location, the instantaneous reaction force in the bearings, or the instantaneous acceleration or velocity of some sensitive member of the stator. For a machine that is to be balanced in its fully assembled state, these sensors may be embedded in the machine and designed as a permanent part of its assembly. As such, they are also potentially very useful for on-line continuous or periodic condition monitoring or for diagnostic testing in the event of problems in service. Such sensors are described in detail in Sec. 4.2.1.

1.5 STATOR MOUNTING AND ISOLATION

In many situations, the machine designer is required to isolate the rotating machinery from the environment. The transmission of vibration from the rotating element to the surrounding environment or from the environment to the rotating element is influenced by the stator mounting, stator mass, and dynamic characteristics of the bearings. For the purpose of demonstrating basic concepts, a single-span two-bearing machine as depicted in Fig. 1.11 has been selected. To simplify the analysis, this generic model with distributed mass along the spans of both the rotor and the stator is replaced by the model with massless shafting and lumped masses, as shown in Fig. 1.46. Dealing with a rotor with a concentrated mass (i.e., a Jeffcott rotor) allows us to use the equations developed in Sec. 1.2.1 for the evaluation of transmissibility. Lumping the stator mass with half the total mass concentrated at each of the two bearings yields an idealized rotor support structure identical to that of Sec. 1.3.1, so that relationships from that section also may be employed. The rotor support system, shown in Fig. 1.46 in the vertical plane only, is assumed to be axisymmetric, so both the rotor and the stator will whirl in circular orbits, and consideration of any single plane is representative of any other plane. Because of the similarity in the modeling, the equations developed in Secs. 1.2.1 and 1.4.1 may then be applied directly to evaluate the transmissibility of the vibration energy.

In this simple idealized model, with the only exciting force due to a single ("static") unbalance in the rotor mass, there are two response modes—a lower-speed mode characterized by in-phase whirling of the rotor and stator masses and a higher-speed mode characterized by out-of-phase whirling. In the more complex case where there may be a moment unbalance caused by two or more dissimilar point unbalances displaced from one another on the shaft (i.e., a "moment" unbalance), there will be a third mode (not considered here) where the

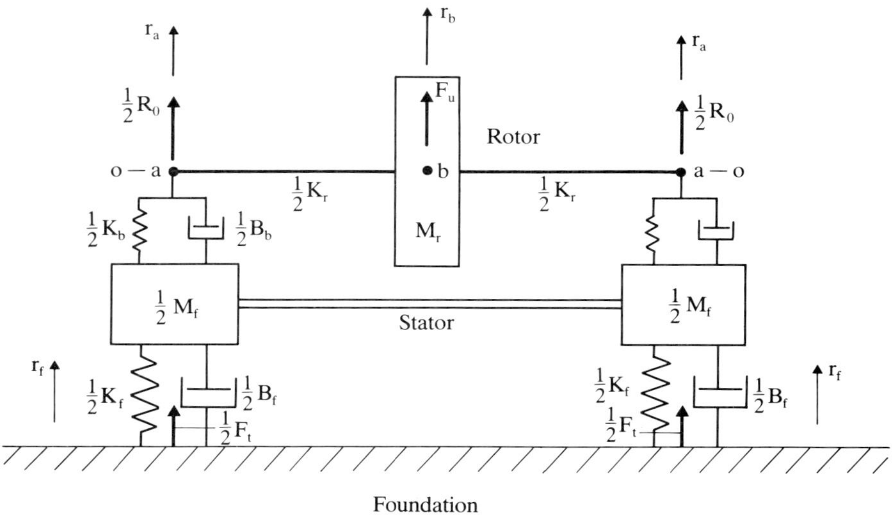

FIGURE 1.46 Simplified model of a single-span two-bearing machine.

two stator masses will vibrate out of phase with one another, as described in Sec. 1.3.1. If the undamped natural frequency of the stator mass on its supporting spring is low relative to the rigid bearing critical speed of the rotor, then the rotor will behave as a rigid body in the lower-speed mode.

1.5.1 Isolation of the Environment from the Machine

The Jeffcott Rotor on Rigid Bearing Supports

We first consider the special case of the system shown in Fig. 1.46 with rigid bearing supports, dealt with previously in Fig. 1.1*a*. The *transmissibility* T is defined as the ratio of the vibratory force R_0 transmitted to the foundation to the exciting force on the rotor F_u. A convenient starting point for evaluating T is the vector diagram of Fig. 1.3. In evaluating vibration isolation, only the amplitudes are of interest, and phase angles are usually ignored. Since R_0 is equal to the vector sum of the spring force F_s and the damping force F_d, the transmissibility is

$$\begin{aligned} T &= \frac{R_0}{F_u} = \frac{(F_s^2 + F_d^2)^{1/2}}{F_u} \\ &= \frac{F_s}{F_u}\left[1 + \left(\frac{F_d}{F_s}\right)^2\right]^{1/2} \end{aligned} \tag{1.55}$$

Making use of Eqs. 1.1, 1.3*a*, 1.3*b*, 1.6, 1.8, and 1.10 yields

$$\begin{aligned} T &= \left(\frac{w_{b1}}{\tau^2}\right)[1 + (2\zeta\tau)^2]^{1/2} \\ &= \left[\frac{1 + (2\zeta\tau)^2}{(1 - \tau^2)^2 + (2\zeta\tau)^2}\right]^{1/2} \end{aligned} \tag{1.56}$$

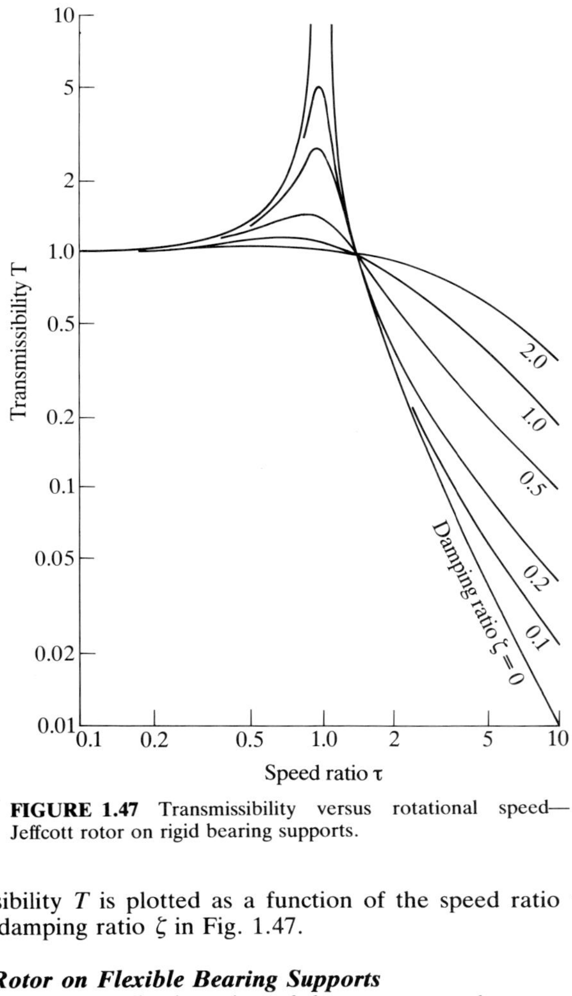

FIGURE 1.47 Transmissibility versus rotational speed—Jeffcott rotor on rigid bearing supports.

The transmissibility T is plotted as a function of the speed ratio τ for selected values of the damping ratio ζ in Fig. 1.47.

The Jeffcott Rotor on Flexible Bearing Supports

A somewhat more generalized version of the most general system shown in Fig. 1.46 is had by the addition of flexible bearing supports, as shown in Fig. 1.1*b*. As in the rigid bearing case, the transmissibility T is given by Eq. 1.55. Making reference to the vector diagram of Fig. 1.5 and using the expressions given in Eqs. 1.6, 1.13, 1.14, 1.16, 1.23, and 1.25, we find for the transmissibility T

$$\begin{aligned} T &= \left(\frac{w_a}{\sigma^2}\right)[1 + (2\eta\sigma)^2]^{1/2} \\ &= \left\{\frac{1 + (2\eta\sigma)^2}{[1 - (1+\kappa)\sigma^2]^2 + 2\eta\sigma(1 - \sigma^2\kappa)}\right\}^{1/2} \end{aligned} \tag{1.57}$$

The speed ratio τ is generally the more convenient ratio for the designer to deal with since the rotational speed is related to the rigid bearing critical speed of the rotor v. Equation 1.57 may be expressed in terms of the speed ratio τ by making use of the relationship between τ and σ as given in Eq. 1.22. We find then for the transmissibility

$$T = \left\{ \frac{\kappa^2 + \kappa(2\eta\tau)^2}{[\kappa - (1+\kappa)\tau^2]^2 + \kappa[2\eta\tau(1-\tau^2)]^2} \right\}^{1/2} \tag{1.58}$$

Because of the presence of the extra parameter (the stiffness ratio κ), the transmissibility T can no longer be represented by a single family of curves, as was the case for the rigid bearing rotor. Typical plots of T versus the speed ratio τ are presented in Fig. 1.48 for the stiffness ratio κ equal to $\frac{1}{8}$, 1, and 8. For each value of κ, the optimum value of the damping ratio η', as determined from Fig. 1.6, is used. As the rotor becomes more flexible relative to the support (i.e., as the value of κ increases), the system becomes more responsive to unbalance and the transmissibility increases markedly. At the same time, the rotational speed for

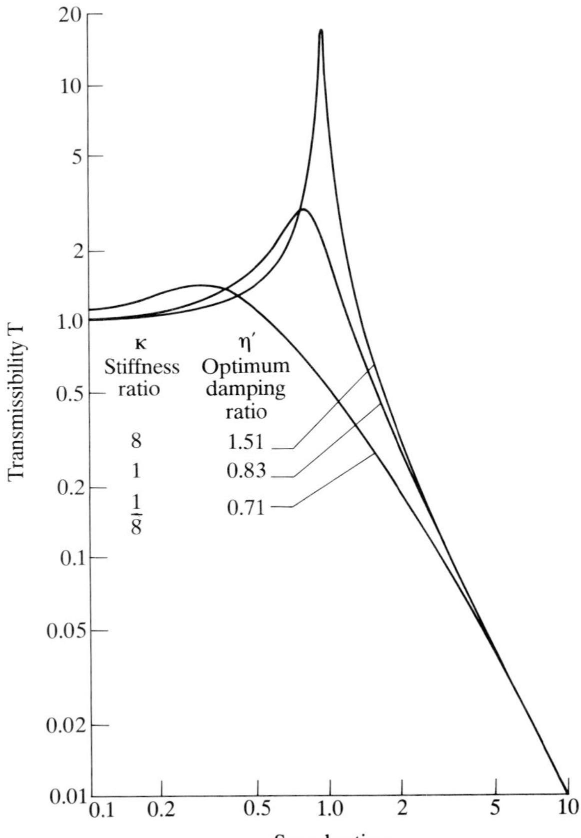

FIGURE 1.48 Transmissibility versus rotational speed—Jeffcott rotor on flexible bearing supports.

maximum transmissibility increases and approaches as a limit the rigid bearing critical speed where τ equals unity.

The Jeffcott Rotor on Flexible Bearings and a Flexibly Mounted Stator Mass The overall transmissibility T for the complete system of Fig. 1.46 can be determined by a two-step procedure using the results already obtained for the simpler Jeffcott rotor systems. The transmissibility T is equal to the product of the two transmissibility factors T_1 and T_2:

$$T = T_1 T_2 \tag{1.59}$$

where T_1 is equal to the bearing reaction R_0 acting on the stator mass M_f divided by the rotor unbalance force F_u and may be evaluated by Eq. 1.58 (derived for the Jeffcott rotor system of Fig. 1.1*b*). However, for the case at hand, "effective" values for the support stiffness K_s and the support damping B_s as derived in Sec. 1.4.1 must be used to properly account for the dynamics of the stator mass M_f. By substituting the identities $\kappa = K_s/K_r$ and $\kappa^{1/2}(2\eta\tau) = \omega B_s/K_r$ in Eq. 1.58, the following expression is obtained for the transmissibility factor T_1:

$$T_1 = \frac{R_0}{F_u} = \left\{\frac{(J_1J_2)^2 + (J_1J_3)^2}{[J_1J_2 - (1 + J_1J_2)\tau^2]^2 + [J_1J_3(1 - \tau^2)]^2}\right\}^{1/2} \tag{1.60a}$$

where

$$J_1 = \frac{K_b}{K_r} \tag{1.60b}$$

$$J_2 = \frac{K_s}{K_b} \qquad \text{(see Eq. 1.44)} \tag{1.60c}$$

$$J_3 = \frac{\omega B_s}{K_b} \qquad \text{(see Eq. 1.45)} \tag{1.60d}$$

And T_2 is equal to the force F_t transmitted to the foundation divided by the bearing reaction R_0 and may be evaluated by Eq. 1.56 which is applicable to the Jeffcott rotor system of Fig. 1.11*a*. Note that this Jeffcott rotor system is dynamically equivalent to that of the stator mass on its spring and dashpot. With the appropriate changes in nomenclature, Eq. 1.56 becomes

$$T_2 = \frac{F_t}{R_0} = \left\{\frac{1 + (2\eta_f\omega/\omega_f)^2}{[1 - (\omega/\omega_f)^2]^2 + (2\eta_f/\omega_f)^2}\right\}^{1/2} \tag{1.61a}$$

where

$$\omega_f = \left(\frac{K_f}{M_f}\right)^{1/2} \tag{1.61b}$$

and

$$\eta_f = \frac{B_f}{2M_f\omega_f} \tag{1.61c}$$

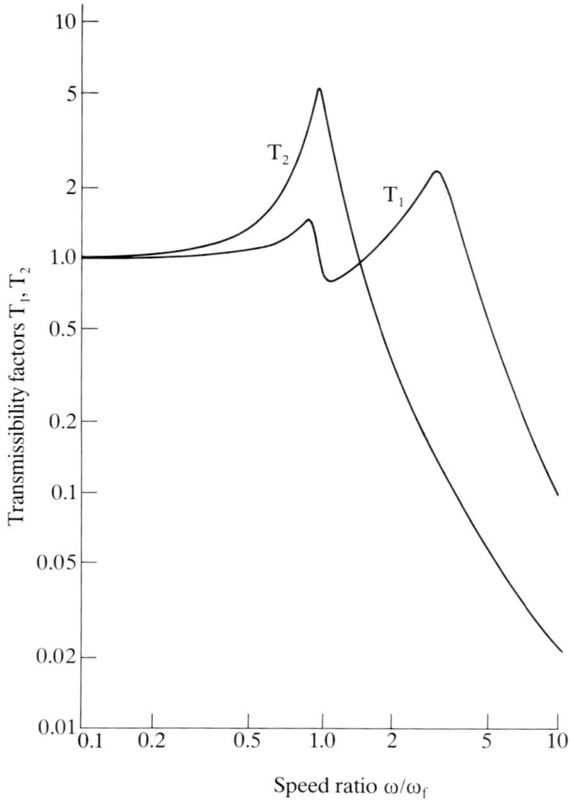

(a)

FIGURE 1.49 (*a*) Transmissibility factors versus rotational speed—high-pressure steam turbine using simplified model; (*b*) transmissibility versus rotational speed—high-pressure steam turbine using simplified model.

Calculated values of T_1, T_2, and T (from Eqs. 1.60, 1.61, and 1.59) are plotted in Fig 1.49*a* and *b* as a function of the speed ratio ω/ω_f for a specific typical system—the high-pressure steam turbine of case 1 treated in Sec. 1.3.1—with system constants taken as

$$M_r g = 2195 \text{ lb} \qquad M_f g = 17{,}560 \text{ lb}$$

$$\frac{60\nu}{2\pi} = 5291 \text{ rpm} \qquad \frac{60\omega_f}{2\pi} = 1500 \text{ cycles/min}$$

$$K_b = 1.461 \times 10^6 \text{ lb/in} \qquad g = 386 \text{ in/s}^2$$

$$G = 1.7 \qquad \eta_f = 0.1$$

and in Eq. 1.60*a*, $\tau = (\omega_f/\nu)(\omega/\omega_f) = (\omega/\omega_f)/3.53$.

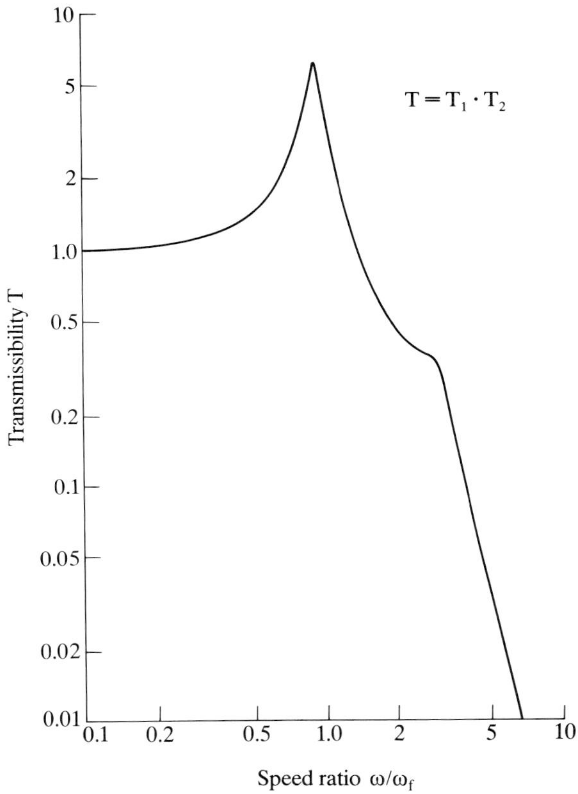

(b)

FIGURE 1.49 (*Continued.*)

Of particular interest is the influence of the critical speeds of the system on the overall transmissibility T. At the speed of the low-frequency translational mode where the speed ratio ω/ω_f is approximately equal to unity, the transmissibility is relatively high. However, the effect on the system's environment will not pose a problem in situations of which this system is typical, because the rotor is a transcritical design and the translational mode in question occurs at a low fraction of the operating speed. At the flexural mode of the rotor where the speed ratio ω/ω_f is about 3.1 and at the maximum operating speed where ω/ω_f is about 4.4, the transmissibility for this particular system is acceptably low. The large value of stator mass to rotor mass and the bearing damping are the important factors in producing this result.

1.5.2 Isolation of the Machine from the Environment

In some situations, such as in rotating equipment mounted aboard vehicles subject to large dynamic loads, the designer is required to isolate the rotating machinery from the environment. In dealing with the transmission of vibration from the machine to the environment in the previous section, the transmissibility was expressed as a ratio of transmitted force to exciting force. For the reverse problem of transmission of vibration from the environment to the machine's interior, the transmissibility is defined by the ratio of the vibration amplitude of the machine element to the vibration amplitude of the support or foundation on which the machine is mounted. For linear systems, the two transmissibilities are numerically equal (see, e.g., Crede and Ruzicka, 1988). In terms of the nomenclature of the forces and amplitudes shown in Fig. 1.46, this equality of transmissibilities can be expressed as

$$T = \frac{F_t}{F_u} = \frac{r_b}{r_f} \tag{1.62}$$

where

$$r_b = \text{whirl radius or vibration amplitude of the rotor mass} \tag{1.63}$$

and

$$r_f = \text{whirl radius or vibration amplitude of the foundation} \tag{1.64}$$

The derivation of the expression for the transmissibility from the machine to the environment is based on the assumption that the system is axisymmetric and that the motion at all locations is a circular orbit. We might expect this limitation to also limit its application to the case of transmissibility of circular whirling excitation from the environment to the machine, which would preclude its applicability to the very real and important cases of planar excitation from the environment. Fortunately, there is a mathematical correspondence between planar beam vibration and the whirling motion of an unbalanced rotor (within certain limitations, which are discussed in Sec. 1.2.1). We then find that if the vibration of the foundation is planar, the vibration of the spinning rotor (in good balance) will also be planar, and so the transmissibility information in Sec. 1.5.1 may also be applied to this situation to a first approximation.

1.6 ROTORDYNAMIC INSTABILITY AND SELF-EXCITED VIBRATIONS*

1.6.1 Introduction

The primary consideration in the design of high-speed rotating machinery in the context of rotordynamics is to control and minimize response to forced vibration (and, in particular, rotor unbalance), as described in preceding sections. But there exists another class of vibration—rotordynamic instability and self-excited vibration—which involves an additional set of design approaches, requirements, and constraints to ensure trouble-free, quiet, and durable rotating machinery (Booth, 1975; Childs, 1978; Ek, 1980). Rotordynamic instabilities and self-excited

* Portions of this section are derived from Ehrich (1987), Ehrich and Childs (1984), and Childs (1987).

vibrations generally take the form of lateral flexural vibrations at a rotor's natural or critical frequency different from, and most often below, the running speed. This "subsynchronous" or "supercritical" vibration generally increases so sharply with increasing running speed or power that if the vibration is monitored, additional increases in running speed are deemed impossible; or if the vibration is not monitored, the equipment is damaged severely or destroyed. Instabilities are expensive in terms of both the delays which they imposed on major engineering projects and redesigns and modifications which are necessary to solve the problems. Instabilities impose a continuing restraint on the performance capabilities of rotating machinery and continue to create difficulties in the design and operation of high-performance rotating machinery. Although major uncertainties persist with regard to the forces which cause unstable motion, a considerable degree of understanding has been achieved concerning this type of motion.

Self-excited systems begin to vibrate of their own accord spontaneously, with the amplitude increasing until some nonlinear effect limits further increase. The energy supplying these vibrations is obtained from a uniform source of power associated with the system which, due to some mechanism inherent in the system, gives rise to oscillating motion. The nature of self-excited vibration compared to forced vibration is as follows: In self-excited vibration, the destabilizing force that sustains the motion is created or controlled by the motion itself; when the motion stops, the alternating force disappears. The self-excited system will vibrate at its own natural frequency, independent of the frequency of any external stimulus. In forced vibration, the sustaining alternating force exists independent of the motion and persists when the vibratory motion is stopped. The system in forced vibration will vibrate at the frequency of the alternating force or at a frequency directly related to that of the alternating force.

The principal concern of this section is rotating machinery, specifically the self-excitation of lateral, or flexural, vibration of rotating shafts (as distinct from torsional, or longitudinal, vibration).

In addition to the description of a large number of such phenomena in standard vibrations textbooks (most typically and prominently by Den Hartog, 1956), the field has been subject to several generalized surveys (Ehrich, 1987; Ehrich and Childs, 1984; Childs, 1987; Kramer, 1972; Vance, 1974). The mechanisms of self-excitation which have been identified can be categorized as follows:

- Whipping and whirling
- Parametric instability
- Stick-slip rubs and chatter
- Instabilities in forced vibrations

In each instance, the physical mechanism is described and aspects of its prevention or its diagnosis and correction are given.

1.6.2 Whirling and Whipping

General Description

In the most important subcategory of instabilities (generally termed *whirling* or *whipping*), the unifying generality is the generation of a tangential force, normal to an arbitrary radial deflection of a rotating shaft, whose magnitude is proportional to (or varies monotonically with) that deflection. At some "onset" rotational speed, such a force system will overcome the stabilizing external damping forces which are generally present and induce a whirling motion of ever-increasing amplitude, limited only by nonlinearities in stiffness and in damping which ultimately limit deflections.

In the case of a whirling or whipping shaft, the equations of motion (for an idealized shaft with a single lumped mass m) are most appropriately written in polar coordinates for the force balance shown in Fig. 1.50 for the radial force

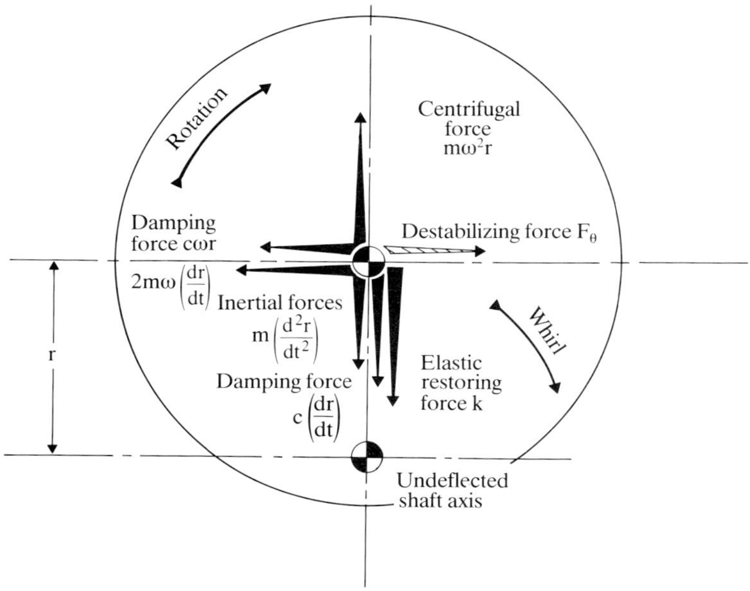

FIGURE 1.50 Force balance for a whirling or whipping shaft.

balance

$$-m\omega^2 r + m\frac{d^2r}{dt^2} + c\frac{dr}{dt} + kr = 0 \tag{1.65}$$

and for the tangential force balance

$$2m\omega\frac{dr}{dt} + c\omega r - F_\theta = 0 \tag{1.66}$$

where we presume a constant rate of whirl ω.

In general, the whirling is predicated on the existence of some physical phenomenon which will induce a force F_θ that is normal to the radial deflection r and in the direction of the whirling motion, i.e., in opposition to the damping force, which tends to inhibit the whirling motion. Often this normal force can be characterized or approximated as being proportional to the radial deflection:

$$F_\theta = k_{r\theta} r \tag{1.67}$$

where the constant is often termed the *cross-coupled stiffness coefficient* since it related the magnitude of the force to a deflection normal to that force. The solution then takes the form

$$r = r_0 e^{at} \tag{1.68}$$

For the system to be stable, the coefficient of the exponent

$$a = \frac{k_{r\theta} - c\omega}{2m\omega} \tag{1.69}$$

must be negative, giving the requirement for stable operation as

$$k_{r\theta} \leq c\omega \tag{1.70}$$

or, in dimensionless form,

$$\frac{k_{r\theta}}{\omega^2 m} \leq \frac{c}{\omega m} = 2\zeta \tag{1.71}$$

As a rotating machine increases its rotational speed, the left-hand side of this inequality (which is generally also a function of shaft rotation speed) may exceed the right-hand side, indicative of the onset of instability. The implication of this relationship is that increasing the stator system's damping ratio is fundamental to pushing the onset of instability to a higher speed. It has also been shown that introduction of anisotropy (differing values of stiffness k_{xx} and k_{yy} between two orthogonal planes) is also effective in stabilizing the system, as illustrated in Fig. 1.51. At the onset of instability

$$a = 0^+ \tag{1.72}$$

so that the whirl speed at onset is found, from Eq. 1.65, to be

$$\omega = \left(\frac{k}{m}\right)^{1/2} \tag{1.73}$$

That is, the whirling speed at the onset of instability is the shaft's natural or critical frequency, irrespective of the shaft's rotational speed. The direction of whirl may be in the same rotational direction as the shaft rotation (forward whirl) or opposite to the direction of shaft rotation (backward whirl) depending on the direction of the destabilizing force F_θ.

When the system is unstable, the solution for the trajectory of the shaft's mass is from Eq. 1.58, an exponential spiral as in Fig. 1.52*a*. Any planar component of this two-dimensional trajectory (also shown in Fig. 1.52*a*) takes the same form as the unstable planar vibration.

This unstable subsynchronous rotor motion is frequently referred to as rotor *whirling* or *whipping*. Whirling motion in rotating machinery does not follow the exponential growth rate predicted by Eq. 1.58. Nonlinearities in the system dissipate energy more rapidly with increasing amplitudes than predicted by a linear model. Hence, whirl amplitudes increase sharply with time but generally achieve a steady-state limit cycle as shown in Fig. 1.52*b*. Additionally, the

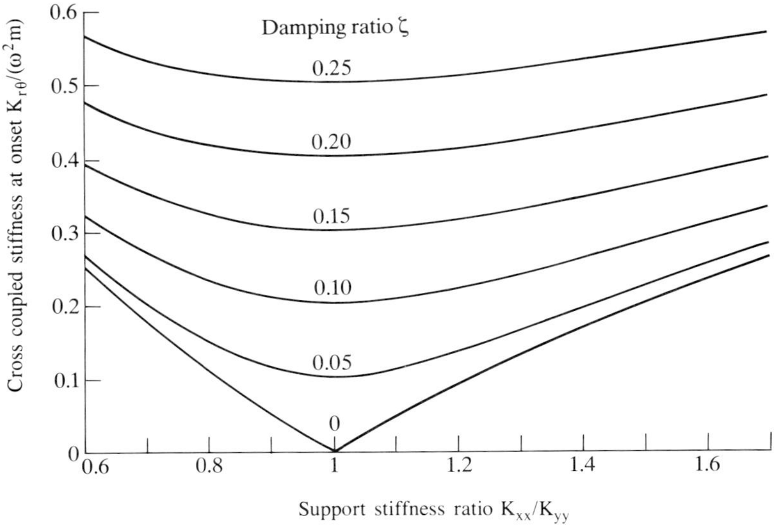

FIGURE 1.51 Equivalence of bearing support anisotropy to damping in suppression of whirl instability (Ehrich, 1989a).

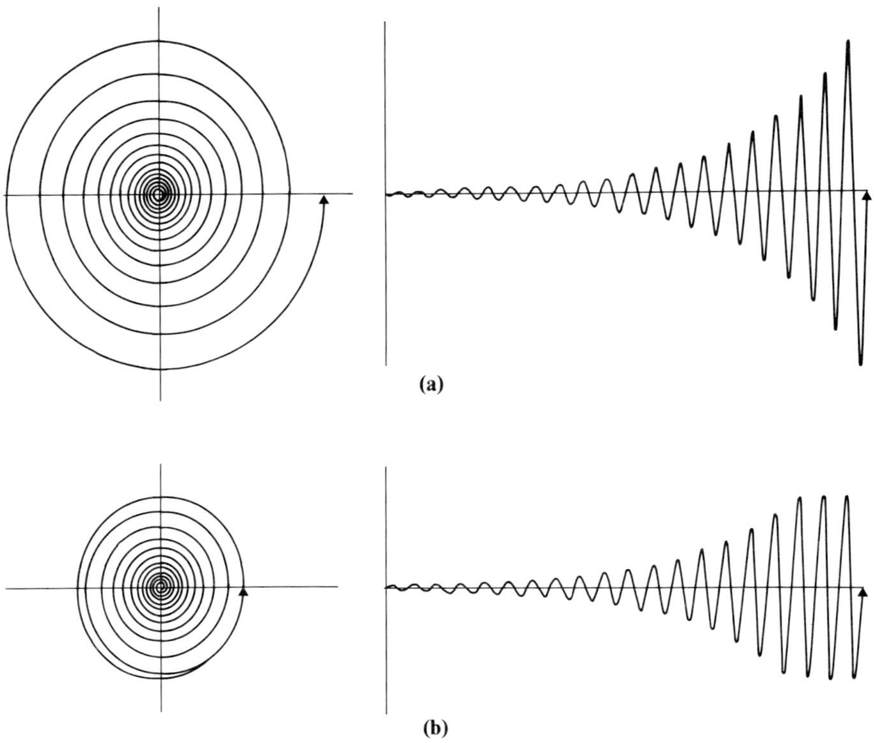

FIGURE 1.52 Shaft motion in self-excited vibration: (*a*) linear model—exponential amplitude increase; (*b*) nonlinear model—limit cycle amplitude.

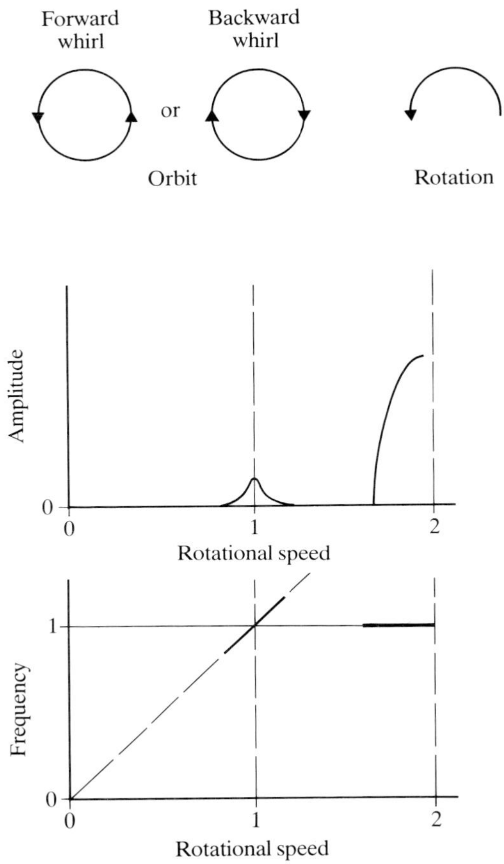

FIGURE 1.53 General attributes of unstable response (whirling or whipping).

response increases sharply with speed, with large but finite amplitudes, as shown in Fig. 1.53. Nevertheless, energy dissipation due to large-amplitude whirl orbits generally results in damaged or destroyed equipment.

The most important sources of whirling and whipping instabilities are

- Internal rotor damping—hysteretic whirl
- Hydrodynamic bearings and seals

  Hydrodynamic bearings

  Oil seals

  Labyrinth seals
- Turbomachinery aerodynamic cross-coupling

  Tip-clearance excitation

Impeller-diffuser interaction forces

Propeller whirl

- Fluid trapped in rotors
- Dry friction rubs
- Torque deflection whirl

All these self-excitation systems involve friction or fluid energy mechanisms to generate the destabilizing force.

These phenomena are rarer than forced vibration due to unbalance or shaft misalignment, and they are difficult to anticipate before the fact or diagnose after the fact because of their subtlety. Also self-excited vibrations are potentially more destructive, since the asynchronous whirling of self-excited vibration induces alternating stresses in the rotor and can lead to fatigue failures of rotating components.

Internal Rotor Damping—Hysteretic Whirl

The mechanism of hysteretic whirl, as observed experimentally (Newkirk, 1924) and defined analytically (Kimball, 1924), may be understood from the schematic representation of Fig. 1.54. With some nominal radial deflection of the shaft, the flexure of the shaft would induce a neutral strain axis normal to the deflection direction. From first-order considerations of elastic beam theory, the neutral axis of stress would be coincident with the neutral axis of strain. The net elastic restoring force would then be perpendicular to the neutral stress axis, i.e.,

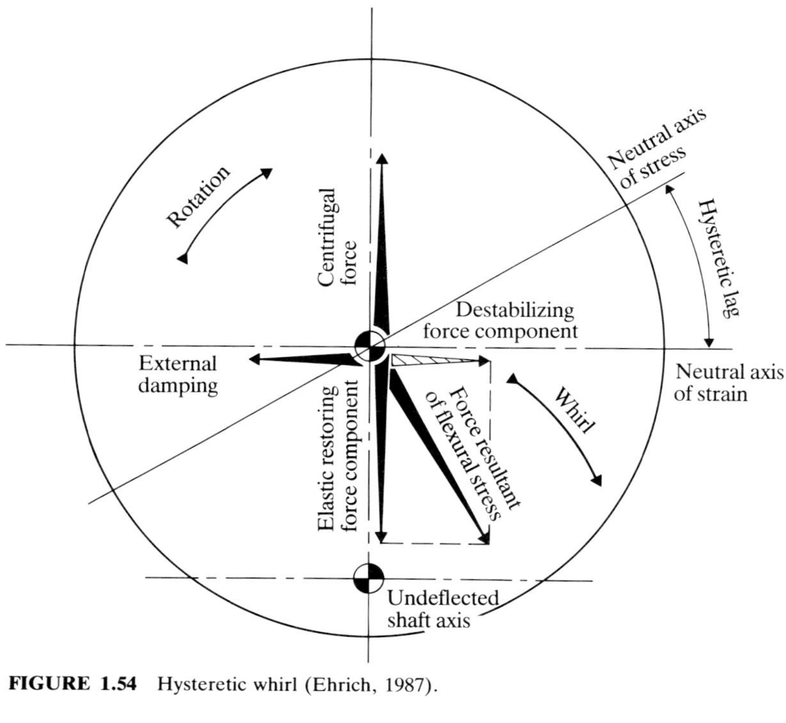

FIGURE 1.54 Hysteretic whirl (Ehrich, 1987).

parallel to and opposing the deflection. In actual fact, hysteresis, or internal friction, in the rotating shaft causes a phase shift in the development of stress as the shaft fibers rotate around through peak strain to the neutral strain axis. The net effect is that the neutral stress axis is displaced in angle orientation from the neutral strain axis, and the resultant force is not parallel to the deflection. In particular, the resultant force has a tangential component normal to deflection, which is the fundamental precondition for whirl. This tangential force component is in the direction of rotation and induces a forward whirling motion, which increases the centrifugal force on the deflected rotor, thereby increasing its deflection. As a consequence, induced stresses are increased, thereby increasing the whirl-inducing force component.

Several surveys and contributions to the understanding of the phenomenon have been published (e.g., Gunter, 1966; Bolotin, 1964; Bently, 1972; Vance and Lee, 1973). It has generally been recognized that hysteretic whirl can occur only at rotational speeds above the first-shaft critical speed (the lower the hysteretic effect, the higher the attainable whirl-free operating rotational speed). It has been shown (Ehrich, 1964) that once whirl has started, the critical whirl speed that is induced (from the spectrum of critical speeds of any given shaft) will have a frequency approximately one half the onset rotational speed.

A straightforward method for hysteretic whirl avoidance is to limit shafts to subcritical operation, but this is unnecessarily and undesirably restrictive. A more effective measure is to limit the hysteretic characteristic of the rotor. Most investigators (e.g., Newkirk, 1924) have suggested that the essential hysteretic effect is caused by damping associated with working at the interfaces of joints in a rotor rather than damping within the material of that rotor's components. Success has been achieved in avoiding hysteretic whirl by minimizing the number of separate elements, restricting the span of concentric rabbets and shrink-fitted parts, and providing secure lockup of assembled elements held together by tie bolts and other compression elements. General recognition of this simple fact has greatly reduced the incidence of classical internal friction as a mechanism for rotordynamic instability in modern turbomachinery, although the application of spline couplings with relatively loose fits in some applications provides a Coulomb damping internal friction destabilizing force which has been known to cause this same class of rotordynamic instability problems. A spline coupling is designed to accept misalignment; however, a misaligned precessing coupling yields internal rotor damping due to alternating Coulomb friction forces. Williams and Trent (1970) identified an instability problem connected with the turbine assembly of a gas turbine, while Marmol et al. (1979) have demonstrated forward whirl at approximately 64 percent of running speed in a spline-coupling apparatus. While lubrication can reduce the friction forces in a spline coupling, this type of unit represents a continuing source of concern from a stability viewpoint.

Hydrodynamic Bearings and Seals

Hydrodynamic Bearings. Historically, journal bearings were the next major cause of rotordynamic instability, as reported by Newkirk and Taylor (1925). This phenomenon can be understood by referring to Fig. 1.55 which illustrates a precessing journal. Consider some nominal radial deflection of a shaft rotating in a clearance filled with fluid (gas or liquid). The entrained, viscous fluid circulates with an average velocity of about one-half the shaft's surface speed. The bearing pressures developed in the fluid is not symmetric about the radial deflection line. Because of viscous losses of the bearing fluid circulating

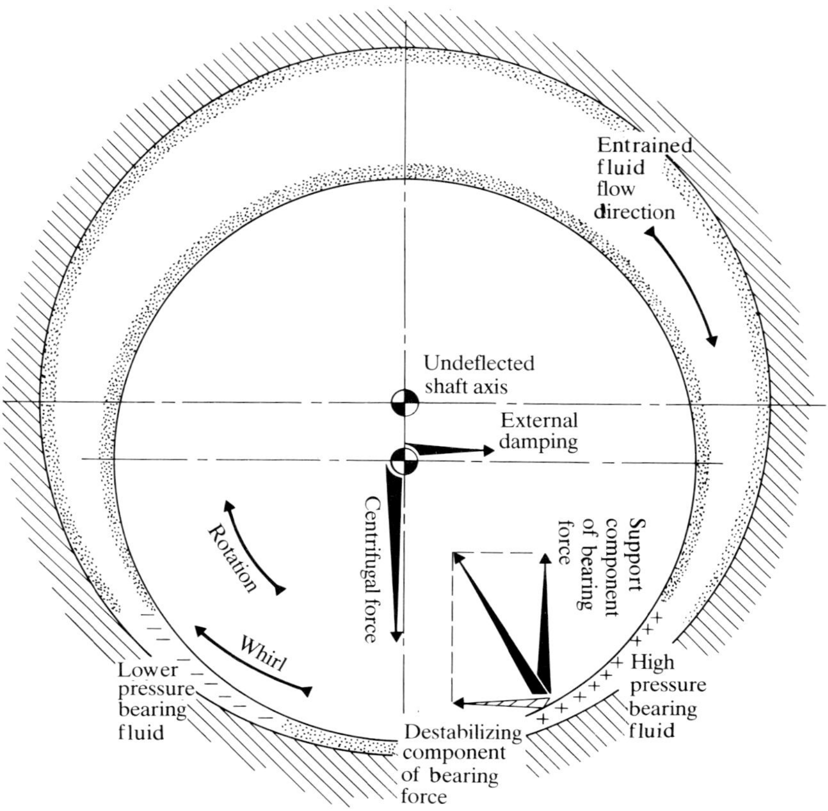

FIGURE 1.55 Fluid bearing whip (Ehrich, 1987).

through the close clearance, the pressure on the upstream side of the close clearance is higher than that on the downstream side. Thus, the resultant bearing force includes a tangential force component in the direction of rotation which tends to induce forward whirl in the rotor. The tendency to instability is evident when this tangential force exceeds inherent stabilizing damping forces.

In more recent years, computational advances allow prediction of the static and dynamic characteristics of journal bearings. The load deflection characteristics of plain journal bearings are inherently nonlinear and are customarily parameterized in terms of the nondimensional Sommerfeld number

$$S = \mu\left(\frac{\omega}{2\pi}\right)\left(\frac{R}{C}\right)^2\left(\frac{DL}{F_0}\right) > 0 \tag{1.74}$$

where

μ = fluid viscosity $\qquad D$ = bearing diameter

ω = shaft running speed $\qquad L$ = bearing length

R = bearing radius $\qquad F_0$ = static load

C = bearing radial clearance

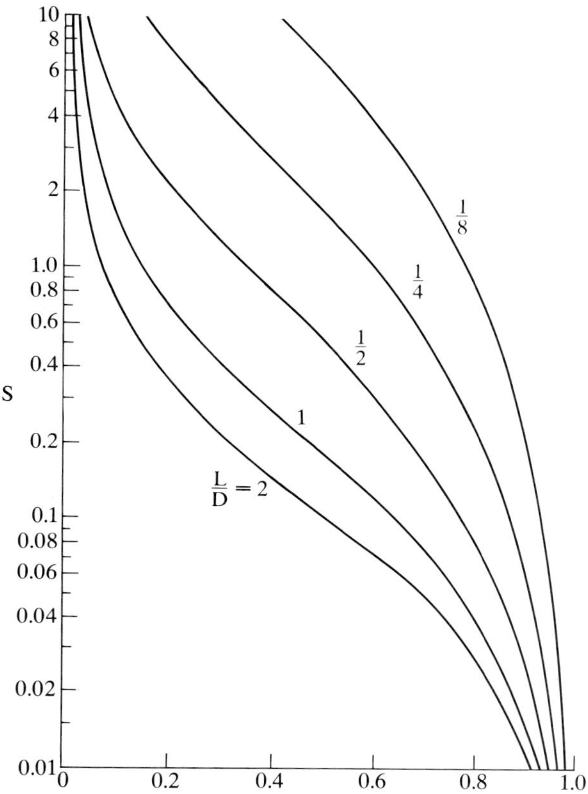

FIGURE 1.56 Equilibrium eccentricity ratio versus Sommerfeld number for a plain journal bearing.

Figure 1.56 illustrates the bearing eccentricity ratio ϵ_0 for various length-to-diameter (L/D) ratios as a function of the Sommerfeld number S. Zero and unity values for ϵ_0 correspond, respectively, to a centered bearing and a bearing in contact with the wall. As the speed increases, S increases and ϵ_0 decreases; i.e., a rotor rises in its bearings as the speed increases. Figure 1.57 illustrates the attitude angle dependency of a bearing on its eccentricity ratio. For a downward static load, the bearing equilibrium point starts at the bottom with zero running speed and "climbs" up the equilibrium locus as the running speed increases.

For an assumed small orbital motion about an equilibrium position, we can calculate the stiffness and damping coefficients for the load deflection relationship

$$-F_x = k_{xx}x + k_{xy}y + c_{xx}\frac{dx}{dt} + c_{xy}\frac{dy}{dt} \tag{1.75}$$

$$-F_y = k_{yy}y + k_{yx}x + c_{yy}\frac{dy}{dt} + c_{yx}\frac{dx}{dt} \tag{1.76}$$

and most bearing manufacturers routinely supply these data. Figure 1.58

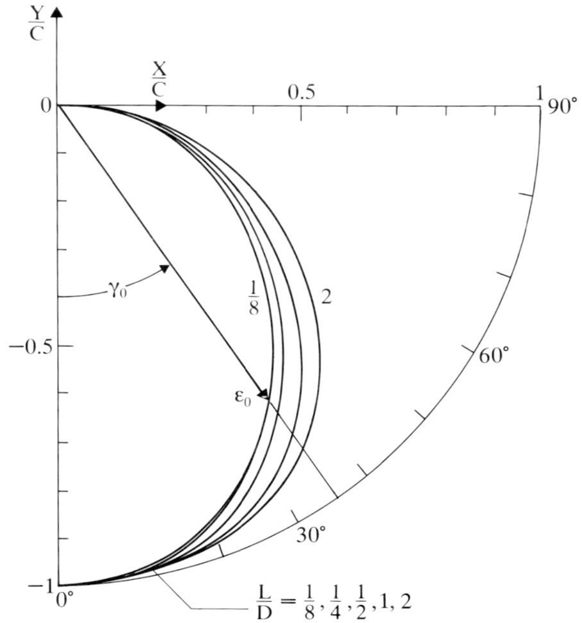

FIGURE 1.57 Attitude angle versus eccentricity ratio for a plain journal bearing.

illustrates the dependency of these coefficients on the static eccentricity ratio. As the running speed increases, the eccentricity ratio becomes progressively smaller and, for values of approximately 0.2 or less, k_{xy} and $-k_{yx}$ are approximately equal and the principal damping coefficients c_{xx} and c_{yy} are approximately equal. In polar coordinates, they then become the single parameters $k_{r\theta}$ and c, and the cross-coupled stiffness mechanism of whirl of Eq. 1.67 is developed.

The *whirl ratio* of a bearing Ω_w is defined as

$$\Omega_w = \frac{k_{r\theta}}{c\omega} \tag{1.77}$$

and for a plain journal bearing is approximately one-half. This ratio fundamentally characterizes the stability performance of a journal bearing in supporting flexible rotors, i.e., rotors which are much more flexible than the bearings by which they are supported. For a flexible rotor the *onset speed of instability* (OSI) is closely approximated by

$$\text{OSI} = \frac{\omega_n}{\Omega_w} \tag{1.78}$$

where ω_n is the rotor critical speed. Hence, a flexible rotor supported in plain journal bearings generally exhibits subsynchronous unstable motion for running speeds greater than twice the lowest rotor critical speed.

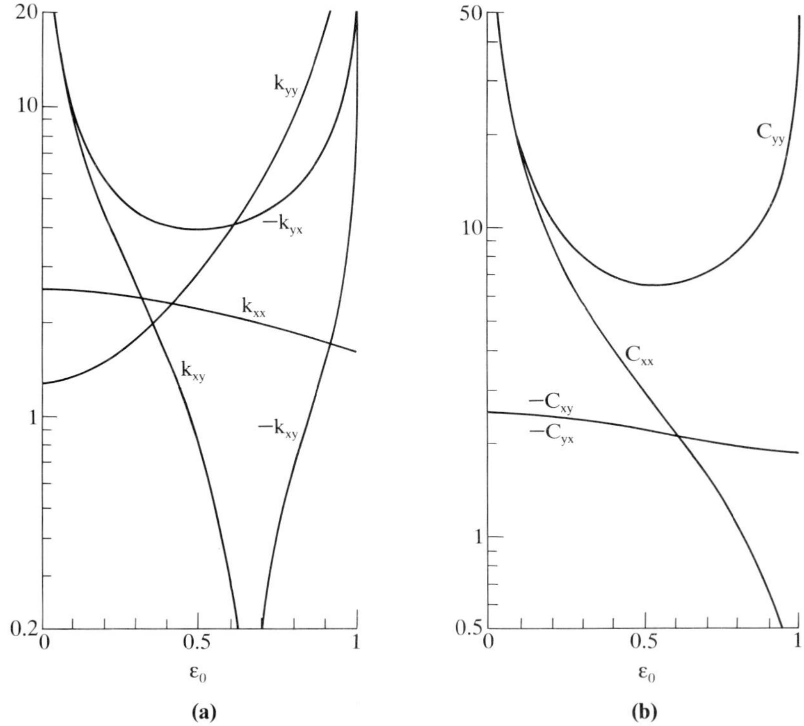

FIGURE 1.58 (*a*) Stiffness and (*b*) damping and coefficients for a plain journal bearing with $L/D = 1$.

Various multilobed bearing configurations have been developed to improve the rotordynamic characteristics of a plain journal bearing. Figure 1.21*c* and *d* illustrates some configurations which provide marginal improvements over plain journal bearings for some applications. However, the whirl ratio of these bearings is only slightly less than one-half. They accordingly provide only a slight improvement in stability margins for flexible rotors over that achieved by plain journal bearings.

The tilting-pad journal bearing of Fig. 1.21*f* represented a real breakthrough in the stability of flexible rotors, since this design eliminates the destabilizing cross-coupling stiffness coefficients. Illustrated are the "load-on-pad" and "load-between-pad" arrangements. Tilting-pad bearings provide less damping at low speeds than plain journal bearings or fixed-arc bearings; however, they are inherently stable and do not excite rotordynamic instability.

The whirl frequency of unstable rotor motion which is driven by hydrodynamic bearings is occasionally referred to as "one-half" frequency whirl. However, this is only approximately correct. The frequency of whirling motion which is caused by oil-film bearings is always slightly less than one-half the running speed frequency, generally 45 to 48 percent of the running speed.

Discussions of the finite difference and finite element methods for calculating stiffness and damping coefficients of hydrodynamic bearings are provided by Lund and Thomsen (1978) and Booker and Heubner (1972).

Stability problems vary considerably in intensity. Frequently an operating unit can be marginally stable and operate for years without a shutdown due to excessive vibrations. Typically only one or two units out of a family of nominally identical units will experience stability problems at any given time. In such marginally stable units supported by hydrodynamic bearings, fairly small changes may be sufficient to permit operation, such as merely changing the operating temperature (Wachel, 1975). Changing from one type of fixed-arc bearing to another or changing bearing preloads may also suffice. Changing from a fixed-arc to a tilting-pad bearing design will eliminate instabilities due to hydrodynamic bearings. Stability improvements involved in changes of hydrodynamic bearings generally rely on a reduction in the destabilizing forces developed by the bearing itself rather than an increase in damping.

Although the terms *whip* and *whirl* are often used interchangeably for the instabilities discussed in the previous section, Muszynska (1986) and others make a careful distinction between this phenomenon, which they refer to as *whip* and a different, but closely related phenomenon found at lower rotational speeds in rotors with fluid-film bearings, which they refer to as *whirl*. While the rotor precessional speed in whip is generally constant and equal to the rotor's natural frequency at any rotational speed above onset speed, the precessional speed in whirl varies with rotational speed, at a constant multiple (approximately 0.5) of the rotational speed. Such whirl has been noticed at subcritical speeds and at supercritical speeds, below the onset of whip on rotors with lightly loaded, 360° lubricated, plain journal bearings. Where such bearings are used, general design practice involves incorporation of some small deviation from precise circularity of the bearing journal to a two-lobed (lenticular) shape or multilobed configuration. Because of this practice, the phenomenon is not generally seen in modern rotating machines.

Floating Contact Seals. In floating contact seals, as described in Sec. 1.4.5 and pictured in Figs. 1.37 and 1.38, the seal rings are designed to "float" in the sense that they rise with the shaft as the running speed and pressure differential increase; however, the friction force and weight eventually exceed the lifting hydrodynamic force developed by the lockup of the seal and the seal rings. The static eccentricity position of the seal with respect to the shaft at lockup is critical to the stability characteristics of the compressor; specifically, destabilizing forces increase sharply with the static eccentricity ratio.

Seal rings in liquid applications are short ($0.05 < L/D < 0.2$), and their flow is generally laminar; hence the Ocvirk (1952) "short-bearing" model has been used in calculating the load capacity and rotordynamic coefficients. The relationship of the static eccentricity ratio to the Sommerfeld number for a short bearing is illustrated in Fig. 1.59. Experience has shown that seal rings do not generally cavitate; hence, the solid line of Fig. 1.59 applies. Because the static attitude angle depends on the friction force, it is, strictly speaking, indeterminate. Kirk (1986b) and Kirk and Miller (1979) have performed transient simulations to determine the lockup position of seals, while Allaire and Kocur (1985) use a quasi-static solution procedure to define the static eccentricity vector. Figure 1.60 illustrates the rotordynamic coefficients for a short seal with an assumed attitude angle of 90°. The seal has no direct stiffness for this attitude angle, and the cross-coupled stiffness and direct damping terms rise sharply with decreasing S_s (increasing static eccentricity ratio). The increase of c_{xx} and c_{yy} with the static eccentricity ratio can be advantageous at low speeds when the rotor is traversing a

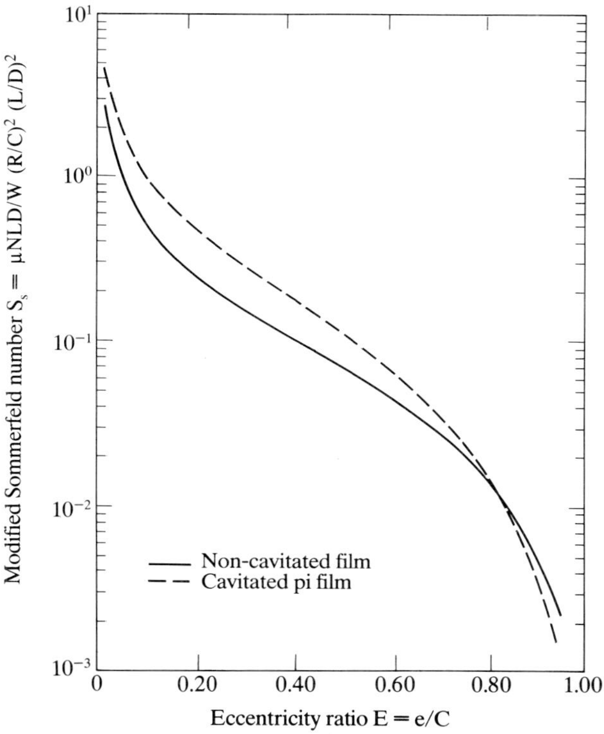

FIGURE 1.59 Modified Sommerfeld number versus eccentricity for oil seals (Kirk and Miller, 1979).

critical speed; however, the increase in k_{xy} and k_{yx} can lead to rotor instability at running speeds on the order of twice the critical speed.

One obvious question that arises with seals is, Why is the stability of a compressor improved by reducing the static eccentricity of an oil seal, while the stability of a rotor supported by journal bearings is enhanced by increasing the static eccentricity? The explanation provided by Allaire and Kocur (1985) is direct and simple, and if goes as follows. The compressors of interest are high-pressure units operating at speeds on the order of twice the critical speed; hence, to avoid stability problems, tilting-pad bearings are used. The seals and bearings are located near each other, and, for the purpose of this discussion, their rotordynamic coefficients can be assumed to be additive. If one examines the seal alone, then stability as defined by the whirl ratio

$$\Omega_{ws} = \frac{k_s}{C_s \omega} \tag{1.79}$$

actually improves slightly as the seal eccentricity ratio increases, since the damping C_s increases slightly faster than the cross-coupled stiffness coefficient k_s. However, the cross-coupled stiffness coefficient for the bearing is zero, and its

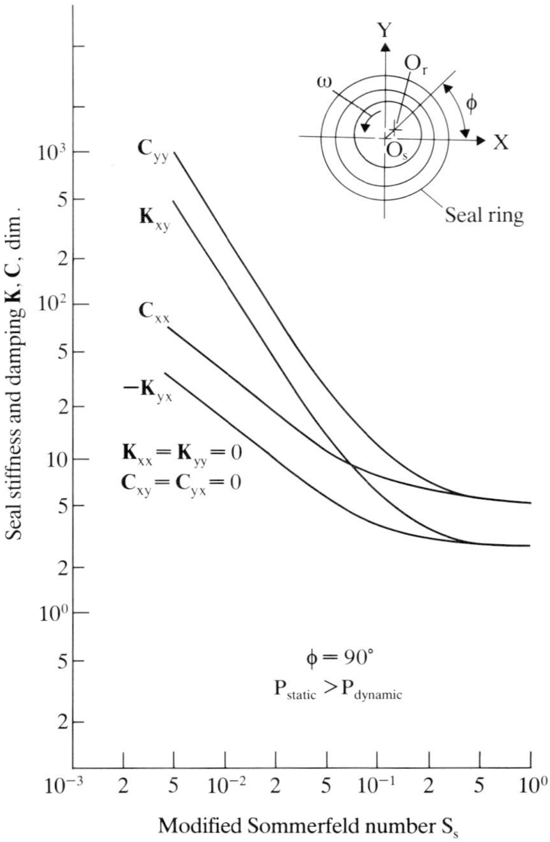

FIGURE 1.60 Stiffness and damping properties for high-pressure oil seals (Kirk, 1986b).

direct damping coefficient C_b is *much* larger than the seal's. Locking up the seal at a large eccentricity increases the bearing load but does not appreciably change the bearing rotordynamic coefficients. The whirl ratio for the seal and bearing together looks like

$$\Omega_w = \frac{k_s}{(C_s + C_b)\omega} \tag{1.80}$$

This parameter is small because of the large value for C_b, but increases rapidly with increasing static eccentricity ratio, because the denominator is approximately constant, while k_s increases sharply with increasing static eccentricity. When a compressor rotor is unstable due to an eccentric or jammed oil seal, the following steps can be taken:

1. The compressor can be "soft started," which means that speed is brought up at reduced suction pressure. Thus the rotor lifts the seal to a smaller eccentricity position before lockup occurs.

2. A better axial pressure balance can be developed which again yields lockup at reduced eccentricities. This approach was taken by Emerick (1982) who modified an existing design to eliminate high axial loads arising from wear. Kirk (1986b) cautions against a balance which is too low, since the seal rings themselves can become unstable.
3. The seal surface can be grooved so the geometry actually reduces the lift capability of the seal, but sharply reduces the cross-coupled stiffness coefficient.

This approach has been successfully taken by Tanaka et al. (1986), Kirk and Miller (1979), and Allaire and Kocur (1985), and it has the additional benefit of not increasing leakage. One compressor manufacturer (Shemeld, 1986), in a patented development, has solved the seal-centering problem by incorporating a tilting-pad bearing element in the high-pressure seal ring. This arrangement is schematically illustrated in Fig. 1.61. Shemeld shows five short-arc pads which

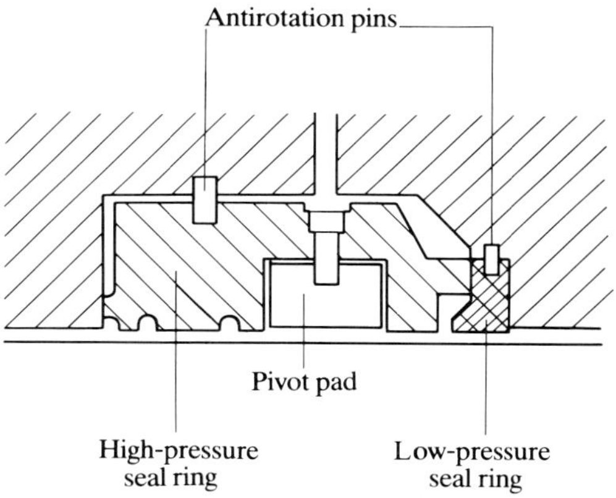

FIGURE 1.61 Pivot-pad oil seal (Shemeld, 1986).

would not restrict seal leakage across the high-pressure ring. Aside from stability considerations, the centering effect of the pads also reduces seal leakage.

Spiral-grooved face seals, described in Sec. 1.4.5 and pictured in Fig. 1.37, generate negligible rotordynamic forces and have been used in conjunction with magnetic bearings to create an oil-free compressor. However, the decision to use a spiral-grooved face seal, instead of a bushing-type radial seal, requires careful consideration. On at least one occasion, a compressor retrofit with spiral-grooved seals was unstable, suggesting that the damping from the oil-bushings seal was contributing to the stability of the compressor. On the other hand, some recent purchasers of compressors with high ratios of running speed to critical speed have insisted on gas seals because of instability predictions associated with oil seals.

Labyrinth Seals. Fluid rotation within labyrinth seals, as described in Sec. 1.4.5 and pictured in Fig. 1.40, causes the destabilizing forces in a manner that is basically similar to the hydrodynamic bearing. The displaced rotor of Fig.

1.55 causes a positive pressure region in the lower half of the seal and a reduced pressure region in the upper half. The net result of this pressure distribution is a transverse force in the direction of rotation.

Test results of Leie and Thomas (1980), Hauck (1982), and Benckert and Wachter (1980a, b) initially demonstrated that substantial destabilizing forces can be developed by labyrinth seals and that these circumferential forces increase with increasing fluid circumferential velocity due to either gas prerotation or shaft rotation. The associated test programs measured reaction force components parallel and perpendicular to the static displacement vector of a labyrinth; accordingly, only direct and cross-coupled stiffness data were obtained.

Childs and Scharrer (1986b) and Scharrer (1986) have recently reported test results for direct- and cross-coupled stiffnesses and direct damping values for "see-through" labyrinth seals. Tests were reported for teeth-on-stator and teeth-on-rotor labyrinths at speeds to 16,000 rpm for a range of pressure ratio and inlet tangential velocity for three different clearances, and the tests support the following conclusions:

1. Cross-coupled stiffness coefficients increase directly with increasing inlet tangential velocities.
2. Seal damping is small but must be accounted for to obtain reasonable rotordynamic predictions.
3. As clearances decrease, teeth-on-rotor seals become less stable and teeth-on-stator seals become more stable.

The prediction capability for labyrinth seals is not as well advanced as for hydrodynamic bearings. Flow in labyrinths is complicated, with a vortex flow developed in the labyrinth cavities. The first "reasonable" model for predicting rotordynamic coefficients of labyrinth seals was proposed by Iwatsubo et al. (1982). Subsequent and increasingly representative models have been developed by Childs and Scharrer (1986a), Wyssman et al. (1984), and Scharrer (1986).

The demonstrated fact that fluid rotation causes the destabilizing cross-coupling coefficients led to the development of *swirl brakes* or *swirl webs* at the labyrinth inlets. As illustrated in Fig. 1.62, this device uses an array of axial vanes

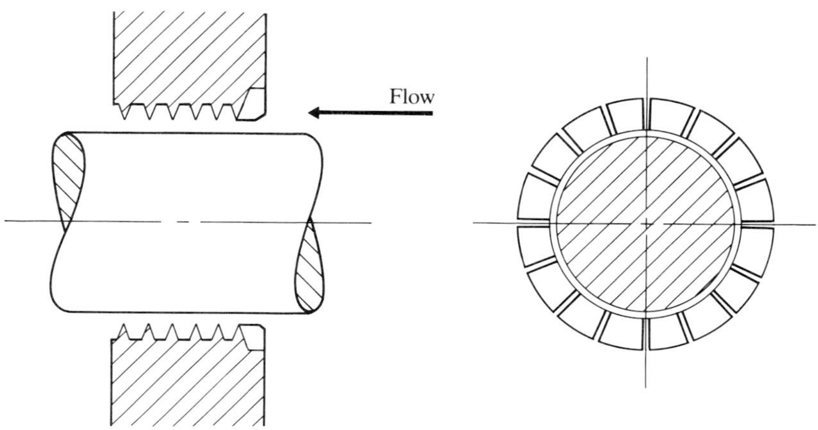

FIGURE 1.62 Swirl brakes at entrance to labyrinth seals.

at the seal inlet to destroy or reduce the inlet tangential velocity. Their effectiveness has been demonstrated in test results (Benckert and Wachter, 1980a), and they have been used successfully in commercial compressors and the turbine interstage seals of the *high-pressure oxygen turbopump* (HPOTP) of the *space shuttle main engine* (SSME).

Another approach for stabilizing compressors involves a "shunt line" and entails rerouting gas from the compressor discharge and injecting flow into one of the early cavities in the balance-piston labyrinth. This approach is illustrated in Fig. 1.63. The natural flow path for leakage into this labyrinth is down the back

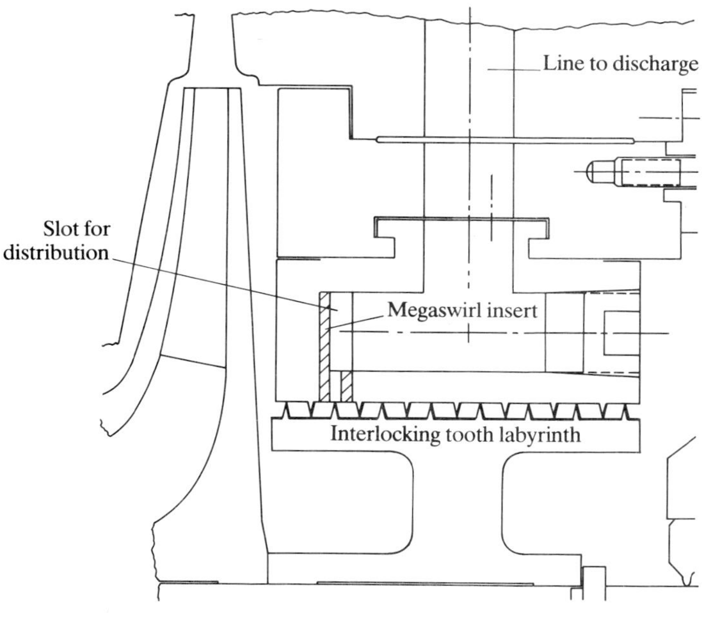

FIGURE 1.63 Shunt diversion from compressor discharge for injection in the balance-drum labyrinth (Kirk, 1986a).

side of the last-stage impeller. The normal leakage flow leaves the impeller with a high tangential velocity on the order of 0.5 times the impeller surface velocity. As it moves radially inward, conservation of moment of momentum requires a substantial tangential acceleration; hence, when the flow enters the labyrinth, it has a *very high tangential* velocity which yields a high destabilizing force. Enough shunt flow is provided to force gas outward along the back face of the impeller. In a conventional shunt, the gas enters the seal with zero tangential velocity and does not acquire much tangential velocity as it proceeds through the labyrinth. A conventional shunt works by sharply reducing the seal tangential velocity and accordingly the destabilizing cross-coupled stiffness coefficients. Of all the labyrinth seals, the destabilizing forces are maximum for the balance-piston labyrinth because it is the longest seal and has the highest pressure drop. For a straight-through machine it also has the highest density. The *shunt-line fix* has

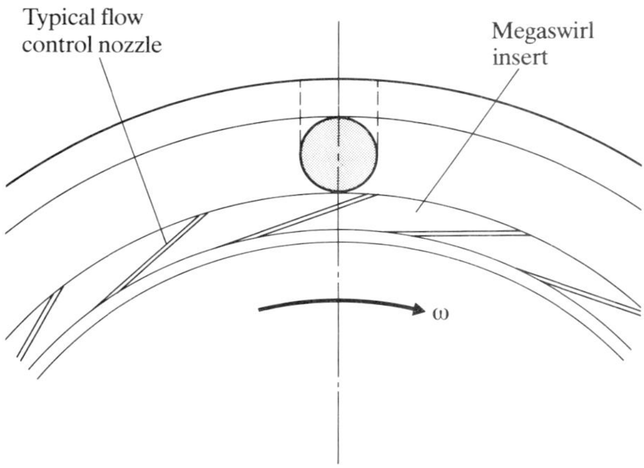

FIGURE 1.64 Cross section of megaswirl insert (Kirk, 1986a).

been used successfully on both straight-through and back-to-back (Zhou, 1986) machines. There are obvious performance penalties involved in the shunt-line approach.

Figure 1.64 illustrates a recent refinement of the shunt-line approach which injects flow against shaft rotation. This approach creates cross-coupled stiffness coefficients which oppose forward whirl of the shaft (Childs and Scharrer, 1986b). Kirk (1986a) reports that a compressor was converted from unstable to stable operation by replacing the conventional (inward-radial-flow) gas injection by the against-shaft-rotation injection.

Some success has been obtained in improving the stability of turbomachinery units by replacing labyrinth seals with honeycomb seals; e.g., the stability characteristics of the HPOTP have been materially improved by replacing a stepped-labyrinth turbine interstage seal with a honeycomb seal. Test results (Benckert and Wachter, 1980b) show that honeycomb seals develop larger destabilizing coefficients than labyrinth seals; however, they seem to develop much higher direct damping values.

Turbomachinery Aerodynamic Cross-Coupling

Tip-Clearance Excitation. Turbomachinery designers are always looking for ways to achieve higher efficiency and performance, and historically they have pursued these objectives by running units faster and adding more stages. Both approaches degrade stability. Adding stages increases the bearing span of a unit and decreases the critical speeds. Increasing the running speed increases the ratio of running speed to critical speed. However, prior to the introduction of tilting-pad bearings, rotating machines in hydrodynamic bearings were generally constrained by "oil whirl" to operate at speeds below twice the first or lowest critical speed. With tilting-pad bearings, designers of turbomachinery are now free to build lightweight, higher-speed units, with the ratios of operating speed to critical speed increased substantially above 2.

Unfortunately, many turbomachines built to these specifications proved to be violently unstable at elevated power levels and at elevated speed ranges which are otherwise stable at lower power levels. Specifically, units have been manufactured and tested by manufacturers at their design operating speed without any problems; however, when these units were put into service, they became unstable at a threshold *power* level and could not be operated satisfactorily at higher power levels. The unknown forces which caused these instabilities were called *aerodynamic* since they apparently were a result of fluid forces acting on the rotor.

A destabilizing mechanism associated with tip clearance with a load-dependent character had been postulated by Thomas (1958) in connection with instability problems of steam turbines. As shown schematically in Fig. 1.65, Thomas

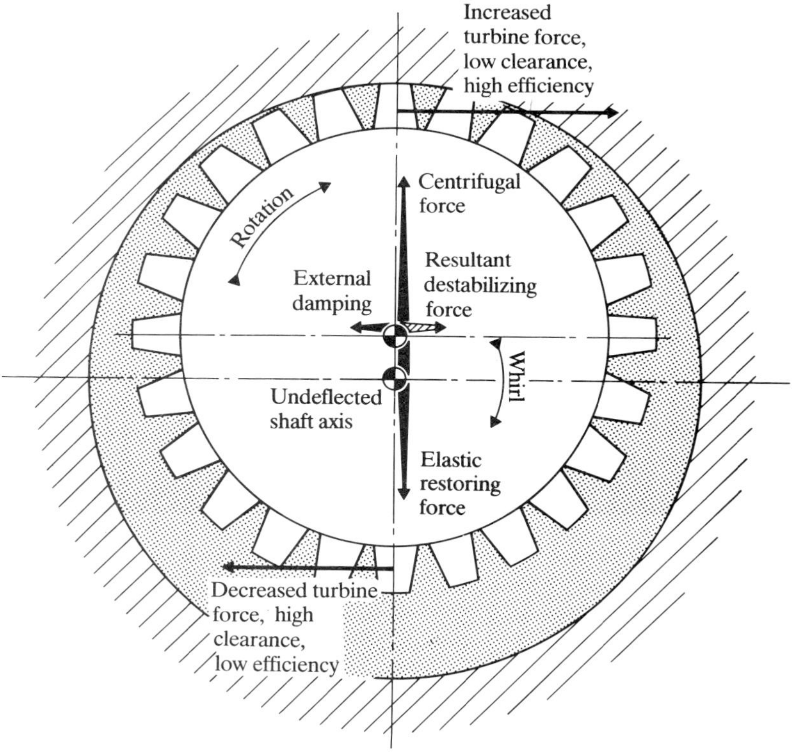

FIGURE 1.65 Tip-clearance excitation (Childs, 1987).

suggested that some nominal radial deflection will close the radial clearance on one side of the turbomachinery component and open the clearance 180° away on the opposite side. We would expect the closer clearance zone to operate more efficiently and to be more highly loaded than the open clearance zone. A resultant net tangential force may be generated to induce a whirl in the direction of rotor rotation (i.e., forward whirl).

Thomas postulated a purely destabilizing reaction-force model of the form

$$K_{r\theta} = \frac{T\beta}{D_m L} \tag{1.81}$$

where T = stage torque
D_m = mean blade diameter
L = blade length
β = change in thermodynamic efficiency per unit change in clearance

In practice, β becomes an empirical adjustment factor which has been used to bring computational predictions of instability into accordance with field experience.

Tests have been made (Urlichs, 1975, 1977) to measue transverse destabilizing forces in unshrouded turbines to determine values for β. For the configurations tested, values of β on the order of 4 to 5 were obtained. However, experienced designers of turbines normally estimate β to be on the order of 1 to 1.6. Alford (1965) subsequently hypothesized the same phenomenon in connection with stability problems of compressors. He reasoned that for the deflected rotor, the blading with the smallest clearance would be most efficient and would be more *lightly* loaded than the blading with larger clearance, 180° away. He then concluded that the net destabilizing force would be in the direction of rotation, resulting in a tendency to *forward* whirl. But if one considers the fact that the blades with the smaller clearance will sustain a *higher* static pressure differential across their tips and therefore be more *highly* loaded than the blades with larger clearance, 180° away, then one must conclude that the net destabilizing force would be counter to the direction of rotation, resulting in a tendency to *backward* whirl.

Horlock and Greitzer (1983) have studied analytically the effect of circumferential efficiency variation associated with tip-clearance variation, and they deduced that the effect of rotation in a whirling rotor would be much more severe than that measured in a static rotor. Vance and Laudadio (1984) have measured the destabilizing forces statically in the deflected rotor of a blower. Their data survey over the whole range of speed and torque, reproduced in Fig. 1.66, shows a complex dependency on rotor speed as well as torque, without any simple linear force-torque relationship. It also indicated the possiblity of both forward and backward whirl.

Impeller-Diffuser Interaction Forces. The clearance excitation force model of Thomas and Alford provides a logical load-dependent model for a turbine or an axial compressor stage but a questionable model for an impeller of a centrifugal pump or compressor. Test experiences with unstable compressors provide strong evidence that impellers are involved in the development of large destabilizing forces; however, to date no satisfactory procedure has been developed to predict impeller forces. While test results have not yet been developed for compressor impellers, they have been reported (Jery et al., 1984; Adkins, 1985; Bolleter et al., 1985; and Ohashi et al., 1986) for centrifugal pump impellers. These test programs have concentrated on the forces at the impeller exit and used very large clearances between the impeller shroud and housing, to minimize shroud forces. The destabilizing force measurements reported from these programs have been relatively benign, requiring running speeds in excess of 2.5 times the lowest natural frequency to yield an instability. However, test results by Bolleter et al. (1985) with more realistic shroud-to-housing clearances yielded substantially larger destabilizing forces.

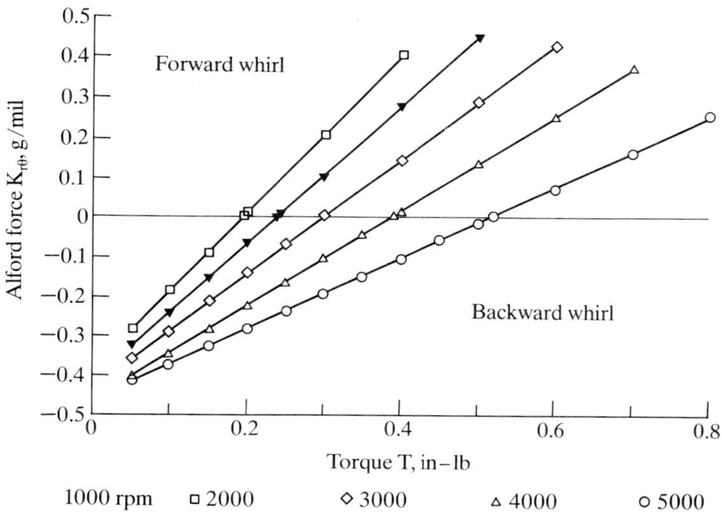

FIGURE 1.66 Alford force measurements, physical data (Vance and Laudadio, 1984).

Childs (1986) has recently completed analysis of the forces developed on a pump impeller shroud which predicts that (1) impeller shrouds develop significant destabilizing forces, (2) conservation-of-momentum considerations for leakage flow inward along the impeller shroud can generate high tangential velocities for the flow entering wear-ring and balance-piston seals, and (3) enlarged clearances of the exit seal due to wear or damage will increase the tangential velocity over the shroud and into the seal. Given that the Mach number for leakage path flow along an impeller shroud tends to be fairly low, the same conclusions are assumed to hold for impellers of compressors.

Kirk (1986a) has recently utilized an analysis due to Jimbo (1956) to calculate the tangential velocity entering a wear-ring seal following flow down the front face of a compressor impeller. His results agree with the results of Childs in predicting comparatively high inlet tangential velocities. The following conclusions can be drawn from these results:

1. Impeller shroud forces themselves are important in determining the rotordynamic stability of compressors.
2. Leakage flow that proceeds radially inward along an impeller shroud creates high tangential velocities at the entrance to exit wear-ring and balance-piston seals.

Analyses for impellers or pumps or compressors remain deficient, and no applicable data have been published for forces developed by impellers of centrifugal compressors.

To circumvent uncertainties with respect to destabilizing forces, Kirk and Donald (1983) proposed an empirical approach for *first-cut* prediction of the stability of multistage centrifugal compressors. The approach is based on the following observations:

1. Units become unstable when the ratio of running speed to first critical speed increases.
2. Units become unstable as the horsepower and density increase.

Figure 1.67 illustrates their suggested definition of parameters to define the problem with the product of inlet pressure P_2 and differential pressure ΔP on the ordinate axis and the ratio of running speed ω to rigid bearing critical speed ω_{cr} on the abscissa. Stable and unstable conditions for the compressors reported by Fowlie and Miles (1975) and Wachel (1975) are entered on this chart and suggest the straight-line boundaries between stable and unstable designs. The results of Fig. 1.67 apply for 'back-to-back' designs. The back-to-back design reduces the axial thrust in a unit but has been more susceptible to rotordynamic instability problems than straight-through designs, presumably because of the center

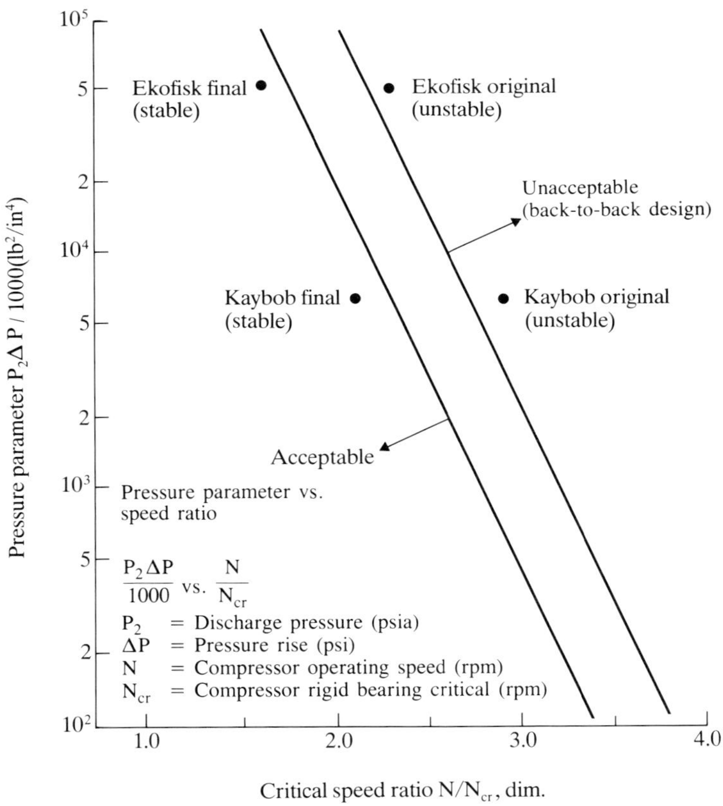

FIGURE 1.67 Suggested stability map for back-to-back centrifugal compressors (Kirk and Donald, 1983).

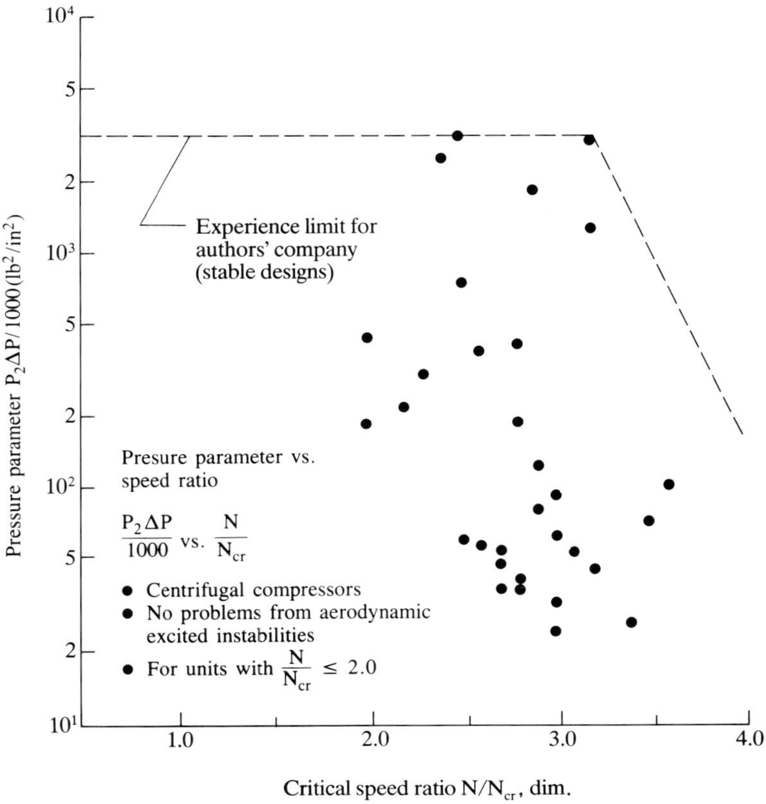

FIGURE 1.68 Suggested stability map for flow-through centrifugal compressors (Kirk and Donald, 1983).

labyrinth. Figure 1.68 shows their experience with straight-through compressors. These two figures can be used as a guide in designing compressors to determine whether *similar* machines are likely to experience stability problems.

Propeller and Turbomachinery Rotor Whirl. Propeller whirl has been identified both analytically by Taylor and Browne (1938) and experimentally by Houbolt and Reed (1961). In this instance of shaft whirling, a small deflection of the shaft is generally accompanied by incremental velocities of the propeller blades. Where the plane of the propeller rotates as a result of the slope deflection of the shaft centerline, incremental axial velocities are induced, as shown in Fig. 1.69. The incremental velocities result in changes in the relative velocities between the blades and the airstream which alter the aerodynamic forces on the blades. At high airspeeds, the forces are destabilizing and can overcome damping forces, to cause destructive whirling of the entire system. The whirling is generally found to be counter to the shaft rotation direction. It has been suggested by Trent and Lull (1972) that equivalent stimulation is possible in turbomachinery in the process of diagnosing and correcting whirl in

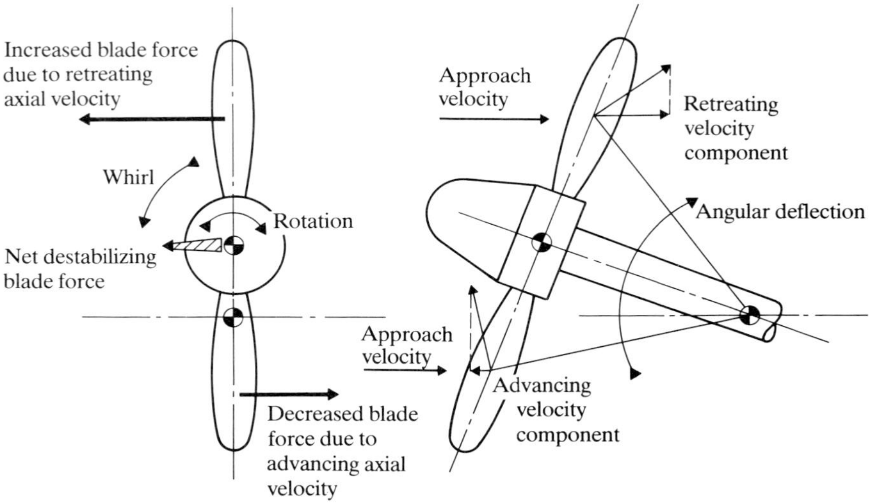

FIGURE 1.69 Propeller whirl (Ehrich, 1987).

an aircraft engine compressor. An attempt has been made by Ehrich (1973) to generalize the analysis for axial-flow turbomachinery. Although it has been shown that this analysis is not accurate, the general deduction seems appropriate that forward whirl may also be possible if the virtual pivot point of the deflected rotor is forward of the rotor (i.e., on the side of the approaching fluid).

Instability is found to be load-sensitive in the sense that it is a function of the velocity and density of the impinging flow. It is not thought to be sensitive to the torque level of the turbomachine since, e.g., experimental work (Houbolt and Reed, 1961) was on an unloaded windmilling rotor. Corrective action is generally recognized to involve stiffening the entire system and manipulating the effective pivot center of the whirling mode to inhibit angular motion of the propeller (or turbomachinery) disk as well as system damping.

Fluid Trapped in the Rotor

There has always been a general awareness that high-speed centrifuges are subject to a special form of instability. It is now appreciated that the same self-excitation may be experienced more generally in high-speed rotating machinery where liquids (e.g., oil from bearing sumps, steam condensate, etc.) may be inadvertently trapped in the internal cavity of hollow rotors. One perspective of the mechanism of instability is shown schematically in Fig. 1.70. For some nominal deflection of the rotor, the fluid is flung out radially in the direction of deflection. But the fluid does not remain in simple radial orientation. The spinning surface of the cavity drags the fluid (which has some finite viscosity) in the direction of rotation. This angle of advance results in the centrifugal force on the fluid having a component in the tangential direction in the direction of rotation. This force then is the basis of instability, since it induces forward whirl, which increases centrifugal force on the fluid and thereby increases the whirl-inducing force.

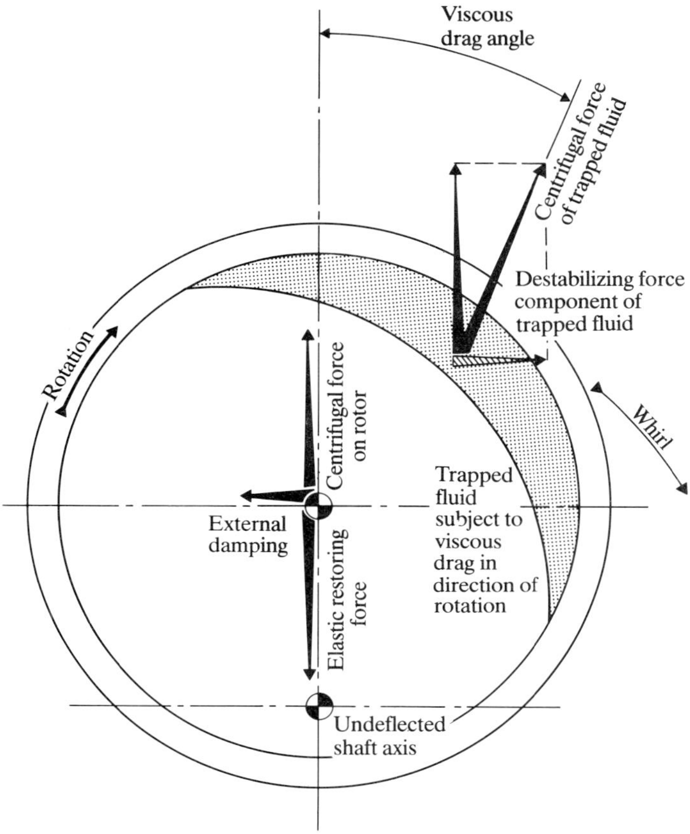

FIGURE 1.70 Whirl due to fluid trapped in rotor (Ehrich, 1987).

As shown in Fig. 1.71, the onset speed for instability is always above the critical rotational speed and below twice the critical rotational speed. Since the whirl is at the shaft's critical frequency, the ratio of whirl frequency to rotational speed is 0.5 to 1.0. Other contributions to the understanding of the phenomenon as well as a complete history of the phenomenon's study have been recorded by Wolf (1968).

In compressors, trapped fluid is only likely to be developed in couplings (Kirk et al, 1983) where subsynchronous motion ranging from 94 to 83 percent of running speed has been reported, depending on the quantity of trapped oil.

This self-excitation can be avoided by running the shafting subcritically, although this is generally undesirable in centrifuge-type applications when further consideration is given to the role of trapped fluids as unbalance in forced vibration of rotating shafts, as described by Ehrich (1967). Where the trapped fluid is not fundamental to the machine's function, the appropriate avoidance measure, if the particular application permits, is to provide drain holes at the outermost radius of all hollow cavities where fluid might be trapped.

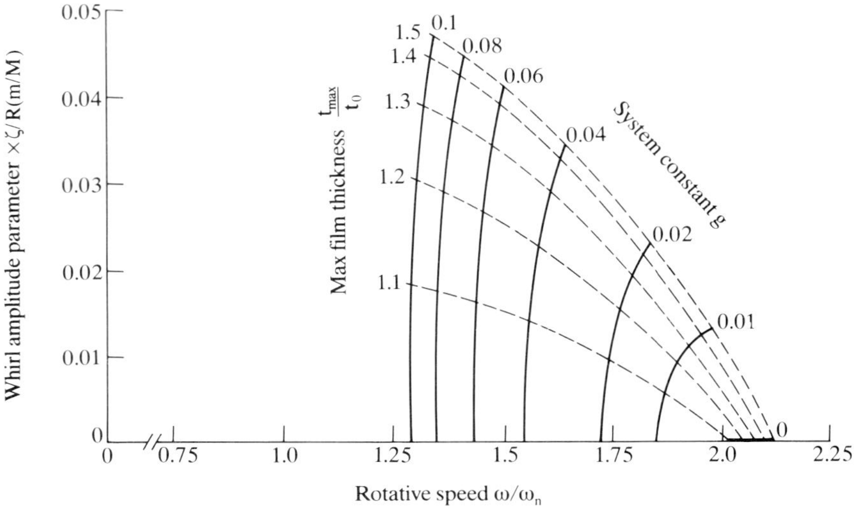

FIGURE 1.71 System response in asynchronous whirl due to trapped fluids (Ehrich, 1967).

Dry Friction Rubs

Dry friction whip is experienced when the surface of a rotating shaft comes in contact with an unlubricated stationary guide or shroud or stator system. This can occur in an unlubricated bearing or with the loss of clearance in a hydrodynamic bearing or inadvertent closure and contact in the radial clearance of labyrinth seals or turbomachinery blading or power screws (Sapetta and Harker, 1967).

The phenomenon may be understood by reference to Fig. 1.72. When radial contact is made between the surface of the rotating shaft and a static part, Coulomb friction induces a tangential force on the rotor since the friction force is approximately proportional to the radial component of the contact force, we have the preconditions for instability. The tangential force induces a whirling motion which induces a larger centrifugal force on the rotor, which in turn induces a large radial contact and hence a larger whirl-inducing friction force.

It is interesting that this whirl system is counter to the shaft rotation direction (i.e., backward whirl). One may envision the whirling system as the rolling (accompanied by appreciable slipping) of the shaft in the stator system.

The same situation can be produced by a thrust bearing where angular deflection is combined with lateral deflection. If contact occurs on the same side of the disk as the virtual pivot point of the deflected disk, then backward whirl results. Conversely, if contact occurs on the side of the disk opposite to where the virtual pivot point of the disk is located, then forward whirl results.

It has been suggested (but not concluded) by Begg (1973) that the whirling frequency is generally less than the critical speed.

The vibration is subject to various types of control. If contact between rotor and stator can be avoided or the contact area can be kept well lubricated, no whipping will occur. Where contact must be accommodated and lubrication is not feasible, whipping may be avoided by providing abradability of the rotor or stator element to allow disengagement before whirl. When dry friction is considered in the context of the dynamics of the stator system in combination with that of the

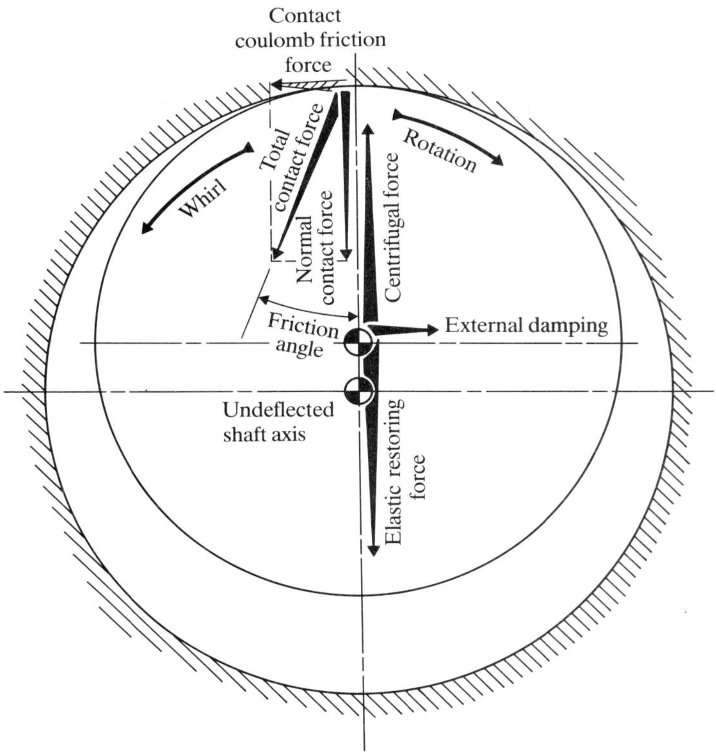

FIGURE 1.72 Dry friction whip (Ehrich, 1987).

rotor system (Ehrich, 1969), whirl can be inhibited if the independent frequencies of the rotor and stator are kept dissimilar, that is, a very stiff rotor should be designed with a very soft-mounted stator element that may be subject to rubs. No first-order interdependence of whirl speed and rotational speed has been established.

Torque Deflection Whirl

Vance (1978) has suggested still another possible mechanism of whirling excitation. The torque-generating and/or torque-absorbing component(s) (e.g., turbine or compressor rotor, motor or generator rotor) typically carried on high-speed rotors may be subject to angular deflection as part of any of the rotors' natural mode of vibration. If that angular deflection also causes an angular deflection of the torque vector associated with the component, then the torque vector will have a small radial component, approximately proportional to the radial deflection. That radial torque vector will tend to bend the shaft and cause a shaft deflection normal to the radial deflection very similar to that described in Sec. 1.6.2 on propeller and turbomachinery rotor whirl.

When this torque system inducing tangential deflection is strong enough to overcome the external damping forces tending to suppress tangential motion, the rotor will become unstable and start to whirl. It appears that extreme ranges of parameters (i.e., very long slender shafts with very high torque) are required for instability. No specific cases of instability in real apparatus are known to have torque whirl instability proved as the specific cause, but it is possible that the identified destabilizing forces may contribute to destabilization where other destabilizing forces are present.

1.6.3 Parametric Instability

General Description

There are systems in engineering and physics which are described by linear differential equations having periodic coefficients

$$\frac{d^2y}{dt^2} + p(t)\frac{dy}{dt} + q(t)y = 0 \tag{1.82}$$

where $p(t)$ and $q(t)$ are periodic in t. These systems also may exhibit self-excited vibrations, but the stability of the system cannot be evaluated by finding the roots of a characteristic equation. A specialized form of this equation, which is representative of a variety of real physical problems in rotating machinery, is *Mathieu's equation*:

$$\frac{d^2y}{dt^2} + (a - 2q \cos 2t)y = 0 \tag{1.83}$$

Mathematical treatment and applications of Mathieu's equation are given by McLachlan (1947).

This general subcategory of self-excited vibrations is termed *parametric instability,* since instability is induced by the effective periodic variation of the system parameters (stiffness, inertia, natural frequency, etc.). Three particular situations of interest in the field of rotating machinery are

- Lateral instability due to asymmetric shafting and/or bearing characteristics
- Lateral instability due to pulsating torque
- Lateral instabilities due to pulsating longitudinal compression

Lateral Instability due to Asymmetric Shafting. If a rotor or its stator contains sufficient levels of asymmetry in the flexibility associated with its two principal axes of flexure, as illustrated in Fig. 1.73, self-excited vibration may occur.

Presupposing a nominal whirl amplitude of the shaft at some whirl frequency, the rotation of the asymmetric shaft at a rotational speed different from the whirling speed will appear as periodic change in flexibility in the plane of the whirling shaft's radial deflection. This will result in an instability in certain specific ranges of rotational speed as a function of the degree of asymmetry. In general, instability is experienced when the rotational speed is approximately one-third and one-half the critical rotational speed and approximately equal to the critical rotational speed (where the critical rotational speed is defined with the average value of shaft stiffness), as in Fig. 1.74. The ratios of whirl frequency to rotational speed will then be approximately 3.0, 2.0, and 1.0. But with gross asymmetries and with the additional complication of asymmetric inertias with principal axes in

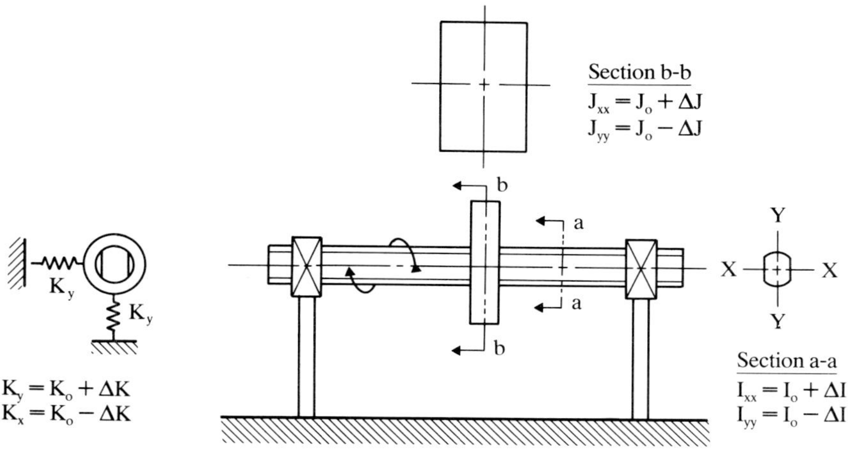

FIGURE 1.73 Lateral instability due to asymmetric shaft stiffness and/or inertia characteristics.

arbitrary orientation to the shaft's principal axes' flexibility, no simple generalization is possible.

There is a considerable body of literature dealing with many aspects of the problem and offering substantial bibliographies (Taylor, 1940; Crandall and Brosens, 1961; Messal and Bronthon, 1971; and Arnold and Haft, 1971).

A closely related phenomenon, usually characterized as a forced vibration, is the vibration of a horizontal shaft with asymmetric stiffness, rotating in a

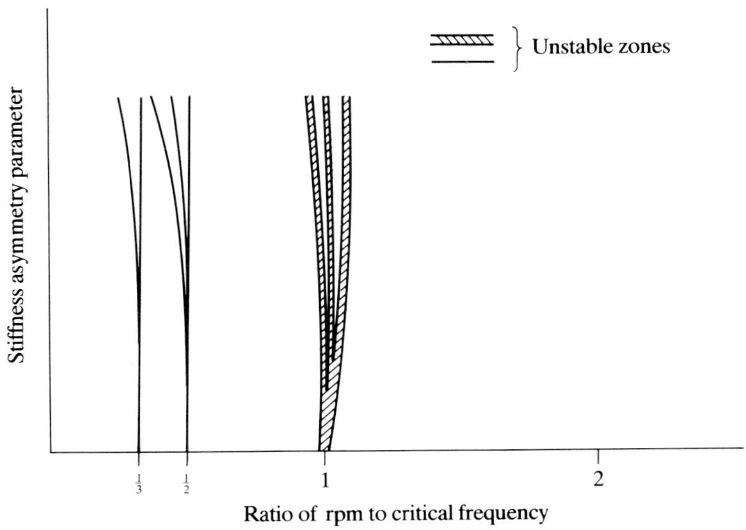

FIGURE 1.74 Instability regimes of rotor system induced by asymmetric stiffness (Messal and Bronthon, 1971).

gravitational field. The shaft will develop a certain amount of gravitational sag when the minor stiffness axis is parallel to the ground. It will develop less sag when the shaft has rotated a quarter turn and the major stiffness axis is parallel to the ground, and so on, completing two full cycles of large and small sag amplitude in a full rotation. This excitation will result in a peak resonant response when the shaft is rotating at half its natural frequency.

Stability is accomplished by minimizing shaft asymmetries and avoiding rotational speed ranges of instability. In circumstances where the functionality of the rotor requires asymmetry (e.g., two-pole electric motors and generators, two-bladed propellers and helicopter rotors), particular care must be taken to satisfy the quantitative parametric limits cited in the references.

Lateral Instability due to Pulsating Torque. Experimental confirmation (Eshleman and Eubanks, 1970) has been achieved that establishes the possibility of inducing first-order lateral instability in a rotor-disk system by the application of a proper combination of constant and pulsating torque. The application of torque to a shaft affects its natural frequency in lateral vibration so that the instability may also be characterized as *parametric.* Analytic formulation and description of the phenomenon are available (Wehrli, 1963). The experimental work by Eshleman and Eubanks (1970) explored regions of shaft speed where the disk always whirled at the first critical speed of the rotor-disk system, regardless of the torsional forcing frequency or the rotor speed within the unstable region.

It therefore appears that combinations of ranges of steady and pulsating torque, which have been identified (Wehrli, 1963) as being sufficient to cause instability, should be avoided in the narrow speed bands where instability is possible in the vicinity of twice the critical speed and lesser instabilities at 2/2, 2/3, 2/4, 2/5, . . . times the critical frequency, as in Fig. 1.75, implying frequency-speed ratios of approximately 0.5, 1.0, 1.5, 2.0, 2.5, ...

Lateral Instability due to Pulsating Longitudinal Loads. Longitudinal loads on a shaft which are on an order of magnitude of the buckling will tend to reduce the natural frequency of that lateral flexural vibration of the shaft. Indeed, when the compressive buckling load is reached, the natural frequency goes to zero. Therefore, pulsating longitudinal loads effectively cause a periodic variation in stiffness, and they are capable of inducing *parametric instability* in rotating as well as stationary shafts (McLachlan, 1947), as noted in Fig. 1.76.

1.6.4 Stick-Slip Rubs and Chatter

We should mention another family of instability phenomena—stick-slip or chatter. Although the instability mechanism is associated with the dry friction contact force at the point of rubbing between a rotating shaft and a stationary element, it must not be confused with dry friction whip, previously discussed. In the case of stick-slip, the instability is caused by the irregular nature of the friction force developed at very low rubbing speeds.

At high velocities, the friction force is essentially independent of the contact speed. But at very low contact speeds we encounter the phenomenon of "stiction," or breakaway friction, where higher levels of friction force are encountered, as in Fig. 1.77. Any periodic motion of the rotor's point of contact, superimposed on the basic relative contact velocity, will be self-excited. In effect, there is negative damping since motion of the rotor's contact point in the

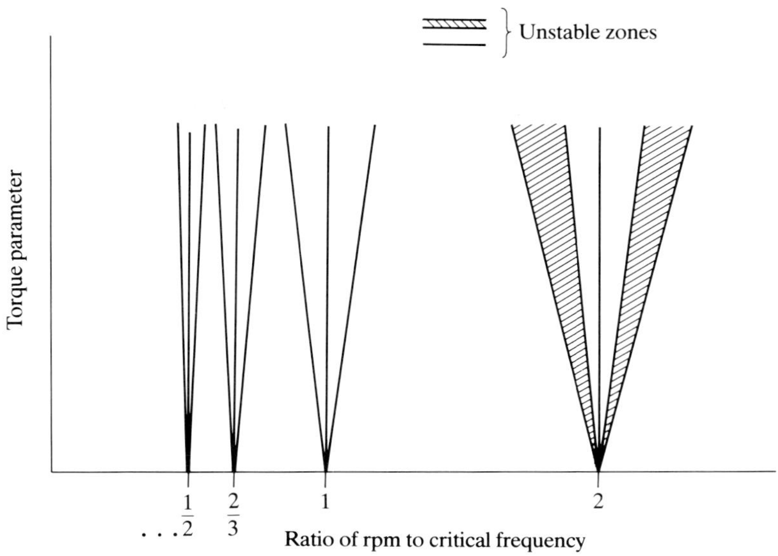

FIGURE 1.75 Instability regimes of rotor system induced by pulsating torque (Wehrli, 1963).

direction of rotation will increase the relative contact velocity and reduce breakaway friction and the net force resisting motion. Rotor motion counter to the contact velocity will reduce relative velocity and increase friction force, again reinforcing the periodic motion. The same generality of reaction force versus rotor surface speed is encountered in machine tools at the cutting edge of the machinery element. In this case, the resultant unstable motion is referred to as *chatter*. The ratio of vibration frequency to rotation speed will be much larger than unity.

While the vibration associated with stick-slip or chatter is often reported to be torsional, planar lateral vibrations can also occur. Surveys of the phenomenon are included in Conn (1960) and Sadowy (1959).

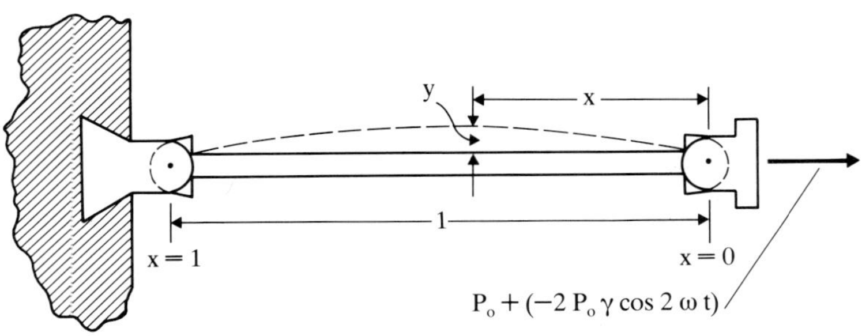

FIGURE 1.76 Lateral instability due to pulsating longitudinal compression (McLachlan, 1947).

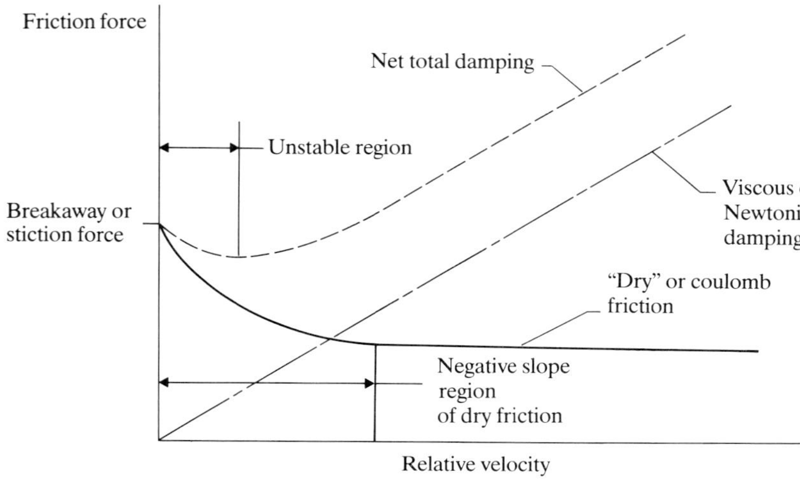

FIGURE 1.77 Dry friction characteristic giving rise to stick-slip rubs or chatter.

Measures for avoidance are similar to those prescribed for dry friction whip: Avoid contact where feasible and lubricate the contact point where contact is essential to the function of the apparatus.

1.6.5 Instabilities in Forced Vibrations

General Description

In a middle ground between the generic categories of force vibrations and self-excited vibrations is the category of instabilities in force vibrations. These instabilities are characterized by forced vibration at a frequency equal to the rotor rotation (generally induced by unbalance), but with the amplitude of that vibration being unsteady or unstable. Such unsteadiness or instability is induced by the interaction of the forced vibration on the mechanics of the system's response, or on the unbalance itself. Two manifestations of such instabilities and unsteadiness have been identified in the literature—bistable vibration and unstable imbalance.

Bistable Vibration

A classic model of one type of unstable motion is the *relaxation oscillator,* or *multivibrator.* A system subject to relaxation oscillation has two fairly stable states, separated by a zone where stable operation is impossible. Furthermore, in each of the stable states, a mechanism exists which will induce the system to drift toward the unstable state. The system will develop a periodic motion with the general shape of a square wave—dwelling at some high amplitude for a time, then suddenly changing to a low or negative amplitude and dwelling there for a time, and finally completing the cycle by suddenly returning to the former high-amplitude state.

While such systems are common in electronic circuitry, they are rather rare in the field of rotating machinery. One instance has been observed (Ehrich, 1965) in a rotor system supported by rolling-element bearings with finite internal clearance. In this situation, the effective stiffness of the rotor is small for small deflections (within the clearance) but large for large deflections (when full contact is made between the rollers and the rotor and stator). Such a nonlinearity in stiffness causes a "rightward-leaning" peak in the response curve when the rotor is operating in the vicinity of its critical speed and is being stimulated by unbalance. In this region, two stable modes of operation are possible, as seen in Fig. 1.78. In region *A-B,* the rotor and stator are in solid contact through the

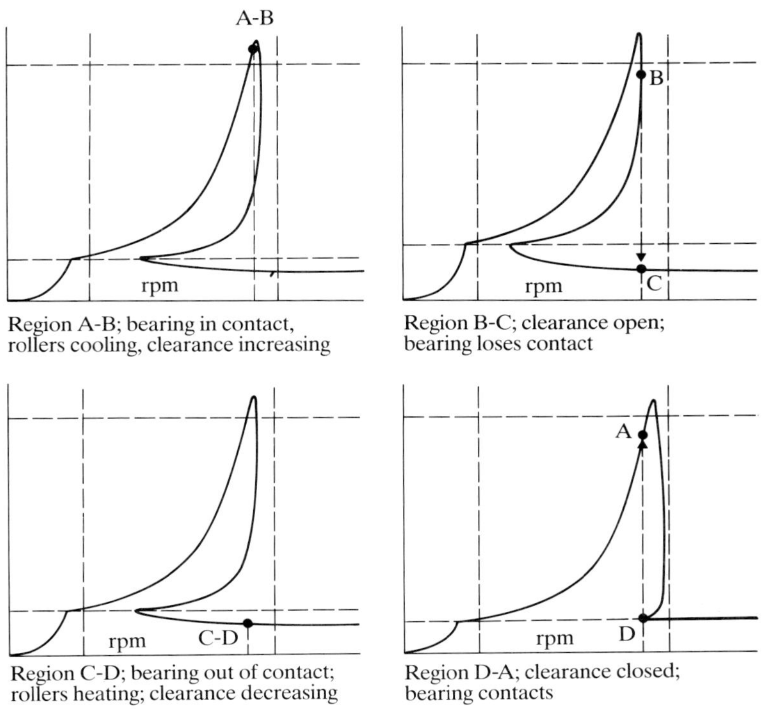

FIGURE 1.78 Rotor response, bistable vibration.

rollers. In region *C-D*, the rotor is whirling within the clearance, out of contact. When operation is at constant speed, either of the nominally stable states can drift toward instability by virtue of thermal effects on the rollers. When the rollers are out of contact, they are unloaded and will skid and heat up, thereby reducing the clearance and leading to closure of contact and a jump in amplitude from *D* to *A*. When the rollers are loaded, they will be cooled by lubrication and will tend to contract and increase the clearance, leading to loss of contact and a jump in amplitude from *B* to *C*. In sequence, these mechanisms are sufficient to cause a relaxation oscillation in the amplitude of the force vibration.

The remedy for this type of self-excited vibration is to eliminate the precondition of skidding rollers by reducing bearing geometric clearance, by preloading the bearing, or by increasing the temperature of any recirculating lubricant.

Unstable Unbalance

Den Hartog (1956) describes the occurrence of unstable vibration of steam turbines where the rotor "would vibrate with the frequency of its rotation, obviously caused by unbalance, but the intensity of the vibration would vary periodically and extremely slowly." The instability in the vibration amplitude is attributable to thermal bowing of the shaft, which is caused by the heat input associated with rubbing at the rotor's deflected "high spot" or by the mass of accumulated steam condensate in the inside of a hollow rotor at the rotor's deflected high spot. In either case, there is a basis for the continuous variation of amplitude, since unbalance gives rise to deflection and the deflection is, in turn, a function of that unbalance.

The phenomenon is sometimes referred to as the *Newkirk effect* in reference to its early recorded experimental observation (Newkirk, 1926). A manifestation of the phenomenon in a steam turbine has been diagnosed and reported by Kroon and Williams (1939), and a bibliography is available (Dimarogonas and Sandor, 1969). An analytic study (Dimarogonas, 1973) shows the possibility of spiraling, oscillating, and constant modes of amplitude variability.

1.6.6 Design of Stable Rotating Machinery

Three general aspects of the generation of rotordynamic instabilities and self-excited vibrations are recognized, and each one suggests an avenue of avoidance or suppression.

1. Rotors are destabilized by cross-coupled stiffness coefficients that yield tangential reaction forces which are both normal to a radial displacement and in the direction of a shaft rotation. This suggests that avoidance can be achieved by reduction or elimination of the source of the destabilizing forces.
2. The sensitivity of rotors to destabilizing forces increases with rotor flexibility. Specifically, rotors become more sensitive to instability problems as the ratio of running speed to critical speed (natural frequency) ω/ω_n increases. This suggests that avoidance can be achieved by increasing system stiffness and critical speeds.
3. Rotor stability is enhanced by increasing external damping. This suggests that avoidance can be achieved by increasing external system damping or, equivalently (as shown in Fig. 1.51), by introduction of anisotropy into the bearing support(s).

As we have seen in the preceding discussions, there are numerous mechanisms by which destabilizing forces are generated, and each one involves unique measures of avoidance, by defeating the destabilizing mechanism.

Increasing the rotor's critical speed is another possible approach for eliminating rotor instability problems. Rotor bearing spans have been reduced and rotor diameters increased to decrease the ratio of the rotor's running speed to critical speed (Kramer, 1972; Vance, 1974). In many instances this tends to be an expensive and time-consuming undertaking, and often it is not feasible without violating other rotordynamic design requirements and constraints needed to avoid forced vibration response.

A very effective and generally applicable avoidance measure, which is also compatible with suppression of forced vibration, is *damping*. Care must be taken to ensure that the damping is *external* to the rotor (i.e., a function of the motion of the rotor with respect to the stator) and not *internal* to the rotor (i.e., a function of rotor strain or deflection of one element of the rotor with respect to another). The latter—internal damping—is actually a potential source of destabilizing forces and is to be avoided. The design of dampers is touched upon in Sec. 1.4.4. As shown in Fig. 1.51, introduction of anisotropy (differing values of stiffness k_{xx} and k_{yy} between orthogonal planes) is also effective in stabilizing systems.

1.7 FORCED VIBRATION RESPONSE IN NONLINEAR AND ASYMMETRIC STATOR SYSTEMS

1.7.1 General Description

Underlying most of the previous considerations of forced vibrations in the design of rotating machinery has been the assumption of linearity and axial symmetry in the elements of the sytem.

- Rotor and stator system stiffnesses are linear and axisymmetric; the reaction force is simply proportional to deflection and independent of angular orientation.
- Rotor and stator damping are linear and axisymmetric; the reaction force is simply proportional to the velocity and independent of angular orientation.
- The rotor principal moments of inertia are equal.

We have seen how certain deviations from these assumptions can play a significant role in causing self-excited vibration (e.g., rotor stiffness asymmetry or rotor moment-of-inertia asymmetry causing parametric instability) or modifying the behavior of self-excited vibration (e.g., increasing stiffness and damping at large amplitudes, limiting the whirl amplitude of unstable systems; anisotropic stator stiffness affecting the onset speed of certain types of whirl).

Nonlinearities and asymmetries can also have a profound effect on the forced vibration behavior of rotating machinery in causing significant shifts in the amplitude and speed or frequency of critical speed amplitude peaks and in causing the appearance of pseudo-critical-speed vibration peaks at speeds markedly different from those anticipated from linear analysis and at other than synchronous ("one per revolution") frequencies. They are therefore of interest to the designer so that critical vibration peaks will not intrude on the running range of a design that might otherwise be free of them.

Figure 1.79 shows three of the important classes of nonlinearity and asymmetry in stator stiffness, compared to the reference linear axisymmetric case, in terms of the force-versus-deflection relationship for the two principal stiffness axes of the stator. All are compared to a reference isotropic (axisymmetric) linear system where both axes have linear and equal stator stiffnesses.

- *Anisotropic linear system*. Both axes have linear stator stiffness, but one axis is more stiff than the other.

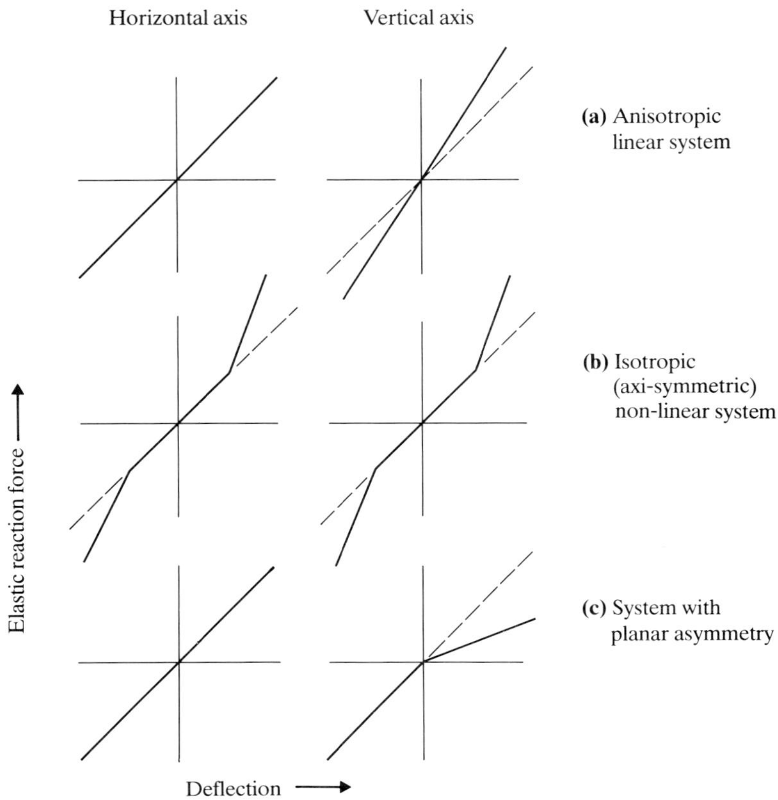

FIGURE 1.79 Three important classes of nonlinearity and asymmetry in stator stiffness.

- *Isotropic (axisymmetric) nonlinear stiffness.* Both axes have equal stator stiffness characteristics, but that characteristic is nonlinear. Most physical systems get stiffer with increasing deflection, as shown.
- *Planar asymmetry.* One stator axis has an asymmetric stiffness (softer for motion in one direction and harder for motion in the opposite direction). The other has a linear stiffness characteristic.

There are, of course, any number of complex variations and combinations of these archetypes, but consideration of these basic three is sufficient to explain most of the unique vibration characteristics that might be anticipated (and avoided) in real rotating machinery.

1.7.2 Anisotropic Linear Systems

A very important and commonly encountered class of deviation from the simple linear axisymmetric stator support stiffness is the anisotropic linear system delineated in Fig. 1.79*a*. In such a system, the stator support stiffnesses in the two principal and orthogonal directions (usually the vertical and horizontal directions for a shaft mounted horizontally) are both linear but differ in value. Typically, a stator mounted on a rigid bed will have a greater stiffness in the vertical direction than in the horizontal direction.

This deviation from axial symmetry results in a change in the forced vibratory response of the system to unbalance, as shown in general form in Fig. 1.80 for the Jeffcott rotor. It can be shown mathematically that the steady state horizontal and vertical responses at any rotative speed are essentially independent of each other. As with any linear system, the frequency of both the horizontal and the vertical vibratory response will be synchronous with the rotative speed (i.e., one per

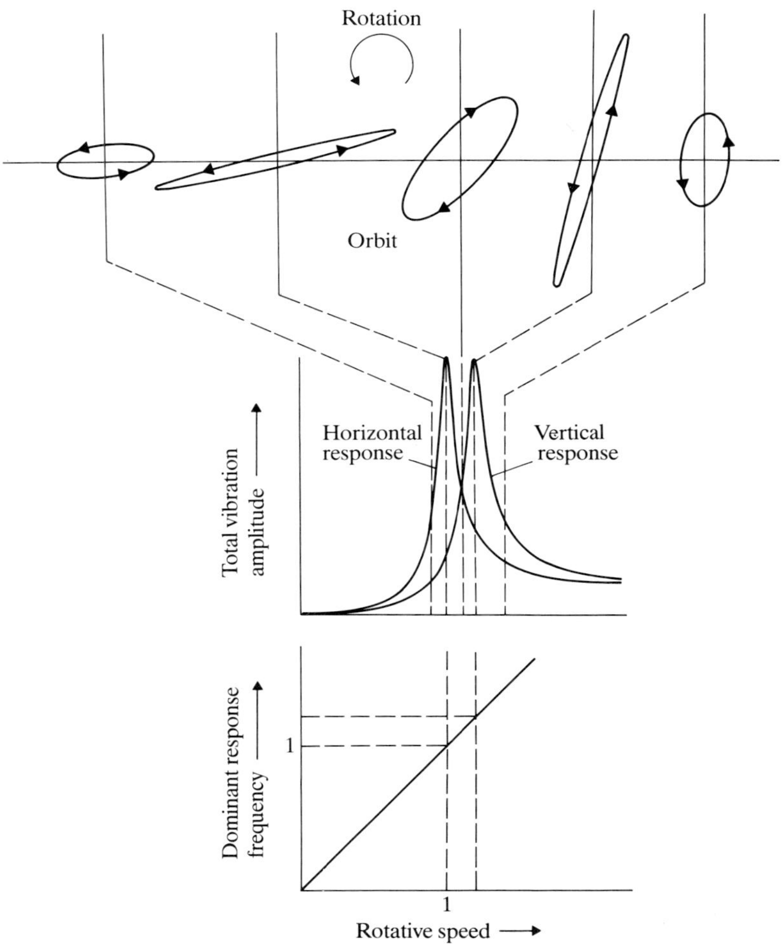

FIGURE 1.80 Response of anisotropic linear system.

revolution). But the critical speed at which the vibratory amplitude peaks will be higher for the axis with the higher stator stiffness (shown in Fig. 1.80 as the vertical axis). As a consequence of this dissimilarity in vertical and horizontal vibration amplitude at any rotative speed, the rotor deflection orbit will be noncircular—elongated in the horizontal direction when passing through the horizontal critical speed and elongated in the vertical direction when passing through the vertical critical speed.

Deviation from a circular orbit does have one consequence of possible significance. When a rotor whirls in a circular orbit synchronously with the rotative speed and in the same direction as rotation, then it moves as a body with frozen deflection in the rotor's frame of coordinates. This means that the internal elastic stresses in the rotor are constant—the rotor fibers feel no vibratory stress and are therefore not subject to fatigue as a consequence of forced vibration due to unbalance. However, when a rotor whirls with an elliptical orbit, it will undergo two maxima and two minima of deflection in each orbit, and therefore the rotor fibers will experience a vibratory stress at a frequency of twice the rotative speed. Care must be taken to ensure that these vibratory stresses are well below the fatigue limits of the rotor material.

There is a second possible consequence of the anisotropy. As has been shown in review of the behavior of the Jeffcott rotor model, the phase angle of the response with respect to the unbalance varies as a function of the rotative speed. At low subcritical speeds, the vibratory deflections are in phase with each other. At the critical speed, the deflection and the unbalance will be 90° out of phase. At high supercritical speeds, the deflection and the unbalance will be 180° out of phase. As long as the vertical and horizontal critical speeds are identical, the phase shifts will be identical at any speed. But with the anisotropic stator, the critical speeds and hence the phase shifts for the two axes will differ at any rotative speed, and particularly so in the speed range between the two critical speeds (see Fig. 1.81). If the difference between the phase angles of the two axes' vibration at any rotative speed is small, the effect is to tilt the major and minor axes of the orbit away from the principal elastic axes of the stator. If the difference between the phase angles is 90°, then the orbit degenerates to a planar motion. If the difference between the phase angles is greater than 90°, then *the whirling motion is in a direction opposite that of the rotor rotation* (i.e., backward whirl), as indicated in the middle orbit shown in Fig. 1.80. While this does not have any major practical consequence and is rarely achieved in real machinery, it is a rather surprising exception to the generality that forced vibration due to unbalance always results in forward whirl.

It has also been shown that anisotropy in bearing support stiffness can substantially enhance the stability of rotors subject to whirl instability (Sec. 1.6.2).

1.7.3 Axisymmetric Nonlinear Systems

A second important class of deviation from simple linear axisymmetric (isotropic) stator support includes systems which are axisymmetric but nonlinear, i.e., systems whose stator reaction force is not simply proportional to the rotor deflection, as shown in Fig. 1.79*b*. Such deviation from linearity is rather common in real physical systems and, when of sufficient magnitude, can have a profound effect on rotor vibratory response. The nonlinearity is generally categorized as being a *hardening* system, where the stiffness increases with

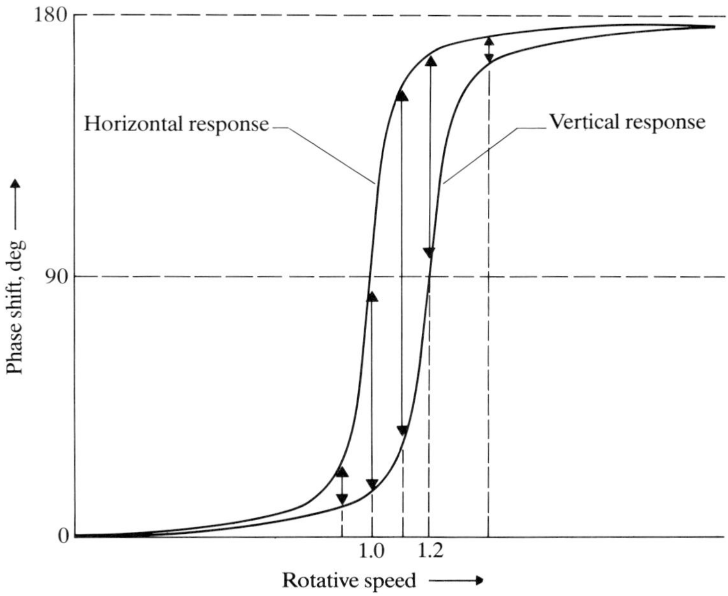

FIGURE 1.81 Phase shift in response for orthogonal axes of system with anisotropic stator.

increasing amplitude, or as a *softening* system, where the stiffness decreases with increasing amplitude. Real physical systems are, for the most part, hardening. Fluid-film bearings have this type of characteristic, as do rolling contact bearings where the contact area of the rolling element increases monotonically with radial load. An even more distinctive manifestation of this type of stiffness characteristic is seen in support elements that operate for low radial deflections within a clearance and for high radial deflections come in contact with a stiff support. This behavior is typical of rolling-element bearings that are designed exclusively for radial loads (i.e., with cylindrical rollers), for squeeze-film dampers, and for any type of bumper or deflection-limiting bearing where actual axisymmetric rub or contact during operation is experienced.

Although the nonlinearity may be manifested by any reaction-force-versus-deflection characteristic which increases in slope with increasing radial deflection, it is most easily characterized, understood, and quantified by using a simple variation consisting of two straight-line (i.e. piecewise linear) segments, as shown in Fig. 1.79*b*. For small deflections, up to the limit of the soft deflection zone, the variation is linear. For any deflection greater than the soft deflection limit, the force reaction increases linearly with increases in deflection, but at a significantly higher rate than in the soft zone. It is convenient to identify two hypothetical limiting natural frequencies for the system: (1) the low natural frequency that would pertain if there were no soft deflection zone upper limit and the system were to operate for any unlimited amplitude in the soft linear system and (2) the higher natural frequency that would pertain if there were no soft deflection zone whatever and the system were to operate for any amplitude in the hard linear system.

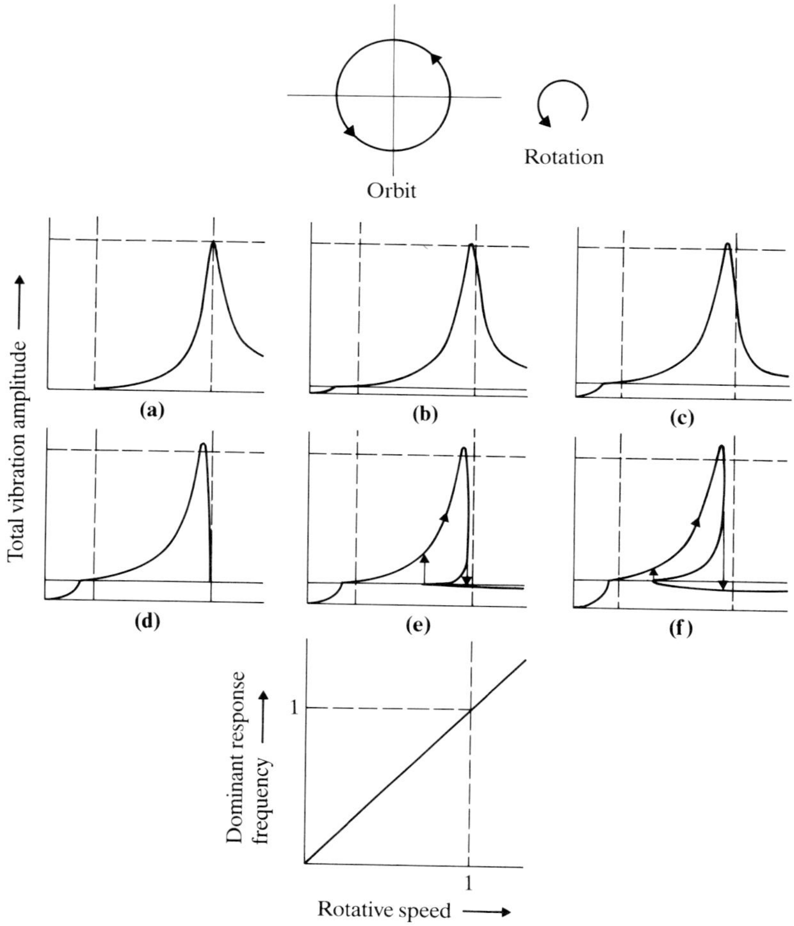

FIGURE 1.82 Response of system with isotropic (axisymmetric) nonlinear stiffness with varying ratio of soft deflection limit to unbalance eccentricity: (*a*) 0, (*b*) 0.5, (*c*) 1, (*d*) 1.36, (*e*) 1.5, (*f*) 2.

Ehrich and O'Connor (1967) and Crandall (1987) have published studies of the forced vibratory behavior of such systems. Typical forced vibratory response to unbalance is shown in Fig. 1.82 for a sequence of selected values of the soft deflection limit, normalized by the system unbalance (expressed as the eccentricity of the rotor's center of gravity) for a Jeffcott rotor. In all cases, the vibration frequency is synchronous with the rotative speed (i.e., one per revolution), the orbit is circular, and the whirling motion is in the same direction as the rotor rotation (i.e., forward whirl), but the nonlinearity has a profound effect on the shape of the response curve, as a function of the dimension of the soft deflection zone.

For the limiting case where the soft deflection zone limit is zero, the response curve is that of the linear Jeffcott rotor system with the hard stator and the high natural frequency by which it is characterized (see Fig. 1.82*a*). As the soft deflection limit is increased, three distinct trends develop.

1. The frequency of the critical speed peak steadily decreases. In the limiting case (not shown in Fig. 1.82) with the soft deflection limit greater than the peak vibration amplitude, the critical frequency reaches the lower of the two limiting frequencies designated above.
2. The maximum amplitude at the critical speed peak increases steadily as the soft deflection limit increases, with the increase being approximately equal to the soft deflection zone limit.
3. The shape of the response curve changes steadily, particularly its right side which tends to grow steeper as the soft deflection limit increases, giving rise to what is usually referred to as a rightward-leaning response peak. When the normalized soft deflection limit reaches a value somewhat above 1, the right-hand side of the response curve is virtually a vertical line, as in Fig. 1.82*d*. At still larger values of the soft deflection limit parameter, an undercut appears under the right-hand side of the response peak.

The undercut in the response curve gives rise to very distinctive vibration behavior. During operation at rotative speeds in the vicinity of the undercut, there appear to be three different possible amplitudes of vibration. In fact, the middle response line represents unstable operation, and its response levels are never seen in practice. That response curve (from the two possible stable branches) which is actually experienced at a given speed depends on the operating path taken to reach that speed. On an acceleration from a low subcritical speed, the system responds with vibration amplitudes characterized by the upper branch and shifts abruptly to the lower branch on the right side of the peak, as shown in Fig. 1.82*e* and *f*. During deceleration from very high supercritical rotative speeds, the vibratory response is characterized by the lower branch and the amplitudes of the critical peak are not experienced. The amplitude exhibits a small jump to the upper response curve when the speed is reduced to a value lower than that of the left-hand end of the undercut. This hysteresis in response—experiencing a high peak with a very steep right-hand side on acceleration and finding virtually no peak on deceleration—is quite unique and characteristic of nonlinear stiffness, so its appearance is useful as a diagnostic indication when aberrant behavior of rotating machinery is being analyzed.

For the less usual case of a softening stator support stiffness, hysteresis in response is also experienced, but in the opposite sequence. On acceleration of the rotor from low subcritical rotative speed, virtually no peak is seen. On deceleration from high speed, a critical speed peak is observed with a very steep left-hand side.

Ehrich and O'Connor (1967) and Crandall (1987) also show other unique behavior of such systems, including supercritical jumps in response amplitude and response branches that are disconnected from the main response curve. But observations of these latter characteristics have not been reported on actual operational rotating machinery and are not considered realistic possibilities.

1.7.4 Systems with Planar Asymmetry

Still another manifestation of nonlinearity and asymmetries in stator system, which can cause significant modification in response, is that categorized as planar asymmetry. The general form of the asymmetry is shown in Fig. 1.79*c*. On one axis, say, the horizontal, the stiffness is linear, as in the standard reference model.

The component of steady-state response behavior at any given rotative speed to forced excitation from unbalance along that axis is essentially unmodified from that of the reference axisymmetric linear model. But on the other orthogonal axis, say, the vertical one, we hypothesize a simple planar asymmetry in the stator's stiffness characteristic—i.e., the stiffness k_1 for motion in one direction (say, up) from the rest position is less than the stiffness k_2 for motion in the other direction (say, down) from the rest position.

Two classes of this type of nonlinearity are found in real apparatus:

1. Systems that are normally supported stiffly (i.e., with stiffness k_2) in both (vertical and horizontal) directions may develop looseness or clearance for motion in one direction which gives rise to a reduced stiffness (k_1) in that direction. This may be a consequence of wear at some interface of the support system or of loosening of the mount fasteners.
2. Some systems that are designed to be normally supported softly (i.e., with stiffness k_1) in both (vertical and horizontal) directions, such as fluid-film bearings, squeeze-film dampers, and rolling-element bearings with cylindrical bearings, may encounter an inadvertent stiff resistance to motion in one direction giving rise to an increased stiffness (k_2) in that direction. This may be the consequence of eccentric misassembly of rotors that are normally intended to be assembled in the center of the clearance and always operate within the clearance, so that the rotor and stator are just making contact at one circumferential location at the rest position. It may also result from the elastic or thermoelastic movement of the centerline of a stator support during operation so that the same sort of point contact or rub is seen.

As a result of this asymmetry in stiffness, the natural periodic motion of the system is effectively *bouncing,* as shown in Fig. 1.83*a,* rather than the simple harmonic oscillation characteristic of linear systems.

One consequence of this unique behavior is a shift in the system's natural and critical frequency. Instead of being $\Omega_1 = (k_1/m)^{1/2}$ for the normally soft system or $\Omega_2 = (k_2/m)^{1/2}$ for the normally stiff system, it lies somewhere between the two at

$$\Omega = \frac{2\Omega_1\Omega_2}{\Omega_1 + \Omega_2} \tag{1.84}$$

But there is another more profound and unexpected effect on the response of the system to forced vibration from unbalance. Ehrich (1966) first noticed and analyzed the appearance of pseudo-critical-speed peak in the response amplitude for a rotating machine operating at twice its fundamental critical rotative speed. The frequency of vibration was the fundamental natural frequency, i.e., at exactly one-half the operating speed. He referred to the phenomenon as *subharmonic vibration.* Bently (1974) did systematic experimental work on the second- and third-order responses of an experimental rotor, observing pseudo-critical-speed peaks at, respectively, 2 times the fundamental critical speed with $\frac{1}{2}$ per revolution frequency and at 3 times the fundamental critical speed with $\frac{1}{3}$ per revolution frequency. Bently referred to the phenomenon as *forced subrotative speed dynamic action.* Childs (1982) analyzed the problem, expanding on Bently's modeling of the problem, and referred to the response as *fractional-frequency rotor motion.* Muszynska (1984) expanded the record with observations of subharmonic pseudo-critical-speed peaks on an experimental rotor up to the fourth order, i.e., occurring at 4 times the fundamental critical speed with

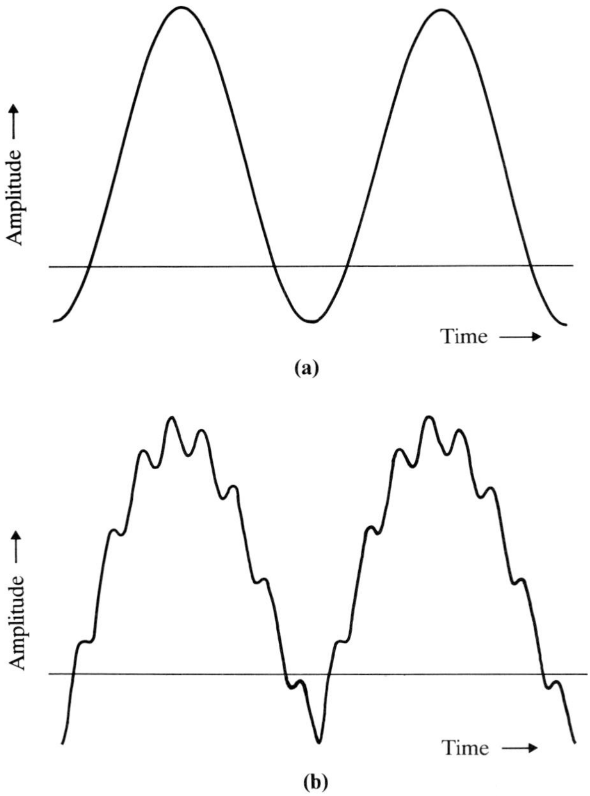

FIGURE 1.83 (*a*) Fundamental waveform at critical speed; (*b*) waveform at 8 times critical speed.

vibration at $\frac{1}{4}$ per revolution and recorded orbits and waveform spectra. Choi and Noah (1987) derived analytically the response for a system up to the third order, and they also noted the existence of a pseudo-critical-speed peak at a rotative speed one-half the fundamental critical rotative speed with the frequency of vibration at the natural or critical frequency (i.e., at twice the rotative speed, giving rise to the nomenclature *superharmonic vibration* for that particular peak). Ehrich (1988) reported observations of eighth- and ninth-order subharmonic pseudo-critical-speed peaks, and he completed an extensive numerical analysis of the phenomenon, including prediction of the complete response curve, prediction of typical orbits, and assessment of the effects of the degree of nonlinearity and damping.

Some of Ehrich's (1988) results are shown in Fig. 1.84 for the response of a typical system with planar asymmetry in the stator stiffness of a high-speed rotor. The most obvious and practically important feature of the response is the appearance of pseudo-critical-speed peaks at 2, 3, 4, and N times the fundamental critical speed with *subharmonic* response frequencies at $\frac{1}{2}$, $\frac{1}{3}$, $\frac{1}{4}$, and $1/N$ of the rotation speed, respectively. The orbits are elongated at the pseudo-critical-speed

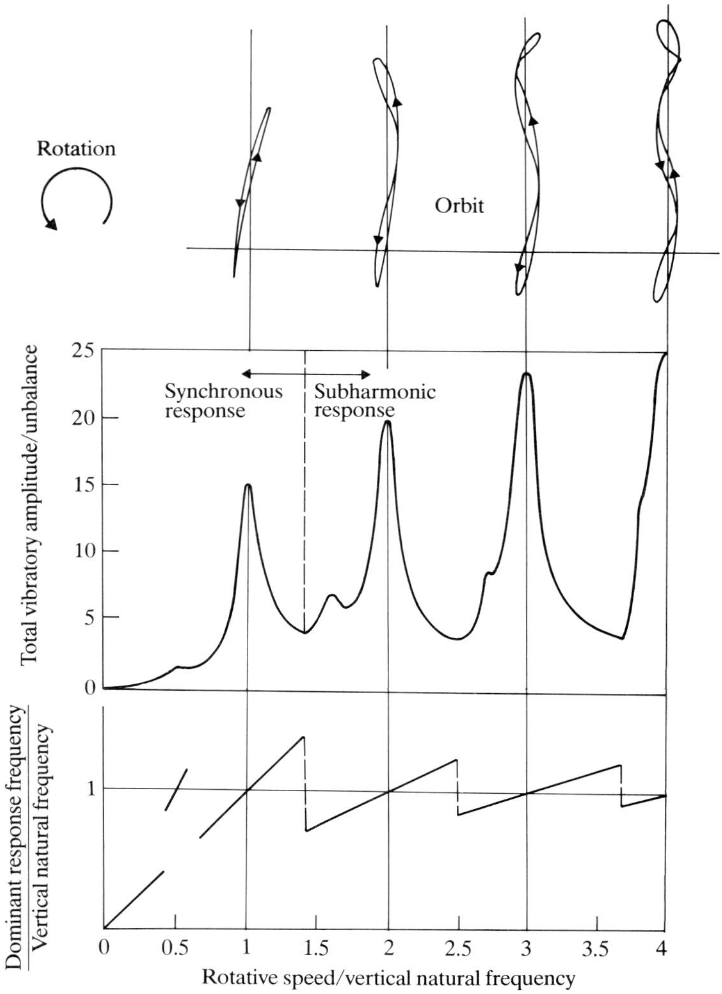

FIGURE 1.84 Response of system with planar asymmetry.

peaks because there is no large response in amplitude on the linear horizontal axis. The orbit is also highly convoluted because of the superimposed one-per-revolution motion that is actually driving the subharmonic response. One may think of the bouncing rotor as being energized at every impact with the hard stator by every Nth impulse of the rotor as it makes contact with the stator, as suggested by the view of the vertical component of the orbit in Fig. 1.83*b*.

Another important feature of the response is the appearance of small regions of erratic response behavior in the zones between the Nth and the $(N + 1)$st pseudo-critical-speed peaks where the transition is being made between the Nth- and the $(N + 1)$st-order response. Ehrich (1989*b*) has shown that the waveform in this zone is generally *chaotic,* varying erratically from behavior typical of

Nth-order response to that typical of $(N+1)$st-order response with a dominant frequency somewhere between $1/N$ and $1/(N+1)$ times the rotative speed and generally equal to the natural frequency of the system.

Figure 1.84 also gives some hint of minor peaking at rotative speeds below the natural frequency of the system (i.e., subcritically). This is termed *superharmonic* response and is manifested by the appearance of minor pseudo-critical-speed peaks at $\frac{1}{2}$, $\frac{1}{3}$, $\frac{1}{4}$, and $1/N$ of the fundamental critical speed with superharmonic response frequencies at 2, 3, 4, and N times the rotative speed, respectively.

The supercritical subharmonic pseudo-critical-speed response is obviously to be avoided, since it introduces asynchronous vibration peaks in what would otherwise be the vibration-free supercritical speed operating range. Avoidance is most directly achieved by eliminating the looseness and rubs that give rise to the nonlinear stiffness, although suppression is also possible by the same measures that suppress synchronous response—increased damping and reduction of residual rotor unbalance. Subharmonic response is so unique and distinctive that its detection and recognition are an excellent tool in the diagnosis of looseness in the stator support and rotor/stator rubs in high-speed rotating machinery.

1.7.5 Nonlinear Transmission Distortion

Nonlinearity in the stiffness of a static rotor support element, even if it does not have a fundamental influence on the response characteristics of a system, can introduce considerable distortion into the otherwise simple harmonic vibrations that are generated in a system which behaves as a linear one. Typically, some clearance or looseness or dead band in the support can cause clipping of the peaks of the transmitted vibration wave. This will be sensed by components mounted on the external static structure of the machine as vibratory stimulus at frequencies other than the synchronous (one per revolution) frequencies for which the components would normally be designed. For clipping of peaks and similar distortions, the additional stimuli will be at frequencies which are whole-number multiples of the fundamental frequency.

This can be particularly troublesome in a system which has two major stimulus frequencies, such as one generated by unbalance in each of two concentric rotors. The waveform generated by superposition of two stimuli of similar amplitude and frequencies Ω_1 and Ω_2 is a *beating wave,* as shown in Fig. 1.85*a*. Ehrich (1972) has shown that clipping this type of waveform can result in significant stimuli at frequencies which are the *sum and difference frequencies* of the two input stimuli, as shown in Fig. 1.85*b*. More generally, the clipped wave form includes a whole spectrum of sum and difference frequencies of the input stimuli and their harmonics, as shown in Table 1.6. These stimuli can be troublesome if they excite resonances in components which are mounted on the machine and which have not been designed to tolerate them.

1.8 TORSIONAL AND LONGITUDINAL VIBRATION

1.8.1 Torsional Vibration

Torsional vibration is an oscillatory angular motion of a system with twisting in the shafting connecting the inertial elements. This oscillatory motion is superim-

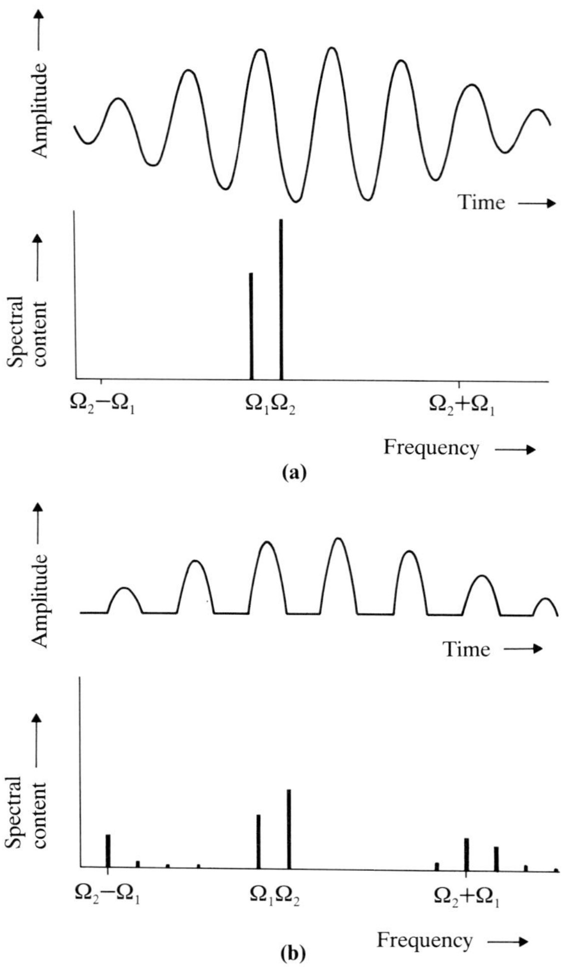

FIGURE 1.85 (*a*) Beating waveform and its spectral analysis; (*b*) distorted (clipped) beating waveform and its spectral analysis.

TABLE 1.6 Development of Sum and Difference Sideband Frequencies

Sideband frequencies			Center frequencies			Sideband frequencies		
		. . .*		0		$\Omega_2 - \Omega_1$		
	$2\Omega_1 - \Omega_2$		Ω_1		Ω_2		$2\Omega_2 - \Omega_1$	
$3\Omega_1 - \Omega_2$		$2\Omega_1$		$\Omega_1 + \Omega_2$		$2\Omega_2$		$3\Omega_2 - \Omega_1$
	etc.				etc.		etc.	

* Symmetry suggests that there should be an entry here, but a unique entry does not exist.

posed on the steady rotational motion of the machine. In modeling the system for analysis, customary practice is to calculate the natural frequencies of torsional vibration of the rotor as a body free in space, not connected to the supporting structure. For many types of machinery, such as a turbine-generator set, this assumption is a very good approximation. While it makes for very simple and accurate analysis, it implies that such systems have a level of damping which is extremely low and essentially limited to internal material damping of the rotor. The torsional damping constant of bearings—the proportionality constant between the bearing torque and rotational speed—is negligibly small for both rolling-element bearings and journal bearings. Thus, unlike lateral vibration of a rotor system where significant damping is derived from the bearings and supporting structure, the simple torsional system is one with very low inherent damping.

Very large amplitudes of torsional vibration can be present in the rotor of a machine with little or no vibration induced in the stationary structure by the torsional vibration. If harmonic exciting torques are present in the system, either in the driving element or in the reactive torque of the driven element, then the possibility exists of a destructive failure of the shafting elements of the rotating system due to the torsional fatigue.

For systems with gearing, the tooth contact force between mating gear elements imposes lateral forces on the bearings and a moment reaction on the gear case. If the system is in a state of torsional vibration, then cyclic force reactions are superimposed on these steady forces. Some vibration can be induced in the stationary structure by the cyclic forces, but generally it is masked by the background machinery noise. However, if the torsional vibration amplitude is large enough to unload the gear teeth, then a clattering noise ensues as a result of the backlash in the gearing and provides a warning of the presence of torsional vibration. This vibration, if allowed to continue, can damage tooth contact surfaces. The cyclic component of the load on the bearings provides some damping as a result of the squeeze-film action.

When the frequency of any harmonic component of the exciting torque is equal to one of the natural torsional frequencies of the rotor, then a condition of resonance exists. The speed at which this coincidence of frequencies occurs is called a *critical speed.* For each natural frequency a series of critical speeds correspond to the various harmonics of the exciting torque. It is important that a dynamic analysis of the system be made and that the forced response of the system be evaluated for those critical speeds in or near the operating range of speed. High levels of response in the operating range of speed should be avoided if at all possible; otherwise special damping devices may be required to limit the resonant response. First we highlight basic concepts in torsional analysis by considering a simple system, two inertias connected by a shaft with torsional flexibility. then we review practical design considerations and traditional calculation procedures as they relate to the two specific machinery systems listed below. Each system is characterized by a different source of torsional excitation.

- *Diesel engine and generator.* For this system, the exciting torque is found in the engine itself. The driving torque is highly irregular with many strong harmonics. Torsional vibration analysis from the very early days has been an essential part of the design and application of all types of multicylinder reciprocating engines. It is frequently necessary to add special damping devices to limit resonant responses.

- *Marine steam-turbine propulsion system.* In this case the exciting torque is found in the driven element, the ship's propeller. Because of the asymmetry of the flow conditions at the stern of the ship in the region of the propeller, torque pulsations are present which excite the entire system. Fortunately the propeller also provides damping. The steam-turbine torque itself is essentially uniform.

A third system (which is of some general interest but is not treated in detail) is a utility steam turbine and generator. When a power system turbine-generator set is operating under normal design conditions, there is no torsional excitation present and hence no torsional vibration problem. Both the turbine torque and the generator air-gap torque are completely uniform. However, abnormal transients in the system to which the generator is attached can, under certain conditions, produce high torsional amplitudes of vibration of limited duration.

The Simple Two-Inertia System: Basic Concepts
The simple two-inertia system is a convenient starting point for introducing some basic concepts for calculation of torsional vibration frequency and forced response. The system is also a useful one for making estimates, particularly where the driving and driven elements of a simple machinery system may each be approximately represented by a lumped inertia. The simple system shown in Fig. 1.86 consists of two disks with polar mass moments of inertia I_1 and I_2 connected

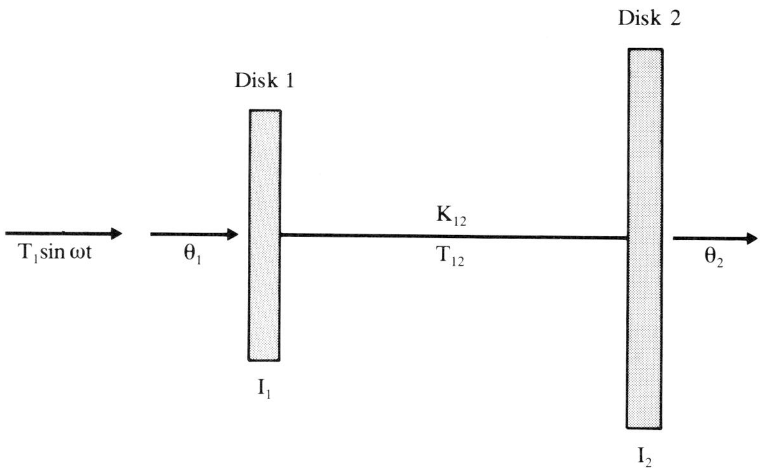

FIGURE 1.86 The simple two-inertia system.

by a massless shaft with a torsional elastic stiffness K_{12}. The angular rotations of the disks are denoted by θ_1 and θ_2. A sinusoidal exciting torque equal to $T_1 \sin \omega t$ is applied to disk 1. Results are presented for forced response derived by two different calculation procedures: exact solutions based on the governing differential equations and resonant responses based on the energy balance method. For each calculation procedure, three different conditions of damping are considered:

1. A viscous damping torque proportional to the rate of change of the angle of twist in the shaft connecting the two disks
2. A viscous damping torque proportional to the angular velocity of disk 1
3. A viscous damping torque proportional to the angular velocity of disk 2.

Exact Solutions. Solutions for the forced response of the system of Fig. 1.86 follow for the three damping conditions.

Damping Condition 1. With the damping torque equal to the viscous damping coefficient B_{12} times the difference in angular velocities of the two disks, a second-order differential equation of motion may be written in terms of the angle of twist in the connecting shaft $\psi = \theta_1 - \theta_2$:

$$\frac{I_1 I_2}{I_1 + I_2}\frac{d^2\psi}{dt^2} + B_{12}\frac{d\psi}{dt} + K_{12}\psi = \frac{I_2}{I_1 + I_2} T_1 \sin \omega t \tag{1.85}$$

For reference, we define the *static angle of twist* ψ_{st} that would exist in the connecting shaft if the exciting torque amplitude T_1 were statically applied to disk 1, producing a uniform acceleration of the rotor, as

$$\psi_{st} = \frac{I_2}{I_1 + I_2}\frac{T_1}{K_{12}} \tag{1.86}$$

A steady-state solution of Eq. 1.85 then yields a relationship for the amplification factor of the angle of twist—the amplitude of the sinusoidally varying angle of twist ψ normalized by the static angle of twist ψ_{st}:

$$\frac{\psi}{\psi_{st}} = [(1 - \tau^2)^2 + (2\zeta\tau)^2]^{-1/2} \tag{1.87}$$

where the frequency ratio τ is equal to the exciting frequency ω divided by the undamped natural frequency ν

$$\tau = \frac{\omega}{\nu} \tag{1.88}$$

with the undamped natural frequency

$$\nu = \left[\frac{K_{12}}{I_1 I_2/(I_1 + I_2)}\right]^{1/2} \tag{1.89}$$

and the damping ratio ζ is equal to the actual system damping B_{12} divided by the critical damping of the system:

$$\zeta = \frac{B_{12}}{2\nu I_1 I_2/(I_1 + I_2)} \tag{1.90}$$

A plot of the amplification factor as a function of τ and ζ is shown in Fig. 1.2*c*.

The twist amplification factor is identical to the amplification factor found in the case of forced linear vibration of a mass on a spring with damping (constant amplitude of exciting force). It is also identical to the whirl amplification factor for the Jeffcott rotor on rigid bearing supports (Sec. 1.2.1, case 2, shaft bow).

The phase angle λ between the sinusoidally varying applied torque with amplitude T_1 and the sinusoidally varying angle of twist with amplitude ψ is given by Eq. 1.2 and plotted in Fig. 1.2*b*.

Equation 1.87 is a special solution for the system shown in Fig. 1.86 which is made possible by the assumed condition of damping. This special solution gives only the difference of the amplitudes of disks 1 and 2. To obtain the individual amplitudes requires the solution of higher-order differential equations. However, this special solution does yield the important design information—the torque T_{12} in the connecting shaft

$$T_{12} = K_{12}(\theta_1 - \theta_2) = K_{12}\psi \tag{1.91}$$

It is often sufficient in design problems to consider only the amplitude of the shaft torque and the resultant torsional stress at resonance ($\tau = 1$). Making use of Eqs. 1.86 and 1.87 yields an expression for the amplitude of the torque T_{12} at resonance

$$|T_{12}| = \frac{T_1[I_2/(I_1 + I_2)]}{2\zeta} \tag{1.92}$$

The assumed form of the damping in the previous analysis is actually representative of material damping in the rotor. Such damping is commonly expressed by the logarithmic decrement δ, which is the natural logarithm of the ratio of the amplitude of any cycle of vibration to the amplitude of the next cycle [$\ln(x_1/x_2)$] when the system is in a state of freely decaying vibration [or approximately equal to the fractional decay in amplitude of any cycle to the next cycle ($\Delta x/x$)]. For the system as defined by Eq. 1.85, the damping ratio is a simple function of the logarithmic decrement

$$\zeta = \frac{\delta}{2\pi} \tag{1.93}$$

This permits us to express the torque amplitude (Eq. 1.92) in the alternative form

$$|T_{12}| = \frac{T_1[I_2/(I_1 + I_2)]\pi}{\delta} \tag{1.94}$$

A representative value for the logarithmic decrement of a rotor steel might be 0.01, which would give a value of the damping ratio equal to 0.0016 and an extremely high amplification factor of 314 at resonance ($\tau = 1$). This level of damping is too low to be of practical significance. It is frequently necessary to add special damping provisions in rotating machinery where a condition of torsional resonance exists.

Damping Condition 2. For the case in which the damper acts on the same disk where the sinusoidal driving torque is applied, the viscous damping torque on disk 1 is equal to $B_1(d\theta_1/dt)$. In this case, the simplicity of the solution for damping condition 1 is not possible. The general solution for the vibratory angular amplitudes θ_1 and θ_2 must be derived separately from the solution of differential equations of order higher than the second. But a more manageable perspective which still gives an insight into the general behavior of the system is available by examination of the response at resonance only ($\tau = 1$):

$$\theta_1 = -\frac{T_1}{B_1 \nu} \cos \nu t \tag{1.95}$$

$$\theta_2 = \frac{T_1}{B_1 \nu} \frac{I_1}{I_2} \cos \nu t \tag{1.96}$$

The ratio θ_2/θ_1 is equal to $-I_2/I_2$ and is independent of the damping. The shaft torque is then found to be

$$|T_{12}| = \frac{T_1}{B_1} \left[\frac{K_{12}(I_1 + I_2)I_1}{I_2} \right]^{1/2} \tag{1.97}$$

Damping Condition 3. For the case where there is a damper on the disk remote from the location where the sinusoidal driving torque is applied, the viscous damping torque acting on disk 2 is equal to $B_2(d\theta_2/dt)$. In this circumstance, the approach to the analysis is similar to that taken for damping condition 2. The angular amplitudes at the resonant condition ($\tau = 1$) are found to be

$$\theta_1 = \frac{-T_1}{K_{12}} \frac{I_2}{I_1} \left(\sin \nu t + \frac{I_2}{I_1} \frac{K_{12}}{B_2 \nu} \cos \nu t \right) \tag{1.98}$$

$$\theta_2 = \frac{T_1}{K_{12}} \frac{I_2}{I_1} \frac{K_{12}}{B_2 \nu} \cos \nu t \tag{1.99}$$

Although there is no simple relationship for the ratio θ_2/θ_1, it approaches the value $-I_1/I_2$ as the value of B_2 approaches zero. The amplitude of the shaft torque is found to be

$$|T_{12}| = T_1 \frac{I_2}{I_1} \left[1 + \left(\frac{K_{12}}{B_2^2} \right) (I_2 + I_1) \left(\frac{I_2}{I_1} \right) \right]^{1/2} \tag{1.100}$$

Energy Balance Method. The energy balance method is often used for calculating the response of a system to a harmonic exciting torque when the frequency of the exciting torque is equal to an undamped natural frequency of the system. The energy input from the exciting torque and the energy dissipated by damping are evaluated on the basis of the undamped mode shape of the system. Equating these two energies yields a solution for the resonant amplitude of the system. The method is dependent on the system having low damping.

For the system of Fig. 1.86, the natural frequency ν is given by Eq. 1.89 and the mode shape by

$$\theta_2 = -\theta_1 \frac{I_1}{I_2} \tag{1.101}$$

For more complicated systems, the natural frequency and mode shape are generally determined by the Holzer method, described in Chap. 2.

For a harmonic exciting torque of amplitude T_1 and frequency ν acting on disk 1, the energy input per cycle E_i for the component of torque in phase with the harmonic angular velocity of amplitude $\nu\theta_1$ is given by

$$E_i = \pi T_1 \theta_1 \tag{1.102}$$

Damping Condition 1. With the viscous damping torque proportional to the rate of change of the angle of twist in the shaft connecting the two disks, the amplitude of the viscous damping torque acting on disk 1 is equal to $\nu B_{12}(\theta_1 - \theta_2)$, and the amplitude of the viscous damping torque acting on disk 2 is equal to $\nu B_{12}(\theta_2 - \theta_1)$. Since these damping torques are in phase with the velocities of angular rotation of disks 1 and 2, the energy dissipated per cycle E_d by each of the two damping torques may be evaluated in the same way as E_i was evaluated (Eq. 1.102) and then combined to give

$$E_d = \pi\nu B_{12}(\theta_2 - \theta_1)^2 \tag{1.103}$$

Setting the input energy per cycle E_i (Eq. 1.102) equal to the dissipated energy per cycle E_d (Eq. 1.103) and expressing the amplitude θ_2 (Eq. 1.101) in terms of the ampltiude θ_1, we find

$$E_i = \pi T_1\theta_1 = E_d = \pi\nu B_{12}\theta_1^2\left(\frac{I_1 + I_2}{I_2}\right)^2 \tag{1.104}$$

From this we derive the angular amplitude of each inertia

$$\theta_1 = \left(\frac{T_1}{\nu B_{12}}\right)\left[\frac{I_2}{I_1 + I_2}\right]^2 \tag{1.105}$$

$$\theta_2 = -\left(\frac{T_1}{\nu B_{12}}\right)\frac{I_1 I_2}{(I_1 + I_2)^2} \tag{1.106}$$

It follows from Eqs. 1.90 and 1.91 that the amplitude of the shaft torque is then

$$|T_{12}| = \frac{T_1[I_2/(I_1 + I_2)]}{2\zeta} \tag{1.107}$$

This result is identical to Eq. 1.92.

If hysteretic damping is assumed in place of viscous damping, the energy E_d dissipated by damping per cycle may, for low values of damping, be taken as twice the logarithmic decrement times the total vibrational energy of the system. This concept, while not essential for the simple two-inertia system, is useful for systems of greater complexity. The total vibrational energy of the system is most conveniently expressed as a summation of kinetic energies for the inertial elements on the system at the position of zero amplitude, where the kinetic energy is a maximum and the potential energy is zero.

$$E_d = \frac{2\delta\nu^2(I_1\theta_1^2 + I_2\theta_2^2)}{2} \tag{1.108}$$

Equating this energy dissipation to the energy input (Eq. 1.102) gives finally for the amplitude of the vibratory shaft torque

$$T_{12} = T_1\frac{I_2}{I_1 + I_2}\frac{\pi}{\delta} \tag{1.109}$$

Because the logarithmic decrement of hysteretic damping δ is relatable to the damping ratio ζ for damping condition 1 by Eq. 1.93, this result is also effectively identical to Eq. 1.107 and to the exact solution Eq. 1.92.

Damping Condition 2. With a damper located at disk 1 where the vibratory torque is applied, the viscous damping torque acting on disk 1 is equal to $\nu B_1\theta_1$.

The energy balance then becomes

$$E_i = \pi T_1 \theta_1 = E_d = \pi \nu B_1 \theta_1^2 \tag{1.110}$$

From this energy balance together with Eq. 1.101, the angular amplitudes of the inertias at resonance are found to be precisely equal to Eqs. 1.95 and 1.96, derived in the exact solution. The amplitude of the shaft torque is therefore defined by Eq. 1.97.

Damping Condition 3. With a damper located at the inertia remote from the location of torque application, the amplitude of the viscous damping torque acting on disk 2 is equal to $\nu B_2 \theta_2$, and the energy balance becomes

$$E_i = \pi T_1 \theta_1 = E_d = \pi \nu B_2 \theta_2^2 \tag{1.111}$$

From this energy balance together with Eq. 1.101, the angular amplitudes of the inertias are found to be

$$\theta_1 = \left(\frac{T_1}{\nu B_2}\right)\left(\frac{I_2}{I_1}\right)^2 \tag{1.112}$$

$$\theta_2 = -\frac{T_1}{\nu B_2}\frac{I_2}{I_1} \tag{1.113}$$

Making use of Eq. 1.91 and eliminating the natural frequency ν give for the amplitude of the shaft torque

$$|T_{12}| = \frac{T_1}{B_2}\frac{I_2}{I_1}\left[K_{12}(I_2 + I_1)\left(\frac{I_2}{I_1}\right)\right]^{1/2} \tag{1.114}$$

The amplitudes given by Eqs. 1.112, 1.113, and 1.114 are not in precise agreement with the corresponding amplitudes given by Eqs. 1.98, 1.99, and 1.100 of the exact solution. Instead, they represent limiting values of the exact-solution amplitudes as the damping B_2 approaches zero. This is consistent with the previously noted limitation of the energy balance method, which is generally applicable only to systems with low damping. Most torsional vibration problems fall in this category.

Diesel Engine and Generator

A lightweight high-speed diesel engine, similar to that referred to by Den Hartog (1956), driving an electric generator has been selected for a case study to demonstrate the procedures for analyzing the torsional vibration of multicylinder reciprocating machinery. Pertinent design data are given in Table 1.7. An

TABLE 1.7 Typical Diesel-Engine Design Data

Engine, V-angle:	4-cycle V-8 diesel; 60°
Crankshaft:	0°, 180°, 180°, 0° with firing order 1L, 1R, 3L, 3R, 4L, 4R, 2L, 2R
Rated power:	50 hp per cylinder at 2000 rpm; 400 hp for engine
Full-load torque:	1580 in · lb per cylinder
Generator:	250 kW; normal speed 2000 rpm

alternative analysis procedure is patterned after that given by Eshleman and Lewis (1988).

The Idealized System. The first step in the analysis is to reduce the system consisting of the engine, flywheel, and generator to an idealized system of lumped-inertia elements connected by massless shaft elements, as shown in Fig. 1.87. The principal problem in this step is the treatment of the reciprocating parts. A single piston, connecting rod and crank, is shown diagrammati-

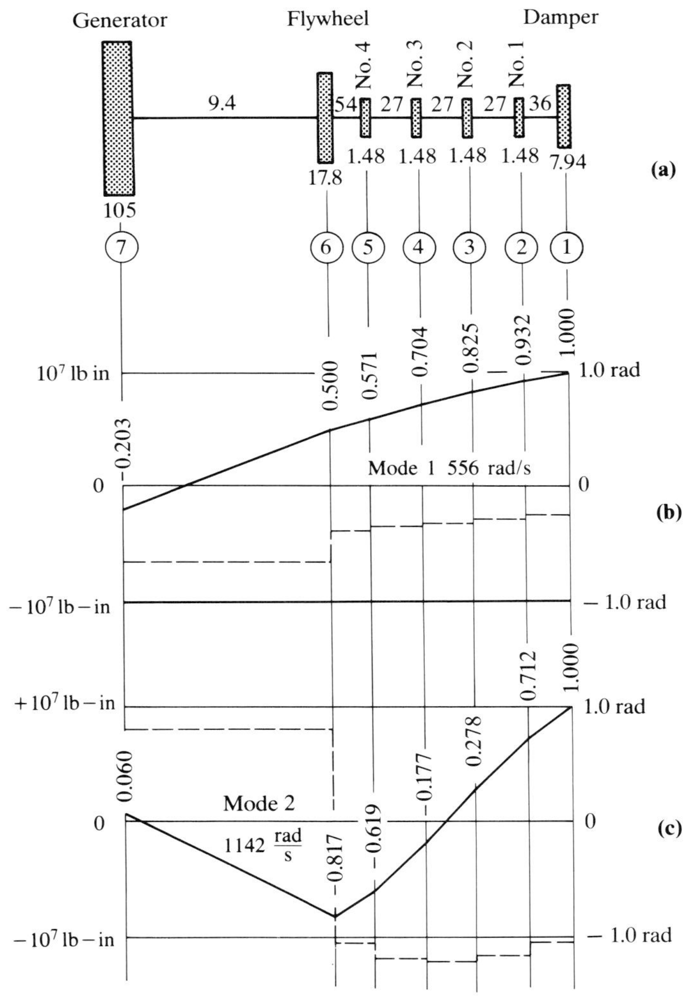

FIGURE 1.87 Idealized representation of a V-8, four-cycle diesel engine driving a generator (inertias are shown as lb · in · s^2 and stiffnesses as 10^6 lb · in/rad); (*b*) rotation amplitude and torque for mode 1; (*c*) rotation amplitude and torque for mode 2.

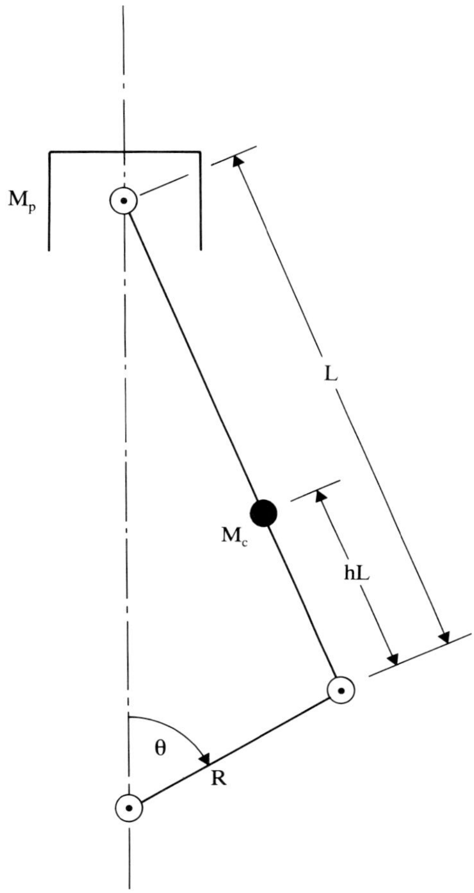

FIGURE 1.88 Schematic of piston, connecting rod, and crank mechanism.

cally in Fig. 1.88. The inertia of this assembly may, in approximate fashion, be split into two parts: an effective mass moment of inertia I_{rec} and an effective mass M_{rec}:

$$I_{\text{rec}} = \left[\frac{M_p}{2} + M_c\left(1 - \frac{h}{2}\right)\right]R^2 \tag{1.115}$$

$$M_{\text{rec}} = M_p + hM_c \tag{1.116}$$

where M_p = mass of piston
M_c = mass of connecting rod
L = length of connecting rod
hL = distance from center of gravity of connecting rod to crankpin
R = radius of crankpin

A lumped inertia consisting of the actual rotating inertia of the crankshaft element plus the effective inertia of the reciprocating parts I_{rec} is placed at each cylinder. For a V-type engine, twice the value of I_{rec} from Eq. 1.115 is added to the inertia of a single rotating crankshaft element. Typical details of the calculation of the stiffnesses of the massless crankshaft elements are given by Eshleman and Lewis (1988).

The undamped natural frequencies and the associated mode shapes of the idealized system are calculated by the Holzer method (see Chap. 2). The amplitudes of angular rotation of the lumped inertias and torque in the various massless shaft elements (constant from one inertia to the next) are shown in Fig. 1.87 for the first mode at its natural frequency of 556 rad/s and for the second mode at 1142 rad/s. Note that all amplitudes of angular rotation and torque are relative to the assumed unit amplitude of angular rotation at position 1 at the right-hand end of the idealized system.

Single-Cylinder Driving Torque. The driving torque of a multicylinder reciprocating engine is highly irregular and has numerous harmonics, each one of which can excite torsional vibration. The evaluation of these harmonics starts with the analysis of the single cylinder. There are two components of the single-cylinder driving torque which must be combined vectorially—the inertia torque and the gas pressure torque.

The Inertia Torque. Due to the reciprocating motion of the pistons and the piston ends of the connecting rods, the moment of inertia of the engine crankshaft varies cyclically with each revolution. A simplified relationship for the harmonics of the inertia torque of a single cylinder T_{rec} is given in normalized form by

$$\frac{T_{rec}}{T_0}=\frac{M_{rec}\Omega^2R^2}{T_0}\sum_{r=1} c_r \sin r\Omega t \tag{1.117}$$

where T_0 = mean driving torque of single cylinder
M_{rec} = effective reciprocating mass (Eq. 1.116)
Ω = rotational speed, rad/s
t = time, s (measured from top-dead-center firing position)
$\lambda = R/L$
r = order of the harmonic = 1, 2, 3, 4, . . .
$r\Omega t$ = angle of rotation of crankshaft from top-dead-center reference position for order r (Figs. 1.88 and 1.89)
$c_1=\lambda/4,\ c_2=-\frac{1}{2},\ c_3=-3\lambda/4;\ c_4=-\lambda^2/4\cdots$

It has been found that satisfactory accuracy is achieved by retaining only the first three harmonics in Eq. 1.117.

The Gas Pressure Torque. The variation of the gas pressure torque as a function of the crank angle is shown in Fig. 1.89*a* for a single cylinder of a four-cycle diesel engine for one complete cycle of firing (2 r). At the four dead-center positions during the 2 r of a firing cycle the torque is zero. When the engine is operated at partial load by a reduced injection of fuel, the gas pressure torque is modified in the firing quarter cycle. The dotted lines 1 and 2 indicate the variation for zero load and half load. At zero load the pressure during the firing period is equal to that during the compression period; so even when there is no average torque, there are alternating torques of considerable magnitude.

Figure 1.89*b*, *c*, and *d* shows the first three harmonics of the gas pressure torque. Since the first harmonic has a frequency of $\frac{1}{2}$ cycle/r, it is commonly referred to as the harmonic of "$\frac{1}{2}$ order." The second and third harmonics are

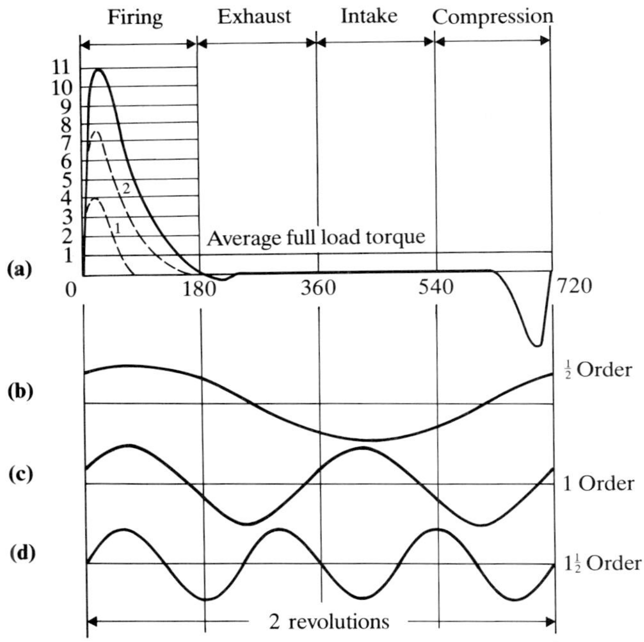

FIGURE 1.89 (*a*) Gas pressure torque of a single cylinder of a four-cycle engine; (*b*) the first harmonic component of the gas pressure torque; (*c*) the second harmonic component of the gas pressure torque; (*d*) the third harmonic component of the gas pressure torque.

then, respectively, referred to as first-order and $1\frac{1}{2}$-order. For a two-cycle engine, which has 1 complete cycle of firing in 1 r, only the integral orders of harmonics are present.

The harmonics of the single-cylinder gas pressure torque are obtained by analyzing families of indicator diagrams taken for a series of values of *mean indicated pressure* (MIP). Such information has been published by Porter (1943) for 11 different engine types where the gas pressure torque T_{gp} is represented by the Fourier series in normalized form

$$\frac{T_{gp}}{T_0} = 1 + \frac{1}{b_0} \sum_{r=1} (a_r \sin r\Omega t + b_r \cos r\Omega t) \tag{1.118}$$

where r = order of harmonic = $\frac{1}{2}$, 1, $1\frac{1}{2}$, 2, $2\frac{1}{2}$, 3, ... A typical set of normalized coefficients a_r/b_0 and b_r/b_0 is given in Table 1.8.

The inertia torque T_{rec}/T_0 and the gas pressure torque T_{gp}/T_0 must be combined vectorially. This is readily accomplished by adding to the sine coefficients a_r/b_0 of Eq. 1.118 for r = 1, 2, 3 (and higher if appropriate) the corresponding sine coefficients $(M_{rec}\Omega^2R^2/T_0)c_r$ of Eq. 1.117. The resulting combined torque T may then be represented in normalized form as

$$\frac{T}{T_0} = 1 + \frac{1}{b_0} \sum_{r=1} (a_r^2 + b_r^2)^{1/2} \cos (r\Omega t + \alpha_r) \tag{1.119}$$

TABLE 1.8 Normalized Harmonic Coefficients of the Gas Pressure Torque of a Single Cylinder of a Four-Cycle Diesel Engine*

r	A_r	a_r/b_0	b_r/b_0
0	1	0	1
$\frac{1}{2}$	2.148	1.357	1.665
1	2.322	2.139	0.903
$1\frac{1}{2}$	2.230	2.221	0.195
2	1.915	1.906	−0.186
$2\frac{1}{2}$	1.569	1.537	−0.315
3	1.290	1.235	−0.373
$3\frac{1}{2}$	1.053		
4	0.827		
$4\frac{1}{2}$	0.630		
5	0.485		
$5\frac{1}{2}$	0.378		
6	0.285		
$6\frac{1}{2}$	0.202		
7	0.142		
$7\frac{1}{2}$	0.105		
8	0.075		
$8\frac{1}{2}$	0.046		
9	0.025		
$9\frac{1}{2}$	0.017		
10	0.013		
$10\frac{1}{2}$	0.008		
11	0.011		

* Set P2 from tables by Porter (1943). MIP = 140 lb/in².

where

$$\alpha_r = \arctan \frac{a_r}{b_r} \tag{1.120}$$

The mean driving torque T_0 applies to the particular MIP for which the set of harmonic coefficients is given, and dividing the harmonic coefficients by b_0 places them on a per-unit basis relative to the mean torque. With

$$A_r = \frac{1}{b_0}(a_r^2 + b_r^2)^{1/2} \tag{1.121}$$

the summary expression for the harmonics of the single-cylinder driving torque is

$$T = T_0 \sum_{r=1} A_r \cos(r\Omega t + \alpha_r) \tag{1.122}$$

Typical values of A_r are also given in Table 1.8.

Engine Torque. We assume that the torque harmonics of all cylinders of the engine are identical and are given by Eq. 1.122. The top-dead-center

(TDC) firing position of the piston in cylinder 1 is taken as the reference point for zero crankshaft angle. The rth-order harmonic of the mth cylinder including a phase adjustment for firing time relative to the cylinder 1 is then

$$T_{mr} = T_0 A_r \cos\left[r\Omega t + \alpha_r - 2\pi r\left(\frac{\gamma_m}{360}\right)\right] \tag{1.123}$$

where $\gamma_m =$ the angle, in degrees, through which the crankshaft must rotate to bring the mth piston to the TDC firing position. The resultant harmonic of rth order for the engine is obtained by summing the contributions of all the cylinders:

$$T_r = \sum_{m=1}^{m=K} T_{mr} \tag{1.124}$$

where $K =$ the total number of cylinders.

Energy Input at a Critical Speed. If we assume that the engine is operating at a critical speed where the frequency of the torque harmonic $r\Omega$ is equal to a torsional natural frequency of the rotor ω_n (with n being the mode number—$1, 2, 3, \ldots$), then the critical speed N_c, in revolutions per minute, is given by

$$N_c = \frac{60}{2\pi}\frac{\omega_n}{r} \tag{1.125}$$

The energy input E_i to the vibrating rotor per cycle of rotor vibration is derived by integrating the work done in each cylinder by the specified torque harmonic over one period of rotor vibration in the nth mode. The sinusoidally varying angular rotation θ_{mn} of the rotor at the mth cylinder position and for the nth mode is

$$\theta_{mn} = \theta_0 \theta_{mn} \sin \omega_{nt} \tag{1.126}$$

where θ_{mn} is the angular rotation value from the Holzer calculation and θ_0 is the actual value of the angular rotation at the starting point, where a unit value was assumed. We integrate the product of the driving torque harmonic T_{mr} (Eq. 1.123) and the angular vibration velocity $d\theta_{mn}/dt$ (from Eq. 1.126) over one period of vibration for the mth cylinder and then sum the resulting energy input for all the cylinders. With the phase angle α_r of Eq. 1.120 replaced by the angle β_{rn} which is to be adjusted to maximize the energy input when all cylinders of an engine design act in the given sequence, we find

$$\frac{E_i}{T_0\theta_0} = \pi A_r (C_{rn} \cos \beta_{rn} + D_{rn} \sin \beta_{rn}) \tag{1.127}$$

where

$$C_{rn} = \sum_{m=1}^{m=K} \theta_{mn} \cos\left[2\pi r\left(\frac{\gamma_m}{360}\right)\right] \tag{1.128}$$

$$D_{rn} = \sum_{m=1}^{m=K} \theta_{mn} \sin\left[2\pi r\left(\frac{\gamma_m}{360}\right)\right] \tag{1.129}$$

The expression $C_{rn} \cos \beta_{rn} + D_{rn} \sin \beta_{rn}$ has a maximum value equal to $(C_{rn}^2 + D_{rn}^2)^{1/2}$ when $\beta_{rn} = \arctan (D_{rn}/C_{rn})$ so that the maximum energy input E_i is given by the simple relationship

$$\frac{E_i}{T_0\theta_0} = \pi A_r (C_{rn}^2 + D_{rn}^2)^{1/2} \tag{1.130}$$

Sample Calculation for Energy Input. To illustrate the use of Eq. 1.130, a system with the basic engine parameters given in Table 1.7 is analyzed. Since the engine is a V type, opposing pairs of cylinders share the same crankpin but are inclined one to the other by the 60° V angle. The TDC firing position of cylinder 1 (left bank) is taken as the zero reference position for the rotation angle of the crankshaft. The rotation angle γ_m required to bring the mth piston into its TDC firing position is determined by the crank angle, the firing order, and, for the right bank, the V angle.

The amplitude of torsional vibration θ_{mn} is taken from Fig. 1.87 for cylinder locations 2 through 5 and for mode 1. The same amplitude of torsional vibration is assigned to each cylinder of an opposing pair. A sample calculation is presented in Table 1.9 for the energy input E_i for the particular critical speed where the frequency of the fourth-order harmonic of the torque ($r = 4$) is equal to the natural frequency of the first mode of vibration ($n = 1$).

TABLE 1.9 Calculation of Energy Input in a V-8 Four Cycle Diesel Engine at the Critical Speed of Order 4 and Vibration Mode #1

Cylinder number m	Fixed crank angle (deg)	Firing order	Crank rotation to firing position γ_m (deg)	Torsional vibration amplitude θ_{m1} (rad)
1 (1L)	0	1	0	0.932
2 (1R)	0	2	60	0.932
3 (2L)	180	7	540	0.826
4 (2R)	180	8	600	0.826
5 (3L)	180	3	180	0.705
6 (3R)	180	4	240	0.705
7 (4L)	0	5	360	0.572
8 (4R)	0	6	420	0.572

with:

$$N_c = \text{critical speed} = \left(\frac{60}{2\pi}\right)\left(\frac{556}{4}\right) = 1327 \text{ rpm} \qquad \text{from (Eq. 1.125)}$$

$$C_{41} = \sum_{m=1}^{m=8} \theta_{m1} \cos 4\gamma_m = 1.516 \qquad \text{from Eq. 1.128}$$

$$D_{41} = \sum_{m=1}^{m=8} \theta_{m1} \sin 4\gamma_m = -2.625 \qquad \text{from Eq. 1.129}$$

$$A_4 = 0.827 \qquad \text{from Table 1.8 for MIP} = 140 \text{ lb/in}^2$$

$$\frac{E_i}{T_0\theta_0} = \pi(0.827)(1.516^2 + 2.625^2)^{-1/2} = 7.87 \qquad \text{from Eq. 1.130}$$

Critical speed spectra covering operating speeds from 800 to 2200 rpm are given in Fig. 1.90 for the first and second modes of vibration. Values of the energy input factor for each critical speed are displayed in bar graphs and are based on full-load operation over the operating speed range. The harmonic order at each critical speed is indicated by the circled number. The maximum value of the energy input occurs at a speed of 1329 rpm and is due to the torque harmonic of order 4 with the system vibrating in the first mode, as in the case considered in the sample calculation of Table 1.9. Energy inputs are generally lower for the second mode of vibration, with a maximum of 3.23 occurring at 1983 rpm (order $5\frac{1}{2}$). There is a cancellation of energy input at the critical speed of order 3 for the first mode, and the critical speeds of orders 2 and 1 lie above the range of rotational speeds of interest. For those reasons, there was no need to include the harmonic coefficients of the inertia torque (orders 1, 2, and 3) in this particular study.

It is reasonable to anticipate that there will be no problem with higher modes of vibration. The third torsional natural frequency is equal to 3420 rad/s, and for operating speeds up to 2200 rpm the only harmonics involved would be orders greater than $14\frac{1}{2}$ for which the energy input is quite small.

While the sample calculation in Table 1.9 is for a V-type engine, the procedure applies equally well to the in-line engine. The idealized system for the in-line engine has an inertia element for each cylinder, and hence each cylinder will have a unique value of angular rotation amplitude.

Resonant Response. The resonant response of the system at any given critical speed is obtained by equating the energy input per cycle of vibration E_i to the energy dissipated in damping per cycle E_d. For the engine being analyzed, damping is supplied to the system by a Houde type of viscous damper located at the outboard end of the engine (position 1 in the idealized system). Other damping effects are present in the system, such as the squeeze-film action in the journal bearings, material damping in the shafting and between fitted parts, and energy absorbed in the engine frame and foundation. Since these effects are highly variable and difficult to evaluate, they are not considered in the calculation of the energy dissipation, and so the calculated response will be conservative by whatever amount these other damping effects may add.

As shown in Fig. 1.91*b*, the Houde damper consists of a loose flywheel *A* free to turn on a bushing *B*. These elements are enclosed in a casing *C* fixed to the shaft. A viscous medium fills the space between the flywheel and the casing.

The differential equation governing the motion of the flywheel is

$$I_f \frac{d^2\theta_f}{dt^2} + B\left(\frac{d\theta_f}{dt} - \frac{d\theta_1}{dt}\right) = 0 \tag{1.131}$$

where θ_1 = angular rotation of rotor at position 1
θ_f = angular rotation of flywheel
I_f = polar mass moment of inertia of flywheel
B = coefficient of viscous friction between flywheel and casing

With $\theta_1 = \theta_0 \sin \omega t$ we find for the steady-state angular rotation of the flywheel

$$\theta_f = \left(\frac{\mu}{1+\mu^2}\right)(\mu\theta_0 \sin \omega t - \theta_0 \cos \omega t) \tag{1.132}$$

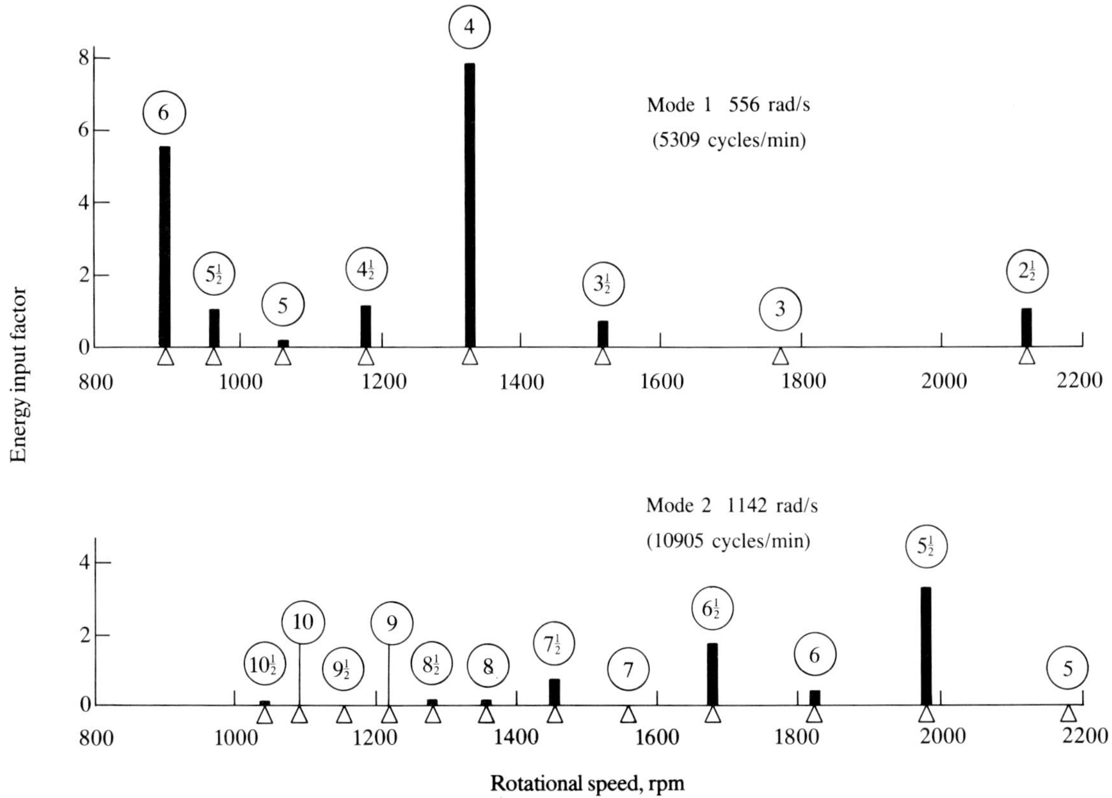

FIGURE 1.90 Critical speeds and energy input factors for modes 1 and 2.

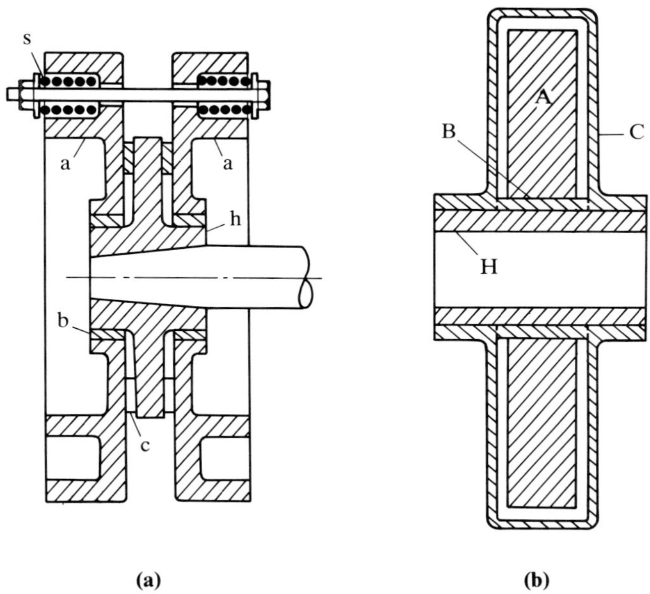

FIGURE 1.91 Torsional dampers for reciprocating engines: (*a*) Lanchester damper with friction element; (*b*) Houde type of viscous Lanchester damper.

where the quantity μ is the dimensionless parameter

$$\mu = \frac{B}{I_f \omega_n} \tag{1.133}$$

The vibratory torque reaction of the flywheel on the rotor T_r is given by

$$T_r = -B\left(\frac{d\theta_f}{dt} - \frac{d\theta_1}{dt}\right) \tag{1.134}$$

and finally

$$T_r = \left(\frac{\mu}{1+\mu^2}\right)(I_f \omega_n^2 \theta_0)(-\mu \sin \omega_n t + \cos \omega_n t) \tag{1.135}$$

The cosine term in Eq. 1.135 is in phase with the vibratory angular velocity of the rotor $d\theta_1/dt$ and is the damping component of the torque T_d. The sine term is in phase with the acceleration of the rotor $d^2\theta_1/dt^2$ and is the acceleration component of the torque T_a.

The energy dissipated in damping per cycle of vibration is found by integrating the torque times the vibratory angular deflection over one full cycle, giving

$$E_d = \frac{\pi I_f \omega_n^2 \theta_0^2 \mu}{1+\mu^2} \tag{1.136}$$

An examination of the acceleration component T_a of Eq. 1.135 reveals that the effective inertia of the loose flywheel relative to the rotor is equal to $I_f\mu^2/(1+\mu^2)$. Thus the polar moment of inertia I_1 to be used in the system is

$$I_1 = I_c + \frac{I_f\mu^2}{1+\mu^2} \tag{1.137}$$

where I_c = polar mass moment of inertia of the casing and other associated rotor mass.

The quantity $\mu/(1+\mu^2)$ has a maximum value of 0.5 when the parameter μ is equal to unity. Thus the optimum damping coefficient B for the Houde damper is simply equal to the product $I_f\omega_n$. For the sample calculation of the system in Fig. 1.87, the polar mass moment of inertia of the flywheel I_f is equal to 8.72 lb · in · s^2. As shown in the critical speed spectra of Fig. 1.90, the highest values of energy input are present for the first mode of vibration whose frequency is 556 rad/s. Designing the Houde damper for optimum damping ($\mu = 1$) at this frequency requires a damping coefficient B which, from Eq. 1.133, must be equal to (1)(8.72)(556) = 4850 lb · in · s. This value of damping is built into the damper by proper choice of design parameters, such as the clearance between the flywheel and the casing and the type of damping fluid.

With the optimum damping, then, the polar moment of inertia I_1 to be used in the system is, from Eq. 1.137,

$$I_1 = I_c + \frac{I_f}{2} \tag{1.138}$$

The polar moment of inertia of the damper casing and associated rotor mass at position 1 is 3.58 lb · in · s^2, so that the resultant polar moment of inertia I_1 = 7.94 lb · in · s^2, as shown in Fig. 1.87. The dissipated energy per cycle is, from Eq. 1.136,

$$E_d = \frac{\pi I_f \omega_n^2 \theta_0^2}{2} \tag{1.139}$$

Equating the energy input per cycle E_i (as found in Table 1.9) with the energy dissipated per cycle for this optimum damping condition (as found in Eq. 1.139) gives the final energy balance

$$E_i = 7.87 T_0 \theta_0 = E_d = \frac{\pi(8.72)(556)^2\theta_0^2}{2} \tag{1.140}$$

With the single-cylinder mean torque T_0 = 1580 lb · in from Table 1.7 we find that θ_0 = 0.00294 rad. This absolute amplitude at position 1 may be used to modify all the normalized values of angular vibratory amplitude in the Holzer calculation shown in Fig. 1.87 to obtain the actual absolute resonant amplitudes for fourth-order engine excitation with the system vibrating in the first mode.

The torque in the ith section of shaft between inertial elements i and $i+1$ is given by

$$T_i = K_i(\theta_{i+1} - \theta_i) \tag{1.141}$$

where K_i is the stiffness of the ith shaft section. The magnitudes of the torques in each massless section are plotted as dashed lines in Fig. 1.87. For the first mode the maximum torque occurs in the section between the flywheel and the generator. Multiplying the normalized Holzer torque by the actual amplitude of

0.00294 rad gives the value of maximum torque of 19,428 lb · in. Some feeling for this magnitude may be gained by comparing it with the mean torque of the engine of 12,640 lb · in, obtained from a simple sum of the mean torques of the eight cylinders. The cyclic torque is 54 percent higher than the mean driving torque. Since diesel-engine systems are normally designed with very conservative mean stresses, this may be acceptable. Final judgment must be based on an evaluation of the cyclic torsional stresses in shafting and coupling elements.

The Houde viscous damper that has been optimized for the first mode of vibration would not be optimum for any other frequency. For the second mode of vibration, the dimensionless damping parameter μ would have a value of 0.49 rather than the optimum value of unity. Energy dissipation E_d, as calculated by Eq. 1.136, would be 80 percent of that obtained for a damper optimized at this particular critical speed. A direct comparison of other parameters requires a rerun of the Holzer calculation for the second mode since the effective inertia I_1 at position 1 is a function of the damping parameter μ, as indicated in Eq. 1.137, and becomes 5.27 instead of 7.94 lb · in · s^2.

Marine Steam Turbine Propulsion System

A 30,000-hp cross-compound marine steam-turbine propulsion system has been selected for a case study. The general arrangement of the machinery is shown in Fig. 1.92*a*. The two steam turbines drive a propeller through double reduction gearing. Design information is listed in Table 1.10. The machinery system has three branches: a propeller branch with a rated speed of 85 rpm, a high-pressure turbine branch with a rated turbine speed of 6650.1 rpm, and a low-pressure turbine branch with a rated turbine speed of 3403.6 rpm. For the given system, the high- and low-pressure turbine branches have been deliberately proportioned to have equal frequencies (as indicated in Table 1.10), a common but not absolutely necessary design practice. This results in the second reduction gear (point 2 in Fig. 1.92*b*) being a torsional node or fixed point. Such a system is then termed a *nodal drive*. The advantage of this arrangement will be made clear in the results which follow.

TABLE 1.10 Basic Design Information for the Cross-Compound Marine Steam-Turbine Propulsion System

Rated power	30,000 hp	
Rated propeller torque	22.24×10^6 in · lb	
Number of propeller blades	5	
Branch frequency		
Propeller	175.17 cycles/min	
High pressure	220.17 cycles/min	
Low pressure	220.17 cycles/min	
	Rated speed (rpm)	**Speed ratio n**
Propeller	85.0	1.000
High-pressure intermediate shaft	799.8	9.409
High-pressure turbine	6650.1	78.236
Low-pressure intermediate shaft	799.8	9.409
Low-pressure turbine	3403.6	40.042

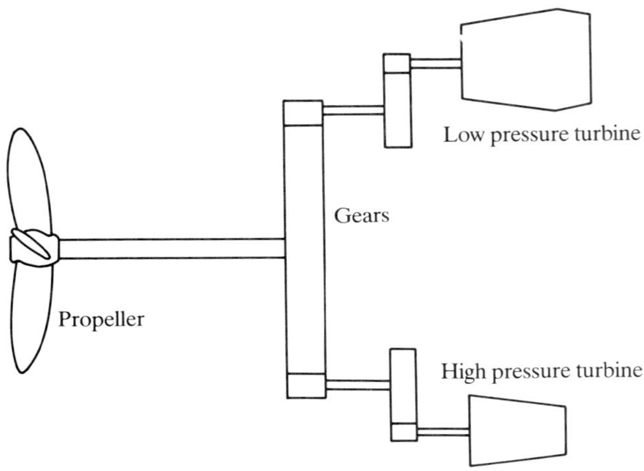

(a)

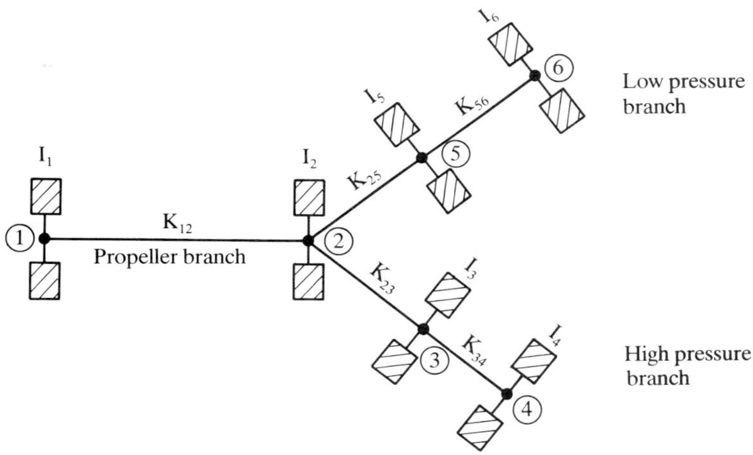

(b)

FIGURE 1.92 Cross-compound marine steam turbine propulsion system: (*a*) actual system configuration, (*b*) equivalent system.

For the purposes of analysis, customary design practice is to convert the actual system of Fig. 1.92*a* to an equivalent system as depicted in Fig. 1.92*b*, where the rotative speeds of the elements in the turbine branches are set equal to the speed of the propeller branch. If each value of polar moment of inertia and each value of torsional stiffness of the actual system is multiplied by the square of the applicable speed ratio n as given in Table 1.10, the equivalent system will have the same natural frequencies as the actual system. The values of the inertias I and the stiffnesses K for the equivalent system are listed in Table 1.11.

TABLE 1.11 Values of Polar Moment of Inertia and Torsional Stiffness for the Equivalent System

Branch	Equivalent inertia In^2 (10^6 lb · in · s^2)	Equivalent stiffness Kn^2 (10^6 lb · in/rad)
Propeller	$I_1 = 2.454$	—
	$I_2 = 0.826$	$K_{12} = 826$
High pressure	$I_3 = 2.411$	$K_{23} = 2140$
	$I_4 = 1.599$	$K_{34} = 87{,}301$
Low pressure	$I_5 = 1.136$	$K_{25} = 18{,}056$
	$I_6 = 24.200$	$K_{56} = 48{,}925$

The complete system has been reduced to six lumped-inertia elements with two inertias in each branch. This modeling should give reasonable accuracy for this system. The Holzer method (Chap. 2) may be used to calculate the natural frequencies and mode shapes of the equivalent system. At a given trial frequency, separate calculations are made for each branch, starting with unit values of rotation amplitude at the outboard ends of each branch (points 1, 4, and 6 in Fig. 1.92*b*). The actual amplitudes at points 4 and 6 are then related to the assumed value at point 1 by the requirement that the rotation amplitudes of the high- and low-pressure branches be equal to that of the propeller branch at the junction point of the branches (point 2). The torque of each branch is then calculated at point 2. Any trial frequency which makes the sum of the three torques at point 2 equal to zero is a natural frequency.

The amplitudes of angular rotation of the inertial elements of the equivalent system are plotted in Fig. 1.93 for the first three modes of vibration, which have natural frequencies of 177.7, 220.2, and 1282.8 cycles/min. The fundamental torsional excitation from the propeller is at blade passing frequency. The harmonics of the propeller excitation at multiples of the blade passing frequency are generally negligible and will be ignored. The critical speeds of rotation are then simply equal to the natural frequencies divided by the number of blades (that is, 5 for the system under consideration). The critical speed for the first mode is 35.5 rpm; for the second, 44.0 rpm; and for the third, 256.6 rpm. At the first critical speed, the maximum amplitude of vibration occurs at the propeller, and a high level of resonant response to excitation from the propeller can be expected. At the second critical speed, the high- and low-pressure branches vibrate in opposition to one another with a node at the low-speed gear (point 2) and zero amplitude at the propeller. This is a consequence of the exact equality of the turbine branch frequencies (220.2 cycles/min, as given in Table 1.10). There will be no resonant response in the turbine branches to excitation from the propeller. In making the Holzer calculation for this mode, only the turbine branches need to be considered.

Before we proceed with the calculation of the resonant response of the system, a review of marine propeller characteristics is required. The steady-state torque curve of Fig. 1.94 applies when all acceleration of the ship has ceased for a given torque T_p and a balance exists between the propeller thrust and the forces resisting the ship's forward motion. This steady-state torque varies with the

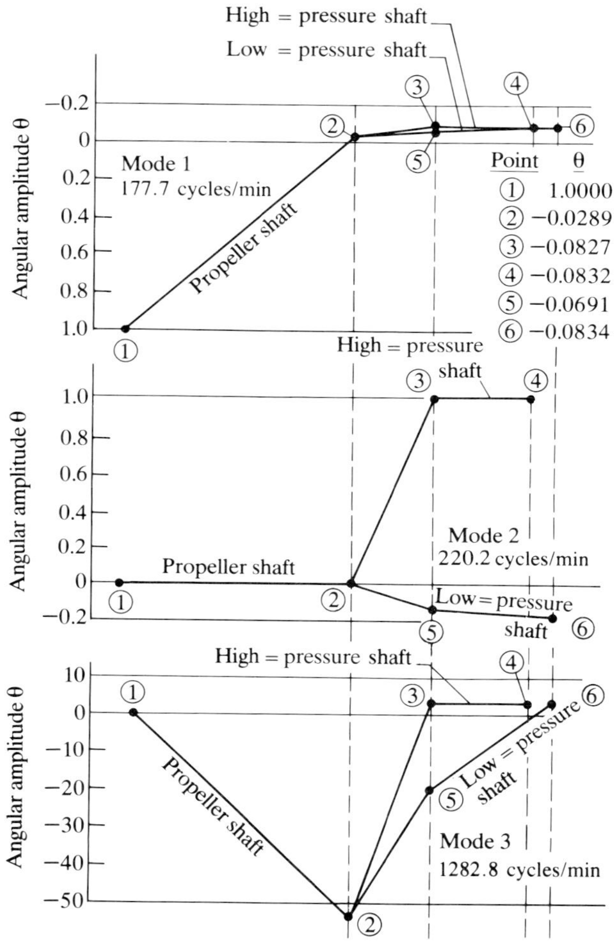

FIGURE 1.93 Mode shapes for the equivalent system—the first three natural frequencies of vibration.

square of the propeller speed Ω_p

$$\frac{T_p}{T_0} = \left(\frac{\Omega_p}{\Omega_0}\right)^2 \tag{1.142}$$

where T_0 = propeller torque at rated speed = 22.24×10^6 lb · in
Ω_0 = rated speed = 8.901 rad/s

If we assume that steady-state operation at point p in Fig. 1.94 applies and that steam flow to the turbines is suddenly increased, then the torque on the propeller immediately increases along the curve of constant ship speed (path A-A) and

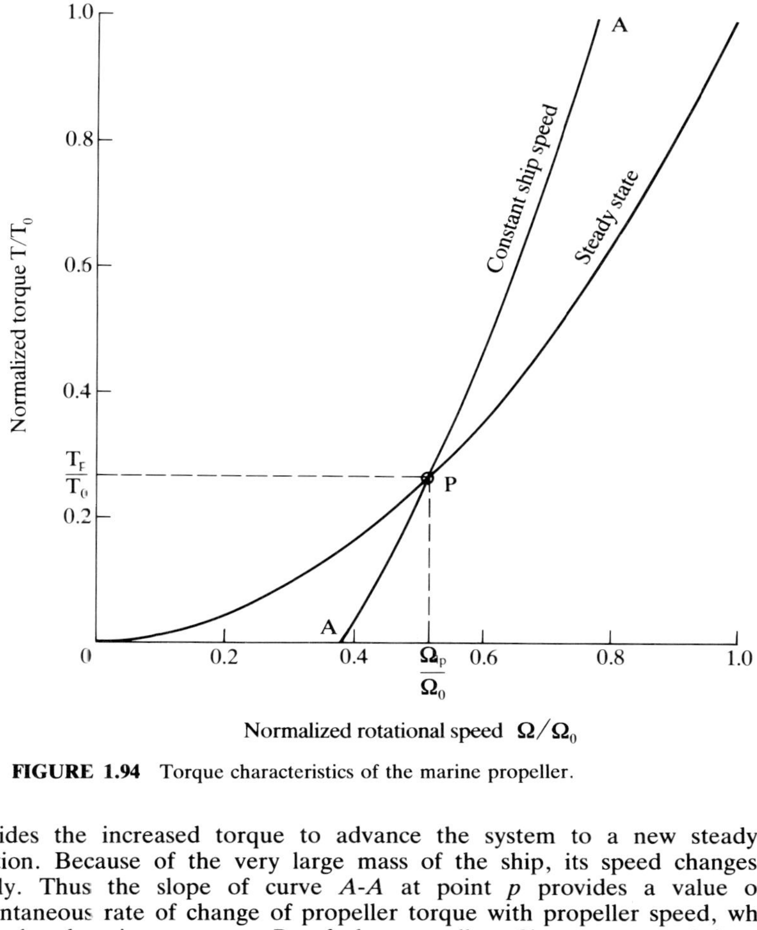

FIGURE 1.94 Torque characteristics of the marine propeller.

provides the increased torque to advance the system to a new steady-state position. Because of the very large mass of the ship, its speed changes very slowly. Thus the slope of curve *A-A* at point *p* provides a value of the instantaneous rate of change of propeller torque with propeller speed, which is then the damping constant B_1 of the propeller. Since accurate information applying to a given ship and propeller combination is usually lacking, the slope of curve *A-A* is customarily related to the slope of the steady-state torque-versus-speed curve by a multiplicative factor *C*. Values for *C* range from 1.5 to 2.0. Determining the slope of the steady-state curve at point *p* from Eq. 1.142, we find

$$B_1 = 2C\left(\frac{T_p}{\Omega_p}\right) \tag{1.143}$$

The resonant response of the system is then evaluated by the energy balance method, described under "The Simple Two-Inertia System," in Sec. 1.8.1. For convenience and ease of scaling, an arbitrary value of the propeller exciting torque T_1 equal to 10 percent of the steady-state torque T_p is assumed. At the first critical speed of 35.5 rpm ($\Omega_p = 3.718$ rad/s), the propeller steady-state exciting torque is

$$T_p = 3.88 \times 10^6 \text{ lb} \cdot \text{in} \qquad \text{from Eq. 1.142}$$

and the energy input is

$$E_i = \pi(0.10 \times 3.88 \times 10^6)\theta_1 = 1.219 \times 10^6 \times \theta_1 \qquad \text{in} \cdot \text{lb} \tag{1.144}$$

per cycle of vibration, from Eq. 1.102. Given that $C = 1.85$, the damping coefficient is

$$B_1 = 3.86 \times 10^6 \text{ lb} \cdot \text{in} \cdot \text{s/rad} \qquad \text{from Eq. 1.143}$$

For mode 1, with $\nu = 18.61$ rad/s (177.7 cycles/min), the energy dissipated in damping by the propeller is

$$\begin{aligned} E_d(\text{prop.}) &= \pi(18.61 \times 3.86 \times 10^6)\theta_1^2 \\ &= 225.7 \times 10^6 \times \theta_1^2 \text{ in} \cdot \text{lb/cycle} \qquad \text{from Eq. 1.110} \end{aligned} \tag{1.145}$$

The damping due to elastic hysteresis in the shafting, sliding fits, and bearings and supporting structure of the gear elements is less well established than the propeller damping. A value of logarithmic decrement δ equal to 2.5 percent is commonly used as an estimate of this combination of damping effects. The energy dissipation per cycle is calculated by an extension of Eq. 1.110 to cover the six-inertia system of Fig. 1.92*b*:

$$E_d(\text{hyst.}) = 2\delta \sum_{i=1}^{6} \frac{I_i \nu^2 \theta_i^2}{2} \tag{1.146}$$

With values of I_i from Table 1.10 and θ_i from Fig. 1.93 for mode 1, we then find

$$E_d(\text{hyst.}) = 23.0 \times 10^6 \times \theta_1^2 \qquad \text{in} \cdot \text{lb/cycle} \tag{1.147}$$

The negative slope of the torque-versus-speed curve of the prime mover (the steam turbine) provides another source of damping. For an impulse steam turbine which is designed for peak efficiency at rated speed, the torque with rated steam flow varies from approximately double the rated torque at zero speed to zero torque at approximately double the rated speed, as plotted in Fig. 1.95. The parabolic variation of power output is indicated by the dashed curve. The damping coefficient of the turbine B_t is equal to minus the rate of change of torque T with speed of rotation Ω at any given steady-state operating point. For steady-state operation with rated steam flow at the design point (torque T_0 and speed Ω_0), the coefficient B_t is equal to T_0/Ω_0. For operation at reduced steam flow (steady-state point t in Fig. 1.95), the damping coefficient is taken equal to the value for rated steam flow multiplied by the torque ratio T_t/T_0. This torque ratio is equal to the square of the speed ratio Ω_t/Ω_0 since for steady-state operation the turbine follows the propeller characteristic. The damping coefficient is then

$$B_t = \left(\frac{T_0}{\Omega_0}\right)\left(\frac{\Omega_t}{\Omega_0}\right)^2 \tag{1.148}$$

The damping coefficients for the individual turbines are obtained by multiplying the value from Eq. 1.148 by the appropriate fractions of the total turbine torque—α for the high-pressure turbine and β for the low-pressure turbine. At rated conditions the power is usually evenly divided between the two turbines, and for the equivalent system, the torques are equal, i.e., $\alpha = \beta = 0.50$. As steam flow is reduced, the high-pressure turbine becomes more heavily loaded and

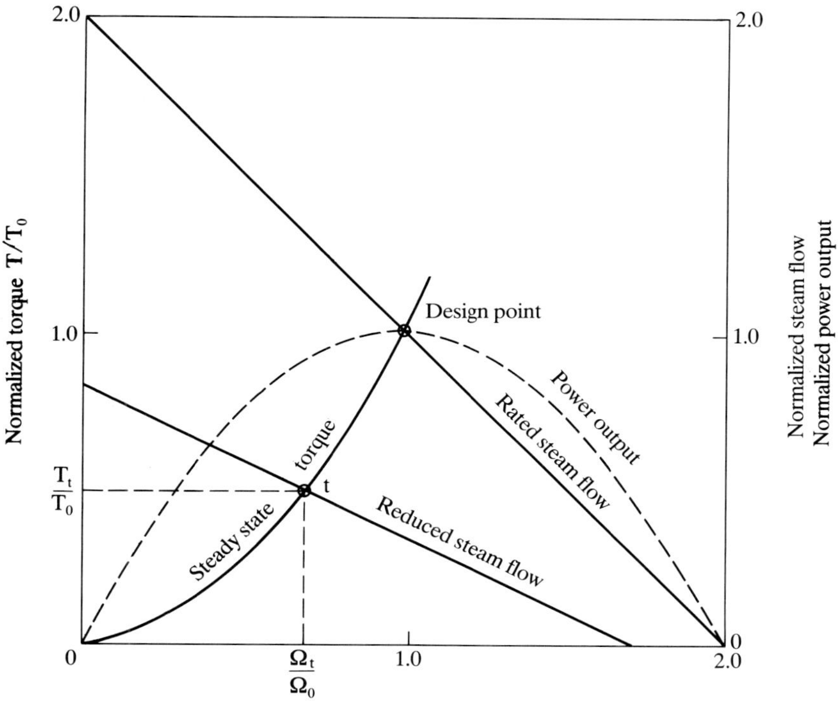

FIGURE 1.95 Torque characteristics of the steam turbine.

values of $\alpha = 0.60$ and $\beta = 0.40$ are more appropriate at the first critical speed. The energy dissipation per cycle of vibration is then

$$E_d(\text{turb.}) = \pi \nu B_t(\alpha\theta_4^2 + \beta\theta_6^2) \tag{1.149}$$

where θ_4 and θ_6 are the amplitudes of angular rotation for the high- and low-pressure turbines, respectively. Substituting numerical values in Eqs. 1.148 and 1.149 gives

$$B_t = 0.436 \times 10^6 \text{ lb} \cdot \text{in} \cdot \text{s/rad}$$

$$E_d(\text{turb.}) = 0.172 \times 10^6 \times \theta_1^2 \qquad \text{in} \cdot \text{lb}$$

A comparison of the energy dissipation in the turbine with the results for the propeller and for hysteresis in the system shows that, for mode 1, the turbine damping is negligible owing to the very small amplitudes of vibration at the turbines.

Equating the energy input from Eq. 1.144 to the sum of the energies dissipated in damping from Eqs. 1.145 and 1.147 and solving for the amplitude of vibration θ_1 at position 1 yield a value of 0.00490 rad for the damped vibratory response at the first critical speed. With the relative amplitudes given in Fig. 1.93, the angles of twist and the torques in the various shaft elements of the equivalent system can now be calculated. The vibratory torque in the propeller shaft is

$$T_{12} = K_{12}(\theta_1 - \theta_2) \tag{1.150}$$

which gives finally

$$T_{12} = 826 \times 1.0289 \times 0.00490 = 4.16 \times 10^6 \text{ lb} \cdot \text{in}$$

Normalizing the vibratory torque T_{12} by the steady-state or mean torque T_p at the first critical speed gives

$$\frac{T_{12}}{T_p} = 1.072$$

Thus the vibratory torque exceeds the mean torque by 7.2 percent, which means that there will be a torque reversal in the propeller shaft during each cycle of vibration—an undesirable condition from the point of view of the gearing. The critical level of propeller excitation at which this torque reversal sets in is 9.3 percent (i.e., 1/1.072, or 93 percent, of the assumed value of normalized exciting torque T_1/T_p, which we had arbitrarily assumed to be 10 percent). Since the expected level of excitation from propellers is actually around 3 percent, the margin with respect to torque reversal is ample. Similar calculations must be made for the shaft elements in the turbine branches. The largest value of the ratio between the vibratory torque and the steady-state or mean torque for the first critical speed is found in the intermediate shaft between the high-speed gear and the low-speed pinion of the low-pressure turbine branch (Sec. 2.5 in Fig. 1.92*b*). With 40 percent of the total turbine torque in this branch, the critical level of propeller excitation is 4.36 percent for torque reversal. Again, there is a design margin over the expected 3 percent level.

In addition to consideration of the response at the first critical speed, we must consider the fact that a certain level of forced vibration will also be present throughout the system over the entire speed range because of the propeller excitation. In Fig. 1.96, the ratio of the vibratory shaft torque to the steady-state torque T_{12}/T_p in the propeller shaft has been plotted as a function of propeller speed with the simplifying assumption of zero damping. Since the moderate level of damping present in the system will have little effect except at speeds close to the first critical speed (which has already been considered), this plot provides a satisfactory though somewhat conservative basis for design. As expected, there is no evidence of any resonant behavior at the second critical speed.

In addition to evaluating the vibratory torques in the connecting shafting, cyclic stresses need to be calculated and compared to fatigue strength. Such calculations are best made by using the geometry of the actual system. Torques from the equivalent system must be divided by the applicable speed ratios n to derive the value of torque for the actual system; and if angular rotations are desired, the equivalent system rotations should be multiplied by the speed ratios n.

Another possible source of torsional excitation in the system is inaccuracy in the gearing geometry which could excite the second mode of vibration. However, with the level of accuracy achieved in modern marine propulsion gears, response to this type of excitation is either absent or negligible. Some propulsion systems are designed without the nodal drive that we have analyzed here. If the turbine branch frequencies are not equal, then there will be a response to propeller

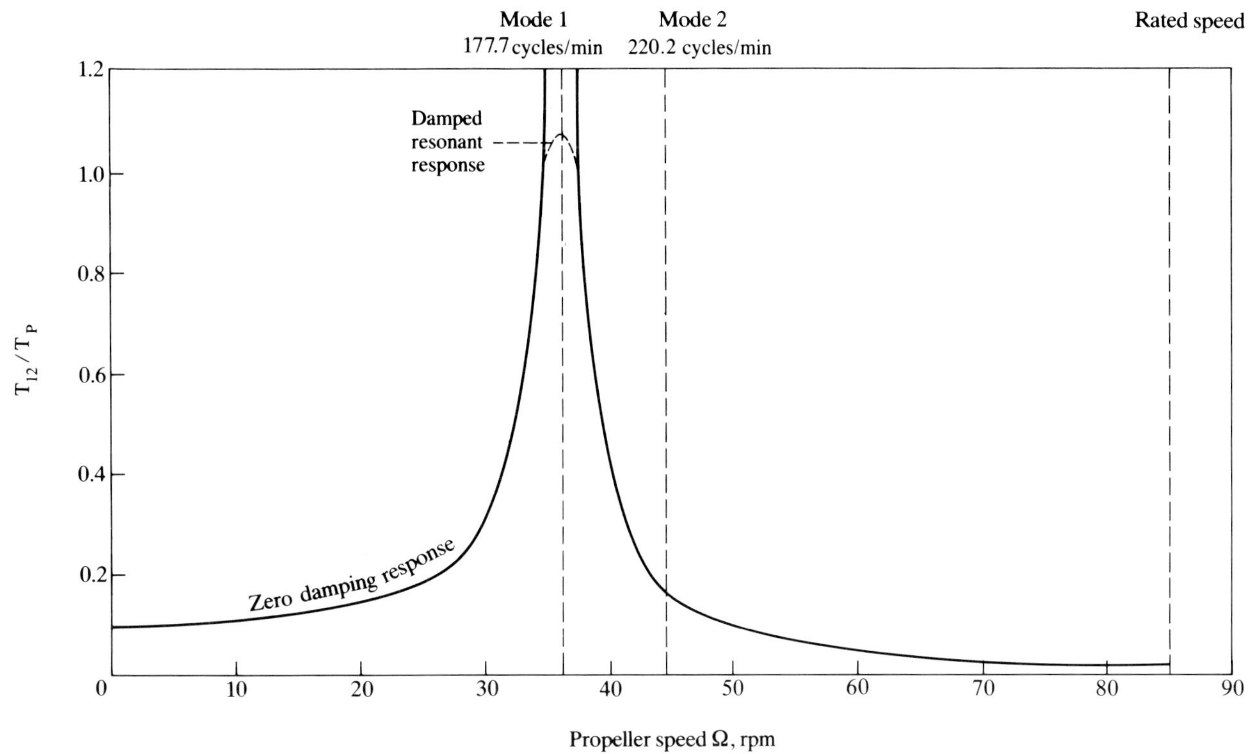

FIGURE 1.96 Vibratory torque T_{12} in the propeller shaft versus propeller speed due to a propeller exciting torque equal to 10 percent of the mean torque T_p (the mean torque varies as the square of the speed).

excitation for mode 2 dependent on the amount of separation of the branch frequencies. This is of no practical consequence provided torque reversals are avoided and cyclic stresses are within established limits. With this relaxation of the nodal drive requirement, restrictions with regard to the proportioning of elements in the turbine branches are also relaxed.

1.8.2 Longitudinal Vibration

Longitudinal vibration involves an alternating longitudinal extension and compression of the rotor body. Longitudinal vibration analysis is similar mathematically to torsional vibration analysis. If mass is substituted for polar mass moment of inertia, longitudinal displacement for angular rotation, and longitudinal force for torque, then analytical procedures for determining the natural frequencies and mode shapes of torsional vibration apply also to longitudinal vibration. However, from a practical point of view, there are significant differences between longitudinal and torsional vibration. Most rotor systems contain a thrust bearing which, for longitudinal vibration, directly couples the rotor to stationary elements of the system, and the calculation model must include these stationary elements. Proper handling of this stationary structure can be the most difficult aspect of the calculation. For torsional vibration, destructive levels of vibration may exist in the rotor with no noticeable activity in the supporting system. On the other hand, when longitudinal vibration is excited in the rotor, the most objectionable action generally occurs in the stationary structure associated with the vibrating system.

In comparison with lateral and torsional vibration, it is relatively rare in practice to encounter systems where a potentially damaging longitudinal vibration exists. Generally, no significant longitudinal exciting force is present, and even if an exciting force does exist, the natural frequency of the rotor body is usually too high relative to the excitation frequency for resonance to develop.

Ship Propulsion System

A ship propulsion system [that of the battleship *North Carolina* (BB55) which went on sea trial in 1941 and was the first in a series of large World War II naval vessels] has been selected for a case study to illustrate some of the major issues involved in analyzing longitudinal vibration in rotors, using data reported by Kane and McGoldrick (1949). The fore-and-aft vibration of the propeller shafting, the attached machinery, and the main steam piping was so severe that it was deemed unsafe to run the rotor up to rated speed. Some pertinent design data together with measured values of the first natural frequency of longitudinal vibration are given in Table 1.12. The ship had four screws with three-bladed propellers on the two outboard shafts and four-bladed propellers on the two inboard shafts. There was a longitudinal exciting force at blade passing rate from the propellers due to the flow dissymmetry at the stern of the ship. Dividing the measured natural frequency by the number of blades on the propeller gives the rotational speed, or critical speed, of each shaft at which the peak resonant response occurred. There was a large variation in shaft length for the various screws since each engine room had a different location along the length of the ship, and the critical speeds covered a range from 75 to 94 percent of the rated speed. It was because these resonances occurred close to rated speed (where the propeller exciting force was a maximum) that the excessively high levels of vibration were experienced.

Although outboard shafts 1 and 4 had the higher values of critical speed, the

TABLE 1.12 Design Information and Measured Frequencies, *U.S.S. North Carolina* (BB55)

Shaft no.	Shaft length (in)	No. of propeller blades	Measured frequency longitudinal mode 1 (cycles/min)	Critical speed: Speed (rpm)	Critical speed: % Rated speed
1	3013	3	520	173	87
2	2341	4	600	150	75
3	1813	4	680	170	85
4	2486	3	560	187	94

maximum resonant response occurred on inboard shaft 3. The inboard shafts were enclosed in deep skegs while the two outboard shafts were carried by double-arm struts, so that the excitation from the inboard propellers was double that of the outboard propellers—8 versus 4 percent as determined by measurement on the ship trials. Therefore shaft 3 has been selected for the sample calculation of the case study. The general arrangement of the geared steam-turbine propulsion machinery is shown in Fig. 1.97, and the model to be used in the Holzer-type calculation is depicted in Fig. 1.98.

The values of parameters relevant to shaft 3 which are required in the model of Fig. 1.98 are given in Table 1.13. Assuming the shafting between the propeller and the reduction gear is of uniform cross section, its dynamic behavior may be adequately approximated by dividing it into four equal parts. The spring constants k_{12}, k_{23}, k_{34}, and k_{45} are each equal to 4 times the overall spring

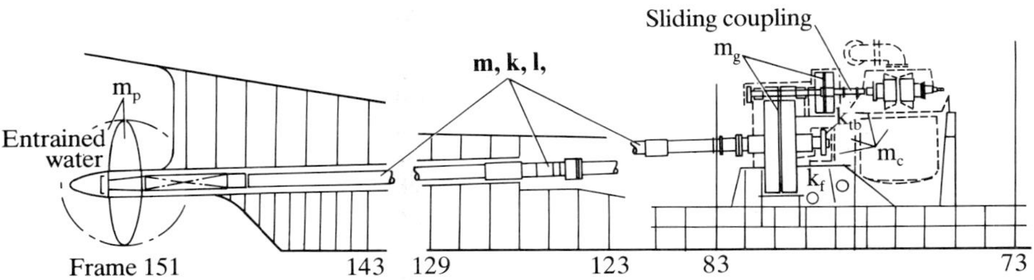

FIGURE 1.97 Geared steam-turbine ship propulsion system with thrust bearing located in forward end of gear case.

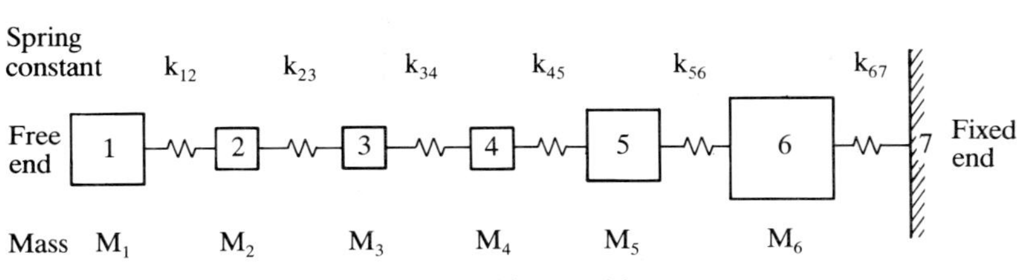

FIGURE 1.98 Calculation model of the ship propulsion system.

TABLE 1.13 System Constants and Holzer Calculation Model of Ship Propulsion System, Shaft 3 (BB55) (Fig. 1.98)

			Holzer calculation at 659 (cycles/min)		Resonant response at 164.8 r/min	
Station no. i	Spring constant k (10^6 lb/in)	Mass M (lb · s²/in)	Force amplitude F (10^6 lb)	Displacement amplitude X (in)	Force amplitude F (10^6 lb)	Displacement amplitude X (in)
1		171.5		1.000		0.1134
1–2	11.44		0.817		0.0926	
2		57.0		0.929		0.1053
2–3	11.44		1.069		0.1212	
3		57.0		0.835		0.0947
3–4	11.44		1.296		0.1470	
4		57.0		0.722		0.0819
4–5	11.44		1.492		0.1692	
5		176.0		0.591		0.0670
5–6	10.00		1.989		0.2256	
6		720.0		0.392		0.0446
6–7	8.50		3.336		0.3783	
7				0		0

constant. Masses M_2, M_3, and M_4 are each equal to one-fourth the total mass of the shafting with the remaining one-fourth of the mass evenly divided between the propeller mass M_1 and the mass of the gear rotors M_5. Mass M_1 also includes an allowance for the virtual mass of the entrained water equal to 50 percent of the propeller mass. It is assumed that the flexible couplings between the turbine rotors and the gear effectively decouple the mass of the turbine rotors from the mass of the gear rotors. The nonrotating mass of the gear and thrust bearing casting and some of the supporting structure in the ship are lumped together as mass M_6 at station 6. The spring constant k_{56} represents the local stiffness of the Kingsbury thrust bearing assembly, as determined by a static test. The spring constant k_{67} is the effective stiffness of the ship's structure between mass M_6 and an assumed fixed point in the hull at station 7.

The results of the Holzer-type calculation are also given in Table 1.13 for the first longitudinal natural mode. The calculated value of the frequency is 659 cycles/min, and with a blade rate excitation from the propeller of 4 cycles/r, the critical speed is equal to 164.8 rpm, which compares well with the measured value of 170 rpm. Vibratory amplitudes of longitudinal force F and longitudinal displacement X from the Holzer calculation are tabulated. All values are relative to the assumed unit value of the displacement X_1 of mass 1 (the propeller mass). The natural frequency of the second mode was calculated to be 1290 cycles/min, which places its critical speed based on blade rate excitation well above the rated speed of 199 rpm. Excitation at 2 times blade rate could excite this mode, but there was no observation of any significant response.

The resonant response of the system in the fundamental mode to blade rate excitation is determined by the energy balance method. The procedure is identical to that used for the torsional analysis under "Marine Steam Propulsion System" in Sec. 1.8.1. The same equations for energy input and energy dissipation can be used by changing torque T to force F, polar mass moment of inertia I to mass M, and rotation angle θ to displacement X. The exciting thrust from the propeller is taken equal to 8 percent of the mean propeller thrust (164,800 lb at the critical speed of 164.8 rpm). The energy input from Eq. 1.102 is

$$E_i = \pi(0.08 \times 164{,}800)X_1 = 41{,}430X_1 \qquad \text{in} \cdot \text{lb/cycle}$$

From information supplied by Kane and McGoldrick (1949) the longitudinal damping coefficient B_1 of the propeller is 923 lb · s/in. With the undamped natural frequency $\nu = 69.0$ rad/s (659 cycles/min), the energy dissipated by the propeller by Eq. 1.110 is

$$E_d(\text{prop.}) = \pi(69.0 \times 923)X_1^2 = 200{,}080X_1^2 \qquad \text{in} \cdot \text{lb/cycle}$$

All distributed damping in the system such as hysteresis in the shafting and structural elements in the ship, friction in bolted connections, and fore-and-aft viscous action in bearings is lumped together and related to the total vibrational energy in the system. A value of log decrement δ equal to 0.075 (3 times the value used for the torsional analysis of Sec. 1.8.1) is used, consistent with the findings of Kane and McGoldrick (1949). This relatively large value of log decrement is undoubtedly due to the higher-level damping to be found in the stationary structure. Substituting this value of δ together with values of M_i and X_i from Table 1.13 in Eq. 1.146 gives for the distributed energy dissipation

$$E_d(\text{hyst.}) = 2(0.075 \times 1.102 \times 10^6)X_1^2 = 165{,}290X_1^2 \qquad \text{in} \cdot \text{lb/cycle}$$

Equating the energy input per cycle E_i to the sum of the energies dissipated by damping per cycle $E_d(\text{prop.}) + E_d(\text{hyst.})$ gives a value for X_1 of 0.1134 in (single amplitude). The resonant amplitudes of the longitudinal force F (for each spring element) and the displacement X (for each mass element) for this level of X_1 are tabulated in Table 1.13.

It is not the cyclic stress in the rotating shafting that is a concern for the case at hand. The maximum amplitude of the cyclic tension-compression stress for line shaft 3 is only on the order of 1000 lb/in^2. Rather the concern involves the effect on the machinery elements—the thrust bearing, the steam piping, the condensers, the turbines and gears, and the supporting structure of the machinery—of the excessive shaking force transmitted by the ship's thrust bearing. The calculated levels of resonant response in Table 1.13 are in reasonable agreement with measurements made on the *U.S.S. North Carolina* on shaft 3. Single amplitudes of vibration of the gear case where observed to be as high as 0.030 to 0.040 in as compared with $X_6 = 0.0446$ in in Table 1.13. The blade frequency thrust variation (again single-amplitude) in the middle region of the shafting was found to be 70 to 80 percent of the mean thrust force of the propeller. The calculated values of F_{23} and F_{34} are equal to 74 and 89 percent, respectively, of the mean thrust force (164,800 lb).

Two other calculated values from Table 1.12 are worthy of mention, although no measurements of these quantities were made. The cyclical force acting across the thrust bearing F_{56} is equal to 137 percent of the mean thrust, which means that the thrust runner crosses the clearance in the bearing, an unacceptable condition from the point of view of bearing life. The cyclical force acting on the hull F_{67} is equal to 229 percent of the mean thrust.

No wholly satisfactory solution to this problem was available which met the practical constraints of the situation. The expedient was adopted of substituting five-bladed propellers for four-bladed propellers on the inboard shafts and four-bladed for three-bladed on the outer shafts. This lowered the critical speeds to speed levels where the reduced propeller excitation did not produce "dangerous vibrations in the propulsion systems."

1.9 REFERENCES

Adkins, D. R., 1985, "Analysis of Hydrodynamic Forces on Centrifugal Pump Impellers, California Institute of Technology, no. 280.23.

Alford, J., 1965, "Protecting Turbomachinery from Self-Excited Rotor Whirl," *Journal of Engineering for Power,* 87(4): 333–334, October.

Allaire, P. E., and J. A. Kocur, 1985, "Oil Seal Effects and Subsynchronous Vibrations in High-Speed Compressors," *Instability in Rotating Machinery,* CP 2409, NASA, Washington, D.C. pp. 205–223, June.

Arnold, R. C., and E. E. Haft, 1971, "Stability of an Unsymmetrical Rotating Cantilever Shaft Carrying an Unsymmetrical Rotor," 71-Vibr-58, ASME, New York, September.

Barrett, L. E., P. E. Allaire, and E. J. Gunter, 1979, *Proceedings of the Conference on the Stability and Dynamic Response of Rotors with Squeeze Film Bearings,* University of Virginia, Charlottesville, May.

Barrett, L. E., E. J. Gunter, and P. E. Allaire, 1978, "Optimum Bearing and Support Damping for Unbalance and Response and Stability of Rotating Machinery," *Journal of Engineering for Power,* 100(1): 89–94.

Begg, I. C., 1973, "Friction Induced Rotor Whirl—A Study in Stability," 73-DET-10, ASME, New York, September.

Benckert, H., and T. Wachter, 1980a, "Flow Induced Spring Coefficients of Labyrinth Seals for Applications in Rotordynamics," CP 2133, NASA, Washington, D.C. pp. 189–212, May.

———, and ———, 1980b, "Flow Induced Spring Constants of Labyrinth Seals," C258/80, Institution of Mechanical Engineers, London, September.

Bently, D. E., 1972, "The Re-Excitation of Balance Resonance Regions by Internal Friction," 72-PET-49, ASME, New York, September.

———, 1974, "Forced Subrotative Speed Dynamic Action of Rotating Machinery," 74-Pet-16, ASME, New York.

Black, H. F., 1975, "The Stabilizing Capacity of Bearings of Flexible Rotors with Hysteresis," 75-DET-55, ASME, New York, September 17–19.

Bolotin, V. V., 1964, *Non-Conservative Problems of the Theory of Elastic Stability,* Pergamon Press, New York.

Bolleter, U., A. Wyss, I. Welte, and R. Sturchler, 1985, "Measurement of Hydrodynamic Matrices of Boiler Feed Pump Impellers," 85-DET-147, ASME, New York, September.

Booker, J. F., and K. N. Heubner, 1972, "Application of Finite Element Methods to Lubrication, An Engineering Approach," *Journal of Lubrication Technology,* 94(4): 313–323, October.

Booser, E. R., 1983, *Handbook of Lubrication,* Vol. 2, CRC Press, Boca Raton, Fla.

Booth, D., 1975, "Phillips' Landmark Injection Project," *Petroleum Engineer,* pp. 105–109, October.

Brown, R. D., and J. A. Hart, 1986, "A Novel Form of Damper for Turbomachinery," CP2443, NASA, Washington, D.C., June.

Cataford, G. F., and R. P. Lancee, 1986, "Oil-Free Compression on a Natural Gas Pipeline," 86-GT-293, ASME, New York, June.

Childs, D. W., 1978, "The Space Shuttle Main Engine High Pressure Fuel Turbopump Instability Problem," *Journal of Engineering for Power,* 100(1): 48–57, January.

———, 1982, "Fractional-Frequency Rotor Motion due to Nonsymmetric Clearance Effects," *Journal of Engineering for Power,* 104(3): 533–541, July.

———, 1986, "Force and Moment Rotordynamic Coefficients for Pump-Impeller Shroud Surfaces," *Proceedings of the Fourth Workshop on Rotordynamic Instability Problems in High Performance Turbomachinery,* NASA, CP, Texas A&M University, College Station, Texas, June 2–4, 1986.

———, 1987, *Identification and Avoidance of Instabilities in High-Performance Turbomachinery,* Von Karman Institute for Fluid Dynamics, lecture, Rhode-Saint-Genese, Belgium, January.

———, and J. K. Scharrer, 1986a, "An Iwatsubo-Based Solution for Labyrinth Seals—Comparison to Experimental Results," *Journal of Engineering for Gas Turbine and Power,* 108(2): 325–331, April.

———, and ———, 1986b, "Experimental Rotordynamic Coefficient Results for Teeth-on-Rotor and Teeth-on-Stator Labyrinth Gas Seals," 86-GT-12, ASME, New York, June.

Choi, Y.-S., and S. T. Noah, 1987, "Non-Linear Steady State Response of a Rotor-Support System," *Journal of Vibration, Acoustics, Stress and Reliability in Design,* 169(3): 255–261, July.

Conn, H., 1960, "Stick-Slip: What It Is, What to Do About It," *Tool Engineering,* 45: 61–65.

Crandall, S. H. and P. J. Brosens, 1961, "Whirling of Unsymmetrical Rotors," *Journal of Applied Mechanics,* 28(4): 567–570.

———, 1987, "Nonlinearities in Rotordynamics," *Proceedings Eleventh International Conference on Nonlinear Oscillations,* Janos Bolyai Mathematical Society, Budapest, Hungary, pp. 44–56.

Crede, C. E., and J. E. Ruzicka, 1988, "Theory of Vibration Isolation," *Shock and Vibration Handbook,* 3d ed., McGraw-Hill, New York.

Den Hartog, J. P., 1956, *Mechanical Vibrations,* 4th ed., McGraw-Hill, New York.

Dimarogonas, A. D., 1973, "Newkirk Effect: Thermally Induced Dynamic Instability of High-Speed Rotors," 73-GT-26, ASME, New York, April.

———, and G. N. Sandor, 1969, "Packing Rub Effect in Rotating Machinery," *Wear,* 14(3): 153–170.

Ehrich, F. F., 1964, "Shaft Whirl Induced by Rotor Internal Damping," *Journal of Applied Mechanics,* 23(1): 109–115.

———, 1965, "Bi-Stable Vibration of Rotors in Bearing Clearance," 65-WA/MD-1, ASME, New York.

———, 1966, "Subharmonic Vibration of Rotors in Bearing Clearance," 66-MD-1, ASME, New York.

———, 1967, "The Influence of Trapped Fluids on High Speed Rotor Vibration," *Journal of Engineering for Industry,* 89(4): 806–812.

———, 1969, "The Dynamic Stability of Rotor/Stator Radial Rubs in Rotating Machinery," 69-Vibr-56, ASME, New York, April.

———, 1972, "Sum and Difference Frequencies in Vibration of High Speed Rotating Machinery," *Journal of Engineering for Industry,* 94(1): 181–184, February.

———, 1973, "An Aeroelastic Whirl Phenomenon in Turbomachinery Rotors," 73-DET-97, ASME, New York, September.

———, 1987, "Self Excited Vibration," *Shock and Vibration Handbook,* 3d ed., McGraw-Hill, sec. 5.

———, 1988, "High Order Subharmonic Response of High Speed Rotors in Bearing Clearance," *Journal of Vibration, Acoustics, Stress and Reliability in Design,* 110(1): 9–16, January.

———, 1989a, "The Role of Bearing Support Stiffness Anisotropy in Suppression of Rotordynamic Instability," DE-Vol. 18–1, ASME, New York.

———, 1989b, "Some Observations of Chaotic Vibration Phenomena in High Speed Rotordynamics," DE-Vol. 18–5, ASME, New York.

———, and D. Childs, 1984, "Self-Excited Vibrations in High-Performance Turbomachinery," *Mechanical Engineering,* 106(5):66–79, May.

———, and J. J. O'Connor, 1967, "Stator Whirl with Rotors in Bearing Clearance," *Journal of Engineering for Industry,* 89(3):381–390, August.

Ek, M. C., 1980, "Solution of the Subsynchronous Whirl Problem in the High-Pressure Hydrogen Turbomachinery of the Space Shuttle Main Engine," *Journal of Spacecraft and Rockets,* 17(3): 208–218, May–June.

Emerick, M. F., 1982, "Vibration and Destabilizing Effects on Floating Ring Seals in Compressors," CP 2250, NASA, Washington, D.C., May.

Eshleman, R. L., and R. A. Eubanks, 1970, "Effects of Axial Torque on Rotor Response: An Experimental Investigation," 70-WA/DE-14, ASME, New York, December.

———, and F. M. Lewis, 1988, "Torsional Vibration in Reciprocating and Rotating Machines," *Shock and Vibration Handbook,* 3d ed., McGraw-Hill, New York, sec. 38.

Foster, E. G., V. Kulle, and R. A. Peterson, 1986, "The Application of Active Magnetic Bearings to a Natural Gas Pipeline Compressor," 86-GT-61, ASME, New York, June.

Fowlie, D. W., and D. D. Miles, 1975, "Vibration Problems with High Pressure Compressors," 75-Pet-28, ASME, New York, September.

Gunter, E. J., 1966, "Dynamic Stability of Rotor-Bearing Systems," SP-113, NASA, chap. 4, Washington, D.C.

———, L. E. Barrett, and P. E. Allaire, 1977, "Design of Nonlinear Squeeze Film Dampers for Aircraft Engines," *(ASME) Journal of Lubrication Technology,* 99(1): 57–64, January.

Hagg, A. C., and G. O. Sankey, 1956, "Some Dynamic Properties of Oil-Film Journal Bearings with Reference to the Unbalance Vibration of Rotors," *Journal of Applied Mechanics,* 23(2): 302–306, June.

Hauck, L., 1982, "Measurement and Evaluation of Swirl-Type Flow in Labyrinth Seals," *Rotordynamic Instability Problems in High-Performance Turbomachinery,* CP 2250, NASA, pp. 242–259, May.

Horlock, J. H., and E. M. Greitzer, 1983, "Non-Uniform Flows in Axial Compressors due to Tip Clearance Variation," *Proceedings Institution of Mechanical Engineers Proceedings,* 197C: 173–178, London, September.

Houbolt, J. C., and W. H. Reed, 1961, "Propeller Nacelle Whirl Flutter," Institute of the Aerospace Sciences 61–34, New York, January.

Iwatsubo, T., N. Motooka, and R. Kawai, 1982, "Flow Induced Force and Flow Pattern of Labyrinth Seal," *Rotordynamic Instability Problems in High-Performance Turbomachinery,* CP 2250, NASA, pp. 205–222, May.

Jery, B., A. Acosta, C. Brennen, and T. Caughey, 1984, "Hydrodynamic Impeller Stiffness, Damping, and Inertia in the Rotordynamics of Centrifugal Pumps," CP 2338, NASA, pp. 137–160, May.

Jimbo, H., 1956, "Investigation of the Interaction of Windage and Leakage Phenomena in a Centrifugal Compressor," 56-A-47, ASME, New York, July.

Kane, J. R., and R. T. McGoldrick, 1949, "Longitudinal Vibrations of Marine Propulsion Systems," *Transactions Society of Naval Architects and Marine Engineers,* 57: 193–252, Jersey City, New Jersey.

Kimball, A. L., 1924, "Internal Friction Theory of Shaft Whirling," *General Electric Review,* 27(4): 244, April.

Kirk, R. G., 1986a, "Labyrinth Seal Analysis for Centrifugal Compressor Design—Theory and Practice," *Proceedings International Federation for the Theory of Machines and*

Mechanisms 1986 International Conference on Rotordynamics, pp. 589–596, Japan Society of Mechanical Engineers, Tokyo, September.

———, 1986b, "Oil Seal Dynamics: Considerations for Analysis of Centrifugal Compressors," *Proceedings of Fifteenth Turbomachinery Symposium,* Texas A&M University, College Station, Texas, November.

———, and G. N. Donald, 1983, "Design Criteria for Improved Stability of Centrifugal Compressors," *Rotor Dynamical Instability,* G00227, ASME, New York, pp. 59–71, June.

———, and W. H. Miller, 1979, "The Influence of High Oil Pressure Seals on Turbo-Rotor Stability," *ASLE Transactions,* 22(1): 14–20, January.

———, R. E. Mondy, and R. C. Murphy, 1983, "Theory and Guidelines to Proper Coupling Design for Rotor Dynamic Considerations," 83-DET-93, ASME, New York, September.

Kramer, E., 1972, "Instabilities of Rotating Shafts," *Proceedings of the Conference: Vibrations Rotating Systems,* Institution of Mechanical Engineers, London, pp. 230–246, February.

Kroon, R. P., and W. A. Williams, 1939, "Spiral Vibration of Rotating Machinery," *Proceedings of the Fifth International Congress of Applied Mechanics,* Wiley, New York, p. 712.

Leie, B., and H. J. Thomas, 1980, "Self-Excited Rotor Whirl due to Tip-Seal Leakage Forces," CP 2133, NASA, Washington, D.C., pp. 303–316, May.

Linn, F. C., and M. A. Prohl, 1951, "The Effect of Flexibility of Support upon the Critical Speeds of High-Speed Rotors," *Transactions Society of Naval Architects and Marine Engineers,* 59: 536–553.

Lund, J. W., 1964, "Spring and Damping Coefficients for the Tilting Pad Journal Bearing," *Transactions American Society of Lubrication Engineers,* 7(4): 342–352.

———, and K. K. Thomsen, 1978, "A Calculation Method and Data for the Dynamic Coefficients of Oil-Lubricated Journal Bearings," *Topics in Fluid Film Bearing and Rotor Bearing System Design and Optimization,* ASME, New York, pp. 1–28.

Marmol, R. A., A. J. Smalley, and Tecza, J. A. 1979, "Spline Coupling Induced Nonsynchronous Rotor Vibrations," 79-DET-60, ASME, New York, September.

———, and J. M. Vance, 1978, "Squeeze Film Damper Characteristics for Gas Turbine Engines," *ASME Journal of Mechanical Design,* 100(1): 139–146, January.

McLachlan, N. W., 1947, *Theory and Applications of Mathieu Functions,* Oxford University Press, New York, p. 40.

Messal, E. E., and R. J. Bronthon, 1971, "Subharmonic Rotor Instability due to Elastic Asymmetry," 71-Vibr-57, ASME, New York, September.

Muszynska, A., 1984, "Partial Lateral Rotor to Stator Rubs," Paper No. C281/84, Institution of Mechanical Engineers, London.

———, 1986, "Whirl and Whip-Rotor/Bearing Stability Problems," *Journal of Sound and Vibration,* 110(3): 443–462.

———, and D. Bentley, 1987, "Rotor Active Anti-Swirl Control," *Rotating Machinery Dynamics,* vol. 2, ASME, New York, pp. 215–222.

Myklestad, N. O., 1956, *Fundamentals of Vibration Analysis,* McGraw-Hill, New York.

Newkirk, B. L., 1924, "Shaft Whipping," *General Electric Review,* 27: 169–178.

———, 1926, "Shaft Rubbing," *Mechanical Engineering,* 48: 830.

———, and H. D. Taylor, 1925, "Shaft Whipping due to Oil Action in Journal Bearing," *General Electric Review,* 28(8): 559–568.

Ocvirk, F. W., 1952, "Short Bearing Approximation for Full Journal Bearings," TN 2808, NASA, Washington, D.C.

Ohashi, H., R. Hatanaka, and A. Sakurai, 1986, "Fluid Force Testing Machine for Whirling Centrifugal Impeller," *Proceedings of the International Federation for Theory of*

Machines and Mechanisms 1986 International Conference on Rotordynamics, Japan Society of Mechanical Engineering, Tokyo, pp. 643–648, September.

Orcutt, F. K., 1967, "The Steady-State and Dynamic Characteristics of the Tilting-Pad Journal Bearing in the Laminar and Turbulent Flow Regimes," *Journal of Lubrication Technology,* 89(3): 392–404.

Pincus, O., and B. Sternlicht, 1961, *Theory of Hydrodynamic Lubrication,* McGraw-Hill, New York.

Porter, F. P., 1943, "Harmonic Coefficients of Engine Torque Curves," *Journal of Applied Mechanics,* 10(1): A33–A48, March.

Roark, R. J., and W. C. Young, 1975, *Formulas for Stress and Strain,* 5th ed., McGraw-Hill, New York, chap. 13.

Sadowy, M., 1959, "Eliminating Chatter in Machine Tools," *Tool Engineering,* 43: 99–103.

San Andres, L. A., and J. M. Vance, 1986a, "Effect of Fluid Inertia on Finite-Length Squeeze-Film Dampers," 86-TC-4D-1, American Society of Lubrication Engineers, Park Ridge, Illinois, October.

———, and ———, 1986b, "Experimental Measurements of the Dynamic Pressure Distribution on a Squeeze Film Bearing Damper Executing Circular Centered Orbits," 86-TC-4D-2, American Society of Lubrication Engineers, Park Ridge, Illinois, October.

Sapetta, L. P., and R. J. Harker, 1967, "Whirl of Power Screws Excited by Boundary Lubrication at the Interface," 67-Vibr-37, ASME, New York, March.

Scharrer, J., 1986, "A Comparison of Experimental and Theoretical Results for Labyrinth Gas Seals," Ph.D. dissertation, Texas A&M Univ., December.

Sedy, J., 1979, "Improved Performance of Film-Riding Gas Seals through Enhancement of Hydrodynamic Effects," *ASLE Transactions,* 23(1): 35–44.

———, 1980, "A New Self-Aligning Mechanism for the Spiral-Groove Gas Seal Stability," *ASLE Transactions,* 36(10): 592–598.

Shaw, M. C., and E. F. Macks, 1949, *Analysis and Lubrication of Bearings,* McGraw-Hill, New York, chap. 10.

Shemeld, D. E., 1986, "A History of Development in Rotordynamics, A Manufacturer's Perspective," CP2443, NASA, Washington, D.C., June.

Tanaka, M., S. Sugimura, Junichi, Mitsui, and A. Nishidui, 1986, "Effect of the Fluid Film Seals on the Stability of Turbocompressor Rotor," *Proceedings of the International Federation for Theory of Machines and Mechanisms 1986 International Conference on Rotordynamics,* Japan Society of Mechanical Engineers, Tokyo, pp. 663–667, September.

Taylor, E. S., and K. A. Browne, 1938, "Vibration Isolation of Aircraft Power Plants," *Journal of the Aeronautical Sciences,* 6(2): 43–49.

Taylor, H. D., 1940, "Critical-Speed Behavior of Unsymmetrical Shafts," *Journal of Applied Mechanics,* 7(2): A-71–A-79, June.

Thomas, J. J. 1958, "Instabile Eigenschwingungen von Turbinenlaufern, Angefacht durch die Spaltstromungen Stopfbuschsen un Beschaufelungen," *AEG-Sonderdruck.*

Trent, R., and W. R. Lull, 1972, "Design for Control of Dynamic Behavior of Rotating Machinery," 72-DE-39, ASME, New York, May.

Urlichs, K., 1975, "Durch Spaltstromungen Hervorgerufene Querkrafte an den Laufern Thermischer Turbomachinen," Ph.D. dissertation, Technical University of Munich, Germany.

———, 1977, "Leakage Flow in Thermal Turbo-Machines as the Origin of Vibration-Exciting Lateral Forces," TT-17409, NASA, Washington, D.C., March.

Vance, J. M., 1974, "High Speed Rotor Dynamics—Assessment of Current Technology for Small Turboshaft Engines," TR-74-66 United States Army Air Mobility Research and Development Laboratory, Fort Eustis, Virginia, July.

———, 1978, "Torque Whirl—A Theory to Explain Nonsynchronous Whirling Failures of Rotors with High Torque Load," *Journal of Engineering for Power,* 100(2): 235–240, April.

———, and F. J. Laudadio, 1984, "Experimental Measurement of Alford's Force in Axial

Flow Turbomachinery," 84-GT-140, ASME, New York, June.

———, and J. Lee, 1973, "Stability of High Speed Rotors with Internal Friction," 73-DET-127, ASME, New York, September.

Wachel, W. D., 1975, "Nonsynchronous Instability of Centrifugal Compressors," 75-Pet-22, ASME, New York, September.

Wehrli, V. C., 1963, "Uber Kritische Drehzahlen unter Pulsierender Torsion," *Ingenieur-Archiv,* 33: 73–84.

Wilcock, D. F., and E. R. Booser, 1957, *Bearing Design and Application,* McGraw-Hill, New York.

Williams, R., Jr., and R. Trent, 1970, "The Effects of Nonlinear Asymmetric Supports on Turbine Engine Rotor Stability," 700320, SAE, Warrendale, Pennsylvania, April.

Wolf, J. A., 1968, "Whirl Dynamics of a Rotor Partially Filled with Liquids," 68-WA/APM-25, ASME, New York, December.

Wyssmann, H. R., T. C. Pham, and R. J. Jenny, 1984, "Prediction of Stiffness and Damping Coefficients for Centrifugal Compressor Labyrinth Seals," *Journal of Engineering for Gas Turbines and Power,* 106(4): 920–926.

Yamamoto, T., 1959, "On Critical Speeds of a Shaft Supported by a Ball Bearing," *Journal of Applied Mechanics,* pp. 199–204, June.

Zhou, Ren-Mu, 1986, "Instability of Multi-Stage Compressor K 1501," CP2443, NASA, Washington, D.C., June.

CHAPTER 2

ANALYTIC PREDICTION OF ROTORDYNAMIC RESPONSE

Dr. Harold D. Nelson
Professor
Department of Engineering
Texas Christian University

Dr. Stephen H. Crandall
Ford Professor of Engineering
Department of Mechanical Engineering
Massachusetts Institute of Technology

2.1 INTRODUCTION

During the design of a new machine, analytical methods are used to predict the operational rotordynamic behavior of proposed configurations, to assess their viability, and to facilitate quantitative design refinement. During the testing of a new machine, and whenever unexpected behavior occurs in operation, analytical models are studied to provide understanding and guidance in remedial redesign or fault detection and correction. The models range from highly idealized models of a single mode which provide qualitative understanding, such as have been reviewed in Chap. 1, to extremely refined and detailed models of total systems which provide rather precise, quantitative predictions of operational behavior. The models simulate the kinematics of important degrees of freedom and permit the determination of the response to the interaction of the dynamic phenomena involved. In this chapter some of the analytical tools for treating the kinematics and dynamics of rotating machinery are reviewed, and the principal analysis techniques are outlined. These are used to analyze free vibrations, forced vibrations, and self-excited vibrations. Models for a wide variety of rotordynamic phenomena are described.

2.1.1 Kinematics of Planar Motion

Elliptic Vibration

The simplest steady whirl motion of a rotor station P is a harmonic motion with whirl frequency ω, as illustrated in Fig. 2.1, and its displacement components in

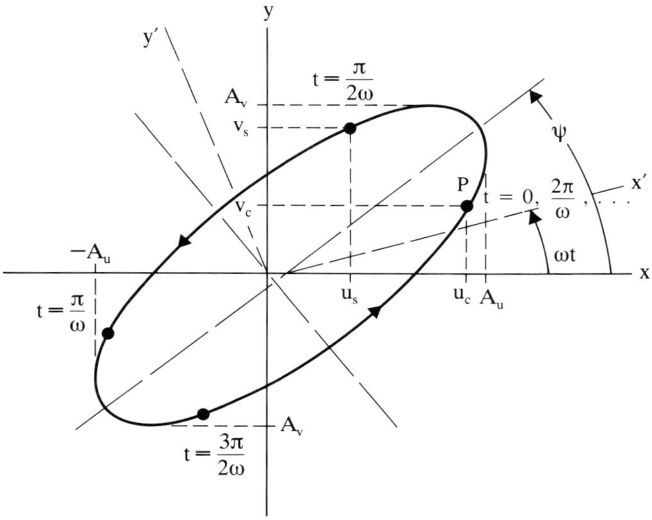

FIGURE 2.1 Elliptic vibration.

the x and y directions are

$$\begin{Bmatrix} u \\ v \end{Bmatrix} = \begin{bmatrix} u_c & u_s \\ v_c & v_s \end{bmatrix} \begin{Bmatrix} \cos \omega t \\ \sin \omega t \end{Bmatrix} = \begin{Bmatrix} A_u \cos(\omega t + \varphi_u) \\ A_v \cos(\omega t + \varphi_v) \end{Bmatrix} \tag{2.1}$$

The point P is said to be vibrating in the x direction with an amplitude $A_u = (u_c^2 + u_s^2)^{1/2}$, circular frequency ω, and phase angle $\varphi_u = \arctan(-u_s/u_c)$. The distance $2A_u$ between the extreme points $x = +A_u$ and $x = -A_u$ is called the *peak-to-peak,* or *double-amplitude, excursion.* Similar statements apply to the motion in the orthogonal y direction. If the parameters A_u, φ_u, A_v, and φ_v are constant, the motion represented by Eq. 2.1 is a single-frequency oscillation. The velocity and acceleration components corresponding to the displacements of (2.1) are

$$\begin{Bmatrix} \dot u \\ \dot v \end{Bmatrix} = \omega \begin{bmatrix} u_s & -u_c \\ v_s & -v_c \end{bmatrix} \begin{Bmatrix} \cos \omega t \\ \sin \omega t \end{Bmatrix} \qquad \begin{Bmatrix} \ddot u \\ \ddot v \end{Bmatrix} = -\omega^2 \begin{Bmatrix} u \\ v \end{Bmatrix} \tag{2.2}$$

and the orbit is called *single harmonic elliptic motion* or *elliptic vibration.* The acceleration of point P is in the opposite direction to the displacement and is proportional to the square of the frequency.

The parameters of the elliptic vibration for a rotor system are generally a function of the spin speed so that the semimajor axis a, semiminor axis b, and attitude angle ψ of the orbit change with this frequency. If $\varphi_v = \varphi_u$ so that the x

and y motions are in phase, the orbit reduces to a straight line, as illustrated in Fig. 2.2*a*. If $u_s = -v_c$ $(+v_c)$ and $u_c = +v_s$ $(-v_s)$ so that the x motion leads (lags) the y motion by 90°, the orbit reduces to a forward (backward) circular motion, as illustrated in Fig. 2.2*b* (2.2*c*).

The preceding elliptical orbits may be represented with the aid of rotating complex vectors by treating the xy plane as a complex plane with real axis along the x axis and imaginary axis along the y axis. If (2.1) is multiplied by the row vector $\{1, i\}$ and cos wt and sin wt are expanded in exponentials with imaginary exponents, the result is

$$\bar{w} = \bar{w}_f e^{+i\omega t} + \bar{w}_b e^{-i\omega t} \tag{2.3}$$

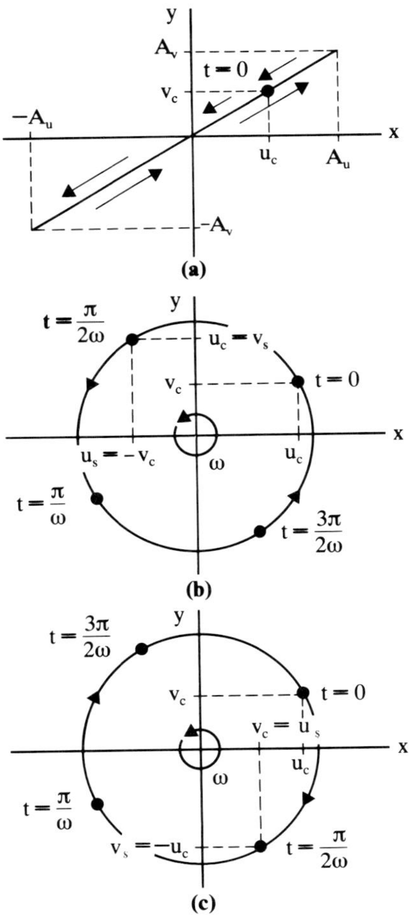

FIGURE 2.2 Special case orbits. (*a*) Straight-line orbit: $\phi_u = \phi_v$; (*b*) forward circular orbit: $u_s = -v_c$, $v_s = +u_c$; (*c*) backward circular orbit: $u_s = +v_c$, $v_s = -u_c$.

where $\bar{w} = u + iv$ is called the *complex displacement* and is the sum of a forward rotating vector with complex amplitude $\bar{w}_f$ and a backward rotating vector with complex amplitude $\bar{w}_b$ given by

$$\begin{aligned}\bar{w}_f &= \tfrac{1}{2}(u_c + v_s) + i\tfrac{1}{2}(-u_s + v_c) = |\bar{w}_f|\, e^{+i\varphi_f} \\ \bar{w}_b &= \tfrac{1}{2}(u_c - v_s) + i\tfrac{1}{2}(u_s + v_c) = |\bar{w}_b|\, e^{+i\varphi_b}\end{aligned} \tag{2.4}$$

where

$$\begin{aligned}|\bar{w}_f| &= \tfrac{1}{2}[(u_c + v_s)^2 + (-u_s + v_c)^2]^{1/2} \\ |\bar{w}_b| &= \tfrac{1}{2}[(u_c - v_s)^2 + (u_s + v_c)^2]^{1/2} \\ \varphi_f &= \arctan\frac{-u_s + v_c}{u_c + v_s} \\ \varphi_b &= \arctan\frac{u_s + v_c}{u_c - v_s}\end{aligned} \tag{2.5}$$

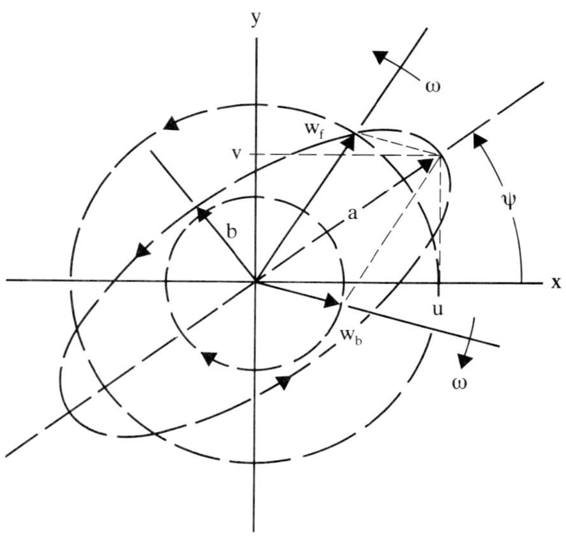

FIGURE 2.3 Elliptic vibration components.

The elliptic vibration illustrated in Fig. 2.3 consists of a forward circular motion with xy components

$$\begin{Bmatrix} u_f \\ v_f \end{Bmatrix} = \frac{1}{2}\begin{bmatrix} u_c + v_s & u_s - v_c \\ -(u_s - v_c) & u_c + v_s \end{bmatrix}\begin{Bmatrix} \cos \omega t \\ \sin \omega t \end{Bmatrix} \tag{2.6}$$

and a backward circular motion with xy components

$$\begin{Bmatrix} u_b \\ v_b \end{Bmatrix} = \frac{1}{2}\begin{bmatrix} u_c - v_s & u_s + v_c \\ u_s + v_c & -u_c + v_s \end{bmatrix}\begin{Bmatrix} \cos \omega t \\ \sin \omega t \end{Bmatrix} \tag{2.7}$$

The semimajor axis a, semiminor axis b, and attitude angle ψ of the ellipse are

$$\begin{aligned} a &= |\bar{w}_f| + |\bar{w}_b| \\ b &= |\bar{w}_f| - |\bar{w}_b| = \frac{1}{a}\det\begin{bmatrix} u_c & u_s \\ v_c & v_s \end{bmatrix} \\ 2\psi &= \varphi_f + \varphi_b \end{aligned} \tag{2.8}$$

A positive (negative) value of the semiminor axis b, as defined by (2.8), indicates a forward (backward) whirling orbit.

The forward and backward components of an elliptic vibration can be measured by making use of the identities

$$\begin{aligned} u_f(\omega t) &= \frac{1}{2}\left[u(\omega t) + v\left(\omega t + \frac{\pi}{2}\right)\right] \\ v_f(\omega t) &= \frac{1}{2}\left[u\left(\omega t - \frac{\pi}{2}\right) + v(\omega t)\right] \\ u_b(\omega t) &= \frac{1}{2}\left[u(\omega t) + v\left(\omega t - \frac{\pi}{2}\right)\right] \\ v_b(\omega t) &= \frac{1}{2}\left[u\left(\omega t + \frac{\pi}{2}\right) + v(\omega t)\right] \end{aligned} \tag{2.9}$$

Superposition of Elliptic Vibrations

Physical phenomena in rotating machinery quite commonly produce transverse shaft motion which consists of the superposition of several elliptic vibrations of typical form (2.3)

$$\bar{w}_k = \bar{w}_{f_k} e^{+i\omega_k t} + \bar{w}_{b_k} e^{-i\omega_k t} \tag{2.10}$$

($k = 1, 2, \ldots, K$) with the total response given by

$$\bar{w} = \sum_{k=1}^{K} (\bar{w}_{f_k} e^{+i\omega_k t} + \bar{w}_{b_k} e^{-i\omega_k t}) \tag{2.11}$$

In general, this response is not periodic, nor is it an elliptic vibration. For the special case when the frequencies ω_k are identical (for example, ω), the total response reduces to

$$\bar{w} = \left(\sum_{k=1}^{K} \bar{w}_{f_k}\right) e^{+i\omega t} + \left(\sum_{k=1}^{K} \bar{w}_{b_k}\right) e^{-i\omega t} \tag{2.12}$$

and is also an elliptic vibration. The total response (2.11) is periodic if the ratios of all possible pairs of frequencies form rational numbers. The period of vibration is given by $2\pi L/\omega_o$, where L represents the lowest common multiple of the ratios ω_k/ω_o, $k = 1, 2, \ldots, K$, and ω_o is the lowest frequency component in (2.11).

A special case of common occurrence is represented by the superposition of a forward elliptic vibration $\bar{w}_f$ at a frequency ω and a second elliptic vibration with a frequency which is an nth-order subharmonic of ω. The total response (Tondl, 1973) is then

$$\bar{w} = \bar{w}_f e^{+i\omega t} + \bar{w}_{1/n} e^{\pm i(1/n)\omega t} \tag{2.13}$$

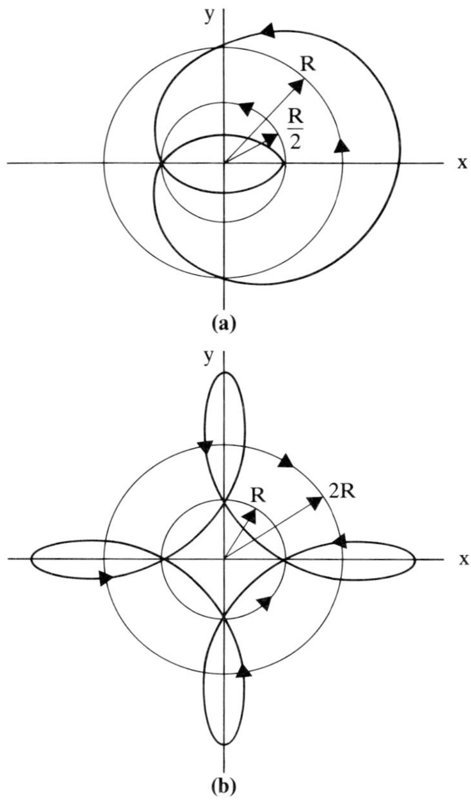

FIGURE 2.4 Superposition of elliptic vibrations (Tondl, 1973). (*a*) Forward subharmonic of order 2: $|w_f| = R$, $|w_{1/2}| = R/2$; (*b*) backward subharmonic of order 3: $|w_f| = R$, $|w_{1/3}| = 2R$.

with a period of $2\pi n/\omega$. For the case of a second-order forward subharmonic $n = 2$ with $\bar{w}_f = R + i0$ and $\bar{w}_{1/2} = R/2 + i0$, the periodic response includes one inner loop, as illustrated in Fig. 2.4*a*, and has a period of $4\pi/\omega$. For the case of a backward subharmonic of order 3 with $\bar{w}_f = R + i0$ and $\bar{w}_{1/3} = 2R + i0$, the periodic response illustrated in Fig. 2.4*b* includes four outer loops and has a period of $6\pi/\omega$.

Coordinate Systems

In the analysis of shaft vibrations it is convenient to utilize several coordinate systems. Consider a stationary rectangular coordinate frame xyz and a rotating rectangular coordinate frame $x'y'z'$ with common origin O and with the z' axis collinear with the z axis. The $x'y'$ axes then lie in the same plane as the xy axes, as illustrated in Fig. 2.5. The orientation of the x' axis is described by $\theta(t)$, and the rate of rotation of the $x'y'z'$ frame relative to the xyz frame is $\omega = d\theta/dt$. Vector OP is simultaneously described by the coordinates (u, v) in the stationary

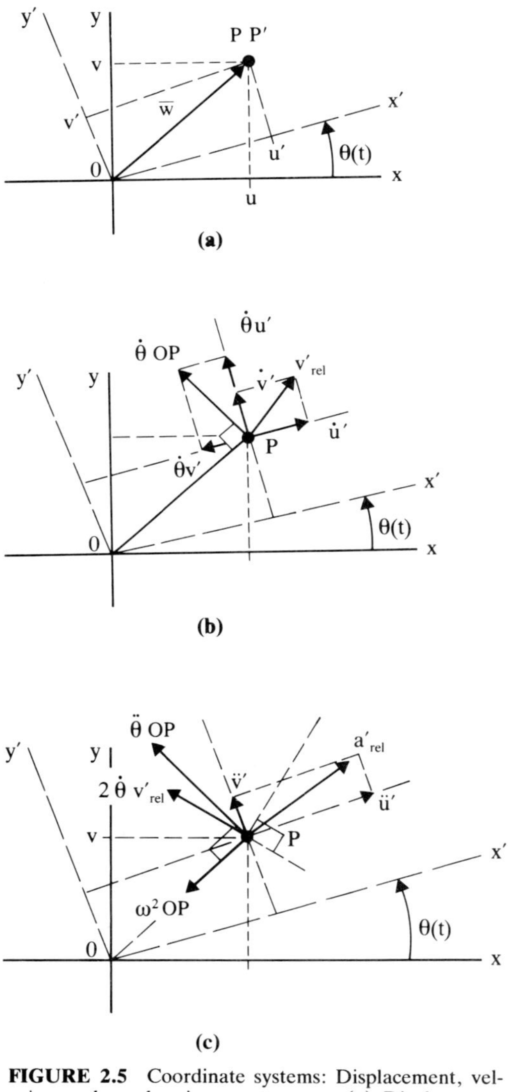

FIGURE 2.5 Coordinate systems: Displacement, velocity, and acceleration components. (*a*) Displacement components of *P*; (*b*) velocity components of *P* for *xyz*; (*c*) acceleration components of *P* for *xyz*.

frame and (u', v') in the rotating frame. The relation between these is represented by the matrix equation

$$\begin{Bmatrix} u \\ v \end{Bmatrix} = [T] \begin{Bmatrix} u' \\ v' \end{Bmatrix} \tag{2.14}$$

where $[T]$ is the rotation transformation matrix

$$[T] = \begin{bmatrix} \cos\theta & -\sin\theta \\ \sin\theta & \cos\theta \end{bmatrix} \qquad [T]^{-1} = [T]^T \tag{2.15}$$

If OP is the displacement of a moving point P in the plane of Fig. 2.5*a*, the corresponding velocity and acceleration components are obtained by differentiating (2.14) with respect to time. These components are illustrated in Fig. 2.5*b* and *c*, respectively. Note that the time derivative of $[T]$ is

$$\frac{d}{dt}[T] = \begin{bmatrix} -\sin\theta & -\cos\theta \\ \cos\theta & -\sin\theta \end{bmatrix}\dot{\theta} = [T]\begin{bmatrix} 0 & -1 \\ 1 & 0 \end{bmatrix}\omega \tag{2.16}$$

and that the result of multiplying a column vector by the unit antisymmetric matrix in (2.16) is a vector of equal length oriented at right angles to the original vector; i.e., the unit antisymmetric matrix represents the operation of rotating a vector through 90° in the counterclockwise sense. The velocity of the point P in Fig. 2.5*a* is

$$\begin{Bmatrix} \dot{u} \\ \dot{v} \end{Bmatrix} = [T]\left(\begin{Bmatrix} \dot{u}' \\ \dot{v}' \end{Bmatrix} + \omega\begin{bmatrix} 0 & -1 \\ 1 & 0 \end{bmatrix}\begin{Bmatrix} u' \\ v' \end{Bmatrix}\right) \tag{2.17}$$

which can be interpreted to state that the velocity of P relative to xyz is the sum of the velocity of P relative to $x'y'z'$ plus the rotational velocity of the fixed point P' in the rotating frame which coincides with P at the instant under consideration. The latter velocity has a magnitude equal to the product of the displacement OP' times the angular velocity ω of $x'y'z'$ and is directed at right angles to OP'. The acceleration of P is

$$\begin{Bmatrix} \ddot{u} \\ \ddot{v} \end{Bmatrix} = [T]\left(\begin{Bmatrix} \ddot{u}' \\ \ddot{v}' \end{Bmatrix} + 2\omega\begin{bmatrix} 0 & -1 \\ 1 & 0 \end{bmatrix}\begin{Bmatrix} \dot{u}' \\ \dot{v}' \end{Bmatrix} + \left(\dot{\omega}\begin{bmatrix} 0 & -1 \\ 1 & 0 \end{bmatrix} - \omega^2\right)\begin{Bmatrix} u' \\ v' \end{Bmatrix}\right) \tag{2.18}$$

and states that the acceleration of P relative to xyz is the sum of four terms: the acceleration of P relative to $x'y'z'$; the centripetal acceleration proportional to ω^2 and opposite in direction to the displacement; the acceleration of P' due to the angular acceleration $d\omega/dt$ of $x'y'z'$ which is directed at right angles to OP'; and the Coriolis acceleration with a magnitude equal to twice the product of the velocity of P relative to $x'y'z'$ and the angular velocity of $x'y'z'$ and is directed at right angles to the relative velocity.

An alternative representation using complex variables can be established by considering the xy and $x'y'$ axes in Fig. 2.5 as the real and imaginary axes of complex planes. The displacement OP of a moving point P from the fixed origin relative to xyz and $x'y'z'$ can be respresented, respectively, by the complex vectors

$$\bar{w} = u + iv \qquad \bar{w}' = u' + iv' \tag{2.19}$$

These complex displacements are related by the transformation

$$\bar{w} = \bar{w}'e^{i\theta} \tag{2.20}$$

and the velocity and acceleration relations are obtained by differentiating (2.20) with respect to time:

$$\begin{aligned} \dot{\bar{w}} &= (\dot{\bar{w}}' + i\dot{\theta}\bar{w}')e^{i\theta} \\ \ddot{\bar{w}} &= [(\ddot{\bar{w}}' - \dot{\theta}^2\bar{w}') + i(\ddot{\theta}\bar{w}' + 2\dot{\theta}\dot{\bar{w}}')]e^{i\theta} \end{aligned} \tag{2.21}$$

Relations (2.20) and (2.21) for the complex vector representation are equivalent to the real representations of (2.14), (2.17), and (2.18). Multiplying a complex vector by i in (2.21) has the same significance as multiplying a column vector by the unit antisymmetric matrix as in (2.17) and (2.18).

2.1.2 Kinematics of Three-Dimensional Motion

When one is observing the motion of a point on a rotating assembly, it is often convenient to utilize several different reference frames. The choice of reference depends usually on the particular characteristics of the system or on the personal preference of the analyst or experimenter. Thus, consider a triaxial coordinate system $\xi\eta\zeta$ which is translating and rotating with respect to a stationary reference frame xyz as shown in Fig. 2.6. The displacement o from the origin O of the

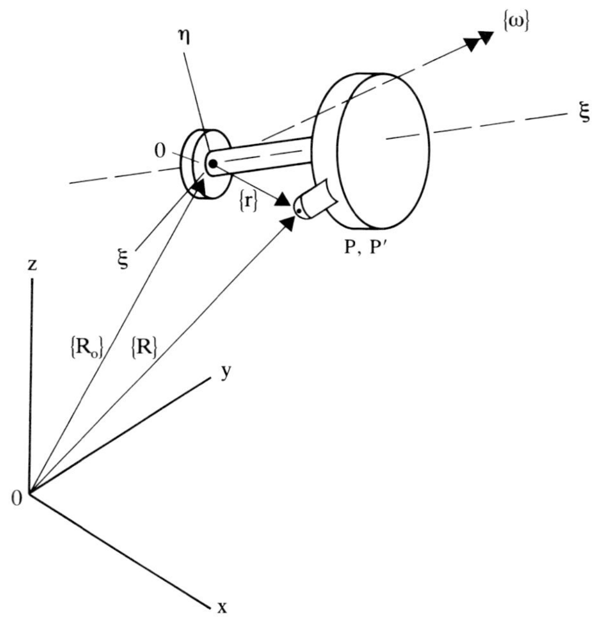

FIGURE 2.6 Stationary xyz and moving $\xi\eta\zeta$ reference frames.

stationary frame is represented by the vector $\{R_o\}$. The rate of rotation of the $\xi\eta\zeta$ frame with respect to the stationary frame is represented by the angular velocity vector $\{\omega\}$. The displacement of a typical moving point P is denoted by $\{R\}$ with respect to O and by $\{r\}$ with respect to o, and they are related by the vecor sum

$$\{R\} = \{R_o\} + \{r\} \tag{2.22}$$

and the velocity and acceleration are obtained by differentiation of (2.22) with respect to time. The relation between a time derivative $(d/dt)_{xyz}$ as observed from the stationary xyz frame and the time derivative $(d/dt)_{\xi\eta\zeta}$ as observed from the rotating $\xi\eta\zeta$ frame can be represented operationally (Crandall et al., 1982) using the vector cross product

$$\left(\frac{d}{dt}\right)_{xyz} = \left[\left(\frac{d}{dt}\right)_{\xi\eta\zeta} + \{\omega\}\times\right] \tag{2.23}$$

When applied to a typical vector $\{A\}$ with components A_ξ, A_η, and A_ζ in the rotating frame, (2.23) takes the form

$$\frac{d}{dt}\begin{Bmatrix} A_\xi \\ A_\eta \\ A_\zeta \end{Bmatrix}_{xyz} = \frac{d}{dt}\begin{Bmatrix} A_\xi \\ A_\eta \\ A_\zeta \end{Bmatrix}_{\xi\eta\zeta} + \begin{bmatrix} 0 & -\omega_\zeta & \omega_\eta \\ \omega_\zeta & 0 & -\omega_\xi \\ -\omega_\eta & \omega_\xi & 0 \end{bmatrix}\begin{Bmatrix} A_\xi \\ A_\eta \\ A_\zeta \end{Bmatrix} \tag{2.24}$$

The first derivative of (2.22) yields the velocity relationship in vector form

$$\{\dot{R}\}_{xyz} = \{\dot{R}_o\}_{xyz} + \{\dot{r}\}_{\xi\eta\zeta} + \{\omega\}\times\{r\} \tag{2.25}$$

which is a three-dimensional generalization of (2.17). The first and third terms on the right represent the absolute velocity of the point P' which is fixed in the $\xi\eta\zeta$ frame and coincides with the moving point P at the instant under consideration. Thus, the absolute velocity of the moving point P is the sum of the absolute velocity of the point P' plus the velocity of P relative to the moving frame. The second derivative of (2.22) yields the absolute acceleration of P

$$\{\ddot{R}\}_{xyz} = \{\ddot{R}_o\} + \{\ddot{r}\}_{\xi\eta\zeta} + 2\{\omega\}\times\{\dot{r}\}_{\xi\eta\zeta} + \{\dot{\omega}\}_{xyz}\times\{r\} + \{\omega\}\times(\{\omega\}\times\{r\}) \tag{2.26}$$

which, in vector notation, is a three-dimensional extension of (2.18). The terms on the right are, respectively, the absolute acceleration of the origin of the moving frame, the relative acceleration of P, the Coriolis acceleration, the contribution due to the angular acceleration of the moving frame, and the centripetal acceleration. Here the sum of the first, fourth, and fifth terms on the right represents the acceleration of point P' which is fixed in the $\xi\eta\zeta$ frame and coincides with the moving point P at the instant under consideration. The second and third terms arise from the motion of P relative to the $\xi\eta\zeta$ frame.

Small Rotations of a Spin Axis

For most applications to rotating machinery, the rotations of the spin axis are small. To address these situations, consider a rigid body spinning about an axis OP. If the axis simultaneously traces out a cone as indicated in Fig. 2.7*a*, the axis is said to be undergoing *precession*. The term *nutation* is used to describe fluctuations in the cone angle, as illustrated in Fig. 2.7*b*, when superposed on precession. These motions are particular cases of rotation of a rigid body about an axis which is itself rotating. To describe these motions quantitatively consider in Fig. 2.8 that the $\xi\eta\zeta$ frame is displaced from the stationary xyz frame by first rotating through angle β about y and then rotating through angle α about ξ'. Angles α and β are convenient Euler angles (Crandall et al., 1982, p. 241) for describing small rotations of the spin axis of a rotor. The positive directions for both β and α are in accord with the right-hand rule for rotations about the y and ξ' axes. In terms of these angles, the absolute angular velocity of the disk in Fig. 2.8 with respect to the xyz frame, when the disk spins about ζ with a spin rate Ω with respect to $\xi\eta\zeta$, is

$$\{\omega\}_d = \dot{\alpha}\{u\}_\xi + \dot{\beta}\cos\alpha\{u\}_\eta + (\Omega - \dot{\beta}\sin\alpha)\{u\}_\zeta \tag{2.27}$$

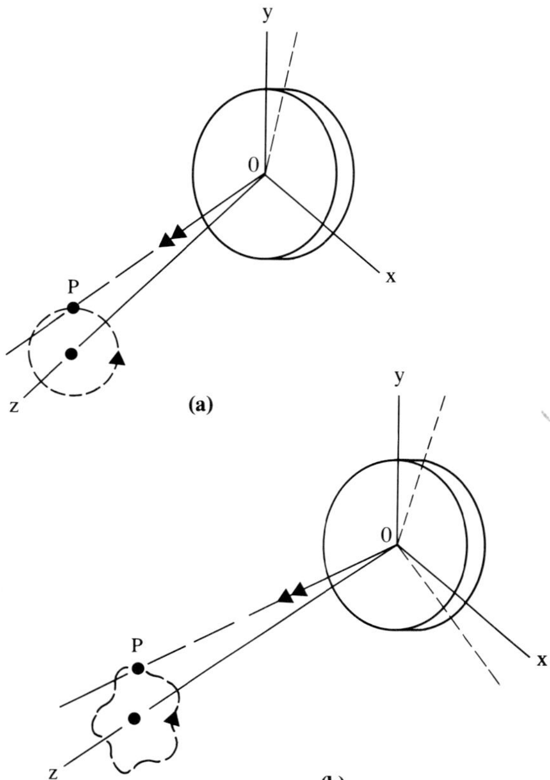

FIGURE 2.7 Rigid-body precession and nutation: (*a*) pure precession; (*b*) precession and nutation.

where $\{u\}_\xi$, $\{u\}_\eta$, and $\{u\}_\zeta$ represent unit vectors in the ξ, η, and ζ directions. When α and β and their derivatives are first-order small quantities, (2.27) reduces to

$$\{\omega\}_d = \dot{\alpha}\{u\}_\xi + \dot{\beta}\{U\}_\eta + (\Omega - \dot{\beta}\alpha)\{u\}_\zeta \tag{2.28}$$

which is correct to second order in the small quantities. Note that the absolute spin component $\Omega - \dot{\beta}\alpha$ about the ζ axis differs by a term of second order from the relative spin Ω.

2.1.3 Vibration Modes of Rotating Systems

The elastic and inertia properties of the rotors of most machines are usually more or less axially symmetric about the spin axis. The major modes of vibration of such rotors are torsional, longitudinal, and transverse. In a torsional vibration of the rotor in Fig. 2.9, the rotation rates $\dot{\theta}_i$ at the various stations fluctuate about

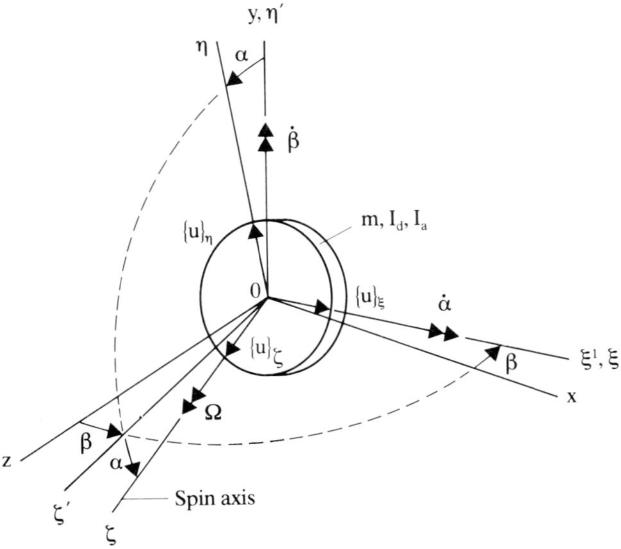

FIGURE 2.8 Small-angle rotation of a spin axis.

the average spin rate Ω, *twisting* the rotor segments between stations. In a longitudinal vibration the axial displacements w_i at the various stations fluctuate, *stretching* and *compressing* the rotor segments between stations. In a transverse or whirling vibration, the rotor center at station i executes an orbital motion in the $x_i y_i$ plane. In general the orbits at the various stations are different, resulting in *bending* of the rotor segments between stations. The spin axes at the various stations are generally tipped at various angles with respect to the undeformed spin axes. If the disks in Fig. 2.9 are flexible or carry flexible blade rows, the disk modes with no nodal diameters are generally coupled to longitudinal modes of the rotor and the disk modes with one nodal diameter are generally coupled to the transverse modes of the rotor. The disk modes with more than one nodal diameter do not couple to the major rotor modes and may be treated separately.

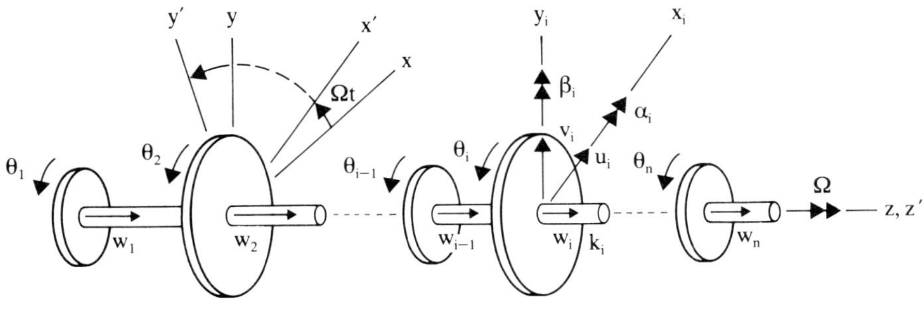

FIGURE 2.9 Longitudinal, torsional, and transverse vibrations of an axisymmetric rotor.

In a single isolated rotor, the torsional, longitudinal, and transverse modes are generally orthogonal to one another. A complete machine, however, has one or more rotors supported by bearings from a stator. The rotors may be in-line, joined by a coupling mechanism, or may be concentric spools separately supported by bearings from the stator or supported by intershaft bearings. The various parts of the machine interact through the bearings, through the working fluids of a fluid-handling machine, or via the magnetic forces of an electrical machine and through gears and couplings. These interconnections may act to couple modes of vibration which would be uncoupled in an isolated rotor. In some machines the inertia and stiffness parameters of the stator are so large in comparison with those of the rotors that very little nonrotating mass is involved in the rotational, longitudinal, or transverse modes of the total system. On the other hand, in lightweight machines such as aircraft gas turbines, the stator inertia and stiffness can be small enough that some of the system torsional, longitudinal, or transverse modes may involve more stator mass in motion than rotor mass.

The torsional modes can be excited by torque fluctuations arising from irregularities in fluid flow due to blade passages or irregularities in magnetic field due to pole structure. Torsional excitation is especially strong in reciprocating machines (Crandall, 1962) where a rotor is connected to pistons in cylinders via connecting rods. Longitudinal vibrations are excited by thrust fluctuations arising from irregularities in fluid flow due to blade passages or irregularities in gears or couplings. Transverse vibrations may be excited by mass unbalance in a rotor, by a bent rotor, or by misalignment of two shafts which are subsequently connected by a coupling. The most insidious form of excitation is self-excitation. This occurs when a certain threshold of some operational parameter such as speed or power is overstepped and a particular system mode becomes unstable. The amplitude of that mode increases rapidly until limited by some nonlinearity such as rub in a seal or even by failure of the machine. Generally the self-excitation may be viewed as caused by the coupling between the rigid-body torsional mode which stores the very large kinetic energy of steady rotation and a particular system mode, usually a whirling mode. When the threshold is exceeded, energy begins to flow from the steady rotation into the system mode which must continue to increase its amplitude to accept additional energy.

2.2 DYNAMICS OF ROTATING SYSTEMS

Equations of motion for dynamic systems are obtained by applying dynamic principles to appropriate models. The distributions of inertia and stiffness properties in the model are essential for determining natural modes and natural frequencies. The distribution of damping properties generally has only minor influence on the natural modes and frequencies but is important for determining response amplitudes and for studying stability. The distributions of dynamic properties can be represented by lumped-parameter models leading to matrix equations or by continuum models leading to partial differential equations. The dynamic principles may be applied directly by selecting appropriate isolated free bodies and enforcing the requirements of the linear and angular momentum principles

$$\{f\} = \frac{d}{dt}\{P\} \qquad \text{and} \qquad \{\tau\}_B = \frac{d}{dt}\{H\}_B \tag{2.29}$$

where $\{f\}$ is the total external force and $\{P\}$ is the total linear momentum of an isolated free body and where $\{\tau\}_B$ is the total external torque and $\{H\}_B$ is the total angular momentum of the isolated free body, both taken about point B which must satisfy certain requirements (Crandall et al., 1982, p. 146). In practice, B is usually the center of mass of the free body or is a fixed point in an inertial reference frame. Alternatively D'Alembert's principle (Crandall et al., 1982, p. 204) can be applied. The D'Alembert forces which are peculiar to rotating systems are the centrifugal force and the Coriolis force. The D'Alembert moment due to turning of the spin axis of the disk is called the *gyroscopic torque*. The dynamic principles may also be applied indirectly to a lumped-parameter model by using Lagrange's equations

$$\frac{d}{dt}\left(\frac{\partial T^*}{\partial \dot{\xi}_j}\right) - \frac{\partial T^*}{\partial \xi_j} + \frac{\partial V}{\partial \xi_j} = \Xi_j \tag{2.30}$$

where T^* and V are the global kinetic coenergy and potential energy, the ξ_j $(j = 1, 2, \ldots, n)$ are the generalized coordinates, and the Ξ_j are the corresponding generalized forces not accounted for in V. For a continuum model Hamilton's principle (Crandall et al., 1982, p. 336) can be be applied to obtain partial differential equations of motion.

2.2.1 Single Rigid Body

Fixed Axis Rotation

Let the disk in Fig. 2.10 represent a rigid body free to rotate about the fixed axis Oz. The angular velocity $\Omega = d\theta/dt$ of the disk can be changed by the application of torque τ_z about axis Oz. Integration of the angular momentum equation in (2.29) yields

$$\int_{t_1}^{t_2} \tau_z(t)\, dt = I_z(\Omega_2 - \Omega_1) \tag{2.31}$$

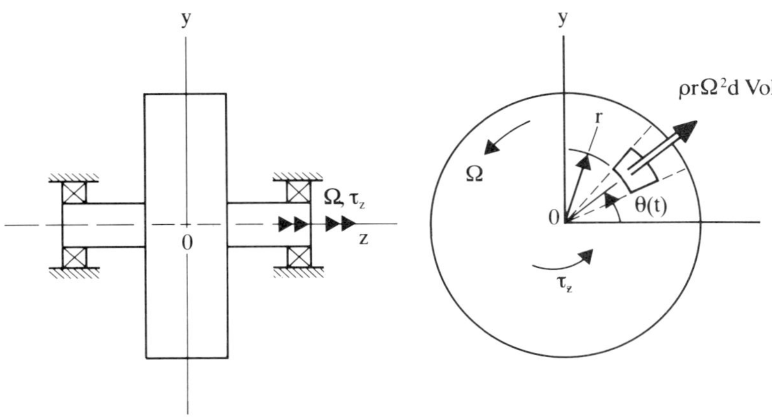

FIGURE 2.10 Rigid disk rotation about a fixed axis.

when the time dependence of the torque is known, or

$$\int_{\theta_1}^{\theta_2} \tau_z(\theta)\, d\theta = \tfrac{1}{2} I_z(\Omega_2^2 - \Omega_1^2) \tag{2.32}$$

when the angular dependence of the torque is known, where I_z is the mass moment of inertia of the body about Oz.

At high speed Ω, the major stress developed in a rotating body is the centrifugal stress required to equilibrate the centrifugal force $\rho r\Omega^2$ per unit volume which acts radially outward at each point in the body, where the radius is r and the mass density is ρ. The distribution of centrifugal stress depends on the geometry of the rotating body. For a thin circular hoop of radius R, the hoop stress is

$$\sigma_\theta = \rho(R\Omega)^2 = \rho V_p^2 \tag{2.33}$$

where $V_p = R\Omega$ is the peripheral speed of the hoop. In other geometries the greatest centrifugal stress is also proportional to the square of the peripheral speed at the rim of the body. For a uniform spoke extending from a hub at $r = 0$ to a free end at $r = R$, the maximum stress (at the hub) is

$$\alpha_r = \tfrac{1}{2}\rho R^2\Omega^2 = \tfrac{1}{2}\rho V_P^2 \tag{2.34}$$

For a uniform disk of radius R of an elastic solid with Poisson's ratio v, the maximum stress condition is the biaxial stress at $r = 0$

$$\sigma_r = \sigma_\theta = \frac{3 + v}{8}\rho V_P^2 \tag{2.35}$$

If a small hole is drilled in the center of the disk, the radial stress drops to zero but the tangential stress is doubled. For other geometries see Roark (1965, pp. 360–366).

The centrifugal force distribution in a rotating rigid body is responsible not only for large internal centrifugal stresses. It may also require external equilibrium forces, at the bearings, which rotate with the body. These undesired rotating bearing reactions will vanish if the centrifugal force distribution is self-equilibrating, i.e., if the center of mass of the body lies on the axis of rotation Oz and if that axis is a principal axis of inertia for the body (Crandall et al., 1982, p. 221). A rigid body with these properties is said to be *balanced* for rotation about the axis in question. Otherwise it is said to be *unbalanced*. The state of balance of a body can be altered by rearranging its mass distribution, by adding or removing mass at particular locations. Practical balancing procedures involve systematic mass redistribution until the vibratory response generated by the rotating bearing reactions falls below some acceptable tolerance level. See Chap. 3.

Rotation about a Rotating Axis

The angular momentum vector $\{H\}$ of a rigid body with respect to a prescribed point B is the product of the inertia matrix $[I]$ corresponding to that point and the angular velocity vector $\{\omega\}$. In general three-dimensional motion, the elements of the inertia matrix in most coordinate frames are time-varying quantities. The elements of the inertia matrix are, however, constant for coordinate frames fixed to the body. Of these the coordinate frame aligned with the principal axes of inertia provides the simplest representation of the inertia matrix. If the principal axes are labeled 1, 2, and 3 and the corresponding principal moments of inertia are I_1, I_2, and I_3, the inertia matrix is represented by a diagonal matrix in this frame, and the angular momentum vector is

$$\begin{Bmatrix} H_1 \\ H_2 \\ H_3 \end{Bmatrix} = \begin{bmatrix} I_1 & 0 & 0 \\ 0 & I_2 & 0 \\ 0 & 0 & I_3 \end{bmatrix} \begin{Bmatrix} \omega_1 \\ \omega_2 \\ \omega_3 \end{Bmatrix} = \begin{Bmatrix} I_1\omega_1 \\ I_2\omega_2 \\ I_3\omega_3 \end{Bmatrix} \tag{2.36}$$

When the angular momentum principle of (2.29) is applied to (2.36) and the derivative evaluated by means of (2.23), the resulting equations

$$\begin{aligned} \tau_1 &= I_1\dot{\omega}_1 - \omega_2\omega_3(I_2 - I_3) \\ \tau_2 &= I_2\dot{\omega}_2 - \omega_3\omega_1(I_3 - I_1) \\ \tau_3 &= I_3\dot{\omega}_3 - \omega_1\omega_2(I_1 - I_2) \end{aligned} \tag{2.37}$$

are called *Euler's equations.* These are relatively simple dynamic relations, but since they apply in a coordinate frame fixed to the moving body, it is usually necessary to integrate kinematic relations simultaneously with (2.37) to determine the resulting motion. One case where Euler's equations provide useful results directly is that of torque-free motion (Crandall et al., 1982, pp. 228–239). A rigid body in a torque-free environment maintains constant angular momentum about its centroid. In general, the angular velocity vector continually changes its magnitude and direction. The angular velocity can remain constant, however, if the rotation is about one of the principal axes of inertia. Euler's equations may be used to show that rotation about a principal axis with smallest or largest moment of inertia is stable, but that rotation about the principal axis with intermediate moment of inertia is unstable. Satellite experiments have demonstrated that when the body is not perfectly rigid and has an internal damping mechanism, then rotation about the axis with smallest moment of inertia is also unstable. Such a quasi-rigid body will thus ultimately end up rotating about the principal axis with largest moment of inertia if left long enough in a torque-free environment.

For an axisymmetric body such as a disk having equal diametral moment of inertia $I_1 = I_2$ with respect to a point on the axis, the moment of inertia I_d with respect to any diameter in the 1-2 plane has the same value $I_d = I_1 = I_2$. This implies that if the 3 axis is aligned with the axis $O\zeta$ in Fig. 2.8, the $\xi\eta\zeta$ frame will have an invariant inertia matrix even though the disk is spinning at rate Ω with respect to the frame. If we call the axial moment of inertia I_3 by the new name I_a, the angular momentum of the disk in Fig. 2.8 is

$$\begin{Bmatrix} H_\xi \\ H_\eta \\ H_\zeta \end{Bmatrix} = \begin{bmatrix} I_d & 0 & 0 \\ 0 & I_d & 0 \\ 0 & 0 & I_a \end{bmatrix} \begin{Bmatrix} \omega_\xi \\ \omega_\eta \\ \omega_\zeta \end{Bmatrix}_d = \begin{Bmatrix} I_d\omega_\xi \\ I_d\omega_\eta \\ I_a\omega_\zeta \end{Bmatrix}_d \tag{2.38}$$

where the angular velocity components are those of $\{\omega\}_d$ from (2.28). When the angular momentum principle of (2.29) is applied to (2.38), using (2.23) and the fact that the angular velocity of the $\xi\eta\zeta$ frame is

$$\begin{Bmatrix} \omega_\xi \\ \omega_\eta \\ \omega_\zeta \end{Bmatrix} = \begin{Bmatrix} \omega_\xi \\ \omega_\eta \\ \omega_\zeta \end{Bmatrix}_d - \begin{Bmatrix} 0 \\ 0 \\ \Omega \end{Bmatrix} = \begin{Bmatrix} \dot{\alpha} \\ \dot{\beta}\cos\alpha \\ -\dot{\beta}\sin\alpha \end{Bmatrix} \tag{2.39}$$

the modified Euler equations are

$$\begin{aligned} \tau_\xi &= I_d\dot{\omega}_\xi + I_a\omega_\eta(\omega_\zeta + \Omega) - I_d\omega_\zeta\omega_\eta \\ \tau_\eta &= I_d\dot{\omega}_\eta - I_a\omega_\xi(\omega_\zeta + \Omega) + I_d\omega_\zeta\omega_\xi \\ \tau_\zeta &= I_a(\dot{\omega}_\zeta + \dot{\Omega}) \end{aligned} \tag{2.40}$$

If α and β and their derivatives are first-order small but Ω and $\dot{\Omega}$ are large, these reduce to

$$\begin{aligned} \begin{Bmatrix} \tau_\xi \\ \tau_\eta \end{Bmatrix} &= \begin{bmatrix} I_d & 0 \\ 0 & I_d \end{bmatrix} \begin{Bmatrix} \ddot{\alpha} \\ \ddot{\beta} \end{Bmatrix} + \Omega \begin{bmatrix} 0 & I_a \\ -I_a & 0 \end{bmatrix} \begin{Bmatrix} \dot{\alpha} \\ \dot{\beta} \end{Bmatrix} \\ \tau_\zeta &= I_a\dot{\Omega} \end{aligned} \tag{2.41}$$

on neglecting second-order terms. These dynamic relations in the $\xi\eta\zeta$ frame can be transformed to the xyz frame by the transformation matrix $[T]$ that represents a rotation of $-\alpha$ about the ξ axis followed by a rotation of $-\beta$ about the y axis which returns the $\xi\eta\zeta$ frame to the orientation of the xyz frame. The individual rotations have the form of (2.15), and the combined transformation is the product

$$[T] = \begin{bmatrix} \cos\beta & 0 & \sin\beta \\ 0 & 1 & 0 \\ -\sin\beta & 0 & \cos\beta \end{bmatrix} \begin{bmatrix} 1 & 0 & 0 \\ 0 & \cos\alpha & -\sin\alpha \\ 0 & \sin\alpha & \cos\alpha \end{bmatrix} \tag{2.42}$$

which for small angles α and β reduces to

$$[T] = \begin{bmatrix} 1 & 0 & \beta \\ 0 & 1 & -\alpha \\ -\beta & \alpha & 1 \end{bmatrix} \tag{2.43}$$

The transformation between torque components in the $\xi\eta\zeta$ frame and xyz frame of Fig. 2.8 thus become

$$\begin{Bmatrix} \tau_x \\ \tau_y \\ \tau_z \end{Bmatrix} = [T] \begin{Bmatrix} \tau_\xi \\ \tau_\eta \\ \tau_\zeta \end{Bmatrix} \tag{2.44}$$

When (2.41) is substituted in (2.44), the result, correct to first order, is

$$\begin{aligned} \begin{Bmatrix} \tau_x \\ \tau_y \end{Bmatrix} &= \begin{bmatrix} I_d & 0 \\ 0 & I_d \end{bmatrix} \begin{Bmatrix} \ddot{\alpha} \\ \ddot{\beta} \end{Bmatrix} + \Omega \begin{bmatrix} 0 & I_a \\ -I_a & 0 \end{bmatrix} \begin{Bmatrix} \dot{\alpha} \\ \dot{\beta} \end{Bmatrix} + \dot{\Omega} \begin{bmatrix} 0 & I_a \\ -I_a & 0 \end{bmatrix} \begin{Bmatrix} \alpha \\ \beta \end{Bmatrix} \\ \tau_z &= I_a\dot{\Omega} \end{aligned} \tag{2.45}$$

and represents the angular momentum requirements in the stationary reference frame for an axisymmetric rigid body undergoing small precession and nutation. Finally, in many applications the spin is constant to at least second order so that to first order $\tau_z = 0$, and

$$\begin{Bmatrix} \tau_x \\ \tau_y \end{Bmatrix} = \begin{bmatrix} I_d & 0 \\ 0 & I_d \end{bmatrix} \begin{Bmatrix} \ddot{\alpha} \\ \ddot{\beta} \end{Bmatrix} + \Omega \begin{bmatrix} 0 & I_a \\ -I_a & 0 \end{bmatrix} \begin{Bmatrix} \dot{\alpha} \\ \dot{\beta} \end{Bmatrix} \tag{2.46}$$

These linearized equations permit the determination of the small motions $\alpha(t)$

and $\beta(t)$ of a spinning disk due to prescribed diametral torques $\tau_x(t)$ and $\tau_y(t)$. Also when the disk is part of a larger system, the system equations of motion can be obtained from D'Alembert's principle if the right-hand sides of (2.46) are introduced with reversed signs as inertia torques acting on the disk. The first terms on the right provide the usual angular acceleration inertia torques. The second terms provide the gyroscopic torques. Their strength is proportional to the axial angular momentum $I_a\Omega$, and they are cross-coupled in the sense that torque about the x axis is proportional to angular velocity about the y axis and vice versa. The applied and inertia torques are shown in Fig. 2.11. The angular

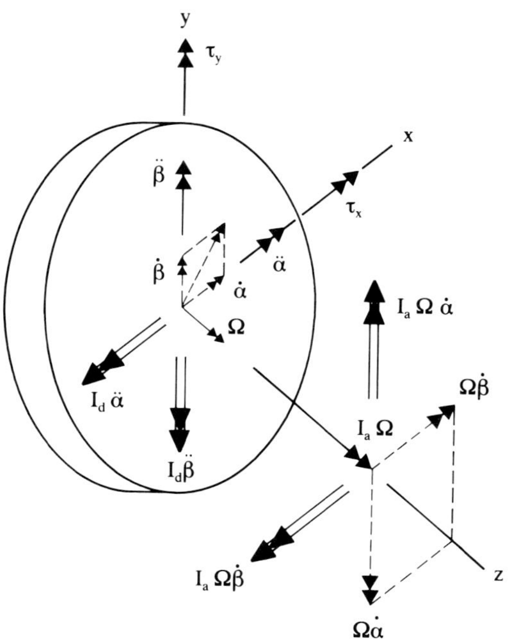

FIGURE 2.11 Inertia torques on a spinning rigid disk.

momentum of the disk changes in the direction of the applied torque. From the gyroscopic cross-coupling a torque τ_x (τ_y) tends to rotate the spin axis toward the x (y) axis.

2.2.2 Flexible Rotor

In this section equations of motion are given for deformable bodies which can twist, stretch, or bend while rotating. Lumped-parameter models lead to matrix equations, and continuum models lead to partial differential equations. The emphasis here is on linear models with stiffness and inertia elements. These are sufficient to predict natural modes and frequencies. Linear damping models which permit prediction of modal damping are also discussed.

Torsional Vibrations

The system of Fig. 2.9 may be used as a lumped-parameter torsional vibration model if all the inertia is assumed to be concentrated in the disks and all the elasticity is assumed to be concentrated in the connecting shafts. Let the axial moment of inertia of the disk at station i be I_i, and let the torsional spring constant for the shaft segment between stations i and $i+1$ be k_i. If an external torque τ_i acts on the ith disk, the angular momentum principle of (2.29) applied to a free body of the ith disk yields

$$I_i\ddot{\theta}_i + k_{i-1}(\theta_i - \theta_{i-1}) - k_i(\theta_{i+1} - \theta_i) = \tau_i \tag{2.47}$$

where θ_i is the rotation of the ith disk relative to the xyz frame. The n equations of this form obtained by letting i go from 1 to n (noting that $k_n = 0$) may be assembled in the single matrix equation

$$\begin{bmatrix} I_1 & 0 & \cdots & 0 & 0 \\ 0 & I_2 & \cdots & 0 & 0 \\ \cdots & \cdots & \cdots & \cdots & \cdots \\ 0 & 0 & \cdots & I_{n-1} & 0 \\ 0 & 0 & \cdots & 0 & I_n \end{bmatrix} \begin{Bmatrix} \ddot{\theta}_1 \\ \ddot{\theta}_2 \\ \vdots \\ \ddot{\theta}_{n-1} \\ \ddot{\theta}_n \end{Bmatrix} + \begin{bmatrix} k_1 & -k_1 & 0 & \cdots & 0 & 0 \\ -k_1 & k_1 + k_2 & -k_2 & \cdots & 0 & 0 \\ \cdots & \cdots & \cdots & \cdots & \cdots & \cdots \\ 0 & 0 & \cdots & -k_{n-2} & k_{n-2} + k_{n-1} & -k_{n-1} \\ 0 & 0 & \cdots & 0 & -k_{n-1} & k_{n-1} \end{bmatrix} \begin{Bmatrix} \theta_1 \\ \theta_2 \\ \vdots \\ \theta_{n-1} \\ \theta_n \end{Bmatrix} = \begin{Bmatrix} \tau_1 \\ \tau_2 \\ \vdots \\ \tau_{n-1} \\ \tau_n \end{Bmatrix} \tag{2.48}$$

which describes the undamped forced torsional vibrations of the system of Fig. 2.9. The free vibration frequencies and modes are obtained from the homogeneous form of (2.48). An important characteristic of systems such as this which are unattached to a stationary reference frame is that the rigid-body mode

$$\theta_1 = \theta_2 = \cdots = \theta_n = A + Bt \tag{2.49}$$

with natural frequency zero is a solution to the free vibration equations for arbitrary A and B. In rotating machinery it is this mode which stores the large kinetic energy of steady rotation. The matrix equation (2.48) may also be obtained by applying Lagrange's equations (2.30) when the energy functions are

$$T^* = \sum_{i=1}^{n} \tfrac{1}{2} I_i \dot{\theta}_i^2 \qquad V = \sum_{i=1}^{n-1} \tfrac{1}{2} k_i (\theta_{i+1} - \theta_i)^2 \tag{2.50}$$

and the generalized forces are $\Xi_i = \tau_i$.

A continuum model for a rotor is shown in Fig. 2.12. Consider a differential element of axial extent dz. Let the axial moment of inertia of the element be $\rho I_a\, dz$. The torque transmitted across a section is $GI_a\, \partial\theta/\partial z$ if $\theta(z, t)$ is the rotation angle at z and GI_a is the torsional modulus. If the section is circular, I_a is the polar moment of area and ρ and G are the mass density and shear modulus, respectively, of the material. For a noncircular section, the mass moment of inertia per unit length ρI_a and the torsional modulus GI_a must be evaluated

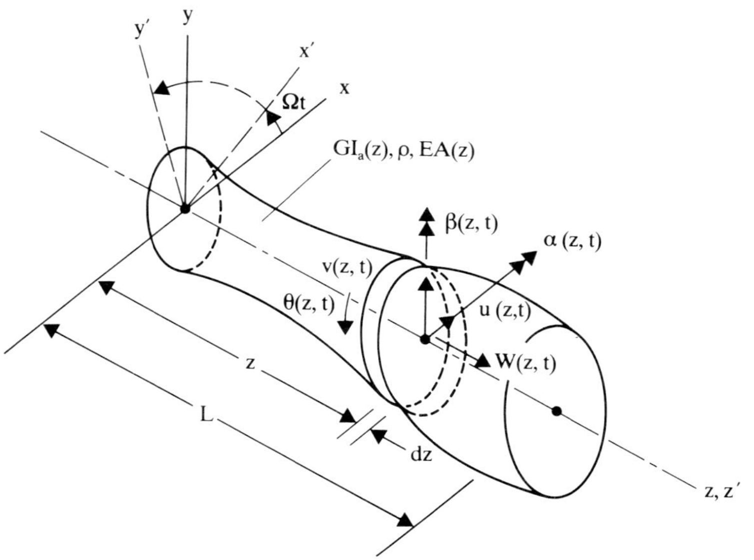

FIGURE 2.12 Continuous rotor torsional and longitudinal vibration model.

separately for the section in question. In general, both ρI_a and GI_a will vary with z. If an external axial torque per unit length $\tau(z, t)$ acts on the rotor, the result of applying the angular momentum principle of (2.29) to the differential element in Fig. 2.12 is the partial differential equation

$$-\frac{\partial}{\partial z}\left(GI_a \frac{\partial \theta}{\partial z}\right) + \rho I_a \frac{\partial^2 \theta}{\partial t^2} = \tau(z, t) \tag{2.51}$$

which applies for $0 < z < L$. If external axial torques $\tau_o(t)$ and $\tau_L(t)$ are applied at the ends, the boundary conditions are

$$\begin{aligned} -\left(GI_a \frac{\partial \theta}{\partial z}\right)_o &= \tau_o(t) \qquad \text{at } z = 0 \\ \left(GI_a \frac{\partial \theta}{\partial z}\right)_L &= \tau_L(t) \qquad \text{at } z = L \end{aligned} \tag{2.52}$$

Equation 2.51 plus the boundary conditions (2.52) describes the undamped forced torsional vibrations of the continuum model of Fig. 2.12. The free vibrations are described by the corresponding equation and boundary conditions when the driving torques $\tau(z, t)$, τ_o and τ_L are all set equal to zero. As in the lumped-parameter model, one of the free vibration modes is the rigid-body mode

$$\theta(t) = A + Bt \qquad \text{for } 0 < z < L \tag{2.53}$$

where A and B are arbitrary constants independent of z.

Longitudinal Vibrations

The treatment of longitudinal vibrations is quite similar to that of torsional vibrations. The major difference is that the rotor is usually attached to a stationary frame by one or more thrust bearings, so that rigid-body modes with unlimited longitudinal displacements are not permitted. The system of Fig. 2.9 can again be used as a lumped-parameter model by assuming that all the mass is concentrated in the disks and all the elasticity is concentrated in the connecting shafts. Let the mass at station i be m_i, and let the extensional spring constant for the shaft segment between stations i and $i+1$ be k_i. If an axial force f_i acts on the ith mass, the linear momentum principle of (2.29) applied to a free body of the ith mass yields

$$m_i\ddot{w}_i + k_{i-1}(w_i - w_{i-1}) - k_i(w_{i+1} - w_i) = f_i \tag{2.54}$$

where w_i is the axial displacement of the ith mass. This equation is completely analogous to (2.47) for torsional vibrations and applies at any station where there is no interconnection with a thrust bearing. To illustrate the inclusion of a thrust bearing, suppose that at station 1 there is such a bearing which modeled as a linear spring with spring constant k_o extending from a stationary reference frame to the first mass. The matrix equation for longitudinal vibration then becomes

$$\begin{bmatrix} m_1 & 0 & \cdots & 0 & 0 \\ 0 & m_2 & \cdots & 0 & 0 \\ \cdots & \cdots & \cdots & \cdots & \cdots \\ 0 & 0 & \cdots & m_{n-1} & 0 \\ 0 & 0 & \cdots & 0 & m_n \end{bmatrix} \begin{Bmatrix} \ddot{w}_1 \\ \ddot{w}_2 \\ \vdots \\ \ddot{w}_{n-1} \\ \ddot{w}_n \end{Bmatrix}$$

$$+ \begin{bmatrix} k_0 + k_1 & -k_1 & 0 & \cdots & 0 & 0 \\ -k_1 & k_1 + k_2 & -k_2 & \cdots & 0 & 0 \\ \cdots & \cdots & \cdots & \cdots & \cdots & \cdots \\ 0 & 0 & \cdots & -k_{n-2} & k_{n-2} + k_{n-1} & -k_{n-1} \\ 0 & 0 & \cdots & 0 & -k_{n-1} & k_{n-1} \end{bmatrix} \begin{Bmatrix} w_1 \\ w_2 \\ \vdots \\ w_{n-1} \\ w_n \end{Bmatrix} = \begin{Bmatrix} f_1 \\ f_2 \\ \vdots \\ f_{n-1} \\ f_n \end{Bmatrix} \tag{2.55}$$

This equation may also be obtained by applying Lagrange's equations (2.30) when the energy functions are

$$T^* = \sum_{i=1}^{n} \tfrac{1}{2} m_i \dot{w}_i^2 \qquad V = \tfrac{1}{2} k_o w_1^2 + \sum_{i=1}^{n-1} \tfrac{1}{2} k_i (w_{i+1} - w_i)^2 \tag{2.56}$$

A continuum model for longitudinal vibration can be constructed analogous to the torsional model for Fig. 2.12. Let $\rho A\,dz$ be the mass of a differential element of axial extent dz, and let $w(z, t)$ be the longitudinal displacement of the plane section originally at z. The longitudinal tensile force transmitted across this section is $EA\,\partial w/\partial z$, where EA is the uniaxial tension modulus. In general, both ρA and EA will vary with z. If an external longitudinal force per unit length $f(z, t)$ acts on the rotor, application of the linear momentum principle of (2.29) to the differential element yields the partial differential equation

$$-\frac{\partial}{\partial z}\left(EA\frac{\partial w}{\partial z}\right) + \rho A\frac{\partial^2 w}{\partial t^2} = f(z, t) \tag{2.57}$$

to be satisfied along the length of the rotor $0<z<L$. If the end $z=0$ is restrained by a longitudinal spring with stiffness k_o and external forces $f_o(t)$ and $f_L(t)$ are applied at the ends, the boundary conditions are

$$
\begin{aligned}
k_o w(0,t) - \left(EA\frac{\partial w}{\partial z}\right)_o &= f_o(t) \qquad \text{at } z=0 \\
\left(EA\frac{\partial w}{\partial z}\right)_L &= f_L(t) \qquad \text{at } z=L
\end{aligned}
\tag{2.58}
$$

Equations 2.57 and 2.58 describe undamped forced longitudinal vibrations. When the forcing terms $f(z,t)$ and f_o and f_L are removed, the corresponding equation and boundary conditions describe the free vibrations.

Transverse Vibrations

The treatment of transverse vibrations of rotors is more involved than that of torsional or longitudinal vibration. There are several reasons for this. First, the deformation of the rotor is due to a combination of bending and shearing which is inherently more complex than simple twisting or stretching. The bending is generally not confined to a single plane, so two orthogonal planes (for example, xz and yz) must be considered. Within each plane it is necessary to characterize the motion of each station point i in Fig. 2.9 by two independent coordinates: a transverse displacement and an angular displacement. Finally, when the rotor is spinning, the motions in two orthogonal planes are coupled by gyroscopic and other asymmetric effects. The location and nature of the bearings play a much more important role in transverse vibrations, which in turn requires the development of more detailed models of bearings and their support structures.

Despite these complications it is possible to derive equations of motion for the transverse vibration of the rotor in Fig. 2.9 which are analogous to (2.47) and (2.54) for torsional and longitudinal vibrations. Again it is assumed that all the deformation takes place in the massless connecting shafts and that the axially symmetric rigid disks have mass m_i, diametral moment of inertia I_{d_i}, axial moment of inertia I_{a_i}, and negligible axial extent. Each disk has 4 degrees of freedom; the center may displace transversely through small displacements u_i and v_i, and the disk may rotate through small angles α_i and β_i, as defined in Fig. 2.8. Figure 2.11 shows the inertia torques acting on the disk. The real forces acting on the disk include the applied forces f_{x_i} and f_{y_i}, the applied torques τ_{x_i} and τ_{y_i}, and internal forces of interconnection with the shaft segments on either side of the disk.

Figure 2.13 shows the forces and moments acting on the deformed ith shaft segment connecting the i and $i+1$ disks. It is assumed that within the segment the shaft is uniform with bending modulus EI_i and effective shear modulus κGA_i, as used in Timoshenko beam theory (Crandall et al., 1982, p. 343). To express the load deformation relations for bending in this ith beam segment in a compact form analogous to the relations for torsion and extension, it is convenient to introduce the vector displacements

$$
\{q_i\} = \begin{Bmatrix} u_i \\ v_i \\ \alpha_i \\ \beta_i \end{Bmatrix} \qquad \text{and} \qquad \{q_{i+1}\} = \begin{Bmatrix} u_{i+1} \\ v_{i+1} \\ \alpha_{i+1} \\ \beta_{i+1} \end{Bmatrix}
\tag{2.59}
$$

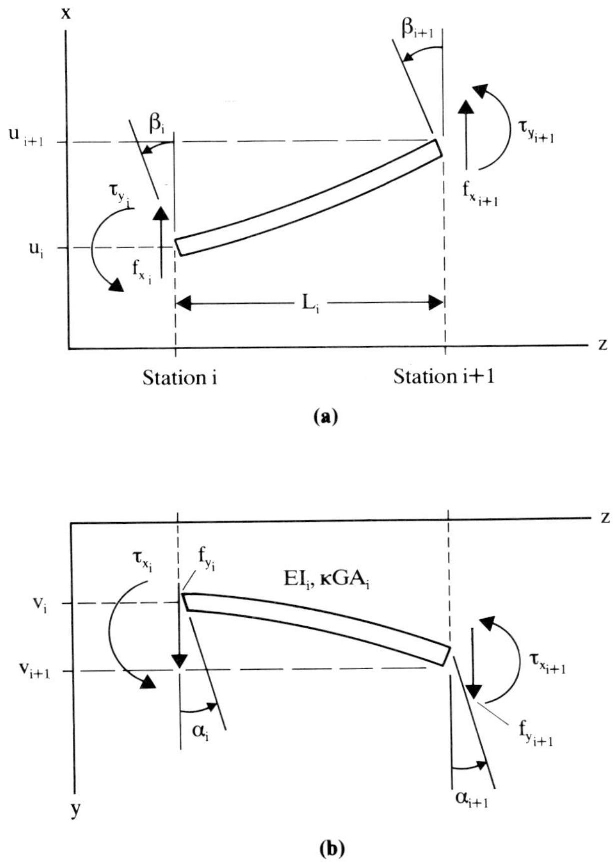

FIGURE 2.13 Transverse deformation of a flexible shaft segment: (*a*) plane of bending *xz*; (*b*) plane of bending *yz*.

at the i and $i+1$ ends of the segment. The transverse deformation in the xz (yz) plane involves the u and β (v and α) displacements. The force-displacement relation for the beam segment is then

$$\begin{Bmatrix} \{Q_i\} \\ \{Q_{i+1}\} \end{Bmatrix} = [k^i] \begin{Bmatrix} \{q_i\} \\ \{q_{i+1}\} \end{Bmatrix} \tag{2.60}$$

where $[k^i]$ is the stiffness matrix for the ith segment. A typical element k_{rs} of this matrix represents the external force (or torque) required at the rth station to produce a unit linear (or angular) displacement at the sth station with all other displacements constrained to zero. The Maxwell–Betti reciprocal theory ensures that the symmetry relation $k_{rs} = k_{sr}$ is satisfied. If both shearing and bending compliance are assumed, as in the Timoshenko beam theory, the stiffness matrix is

$$[k^i] = \begin{bmatrix} [k^i_{i,i}] & [k^i_{i,i+1}] \\ [k^i_{i+1,i}] & [k^i_{i+1,i+1}] \end{bmatrix} \tag{2.61}$$

where

$$[k^i_{i,i}] = \frac{EI_i}{L_i^3(1+12\varepsilon_i)} \begin{bmatrix} 12 & 0 & 0 & 6L_i \\ 0 & 12 & -6L_i & 0 \\ 0 & -6L_i & 4L_i^2(1+3\varepsilon_i) & 0 \\ 6L_i & 0 & 0 & 4L_i^2(1+3\varepsilon_i) \end{bmatrix}$$

$$[k^i_{i,i+1}] = \frac{EI_i}{L_i^3(1+12\varepsilon_i)} \begin{bmatrix} -12 & 0 & 0 & 6L_i \\ 0 & -12 & -6L_i & 0 \\ 0 & 6L_i & 2L_i^2(1-6\varepsilon_i) & 0 \\ -6L_i & 0 & 0 & 2L_i^2(1-6\varepsilon_i) \end{bmatrix} \tag{2.62}$$

$$= [k^i_{i+1,i}]^T$$

The submatrix $[k^i_{i+1,i+1}]$ is equal to the submatrix $[k^i_{i,i}]$ except that the polarity of the $6L_i$ terms is reversed. The factor $\varepsilon_i = EI_i/(\kappa G A_i L_i^2)$ is a dimensionless measure of the shear compliance of the ith segment. When shear compliance is neglected, as in Bernoulli–Euler beam theory, $G_i \to \infty$ and $\varepsilon_i \to 0$. The decision as to which theory to use for a particular rotor can usually be based on evaluating ε_i for a few segments. If $12\varepsilon_i$ is negligible in comparison to unity, then the simpler theory may be used safely.

The equations of motion for the ith disk are obtained by applying the linear and angular momentum principles (2.29) in both the xz and yz planes. A convenient form of the angular momentum requirements applicable for small tip angles is given by (2.46). Alternatively, D'Alembert's principle may be applied, using the inertia torques of Fig. 2.11. The resulting four scalar equations may be written as a matrix equation by introducing the mass matrix $[m_i]$, gyroscopic matrix $[g_i]$, and the external excitation vector $\{Q_i\}$:

$$[m_i] = \begin{bmatrix} m & 0 & 0 & 0 \\ 0 & m & 0 & 0 \\ 0 & 0 & I_d & 0 \\ 0 & 0 & 0 & I_d \end{bmatrix}_i \qquad [g_i] = \begin{bmatrix} 0 & 0 & 0 & 0 \\ 0 & 0 & 0 & 0 \\ 0 & 0 & 0 & I_a \\ 0 & 0 & -I_a & 0 \end{bmatrix}_i$$

$$\{Q_i\} = \begin{Bmatrix} f_x \\ f_y \\ \tau_x \\ \tau_y \end{Bmatrix}_i \tag{2.63}$$

where f_{x_i} and τ_{y_i} are the external force and torque applied to the ith disk in the xz plane, as indicated in Fig. 2.11, and f_{y_i} and τ_{x_i} are the corresponding external excitation components in the yz plane. The resulting equations, including the interconnection forces with the $i-1$ and i beams from (2.60), are

$$[m_i]\{\ddot{q}_i\} + \Omega[g_i]\{\dot{q}_i\} + [k^i_{i,i}]\{q_i\} - [k^i_{i,i+1}]\{q_{i+1}\}$$
$$- [k^{i-1}_{i,i-1}]\{q_{i-1}\} + [k^{i-1}_{i,i}]\{q_i\} = \{Q_i\} \tag{2.64}$$

where Ω is the uniform rotation rate of the rotor. The only coupling between the xz and yz planes of motion is provided by the gyroscopic matrix $[g_i]$. Note that when $\Omega = 0$, (2.64) can be separated into two independent second-order matrix equations associated with deformation in the xz and yz planes, and these equations are similar in form to (2.47) and (2.54) for torsional and longitudinal vibration. Equation 2.64 represents four scalar equations of motion for the ith disk. For a rotor with n disks, the governing equations consist of $4n$ such relations plus the constraint conditions imposed at the bearings and at the ends of the rotor. In the example which follows, these constraints are illustrated for a particular case. If, for the present, detailed consideration of these matters is postponed by assuming that any bearing reactions can be represented by the external excitation vectors, a standard form for the equations of an n-disk rotor can be exhibited. The system response and excitation vectors are defined as

$$\{q\} = \begin{Bmatrix} \{q_1\} \\ \vdots \\ \{q_n\} \end{Bmatrix}_{4n\times 1} \qquad \{Q\} = \begin{Bmatrix} \{Q_1\} \\ \vdots \\ \{Q_n\} \end{Bmatrix}_{4n\times 1} \tag{2.65}$$

and in terms of these the equations of motion of an n-disk rotor are

$$[M]\{\ddot{q}\} + \Omega[G]\{\dot{q}\} + [K]\{q\} = \{Q\} \tag{2.66}$$

where $[M]$ is a diagonal mass matrix, $[G]$ is a banded skew symmetric gyroscopic matrix, and $[K]$ is a banded symmetric stiffness matrix. Equation 2.66 may also be derived by applying Lagrange's equations (2.30) and retaining only linear terms when the energy functions are

$$\begin{aligned} T^* &= \tfrac{1}{2}\sum_{i=1}^{n}[m_i(\dot{u}_i^2 + \dot{v}_i^2) + I_{d_i}(\dot{\alpha}_i^2 + \dot{\beta}_i^2) + I_{a_i}(\Omega - \dot{\beta}_i\alpha_i)^2] \\ V &= \tfrac{1}{2}\{q\}^T[K]\{q\} \end{aligned} \tag{2.67}$$

Example 2.1: Jeffcott Rotor Model. Consider the rotor model (Jeffcott, 1919) shown in Fig. 2.14. The rigid disk has mass m, diametral moment of inertia I_d, and axial moment of inertia I_a. The flexible shaft segments on each side are identical, each having length L, negligible mass, bending modulus EI, and equivalent shear modulus κGA. The ideal bearings constrain the transverse displacements to zero at stations $i = 1$ and $i = 3$ and exert no diametral moments on the shaft. Since the system has only the single disk at station $i = 2$, the equations of motion consist of (2.64) written once for $i = 2$ plus the end conditions

$$u_1 = v_1 = 0 \qquad u_3 = v_3 = 0$$

and

$$\tau_{x_1} = \tau_{y_1} = 0 \qquad \tau_{x_3} = \tau_{y_3} = 0 \tag{2.68}$$

Here the system displacement vector is

$$\{q\} = [u_1 \;\; v_1 \;\; \alpha_1 \;\; \beta_1 \;\; u_2 \;\; v_2 \;\; \alpha_2 \;\; \beta_2 \;\; u_3 \;\; v_3 \;\; \alpha_3 \;\; \beta_3]^T \tag{2.69}$$

In this example the condition of zero end moments and zero inertia properties at the endpoints can be used to relate the endpoint rotations to the motion at station 2. Using this and imposing the endpoint displacement constraints (2.68)

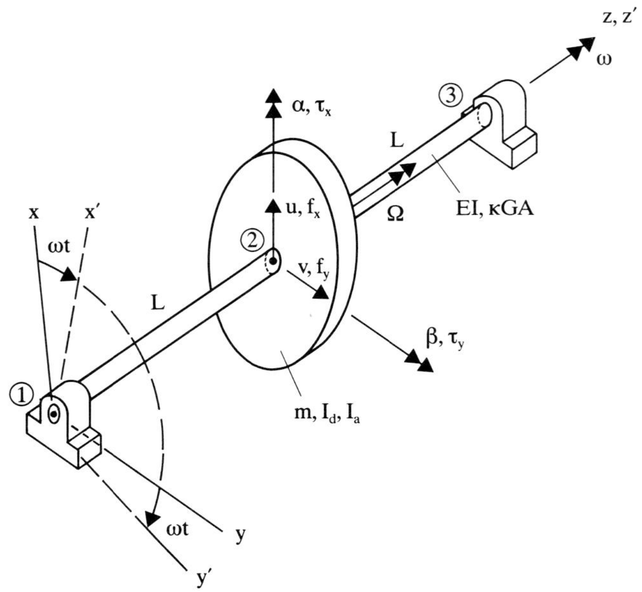

FIGURE 2.14 Jeffcott rotor model.

allow the system equations (2.66) to be written in the form

$$\begin{bmatrix} m & 0 & 0 & 0 \\ 0 & m & 0 & 0 \\ 0 & 0 & I_d & 0 \\ 0 & 0 & 0 & I_d \end{bmatrix}\{\ddot{q}\} + \Omega\begin{bmatrix} 0 & 0 & 0 & 0 \\ 0 & 0 & 0 & 0 \\ 0 & 0 & 0 & I_a \\ 0 & 0 & -I_a & 0 \end{bmatrix}\{\dot{q}\}$$

$$+\begin{bmatrix} k_T & 0 & 0 & 0 \\ 0 & k_T & 0 & 0 \\ 0 & 0 & k_R & 0 \\ 0 & 0 & 0 & k_R \end{bmatrix}\{q\} = \{Q\} \qquad (2.70)$$

where the translational and rotational stiffness coefficients are

$$k_T = \frac{6EI}{(1+3\varepsilon)L^3} \qquad k_R = \frac{6EI}{(1+3\varepsilon)L} \qquad (2.71)$$

Note that $[M]$ and $[K]$ are symmetric (actually diagonal) and $[G]$ is skew symmetric. In addition, the translational and rotational motions of the disk are elastically decoupled since the disk is located at the center of the shaft. The translational equations of motion are

$$\begin{bmatrix} m & 0 \\ 0 & m \end{bmatrix}\begin{Bmatrix} \ddot{u} \\ \ddot{v} \end{Bmatrix} + \begin{bmatrix} k_T & 0 \\ 0 & k_T \end{bmatrix}\begin{Bmatrix} u \\ v \end{Bmatrix} = \begin{Bmatrix} f_x \\ f_y \end{Bmatrix} \qquad (2.72)$$

and the rotational equations of motion are

$$\begin{bmatrix} I_d & 0 \\ 0 & I_d \end{bmatrix} \begin{Bmatrix} \ddot{\alpha} \\ \ddot{\beta} \end{Bmatrix} + \Omega \begin{bmatrix} 0 & I_a \\ -I_a & 0 \end{bmatrix} \begin{Bmatrix} \dot{\alpha} \\ \dot{\beta} \end{Bmatrix} + \begin{bmatrix} k_R & 0 \\ 0 & k_R \end{bmatrix} \begin{Bmatrix} \alpha \\ \beta \end{Bmatrix} = \begin{Bmatrix} \tau_x \\ \tau_y \end{Bmatrix} \tag{2.73}$$

When the rotor is not rotating (that is, $\Omega = 0$), the natural modes and natural frequencies of the Jeffcott rotor can be determined by inspection. From (2.72) there are two independent modes of transverse vibration of equal natural frequency $\omega_T = (k_T/m)^{1/2}$. One of the modes is a straight-line oscillation in the x direction, and the other is a straight-line oscillation in the y direction, as depicted in Fig. 2.15*a*. If these two modes are given the same amplitude, they can be

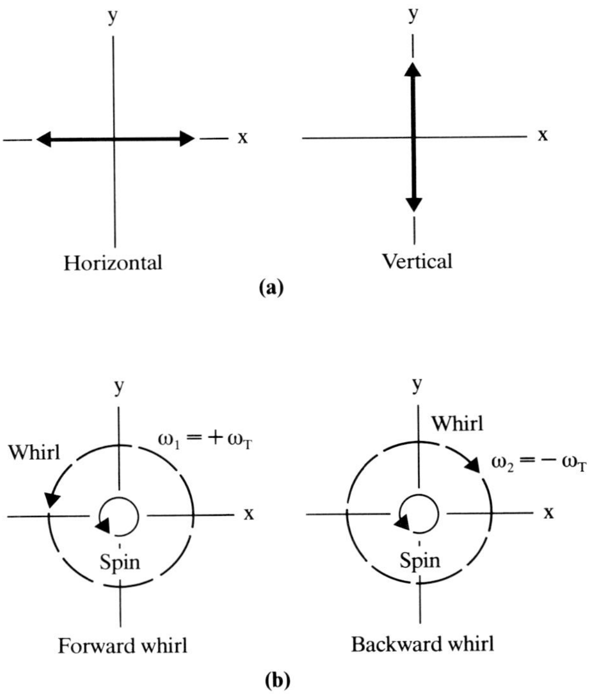

FIGURE 2.15 Jeffcott rotor translational modes: (*a*) straight-line modes; (*b*) circular precession modes.

superposed with the proper phasing to form a circular precessional whirl mode which is either corotating (forward) or counterrotating (backward) with the rotor spin, as illustrated in Fig. 2.15*b*. These frequencies are independent of the rotor spin speed Ω for this model. It is convenient here to introduce the complex rotation $\bar{\gamma} = \alpha + i\beta$ and rewrite the homogeneous form of (2.73) as

$$I_d\ddot{\bar{\gamma}} - i\Omega I_a\dot{\bar{\gamma}} + k_R\bar{\gamma} = 0 \tag{2.74}$$

The rotational natural modes for this system are of the form $\bar{\gamma} = \bar{\gamma}_o e^{i\omega t}$, and the natural frequencies are given by

$$\omega_1,\ \omega_2 = \frac{\Omega I_a}{2I_d} \pm \left[\left(\frac{\Omega I_a}{2I_d}\right)^2 + \frac{k_R}{I_d}\right]^{1/2} \tag{2.75}$$

For a positive spin speed Ω_1, ω_1 is positive and corresponds to a corotating precessional mode and ω_2 is negative and corresponds to a counterrotating precessional mode. For a negative spin speed, ω_1 is positive and corresponds to a counterrotating mode while ω_2 is negative and corresponds to a corrotating mode. The translational and rotational natural frequencies of precession (whirl) are plotted versus rotor spin speed in Fig. 2.16, and the modes of precession are

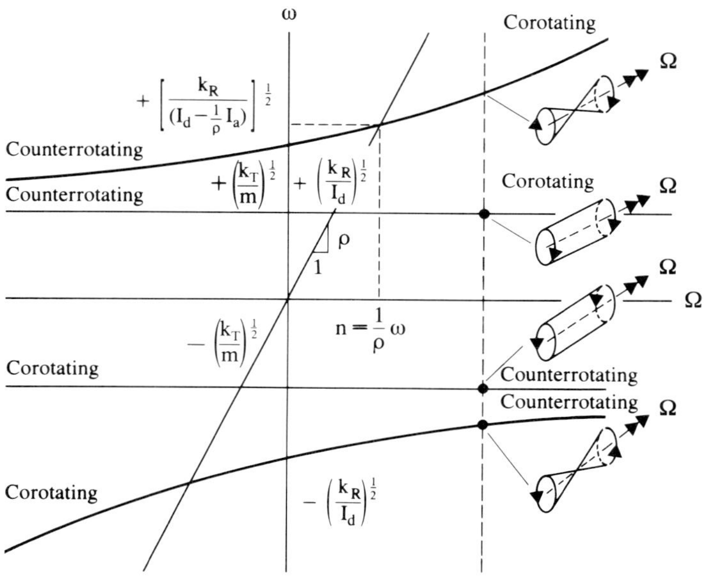

FIGURE 2.16 Jeffcott rotor whirl speeds and modes.

also illustrated. The rotational modes provide perhaps the simplest example of a rotordynamic mode subject to gyroscopic effects. As illustrated in Fig. 2.16, the corotating modes tend to increase in frequency magnitude as the spin speed magnitude increases due to a gyroscopic stiffening effect, while the counterrotating modes tend to decrease in frequency.

It is interesting to consider the rotational motion of the Jeffcott rotor relative to the rotating reference $x'y'z'$ shown in Fig. 2.14. The uniform rotation rate of this reference about the z axis is ω, and by using the transformations (2.20) and (2.21)

$$\bar{\gamma} = \alpha + i\beta = (\alpha' + \beta')e^{i\omega t} = \bar{\gamma}' e^{i\omega t} \tag{2.76}$$

the transformed rotational equation of motion is

$$I_d \ddot{\bar{\gamma}}' + i\omega\left(2I_d - \frac{\Omega}{\omega} I_a\right)\dot{\bar{\gamma}}' + \left[k_R - \omega^2\left(I_d - \frac{\Omega}{\omega} I_a\right)\right]\bar{\gamma}' = 0 \tag{2.77}$$

Along a specified straight line of slope $\omega/\Omega = \rho$ in the $\Omega\omega$ plane of Fig. 2.16 we seek a constant solution to (2.77). A constant solution relative to the $x'y'z'$ reference corresponds to a circular solution in the xyz reference. Such a solution exists if

$$\omega = \left[\frac{k_R}{I_d - (1/\rho)I_a}\right]^{1/2} \tag{2.78}$$

The frequency ω corresponds to a natural frequency of circular whirl relative to the fixed reference frame, and the form of (2.78) clearly illustrates the *gyroscopic stiffening* effect for the corotating modes when ρ is positive. It is also clear that when $I_d - (1/\rho)I_a$ is negative, a corotating mode does not exist. This corresponds to a situation when the gyroscopic stiffening is so high that an intersection does not occur between the line $\omega = \rho\Omega$ and the rotational natural frequency curve. For a negative whirl-to-spin ratio, a counterrotating mode always exists. It is also interesting to note that as $k_R \rightarrow 0$, the corotating mode frequency approaches $\Omega I_a/I_d$ while the frequency of the counterrotating mode approaches zero.

Continuum Model for Transverse Vibration. Consider again the continuous rotor of Fig. 2.12. In addition to the mass per unit length ρA, the diametral moment of inertia per unit length is ρI_d, and the axial moment of inertia per unit length is ρI_a. Let the transverse displacements of the center of the infinitesimal element be defined by $u(z, t)$ and $v(z, t)$, and let the disk rotate through small angles $\alpha(z, t)$ and $\beta(z, t)$. The internal bending moments transmitted across the section at z are

$$M_x = +EI\frac{\partial \alpha}{\partial z} \qquad M_y = +EI\frac{\partial \beta}{\partial z} \tag{2.79}$$

and the shear forces transmitted across the section at z are

$$Q_x = \kappa GA\left(\frac{\partial u}{\partial z} - \beta\right) \qquad Q_y = \kappa GA\left(\frac{\partial v}{\partial z} + \alpha\right) \tag{2.80}$$

The bending modulus EI, effective shear modulus κGA, and inertia parameters vary with position z along the rotor. If external forces $f_x(z, t)$ and $f_y(z, t)$ per unit length and external torques $\tau_y(z, t)$ and $\tau_x(z, t)$ per unit length, in the xz and yz planes, respectively, act on the rotor while it rotates at uniform rate Ω, then application of the linear and angular momentum principles of (2.29) yield the following four partial differential equations:

$$\begin{aligned}
\rho A\frac{\partial^2 u}{\partial t^2} - \frac{\partial}{\partial z}\left[\kappa GA\left(\frac{\partial u}{\partial z} - \beta\right)\right] &= f_x \\
\rho A\frac{\partial^2 v}{\partial t^2} - \frac{\partial}{\partial z}\left[\kappa GA\left(\frac{\partial v}{\partial z} + \alpha\right)\right] &= f_y \\
\rho I_d\frac{\partial^2 \alpha}{\partial t^2} + \Omega\rho I_a\frac{\partial \beta}{\partial t} - \frac{\partial}{\partial z}\left(EI\frac{\partial \alpha}{\partial z}\right) + \kappa GA\left(\frac{\partial v}{\partial z} + \alpha\right) &= \tau_x \\
\rho I_d\frac{\partial^2 \beta}{\partial t^2} - \Omega\rho I_a\frac{\partial \alpha}{\partial t} - \frac{\partial}{\partial z}\left(EI\frac{\partial \beta}{\partial z}\right) - \kappa GA\left(\frac{\partial u}{\partial z} - \beta\right) &= \tau_y
\end{aligned} \tag{2.81}$$

for the displacements (u, v, α, β) to be satisfied along the length of the rotor, $0 < z < L$. The formulation is completed by appending boundary conditions which may be expressed in terms of the terminal values of the displacements by using (2.79) and (2.80).

The above formulation includes the shear compliance of the Timoshenko beam theory. It can be reduced to the simpler theory which neglects shear compliance by first eliminating the shear force term between the first and fourth equations of (2.81) (and similarly between the second and third equations) and then setting $\alpha = -\partial v/\partial z$ and $\beta = +\partial u/\partial z$ to obtain

$$\begin{aligned} \frac{\partial^2}{\partial z^2}\left(EI\frac{\partial^2 u}{\partial z^2}\right) + \rho A\frac{\partial^2 u}{\partial t^2} - \frac{\partial}{\partial z}\left(\rho I_d\frac{\partial^3 u}{\partial z\,\partial t^2}\right) - \Omega\frac{\partial}{\partial z}\left(\rho I_a\frac{\partial^2 v}{\partial z\,\partial t}\right) &= f_x - \frac{\partial \tau_y}{\partial z} \\ \frac{\partial^2}{\partial z^2}\left(EI\frac{\partial^2 v}{\partial z^2}\right) + \rho A\frac{\partial^2 v}{\partial t^2} - \frac{\partial}{\partial z}\left(\rho I_d\frac{\partial^3 v}{\partial z\,\partial t^2}\right) + \Omega\frac{\partial}{\partial z}\left(\rho I_a\frac{\partial^2 u}{\partial z\,\partial t}\right) &= f_y + \frac{\partial \tau_x}{\partial z} \end{aligned} \tag{2.82}$$

to be satisfied by $u(z, t)$ and $v(z, t)$ in the interval $0 < z < L$. Here the boundary conditions, for example in the xz plane are expressed in terms of the terminal values of u, its slope $\partial u/\partial z$, the bending moment M_y, and the shear force Q_x. Since the constitutive equations (2.80) are no longer available in this simpler theory, the shear forces must be obtained from dynamic requirements equivalent to the third and fourth equations of (2.81). Thus the boundary shear forces are

$$\begin{aligned} Q_x &= -\frac{\partial}{\partial z}\left(EI\frac{\partial^2 u}{\partial z^2}\right) + \rho I_d\frac{\partial^3 u}{\partial z\,\partial t^2} + \Omega\rho I_a\frac{\partial^2 v}{\partial z\,\partial t} - \tau_y \\ Q_y &= -\frac{\partial}{\partial z}\left(EI\frac{\partial^2 v}{\partial z^2}\right) + \rho I_d\frac{\partial^3 v}{\partial z\,\partial t^2} - \Omega\rho I_a\frac{\partial^2 u}{\partial z\,\partial t} + \tau_x \end{aligned} \tag{2.83}$$

When the boundary conditions as well as the rotor properties are axially symmetric, the order of systems (2.81) and (2.82) can be halved by introducing complex vector responses and excitations

$$\begin{aligned} \bar{w} &= u + iv \qquad & \bar{\gamma} &= \alpha + i\beta \\ \bar{f} &= f_x + if_y \qquad & \bar{\tau} &= \tau_x + i\tau_y \end{aligned} \tag{2.84}$$

Then the second and fourth equations of (2.81) are multiplied by i and added, respectively, to the first and third to obtain

$$\begin{aligned} &\rho A\frac{\partial^2 \bar{w}}{\partial t^2} - \frac{\partial}{\partial z}\left[\kappa GA\left(\frac{\partial \bar{w}}{\partial z} + i\bar{\gamma}\right)\right] = \bar{f} \\ &\rho I_d\frac{\partial^2 \bar{\gamma}}{\partial t^2} - i\Omega\rho I_a\frac{\partial \bar{\gamma}}{\partial t} - \frac{\partial}{\partial z}\left(EI\frac{\partial \bar{\gamma}}{\partial z}\right) - i\kappa GA\left(\frac{\partial \bar{w}}{\partial z} + i\bar{\gamma}\right) = \bar{\tau} \end{aligned} \tag{2.85}$$

for an axially symmetric Timoshenko rotor. The same procedure applied to (2.82) yields

$$\frac{\partial^2}{\partial z^2}\left(EI\frac{\partial^2 \bar{w}}{\partial z^2}\right) + \rho A\frac{\partial^2 \bar{w}}{\partial t^2} - \frac{\partial}{\partial z}\left(\rho I_d\frac{\partial^3 \bar{w}}{\partial z\,\partial t^2}\right) + i\Omega\frac{\partial}{\partial z}\left(\rho I_a\frac{\partial^2 \bar{w}}{\partial z\,\partial t}\right) = \bar{f} + i\frac{\partial \bar{\tau}}{\partial z} \tag{2.86}$$

2.2.3 Rotating Machinery Systems

Concepts in Modeling

The modeling and analysis of a complicated machine with rotating assemblies involve several different phases of effort. Initially, a judgment needs to be made as to the required complexity of a system model. This choice is based on the objectives of the analysis and is assisted by sound engineering judgment and experience with the particular type of machine under consideration. In general, it is advisable to utilize as simple a model as possible to provide reasonable estimates of the desired system dynamic characteristics. Ultimate verification of the model validity is then made by comparison with experimental results, and additional model complexity and refinement are made as required.

The primary components of the system must be identified and a set of simplifying assumptions made regarding the constitution of each component and how it interacts with connecting components. This set of system components and associated set of simplifying assumptions form a physical model of the system. Figure 2.17 shows a rotating assembly of a gas-turbine aircraft engine, assumed to be axially symmetric, along with a schematic of a physical model of the engine. In

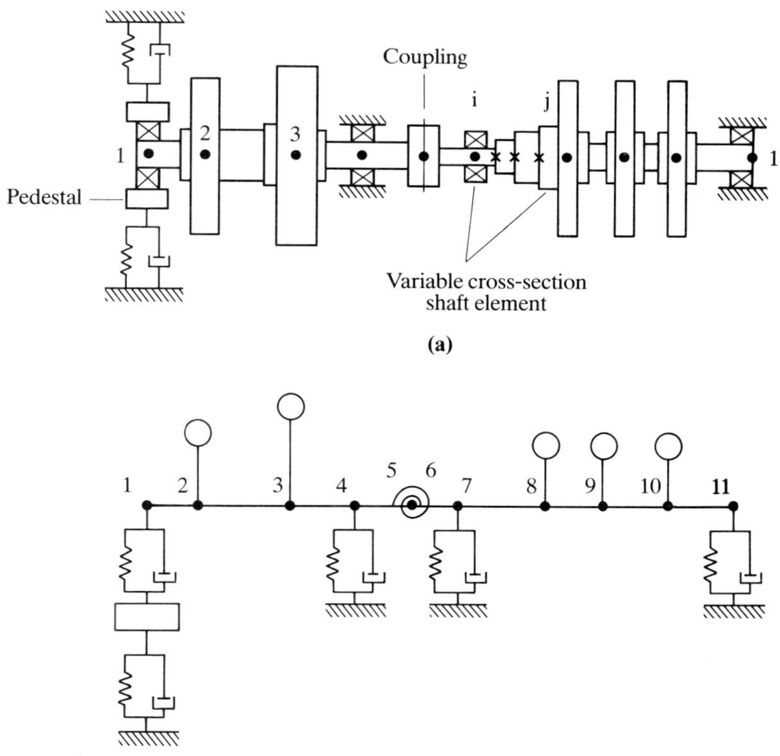

FIGURE 2.17 Gas-turbine rotating assembly: (*a*) physical model; (*b*) schematic.

this illustration, the real system is approximated by five disks to represent the two compressor wheels and three turbine disks, nine beam elements, four bearings, one coupling, and a bearing pedestal to approximate the support structure dynamics at bearing 1. Some of the assumptions associated with this model include neglecting disk flexibility, aerodynamic interconnection between rotor and stator stages, seal forces, higher-frequency support structure dynamics at bearing 1, and support structure dynamics at the other bearings.

Once a physical model is established for a real system, a set of mathematical equations consistent with the modeling assumptions is generated for each component of the system. In the next section, governing equations of motion are presented for several typical rotating machinery components. In later sections, methods of assembly of these component equations to form a set of system equations of motion are presented. Methods of analysis of these equations, the system mathematical model, are presented in Sec. 2.3.

As discussed previously, system models are normally classified as continuous-parameter or discrete-parameter models. Continuous models, which result in partial differential equations of motion, are particularly useful for simple systems with constant or smoothly varying properties. Complicated machines, however, such as illustrated in Fig. 2.17, normally include many abrupt changes in geometry and many highly localized mechanisms. As a result, continuous models are not usually practical, and almost all present modeling and analysis are done by using some form of discrete-parameter models.

Rotor Component Models

Typical components of rotating machinery include disks, beam segments, bearings, dampers, seals, gears, couplings, static support structure, and other fluid interconnection mechanisms. Physical models and the associated equations of motion for a few of these components are discussed and listed below. The emphasis here is on lateral rotating assembly motion rather than longitudinal and torsional motion.

Disks. Turbine and compressor wheels and gears are usually modeled as rigid disks, and it is usually assumed that the inertia coupling between the longitudinal, torsional, and lateral motion of the disk is negligible. In many applications it is also acceptable to consider the width of the disk to be negligible in comparison with the overall rotating assembly length. Thus, the resulting *thin-disk model* is depicted in Fig. 2.18. The motion of the geometric center of the ith typical disk is defined by the two translations (u_i, v_i) and the two rotations (α_i, β_i) in the x and y directions, respectively. The equations of motion are

$$[m_i]\{\ddot{q}_i\} + \Omega[g_i]\{\dot{q}_i\} + \dot{\Omega}[g_i]\{q_i\}$$

$$= m_i\left(\begin{Bmatrix} a_{x'}\Omega^2 + a_{y'}\dot{\Omega} \\ a_{y'}\Omega^2 - a_{x'}\dot{\Omega} \\ 0 \\ 0 \end{Bmatrix}_i \cos\varphi + \begin{Bmatrix} -a_{y'}\Omega^2 + a_{x'}\dot{\Omega} \\ a_{x'}\Omega^2 + a_{y'}\dot{\Omega} \\ 0 \\ 0 \end{Bmatrix}_i \sin\varphi\right)$$

$$+ m_i g\begin{Bmatrix} 0 \\ -1 \\ 0 \\ 0 \end{Bmatrix} + \{Q_i\}_c \qquad (2.87)$$

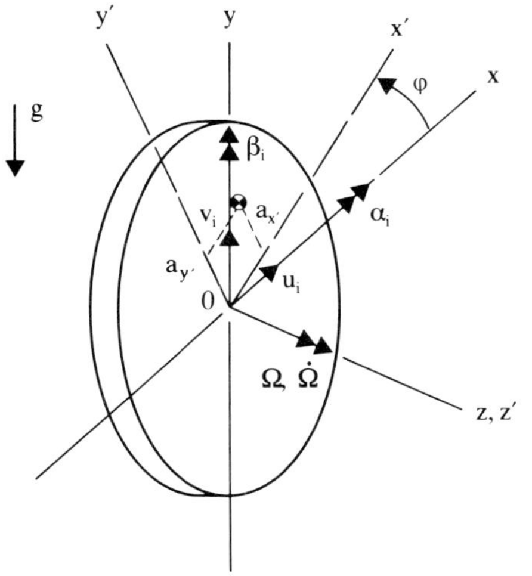

FIGURE 2.18 Thin rigid disk model.

where $[m_i]$ and $[g_i]$ are defined by (2.63) and $\{q_i\}$ by (2.59) and $\{Q_i\}_c$ is the vector of interconnection forces acting on the disk from connecting components. The mass center of the disk is located, as shown in Fig. 2.18, at the point $(a_{x'}, a_{y'})$ with respect to the rotating $x'y'$ frame, and the resulting centrifugal and tangential inertia force components are referred to as *unbalance forces*. The translational and rotational motions are uncoupled for this model.

If the axial extent of the rigid disk is large compared to the shaft length, one approach to model the "thick disk" is to treat it as two thin disks connected by a beam segment of appropriate length, as illustrated in Fig. 2.19. The inertia parameters for the thin disks are selected to provide the correct inertia properties at the thick-disk mass center. This 8-degree-of-freedom model is then reduced to a 4-degree-of-freedom model imposing a rigid-body constraint between the motion of the two thin disks. The coordinates $\{q_i\} = [u_i \; v_i \; \alpha_i \; \beta_i]^T$ and $\{\underline{q}_i\} = [\underline{u}_i \; \underline{v}_i \; \underline{\alpha}_i \; \underline{\beta}_i]^T$ are considered to be active and dependent, respectively, as indicated by the constraint equations

$$\begin{Bmatrix} \{q_i\} \\ \{\underline{q}_i\} \end{Bmatrix} = \begin{bmatrix} 1 & 0 & 0 & 0 \\ 0 & 1 & 0 & 0 \\ 0 & 0 & 1 & 0 \\ 0 & 0 & 0 & 1 \\ 1 & 0 & 0 & L \\ 0 & 1 & -L & 0 \\ 0 & 0 & 1 & 0 \\ 0 & 0 & 0 & 1 \end{bmatrix} \{q_i\} \tag{2.88}$$

The use of constraints, such as given by (2.88), is discussed in more detail in Sec. 2.3.

If the flexibility of the disk is significant in its influence on the lateral dynamics of a rotating assembly, the thin rigid disk of Fig. 2.18 can be generalized to

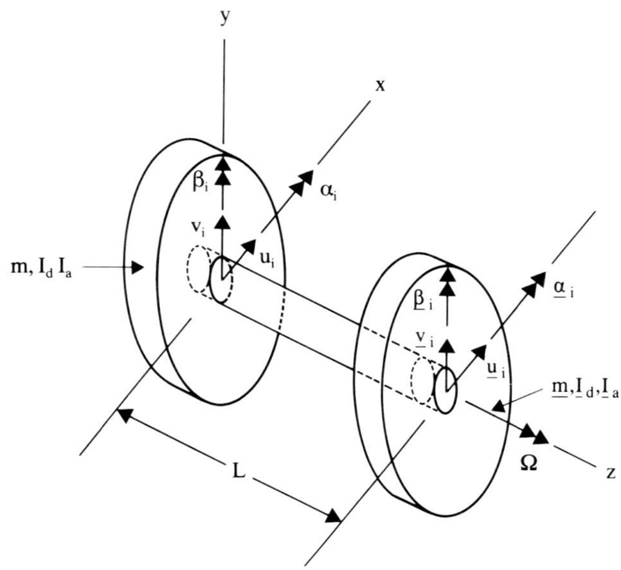

FIGURE 2.19 Thick rigid disk model.

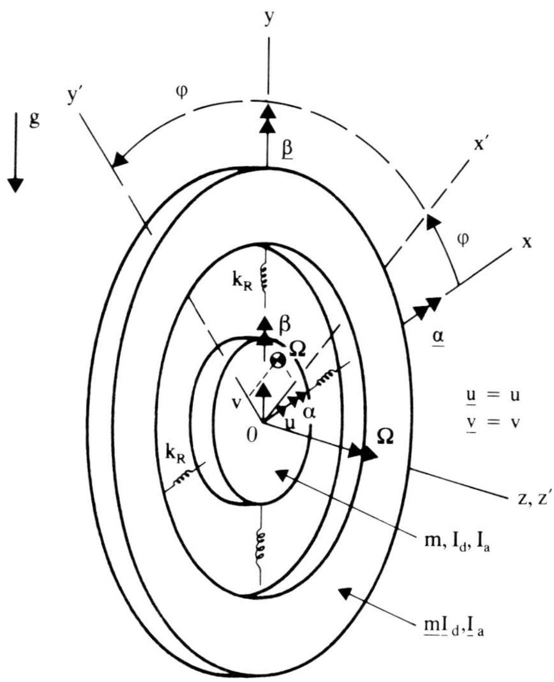

FIGURE 2.20 Six-degree-of-freedom flexible disk model.

account for this flexibility. As stated earlier for axisymmetric disks, the only disk modes that couple to the lateral motion of the rotating assembly are those with a single nodal diameter. The other modes are self-equilibrating in terms of the moment at the connection to the rotating assembly. Thus, one simple generalization is to consider the disk to consist of two concentric thin rigid disks interconnected by a rotational spring k_R with zero translational flexiblity, as illustrated in Fig. 2.20. The resulting model is a 6-degree-of-freedom model defined by the translational motion (u_i, v_i) of the geometric center O and the rotational motions (α_i, β_i) and $(\underline{\alpha}_i, \underline{\beta}_i)$ of the inner and outer disks, respectively. The inertia parameters $(\underline{m}_i, \underline{I}_{d_i})$ of the outer disk can be chosen, along with k_R, so as to approximate the important one-nodal-diameter natural mode of the disk. The translation and rotation coordinates are uncoupled, and their respective equations of motion, with $\bar{m} = m + \underline{m}$, are

$$\begin{bmatrix} \bar{m} & 0 \\ 0 & \bar{m} \end{bmatrix}_i \begin{Bmatrix} \ddot{u}_i \\ \ddot{v}_i \end{Bmatrix} = \bar{m}\left(\begin{Bmatrix} a_{x'}\Omega^2 + a_{y'}\dot{\Omega} \\ a_{y'}\Omega^2 - a_{x'}\dot{\Omega} \end{Bmatrix}_i \cos\varphi + \begin{Bmatrix} -a_{y'}\Omega^2 + a_{x'}\dot{\Omega} \\ a_{x'}\Omega^2 + a_{y'}\dot{\Omega} \end{Bmatrix}_i \sin\varphi\right)$$
$$+ \bar{m}_i g \begin{Bmatrix} 0 \\ -1 \end{Bmatrix} + \begin{Bmatrix} f_{x_i} \\ f_{y_i} \end{Bmatrix}_c \tag{2.89a}$$

and

$$\begin{bmatrix} I_d & 0 & 0 & 0 \\ 0 & I_d & 0 & 0 \\ 0 & 0 & \underline{I}_d & 0 \\ 0 & 0 & 0 & \underline{I}_d \end{bmatrix}_i \{\ddot{q}_R\}_i + \Omega \begin{bmatrix} 0 & I_a & 0 & 0 \\ -I_a & 0 & 0 & 0 \\ 0 & 0 & 0 & \underline{I}_a \\ 0 & 0 & -\underline{I}_a & 0 \end{bmatrix}_i \{\dot{q}_R\}_i$$
$$+ \begin{bmatrix} k_R & \dot{\Omega}I_a & -k_R & 0 \\ -\dot{\Omega}I_a & k_R & 0 & -k_R \\ -k_R & 0 & k_R & \dot{\Omega}\underline{I}_a \\ 0 & -k_R & -\dot{\Omega}\underline{I}_a & k_R \end{bmatrix}_i \{q_R\}_i = \begin{Bmatrix} \tau_{x_i} \\ \tau_{y_i} \\ 0 \\ 0 \end{Bmatrix}_c \tag{2.89b}$$

where $\{q_R\}_i = [\alpha_i \ \beta_i \ \underline{\alpha}_i \ \underline{\beta}_i]^T$. The subscript c on the right-hand side (RHS) force vector denotes interconnection forces from connecting components.

Bearings. The rotating assembly of a machine interacts with other rotating assemblies and/or the static support structure through a variety of mechanisms, the most prevalent being rolling-element bearings, fluid-film bearings and dampers, seals, splines, couplings, and aerodynamic interconnection mechanisms. The characteristics of these components are discussed in other sections of this handbook. In most cases these components are nonlinear in their force-displacement and velocity relations. In this section it is assumed that a rotating machine operates in a small neighborhood of a static configuration. When dynamic loading, such as rotating unbalance or time-varying maneuver loads, is superposed on the static loading, it is assumed that the corresponding interconnection forces acting on the rotating assemblies can be closely approximated by linear force-displacement and velocity relations. In general, the force components from a typical interconnection mechanism acting between stations i and j, as illustrated in Fig. 2.21, can be represented by the form

$$Q_x = Q_x(u, v, \dot{u}, \dot{v}) \qquad Q_y = Q_y(u, v, \dot{u}, \dot{v}) \tag{2.90}$$

where $u = u_j - u_i$, etc., are the relative displacements and velocities between the

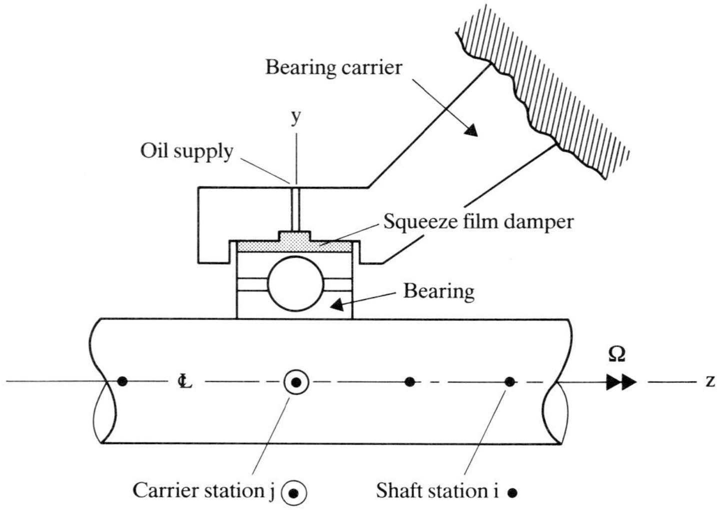

FIGURE 2.21 Bearing/carrier schematic.

two stations. In the neighborhood of an equilibrium configuration (u_o, v_o) the incremental force components obtained by a Taylor series expansion of relations (2.90) about the equilibrium configuration are

$$\begin{aligned} Q_x &= (k_{xx}u + k_{xy}v + c_{xx}\dot{u} + c_{xy}\dot{v}) \\ Q_y &= (k_{yx}u + k_{yy}v + c_{yx}\dot{u} + c_{yy}\dot{v}) \end{aligned} \tag{2.91}$$

with $k_{xx} = (\partial Q_x/\partial u)_o$, $c_{yx} = (\partial Q_y/\partial \dot{u})_o$, etc. The linear constitutive relation for the interconnection between stations i and j may then be written as

$$\begin{Bmatrix} f_{x_i} \\ f_{y_i} \\ f_{x_j} \\ f_{y_j} \end{Bmatrix}_c = \begin{bmatrix} [k^b] & -[k^b] \\ -[k^b] & [k^b] \end{bmatrix} \begin{Bmatrix} u_i \\ v_i \\ u_j \\ v_j \end{Bmatrix} + \begin{bmatrix} [c^b] & -[c^b] \\ -[c^b] & [c^b] \end{bmatrix} \begin{Bmatrix} \dot{u}_i \\ \dot{v}_i \\ \dot{u}_j \\ \dot{v}_j \end{Bmatrix} \tag{2.92}$$

where

$$[k^b] = \begin{bmatrix} k_{xx} & k_{xy} \\ k_{yx} & k_{yy} \end{bmatrix} \quad \text{and} \quad [c^b] = \begin{bmatrix} c_{xx} & c_{xy} \\ c_{yx} & c_{yy} \end{bmatrix}$$

It is assumed here that the interconnection mechanism is dependent only upon translation coordinates and that the inertia effects of the "bearing" are either negligible or included by other means, such as rigid disks at stations i and j.

Shaft Segments. The distribution of mass and elasticity in a rotor system is generally very irregular, such as illustrated in Fig. 2.17. The normal practice with such systems is to divide the rotating assembly into a number of shaft segments which are individually variable and discontinuous in cross section, also depicted in Fig. 2.17. Two common modeling approaches are then used to reduce the distributed parameter component to a finite-degree-of-freedom element. One approach is a lumped-mass model, and the other is a consistent mass model.

Uniform Beam Element. First, consider a uniform-cross-section element as shown in Fig. 2.22. One possible lumped-mass model is shown in Fig. 2.19, and it

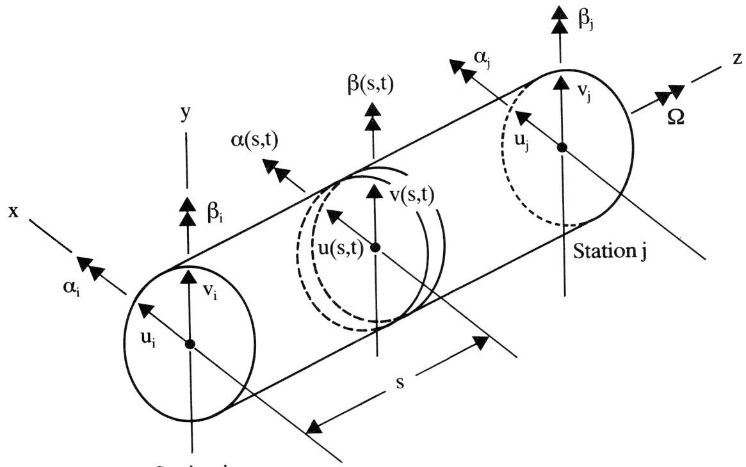

FIGURE 2.22 Cylindrical shaft element.

assumes that the inertia properties are lumped at the two endpoints in the form of identical thin rigid disks which are connected by a massless elastic beam. The inertia properties of this continuous shaft segment, treated as a rigid body, are determined at the mass center and are then equally divided between the two disks. The equations of motion for this lumped-mass component, at constant spin speed Ω, are

$$\begin{bmatrix}[m_i] & [0]\\ [0] & [m_j]\end{bmatrix}\begin{Bmatrix}\{\ddot{q}_i\}\\ \{\ddot{q}_j\}\end{Bmatrix}+\Omega\begin{bmatrix}[g_i] & [0]\\ [0] & [g_j]\end{bmatrix}\begin{Bmatrix}\{\dot{q}_i\}\\ \{\dot{q}_j\}\end{Bmatrix}+\begin{bmatrix}[k^i_{i,i}] & [k^i_{i,j}]\\ [k^i_{j,i}] & [k^i_{j,j}]\end{bmatrix}\begin{Bmatrix}\{q_i\}\\ \{q_j\}\end{Bmatrix}$$
$$=\begin{Bmatrix}\{Q_{uc_i}\}\\ \{Q_{uc_j}\}\end{Bmatrix}\cos\varphi+\begin{Bmatrix}\{Q_{us_i}\}\\ \{Q_{us_j}\}\end{Bmatrix}\sin\varphi+\begin{Bmatrix}\{Q_{g_i}\}\\ \{Q_{g_j}\}\end{Bmatrix}+\{Q\}_c \quad (2.93)$$

where $[m_i]=[m_j]$ and $[g_i]=[g_j]$ are defined by (2.63), $\{q_i\}$ by (2.59), and $m_i=\rho AL/2$, $I_{a_i}=\rho IL$, $I_{d_i}=(\rho IL+\rho AL^3/12)/2$. The stiffness arrays are defined by (2.62), and the force vectors are defined by correspondence with (2.87) and with the forces equal at stations i and j.

A consistent mass model of the uniform-cross-section shaft element is obtained (Nelson and McVaugh, 1976; Nelson, 1980) by choosing to represent the translation of a typical cross section of the element in terms of the endpoint coordinates as stated by

$$\begin{Bmatrix}u(s,t)\\ v(s,t)\end{Bmatrix}=\begin{bmatrix}\varphi_1 & 0 & 0 & \varphi_2 & \varphi_3 & 0 & 0 & \varphi_4\\ 0 & \varphi_1 & -\varphi_2 & 0 & 0 & \varphi_3 & -\varphi_4 & 0\end{bmatrix}\begin{Bmatrix}\{q_i\}\\ \{q_j\}\end{Bmatrix} \quad (2.94)$$

where the φ_i $(i = 1, 2, 3, 4)$ are shape functions that satisfy the boundary conditions of the element. The cross-sectional bending rotations and shear angles are defined similarly, and the energy and work functions for the uniform element are then expressed in terms of integrals over the element. The subsequent use of Lagrange's equations, for a shaft element of length L and radius r, provides the following constant-spin-speed ($\varphi = \Omega t$) equations of motion:

$$\begin{bmatrix} [m^i_{i,i}] & [m^i_{i,j}] \\ [m^i_{j,i}] & [m^i_{j,j}] \end{bmatrix} \begin{Bmatrix} \{\ddot{q}_i\} \\ \{\ddot{q}_j\} \end{Bmatrix} + \Omega \begin{bmatrix} [g^i_{i,i}] & [g^i_{i,j}] \\ [g^i_{j,i}] & [g^i_{j,j}] \end{bmatrix} \begin{Bmatrix} \{\dot{q}_i\} \\ \{\dot{q}_j\} \end{Bmatrix}$$

$$+ \begin{bmatrix} [k^i_{i,i}] & [k^i_{i,j}] \\ [k^i_{j,i}] & [k^i_{j,j}] \end{bmatrix} \begin{Bmatrix} \{q_i\} \\ \{q_j\} \end{Bmatrix} = \{Q^i_{uc}\} \cos \Omega t + \{Q^i_{us}\} \sin \Omega t + \{Q^i_g\} + \{Q^i\}_c \quad (2.95)$$

where

$$[m^i_{i,i}] = \begin{bmatrix} \mu_1 & 0 & 0 & \mu_2 L \\ 0 & \mu_1 & -\mu_2 L & 0 \\ 0 & -\mu_2 L & \mu_3 L^2 & 0 \\ \mu_2 L & 0 & 0 & \mu_3 L^2 \end{bmatrix}$$

$$[m^i_{i,j}] = \begin{bmatrix} \mu_4 & 0 & 0 & -\mu_5 L \\ 0 & \mu_4 & \mu_5 L & 0 \\ 0 & -\mu_5 L & \mu_6 L^2 & 0 \\ \mu_5 L & 0 & 0 & \mu_6 L^2 \end{bmatrix} = [m^j_{j,i}]^T$$

$$[g^i_{i,i}] = \begin{bmatrix} 0 & -\mu_7 & \mu_8 L & 0 \\ \mu_7 & 0 & 0 & \mu_8 L \\ -\mu_8 L & 0 & 0 & -\mu_9 L^2 \\ 0 & -\mu_8 L & \mu_9 L^2 & 0 \end{bmatrix} \quad (2.96)$$

$$[g^i_{i,j}] = \begin{bmatrix} 0 & \mu_7 & \mu_8 L & 0 \\ -\mu_7 & 0 & 0 & \mu_8 L \\ \mu_8 L & 0 & 0 & \mu_{10} L^2 \\ 0 & \mu_8 L & -\mu_{10} L^2 & 0 \end{bmatrix} = -[g^i_{j,i}]^T$$

and $[m^i_{j,j}] = [m^i_{i,i}]$ and $[g^i_{j,j}] = [g^i_{i,i}]$ except that the polarity of the μ_2 and μ_8 terms is reversed. The coefficients of these inertia arrays are

$$\begin{aligned}
\mu_1 &= (156 + 3528\varepsilon + 20{,}160\varepsilon^2)\alpha_T + 36\alpha_R \\
\mu_2 &= (22 + 462\varepsilon + 2520\varepsilon^2)\alpha_T + (3 - 180\varepsilon)\alpha_R \\
\mu_3 &= (4 + 84\varepsilon + 504\varepsilon^2)\alpha_T + (4 + 60\varepsilon + 1440\varepsilon^2)\alpha_R \\
\mu_4 &= (54 + 1512\varepsilon + 10{,}080\varepsilon^2)\alpha_T - 36\alpha_R \\
\mu_5 &= (13 + 378\varepsilon + 2520\varepsilon^2)\alpha_T - (3 - 180\varepsilon)\alpha_R \\
\mu_6 &= -(3 + 84\varepsilon + 504\varepsilon^2)\alpha_T - (1 + 60\varepsilon - 720\varepsilon^2)\alpha_R \\
\mu_7 &= 72\alpha_R \qquad \mu_8 = 2(3 - 180\varepsilon)\alpha_R \\
\mu_9 &= 2(4 + 60\varepsilon + 1440\varepsilon^2)\alpha_R \qquad \mu_{10} = 2(1 + 60\varepsilon)\alpha_R \\
\alpha_T &= \frac{\rho AL}{420(1 + 12\varepsilon)^2} \qquad \alpha_R = \frac{pAr^2}{120L(1 + 12\varepsilon)^2}
\end{aligned} \quad (2.97)$$

where $\varepsilon = EI/(\kappa GAL^2)$. The stiffness arrays of (2.95) are defined by (2.62). The form of the unbalance force vectors of (2.95) is dependent on the center-of-gravity eccentricity distribution of the element. For a linear distribution between $(a_{x'L}, a_{y'L})$ at the left end and $(a_{x'R}, a_{y'R})$ at the right end, these vectors are

$$\{Q^i_{uc}\} = \frac{\rho AL}{120}\Omega^2[v_1 \; v_2 \; v_3L \; v_4L \; v_5 \; v_6 \; v_7L \; v_8L]^T$$
$$\{Q^i_{us}\} = \frac{\rho AL}{120}\Omega^2[-v_2 \; v_1 \; -v_4L \; v_3L \; -v_6 \; v_5 \; -v_8L \; v_7L]^T \tag{2.98}$$

with

$$\begin{aligned}
v_1 &= (42 + 480\varepsilon)a_{x'L} + (18 + 240\varepsilon)a_{x'R} \\
v_2 &= (42 + 480\varepsilon)a_{y'L} + (18 + 240\varepsilon)a_{y'R} \\
v_3 &= -(6 + 60\varepsilon)a_{y'L} - (4 + 60\varepsilon)a_{y'R} \\
v_4 &= (6 + 60\varepsilon)a_{x'L} + (4 + 60\varepsilon)a_{x'R} \\
v_5 &= (18 + 240\varepsilon)a_{x'L} + (42 + 480\varepsilon)a_{x'R} \\
v_6 &= (18 + 240\varepsilon)a_{y'L} + (42 + 480\varepsilon)a_{y'R} \\
v_7 &= (4 + 60\varepsilon)a_{y'L} + (6 + 60\varepsilon)a_{y'R} \\
v_8 &= -(4 + 60\varepsilon)a_{x'L} - (6 + 60\varepsilon)a_{x'R}
\end{aligned} \tag{2.99}$$

The gravity force vector is given by

$$\{Q^i_g\} = (\rho A)_i \frac{g}{12}[0 \; -6L \; L^2 \; 0 \; 0 \; -6L \; -L^2 \; 0]^T_i \tag{2.100}$$

and the vector $\{Q^i\}_c$ consists of interconnection shear and moment forces acting on element i from connecting components.

Conical Beam Element. The uniform-cross-section element discussed above has been generalized to accommodate beams with linearly varying radius. The resulting conical beam rotating element provides increased ease in modeling variable-cross-section rotating assemblies. Conical elements developed by Rouch and Kao (1979), To (1981), and Greenhill et al. (1985) include shear deformation as additional degrees of freedom. These additional endpoint degrees of freedom can then be statically condensed (Guyan, 1965) to form an 8-degree-of-freedom shaft element with two translation and two rotation coordinates at each endpoint. An element developed by Genta and Gugliotta (1988) is an 8-degree-of-freedom element, thereby not requiring condensation, and it exhibits equivalent accuracy to previous elements. The accuracy of these elements deteriorates as the length of the element increases since the displacement shape functions become less accurate. Thus, relatively short segments are recommended.

All the conical and cylindrical elements cited above are beam-type elements and do not account for the ovalizing effect present in shell-type elements. As a result, caution must be exercised when beam-type elements are applied to thin-wall segments. Thin-wall segments should probably be analyzed by using more appropriate assumptions to generate the element equations of motion.

Discontinuous Beam Element. One approach that can be used to obtain a mathematical model for a discontinuous shaft element, such as shown in Fig.

2.17, is to treat it as an assemblage of cylindrical and/or conical beam subelements and then utilize static condensation (Guyan, 1965) to reduce the model to an 8-degree-of-freedom equivalent element in terms of the endpoint coordinates. The assembly process is illustrated in Sec. 2.3, and the condensation process is also presented. This approach is quite accurate so long as the variations in cross section between subelements is not too large.

Couplings. When two collinear shafts are connected by a coupling, as illustrated schematically in Fig. 2.23, often the coupling can be modeled as an

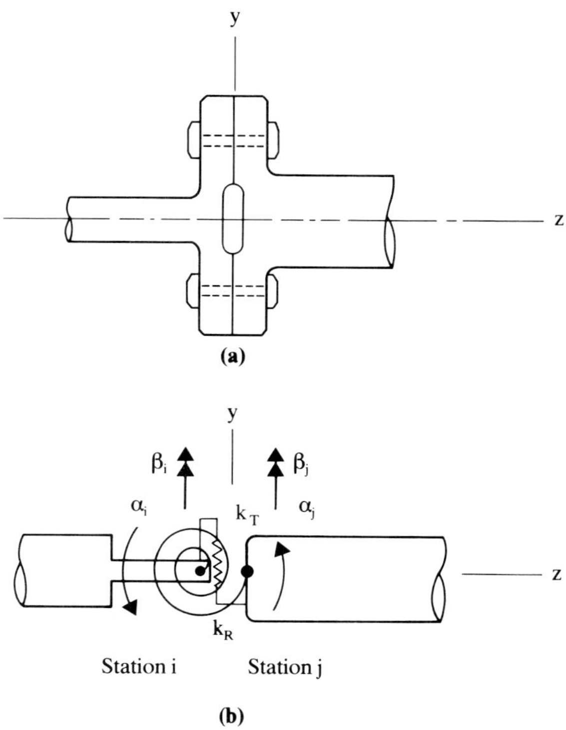

FIGURE 2.23 Flexible coupling model: (*a*) coupling cross section; (*b*) schematic.

elastic component with isotropic translational stiffness k_T and rotational stiffness k_R between station i on one shaft and station j on the other. The equation of motion for the coupling is then

$$\begin{bmatrix} [k^c] & -[k^c] \\ -[k^c] & [k^c] \end{bmatrix} \begin{Bmatrix} \{q_i\} \\ \{q_j\} \end{Bmatrix} = \{Q^c\}_c \tag{2.101}$$

where $[k^c] = \text{diag}\,(k_T, k_T, k_R, k_R)$. Internal damping can be added to this model, and inertia effects can be taken into account by including thin rigid disks at each of the connecting stations. Internal damping is a destablizing mechanism for supercritical rotation, and sufficient external damping must be present in the system to ensure stable operation. Low values of the stiffness coefficients closely simulate releases, and high values simulate rigid connections.

Squeeze-Film Dampers. A squeeze-film damper is a strongly nonlinear hydrodynamic device which, when properly designed, extracts energy from the lateral motion of the rotor system so as to reduce vibration levels and bearing loads. One possible configuration of this type of damper is shown in Fig. 2.24,

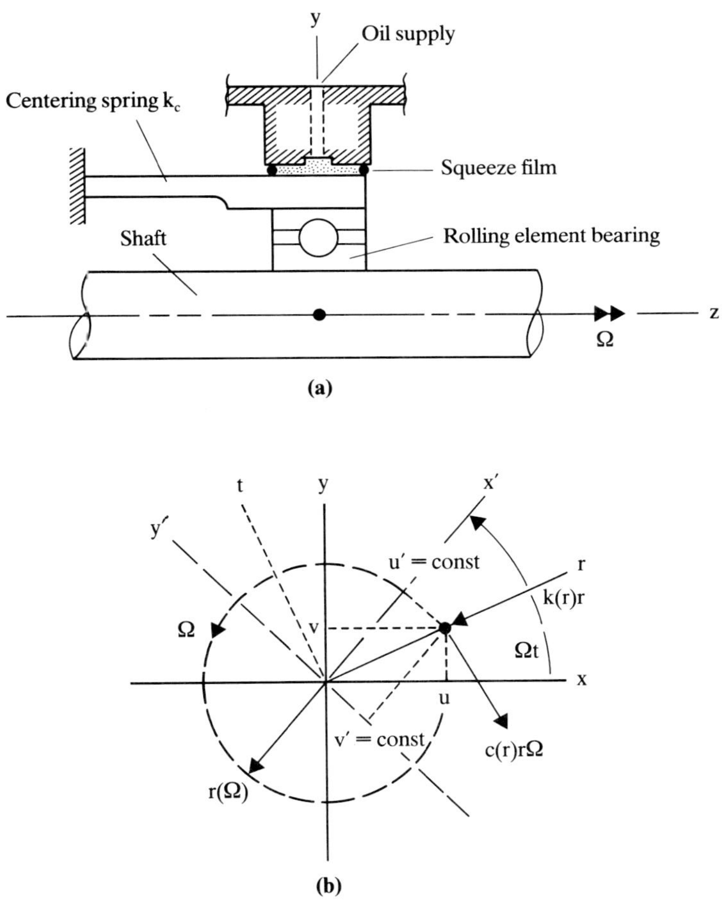

FIGURE 2.24 Squeeze-film damper with centering spring: (*a*) schematic; (*b*) typical circular response.

and it includes a centering spring k_c which can be preloaded to offset gravity or other constant side loads. As a result, the motion of the shaft center is nearly circular, and this centered circular orbit response of radius $r(\Omega)$ is synchronous with the unbalance excitation at shaft speed Ω. The centering spring and fluid film act in parallel so that the force acting on the damper is given by the equations

$$\begin{Bmatrix} f_{x_i} \\ f_{y_i} \end{Bmatrix} = \begin{bmatrix} k_c + k(r) & 0 \\ 0 & k_c + k(r) \end{bmatrix} \begin{Bmatrix} u_i \\ v_i \end{Bmatrix} + \begin{bmatrix} c(r) & 0 \\ 0 & c(r) \end{bmatrix} \begin{Bmatrix} \dot{u}_i \\ \dot{v}_i \end{Bmatrix} \tag{2.102}$$

The amplitude-dependent stiffness coefficient $k(r)$ and damping coefficient $c(r)$ are the secant stiffness and damping coefficients associated with the radial force–radial displacement and tangential force–radial displacement relations, respectively, for the damper. The support structure for the damper is considered to be rigid for the model of (2.102), although this constraint can be relaxed and the model generalized to include isotropic flexibility of the support structure.

2.3 ANALYTICAL PROCEDURES

Two procedures are commonly used for the analysis of free and forced response of rotor dynamic systems. These are the *direct stiffness method* (DSM) and the *transfer matrix method* (TMM). Both methods utilize the component equations of the system such as presented in Sec. 2.2.3. The equations of motion for each of these components were developed by using the appropriate force-displacement and force-velocity relations and the momentum principles of (2.29). The system equations are assembled as required for the DSM and the TMM by utilizing geometric displacement constraints which ensure proper connectivity of the components. The assembly procedures for both methods of analysis are described in more detail in the sections that follow.

Typical analyses for rotordynamic systems include the determination of the free vibration natural frequencies and modes of whirl and the stability characteristics of these modes. The critical speeds of a rotor system are defined as rotating assembly spin speeds which provide excitations that coincide with one of the system's natural frequencies, thereby producing a resonant condition. The direct determination of these speeds is possible for some types of systems and constitutes another analysis associated with free rotor vibration. Typical forced response analyses include the determination of static response due to gravity, side loading, misalignment effects, and constant maneuver loads; steady-state dynamic response due to rotating unbalance, periodic support structure motion, and aerodynamic and hydrodynamic interconnection; and transient response from excitations such as sudden change in unbalance due to blade loss and shock loading from a hard landing or maneuver.

Most analyses utilize linear equations of motion based on small amplitude motion in the neighborhood of an equilibrium configuration. It is normally possible to extract a considerable amount of very useful information about the system with these linear equations. In some cases, however, the strength of the nonlinearities is so large that linearization does not provide sufficiently accurate simulations. In these cases, the most common approach is to integrate the nonlinear equations of motion numerically either directly in physical coordinates or in terms of modal coordinates associated with some form of component mode synthesis. Some of these procedures are discussed in Sec. 2.3.3.

2.3.1 Direct Stiffness Method

Assembly of System Equations

A general rotordynamic system is composed of a large number of components which typically include rigid and flexible disks, bearings, dampers, seals, couplings, and shaft segments. The equations of motion for several of these components are presented in Sec. 2.2.3 in terms of fixed xyz reference coordinates. These equations can easily be transformed to rotating $x'y'z'$ reference coordinates by using transformation (2.14). The choice of fixed reference or rotating reference may be made according to the preference of the analyst in some situations, but is usually due to particular computational advantages that exist for specific types of analyses.

The procedures for assembly of the system equations are conceptually equivalent for the fixed and rotating reference coordinates. The fixed reference is used in the explanations that follow as an arbitrary choice. It is, of course, possible to utilize the transformation of (2.14) between the fixed and rotating reference coordinates to transform the assembled system equations to whichever reference system is desired for a particular analysis.

In modeling the dynamic characteristics of a rotordynamic system, several steps are required in the development of the equations of motion. The first step is to define one or more reference frames which are useful for observing the motion of the system. The fixed and rotating references defined earlier are the usual choices. The second step is to divide the real system into a finite-degree-of-freedom model consisting of an interconnected set of discrete elements such as presented in Sec. 2.2.3. The dual-shaft system, shown schematically in Fig. 2.25, is used here as an example in presenting the various steps in developing a set of system equations of motion. One possible discrete model of the illustrative system is also shown. It consists of two shafts, four rigid disks, three bearings to the support structure, one intershaft bearing, eight rotating shaft segments, and a multiple degree-of-freedom support structure. This discrete model includes 10 rotating assembly stations, each with two orthogonal translations and two orthogonal rotations. Separate reference systems can be used for each component; however, the common choice of the fixed xyz reference for each component greatly simplifies the assembly process.

The third step in the modeling process is to choose a set of system coordinates. This is particularly simple if a common reference is used for all components and if this reference is also used as the system or global reference. For this choice, the system displacement vector consists of all the station displacements and is defined as

$$\{q\}^T_{n\times 1} = [\underbrace{\{q_1\}^T, \ldots, \{q_6\}^T}_{\text{shaft 1}}, \underbrace{\{q_7\}^T, \ldots, \{q_{10}\}^T}_{\text{shaft 2}}] \tag{2.103}$$

The fourth step is to define the connectivity of the discrete elements with the system displacement vector (2.103). The complete set of component coordinates constitutes a dependent set of coordinates, and the connectivity statements essentially represent geometric constraint relations between the component coordinates and the system coordinates. For the rth typical component with n_r degrees of freedom, the connectivity statement is of the form

$$\{q_r\}_{n_r\times 1} = [B_r]\{q\}_{n\times 1} \tag{2.104}$$

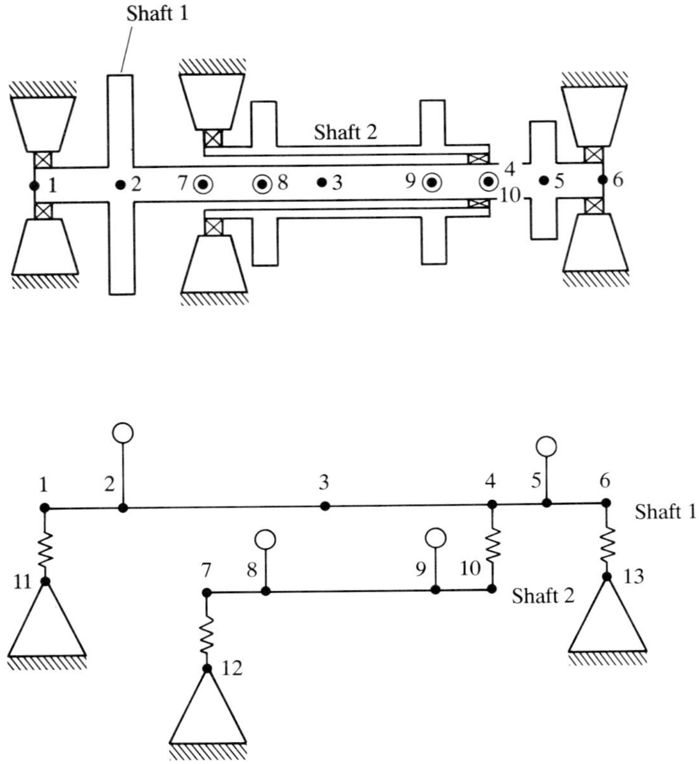

FIGURE 2.25 Dual shaft DSM model and schematic.

The nonzero elements of $[B_r]$ are simply 1s when the component and system reference systems are chosen to be the same. The unit elements of $[B_r]$ identify the coordinates of $\{q\}$ which are common to $\{q_r\}$. The connectivity statements for the rigid disk at station 2, the shaft segment between stations 8 and 9, and the intershaft bearing between stations 4 and 10 are listed below for illustrative purposes.

Rigid disk at station 2:

$$\{q_2^d\} = \{q_2\} = [[0]_{4\times4}[I]_{4\times4}[0]_{4\times32}]\{q\}$$

Shaft segment between stations 8 and 9:

$$\{q_8^e\} = \begin{Bmatrix} \{q_8\} \\ \{q_9\} \end{Bmatrix} = [[0]_{8\times28}[I]_{8\times8}[0]_{8\times4}]\{q\} \tag{2.105}$$

Intershaft bearing between stations 4 and 10:

$$\{q_4^b\} = \begin{Bmatrix} \{q_4\} \\ \{q_{10}\} \end{Bmatrix} = \left[[0]_{8\times12}\begin{bmatrix} [I]_{4\times4} \\ [0]_{4\times4} \end{bmatrix}[0]_{8\times20}\begin{bmatrix} [0]_{4\times4} \\ [I]_{4\times4} \end{bmatrix}\right]\{q\}$$

The n_r equations of motion for the rth typical component of the system, such as presented in Sec. 2.2.3, is

$$[M_r]\{\ddot{q}_r\} + ([D_r] + \Omega_r[G_r])\{\dot{q}_r\} + [K_r]\{q_r\} = \{Q_r\} + \{Q_r\}_c \qquad (2.106a)$$

where $\{Q_r\}$ and $\{Q_r\}_c$, respectively, represent the vector of applied forces and interconnection forces from connecting components. In D'Alembert form, (2.106a) is

$$\{Q_r\} + \{Q_r\}_c - [M_r]\{\ddot{q}_r\} - ([D_r] + \Omega_r[G_r])\{\dot{q}_r\} - [K_r]\{q_r\} = \{0\}_{n_r \times 1} \qquad (2.106b)$$

The system equations are formally assembled by utilizing the principle of virtual displacements which states, "The work done by the external forces acting on the system and the work done by the internal forces must vanish for any virtual displacement" (Oden, 1967), i.e.,

$$\delta W = 0 \qquad (2.107)$$

This virtual work consists of the virtual work of all R components of a system. Thus,

$$\delta W = \sum_{r=1}^{R} \delta W_r = 0 \qquad (2.108)$$

and by using (2.106b) this becomes

$$\sum_{r=1}^{R} \{\delta q_r\}^T(\{Q_r\} + \{Q_r\}_c - [M_r]\{\ddot{q}_r\} - ([D_r] + \Omega_r[G_r])\{\dot{q}_r\} - [K_r]\{q_r\}) = 0 \qquad (2.109)$$

The introduction of the connectivity relations (2.104) into (2.109) yields the following set of system equations of motion:

$$[M]\{\ddot{q}\} + ([D] + [G])\{\dot{q}\} + [K]\{q\} = \{Q\}_{n \times 1} \qquad (2.110)$$

According to Newton's third law, the component interconnection forces $\{Q_r\}_c$ cancel upon assembly of all the component equations.

The system matrices of (2.110) are defined as typified by (2.111) for the mass matrix. It is important to note that the rotating assembly spin speeds need to be included when one is assembling the system gyroscopic matrix and for speed-dependent support properties.

$$[M] = \sum_{r=1}^{R} ([B_r]^T[M_r][B_r]) \qquad (2.111)$$

and so forth. The system force vector is obtained from

$$\{Q\} = \sum_{r=1}^{R} [B_r]^T\{Q_r\} \qquad (2.112)$$

The system mass matrix $[M]$ is symmetric and consists of contributions from the rigid disks, flexible disks, and flexible shaft elements. The system gyroscopic matrix $[G]$ consists of contributions from the same components and is skew symmetric and spin-speed-dependent. The dissipation $[D]$ and stiffness $[K]$ arrays consist of symmetric arrays from the undamped shaft elements and generally asymmetric spin-speed-dependent contributions from the bearings and other components, e.g., seals and aerodynamic mechanisms, which are commonly modeled as pseudobearings. The system applied force vector $\{Q\}$ includes all external forces such as gravity, rotating unbalance, side loads, and maneuver loads.

The forms of the system arrays of (2.110) are shown in Fig. 2.26 for the

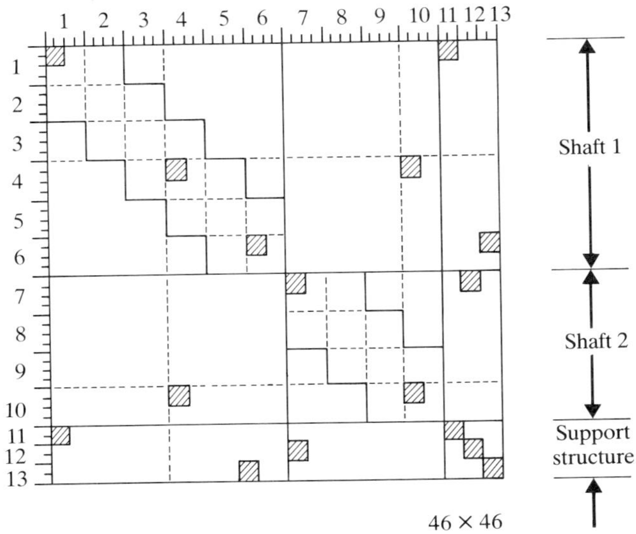

FIGURE 2.26 System arrays for dual-shaft support structure example: (*a*) stiffness matrix $[K]$; (*b*) dissipation matrix $[C]$; (*c*) inertia matrices $[M]$ and $[G]$.

dual-shaft support structure example of Fig. 2.25. In this example the support structure is modeled as a reduced-order model with two translations at each of the three bearing support stations. The coordinate ordering in these system arrays is consistent with the displacement vector defined by (2.103) for the rotating assemblies. The support structure coordinates associated with stations 11, 12, and 13—$[u_{11}\ v_{11}\ u_{12}\ v_{12}\ u_{13}\ v_{13}]^T$—are appended to (2.103) to form a 46th-order vector. The form of the cosine component of the system constant-speed unbalance force vector is illustrated by (2.113) below (the sine component has the same form), and the form of the system stiffness, dissipation, and inertia arrays is illustrated in Fig. 2.26. These arrays are of order 46 with 4 degrees of freedom for each rotor station and 2 degrees of freedom for each support structure station.

$$\{Q_{uc}\}_{46\times 1} = \sum_{r=1}^{6} \Omega_1^2 \begin{Bmatrix} \begin{Bmatrix} \cdots \\ \cdots \end{Bmatrix} \begin{matrix} \leftarrow \text{row } 4r-3 \\ \leftarrow \text{row } 4r \end{matrix} \\ \text{-------} \\ \{0\}_{16\times 1} \\ \text{-------} \\ \{0\}_{6\times 1} \end{Bmatrix} \cos \Omega_1 t + \sum_{r=7}^{10} \Omega_2^2 \begin{Bmatrix} \{0\}_{24\times 1} \\ \text{-------} \\ \begin{Bmatrix} \cdots \\ \cdots \end{Bmatrix} \begin{matrix} \leftarrow \text{row } 4r-3 \\ \leftarrow \text{row } 4r \end{matrix} \\ \text{-------} \\ \{0\} \end{Bmatrix} \cos \Omega_2 t \qquad (2.113)$$

Geometric Constraints

The concept of imposing geometric constraints on the coordinates of a model is introduced in Sec. 2.2.3. In the first illustration a "thick, rigid disk" is formed by imposing a rigid-body relation between two "thin disks" separated by a flexible shaft. The second illustration described in the section on conical beam elements utilized the concept of static condensation to release the internal coordinates for a variable-cross-section shaft element. These are both examples of applying a geometric constraint between coordinates. After a set of system equations is assembled, it is often desirable to impose geometric conditions between the coordinates. The introduction of these constraints to the system equations of motion then results in a reduction in the system degrees of freedom.

In matrix notation the coordinate relationship associated with geometric constraints between the original system displacement vector, $\{q\}$ of order n, and the reduced system displacement vector, $\{\underline{q}\}$ of order $\underline{n} \leq n$, is of the form

$$\{q\} = [\Psi]_{n \times \underline{n}}\{\underline{q}\} \tag{2.114}$$

This relation, when applied to the system equations of motion (2.110), provides the reduced-order system equations

$$[\underline{M}]\{\ddot{\underline{q}}\} + ([\underline{D}] + [\underline{G}])\{\dot{\underline{q}}\} + [\underline{K}]\{\underline{q}\} = \{\underline{Q}\}_{\underline{n} \times 1} \tag{2.115}$$

where

$$[\underline{M}] = [\Psi]^T[M][\Psi] \tag{2.116}$$

and so forth, and

$$\{\underline{Q}\} = [\Psi]^T\{Q\}$$

One common use of a relation such as (2.114) is to "release" selected coordinates from a rotordynamic model. The usual procedure is to reorder the system coordinates into a subset of independent or active coordinates $\{q_a\}$ and a complementary set of dependent coordinates $\{q_d\}$. A dependency relation between these two sets of coordinates can be established from the homogeneous static problem associated with (2.110), partitioned appropriately, to obtain

$$\begin{bmatrix} [K_{aa}] & [K_{ad}] \\ [K_{da}] & [K_{dd}] \end{bmatrix} \begin{Bmatrix} \{q_a\} \\ \{q_d\} \end{Bmatrix} = \{0\} \tag{2.117}$$

From the second row of (2.117), a dependency relation is

$$\{q_d\} = (-[K_{dd}]^{-1}[K_{da}])\{q_a\} = [\psi]\{q_a\} \tag{2.118}$$

and a transformation of coordinates may be formed by combining the constraint of (2.118) with an identity relation to obtain

$$\begin{Bmatrix} \{q_a\} \\ \{q_d\} \end{Bmatrix} = \begin{bmatrix} [I] \\ [\psi] \end{bmatrix} \{q_a\} \tag{2.119}$$

A reordering of the coordinates in (2.119) to the original order yields the constraint relation of coordinates as defined by (2.114). The reduction process then follows the procedure illustrated by (2.116) to provide the reduced-order set of system equations (2.115).

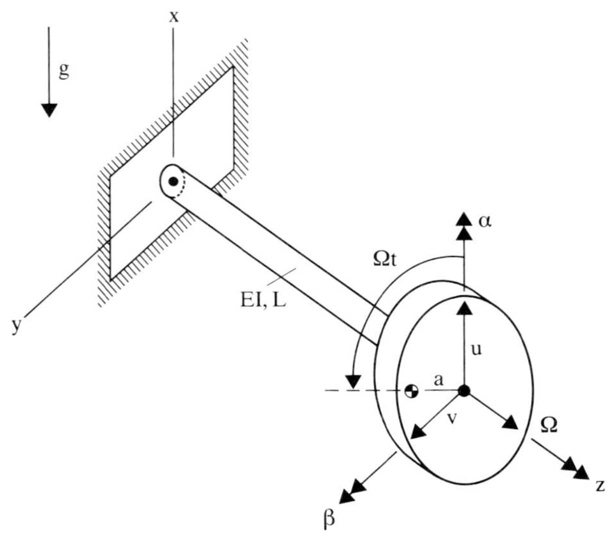

FIGURE 2.27 Cantilevered rotor system.

To illustrate this static condensation process, consider the cantilevered rotor system in Fig. 2.27. This model consists of a thin, rigid disk and a uniform flexible shaft with negligible mass and a negligible shear compliance. The shaft is cantilevered, and the translation and rotation coordinates at $z = 0$ are constrained to zero. The equations of motion for this 4-degree-of-freedom model, in terms of fixed reference coordinates, are

$$\begin{bmatrix} m & 0 & 0 & 0 \\ 0 & m & 0 & 0 \\ 0 & 0 & I_d & 0 \\ 0 & 0 & 0 & I_d \end{bmatrix} \begin{Bmatrix} \ddot{u} \\ \ddot{v} \\ \ddot{\alpha} \\ \ddot{\beta} \end{Bmatrix} + \Omega \begin{bmatrix} 0 & 0 & 0 & 0 \\ 0 & 0 & 0 & 0 \\ 0 & 0 & 0 & I_a \\ 0 & 0 & -I_a & 0 \end{bmatrix} \begin{Bmatrix} \dot{u} \\ \dot{v} \\ \dot{\alpha} \\ \dot{\beta} \end{Bmatrix}$$

$$+ \frac{EI}{L^3} \begin{bmatrix} 12 & 0 & 0 & -6L \\ 0 & 12 & 6L & 0 \\ 0 & 6L & 4L^2 & 0 \\ -6L & 0 & 0 & 4L^2 \end{bmatrix} \begin{Bmatrix} u \\ v \\ \alpha \\ \beta \end{Bmatrix} = ma\Omega^2 \begin{Bmatrix} \cos \Omega t \\ \sin \Omega t \\ 0 \\ 0 \end{Bmatrix} + \begin{Bmatrix} -mg \\ 0 \\ 0 \\ 0 \end{Bmatrix} \qquad (2.120)$$

where rotating unbalance and constant gravity force loading have been included. For dynamic characteristics of this system, at frequencies which are low relative to its natural frequencies of whirl, it may be acceptable to reduce the degrees of freedom from 4 to 2. The loss in model accuracy as a result of such a reduction can, of course, only be assessed in comparison with higher-order model characteristics. Assume, for illustration purposes, that an order reduction is acceptable in this case. To accomplish this, the elastic relation between translation and rotation coordinates is introduced. The intent here is to release the rotation coordinates or, in other words, treat the translation coordinates as active and the rotation coordinates as dependent. From the homogeneous static form of (2.120)

$$\frac{EI}{L^3}\begin{bmatrix} 12 & 0 & 0 & -6L \\ 0 & 12 & 6L & 0 \\ 0 & 6L & 4L^2 & 0 \\ -6L & 0 & 0 & 4L^2 \end{bmatrix}\begin{Bmatrix} u \\ v \\ \alpha \\ \beta \end{Bmatrix} = \{0\} \tag{2.121}$$

Coordinate reordering is not required here, and following (2.119) the desired transformation is

$$\begin{matrix} u \\ v \\ \alpha \\ \beta \end{matrix} = \begin{matrix} 1 & 0 \\ 0 & 1 \\ 0 & -\dfrac{3}{2L} \\ \dfrac{3}{2L} & 0 \end{matrix} \begin{Bmatrix} u \\ v \end{Bmatrix} \tag{2.122}$$

The introduction of this transformation into (2.120) yields the reduced-order system equations

$$\begin{bmatrix} \bar{m} & 0 \\ 0 & \bar{m} \end{bmatrix}\begin{Bmatrix} \ddot{u} \\ \ddot{v} \end{Bmatrix} + \Omega\begin{bmatrix} 0 & \bar{I}_a \\ -\bar{I}_a & 0 \end{bmatrix}\begin{Bmatrix} \dot{u} \\ \dot{v} \end{Bmatrix} + \begin{bmatrix} \bar{k} & 0 \\ 0 & \bar{k} \end{bmatrix}\begin{Bmatrix} u \\ v \end{Bmatrix} = ma\Omega^2\begin{Bmatrix} \cos\Omega t \\ \sin\Omega t \end{Bmatrix} + \begin{Bmatrix} -mg \\ 0 \end{Bmatrix} \tag{2.123}$$

where $\bar{m} = m + 9I_d/(4L^2)$, $\bar{I}_a = 9I_a/(4L^2)$, and $\bar{k} = 3EI/L^3$.

Another frequent use of a geometric constraint relation (2.114) is to impose certain boundary conditions and/or interconnection conditions between various components of a system. To illustrate this, consider the same model as in Fig. 2.27 with the additional requirement that the translation coordinates at the rigid disk are constrained to zero by a spline connection, as illustrated in Fig. 2.28. The original 4-degree-of-freedom system is then reduced to 2 by imposing the conditions $u = v = 0$. The relation equivalent to (2.114) for this case is

$$\begin{Bmatrix} u \\ v \\ \alpha \\ \beta \end{Bmatrix} = \begin{bmatrix} 0 & 0 \\ 0 & 0 \\ 1 & 0 \\ 0 & 1 \end{bmatrix}\begin{Bmatrix} \alpha \\ \beta \end{Bmatrix} \tag{2.124}$$

and the reduced-order equations of motion are

$$\begin{bmatrix} I_d & 0 \\ 0 & I_d \end{bmatrix}\begin{Bmatrix} \ddot{\alpha} \\ \ddot{\beta} \end{Bmatrix} + \Omega\begin{bmatrix} 0 & I_a \\ -I_a & 0 \end{bmatrix}\begin{Bmatrix} \dot{\alpha} \\ \dot{\beta} \end{Bmatrix} + \frac{EI}{L}\begin{bmatrix} 4 & 0 \\ 0 & 4 \end{bmatrix}\begin{Bmatrix} \alpha \\ \beta \end{Bmatrix} = \{0\} \tag{2.125}$$

Additional information on the philosophy and details of static reduction can be found in many references including Craig (1981) and Irons (1965). Methods of dynamic reduction are discussed in a later section in this chapter.

Prescribed Coordinate Motion

The transformation of coordinates discussed in the previous section for imposing zero boundary conditions is a special case of a more general situation where the time history of a subset of the system coordinates is prescribed. For rotor systems, this situation can arise when specific connection points of the support structure move in a prescribed manner independent of the remaining support

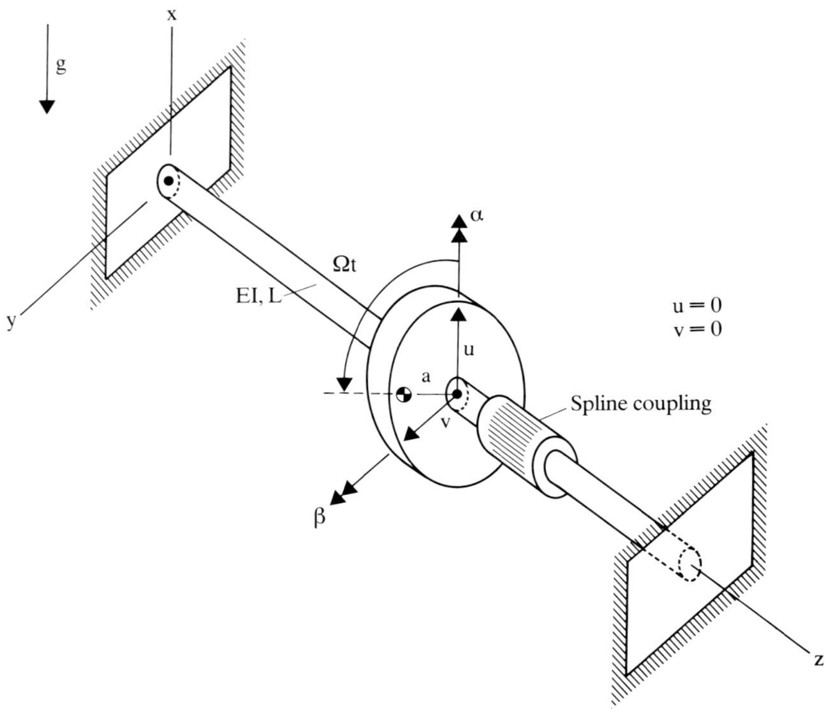

FIGURE 2.28 Cantilever rotor model with spline constraint.

structure/rotating assembly system. Typical applications include the simulation of loads due to vehicle maneuvers such as turns or hard landings.

To illustrate this, consider the system displacement vector to be partitioned into a "prescribed" subset $\{q_p\}$ and a complementary subset of "active" coordinates $\{q_a\}$. The system equation of motion (2.110) can then be written in the form (with $[\Delta] = [D] + [G]$)

$$\begin{bmatrix} [M_{aa}] & [M_{ap}] \\ [M_{pa}] & [M_{pp}] \end{bmatrix} \begin{Bmatrix} \{\ddot{q}_a\} \\ \{\ddot{q}_p\} \end{Bmatrix} + \begin{bmatrix} [\Delta_{aa}] & [\Delta_{ap}] \\ [\Delta_{pa}] & [\Delta_{pp}] \end{bmatrix} \begin{Bmatrix} \{\dot{q}_a\} \\ \{\dot{q}_p\} \end{Bmatrix} + \begin{bmatrix} [K_{aa}] & [K_{ap}] \\ [K_{pa}] & [K_{pp}] \end{bmatrix} \begin{Bmatrix} \{q_a\} \\ \{q_p\} \end{Bmatrix} = \begin{Bmatrix} \{Q_a\} \\ \{Q_p\} \end{Bmatrix} \quad (2.126)$$

The force vector $\{Q_a\}$ associated with the active coordinates must be prescribed; however, the force vector $\{Q_p\}$ must be determined from the equations of motion.

It is convenient to utilize the concept of static constraint modes for situations with prescribed coordinate motion. A static constraint mode is a displacement pattern of the rotor system due to a unit static displacement of a particular prescribed coordinate. The following transformation is introduced, utilizing static constraint modes:

$$\begin{Bmatrix} \{q_a\} \\ \{q_p\} \end{Bmatrix} = \begin{bmatrix} [I] & [\chi] \\ [0] & [I] \end{bmatrix} \begin{Bmatrix} \{r_a\} \\ \{q_p\} \end{Bmatrix} \tag{2.127}$$

The columns of $[[\chi]^T[I]^T]^T$ are static constraint modes for the system. The first row of (2.127) states that the active absolute coordinates $\{q_a\}$ are represented as the sum of static constraint mode motion due to the prescribed coordinates plus a motion $\{r_a\}$ relative to a zero constraint mode configuration. The second row is simply an identity. The introduction of (2.127) into (2.126), upon separation, yields the following matrix equation for the relative active coordinates:

$$[M_{aa}]\{\ddot{r}_a\} + [\Delta_{aa}]\{\dot{r}_a\} + [K_{aa}]\{r_a\} = \{Q_a\} - ([M_{aa}][\chi] + [M_{ap}])\{\ddot{q}_p\} - ([\Delta_{aa}][\chi] + [\Delta_{ap}])\{\dot{q}_p\} \tag{2.128}$$

The solution of (2.128) for $\{r_a\}$ and subsequent use of (2.127) provide the active coordinate time history. The force vector $\{Q_p\}$ which is required to produce the prescribed coordinate motion may then be determined from (2.126).

Free Vibration Analysis

The equations of motion for a typical n-degree-of-freedom discrete rotordynamic system, from (2.110), are of the form

$$[M]\{\ddot{q}\} + [\Delta]\{\dot{q}\} + [K]\{q\} = \{Q\} \tag{2.129}$$

The matrix $[M]$ of acceleration coefficients is symmetric while the matrix $[\Delta]$ of velocity coefficients and the matrix $[K]$ of displacement coefficients are generally asymmetric. Any square matrix may be written as the sum of a symmetric matrix and a skew symmetric matrix. In the case of $[\Delta] = [D] + [G]$, damping matrix $[D]$ is symmetric and gyroscopic matrix $[G]$ is skew symmetric. Similarly, matrix $[K]$ can be decomposed (Adams and Padovan, 1981) into the sum of a symmetric stiffness matrix $[S]$ representing conservative forces plus a skew symmetric circulant matrix $[C]$ representing nonconservative forces.

Matrix $[M]$ is a constant-coefficient array for usual rotordynamic systems, while the matrices $[\Delta]$ and $[K]$ generally depend on the spin velocities and accelerations of the rotating assembly. Important speed-dependent terms include those due to gyroscopic effects and fluid-film bearing reactions. It is assumed here that the rotating assemblies are axisymmetric so that the system arrays are not parametrically excited in the fixed reference coordinates.

Whirl Speeds and Modes The free vibration solutions of (2.129) are of the form

$$\{q\} = \{u\}e^{\lambda t} \tag{2.130}$$

and when substituted into the homogeneous form of (2.129), they provide the eigenvalue problem

$$[H(\lambda)]_{n\times n}\{u\} = [\lambda^2[M] + \lambda[\Delta] + [K]]\{u\} = \{0\} \tag{2.131}$$

for the right eigenvector $\{u\}$ and associated adjoint eigenvalue problem

$$\{v\}^T[H(\lambda)] = \{0\}^T \tag{2.132}$$

for the left eigenvector $\{v\}$. The array $[H(\lambda)]$ is the dynamic stiffness matrix, and (2.131) and (2.132) yield solutions λ_j, $\{u_j\}$, $\{v_j\}$ $(j = 1, 2, \ldots, 2n)$. There are $2n$ eigensolutions, where n is the order of the system arrays, and they are purely real for overdamped modes and appear in complex conjugate pairs for underdamped or undamped modes. In general, the eigenvalues and associated right eigenvectors are of the form

$$\lambda_j = \sigma_j + i\omega_j \qquad \{u_j\} = \text{Re}\,\{u_j\} + i\,\text{Im}\,\{u_j\} \qquad j = 1, 2, \ldots, 2n \tag{2.133}$$

and are a function of the rotating assembly spin speeds. The imaginary part ω_j of the eigenvalue is the damped natural frequency of whirl (whirl speed), and the real part σ_j is the damping coefficient. A stable mode requires a nonpositive value for σ_j.

It is common practice to plot both the whirl frequencies and damping coefficients versus a reference spin speed Ω_{ref} for the system. For a single-shaft system, the reference spin speed is simply the shaft speed. For a multishaft system, however, it is necessary to arbitrarily choose one of the shaft spin speeds as the reference. The other shaft speeds may or may not be dependent on this reference speed. These plots are called *whirl speed maps,* and one is illustrated in Fig. 2.29 with three whirl speeds displayed. In this typical plot the first mode is

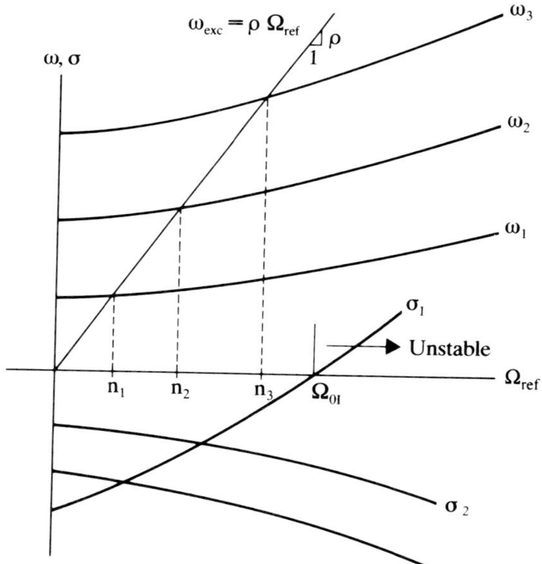

FIGURE 2.29 Whirl speed map.

illustrated to go unstable at an onset of instability speed Ω_{oI}. In most practical applications, if the system becomes unstable, it is usually the first forward mode whirl which yields the instability while the remaining modes remain stable.

Several special cases, which occur quite often, are worth mention (1) For conservative nongyroscopic systems, $[D] = [G] = [C] = [0]$, λ_j^2, $\{u_j\}$, and $\{v_j\}$ are real. (2) For damped nongyroscopic and noncirculatory systems, $[G] = [C] = [0]$,

λ_j and $\{v_j\} = \{u_j\}$ are complex. (3) For conservative gyroscopic systems, $[D] = [C] = [0]$, λ_j are pure imaginary, and $\{v_j\} = \{u_j\}^*$ are complex.

For purely real modes,

$$\lambda_j = \text{Re}\,\lambda_j = \sigma_j \qquad \{u_j\} = \text{Re}\,\{u_j\} \tag{2.134}$$

and the eigensolution, from (2.130), is

$$\{q_j(t)\} = \text{Re}\,\{u_j\}\,e^{\sigma_j t} \tag{2.135}$$

which is illustrated in Fig. 2.30*a* for the translational motion at a typical

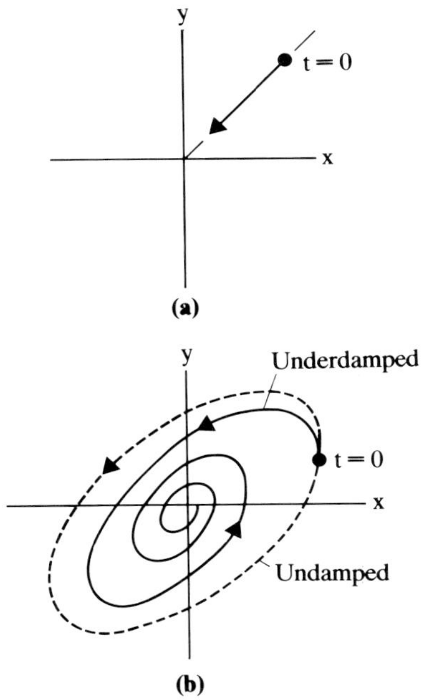

FIGURE 2.30 Rotor precessional modes, typical station orbits: (*a*) overdamped mode; (*b*) undamped and underdamped modes.

station. It is simply a straight-line exponential decay for a stable situation. For underdamped modes, the eigenvalues and vectors appear in complex conjugate pairs ($k = j + n$)

$$\begin{aligned} \lambda_j &= \sigma_j + i\omega_j \qquad & \{u_j\} &= \text{Re}\,\{u_j\} + i\,\text{Im}\,\{u_j\} \\ \lambda_k &= \sigma_j - i\omega_j \qquad & \{u_k\} &= \text{Re}\,\{u_j\} - i\,\text{Im}\,\{u_j\} \end{aligned} \tag{2.136}$$

The sum of the two complex eigensolutions associated with the pair of eigenvalues and right eigenvectors of (2.136) provides a precessional mode solution which is real and has the form

$$\{p_j\} = \{q_j\} + \{q_k\} = 2e^{\sigma_j t}(\text{Re}\,\{u_j\}\cos\omega_j t - \text{Im}\,\{u_j\}\sin\omega_j t) \tag{2.137}$$

At a typical station of the rotor system, the precessional motion of (2.137) is illustrated in Fig. 2.30*b* for the translation coordinates. In the event that the mode is undamped $\sigma_j = 0$, Eq. 2.137 reduces to elliptic vibration, such as given by (2.1).

The right eigenvectors $\{u_j\}$ and left eigenvectors $\{v_j\}$ of $[H(\lambda)]$ satisfy the following biorthogonality conditions (Fawzy and Bishop, 1976) with norms N_j:

$$\{v_k\}^T[(\lambda_j + \lambda_k)[M] + [\Delta]]\{u_j\} = N_j\delta_{jk} \tag{2.138a}$$

$$\{v_k\}^T[[K] - \lambda_j\lambda_k[M]]\{u_j\} = -\lambda_j N_j\delta_{jk} \tag{2.138b}$$

These conditions are utilized in a later section to decouple the system equations when an eigenvector or modal expansion approach is used for response analyses.

The eigenvalue problems (2.131) and (2.132) are second-order in λ, and there are limited numerical procedures available (Leung, 1988) for solving these quadratic problems. As a result, it is generally advisable to pose the system equations (2.129) in first-order form by defining the state vector $\{x\} = [\{\dot{q}\}^T \ \{q\}^T]^T$ and associated first-order set of system equations

$$[A]\{\dot{x}\} + [B]\{x\} = \{X\} \tag{2.139}$$

where

$$[A] = \begin{bmatrix} [0] & [M] \\ [M] & [\Delta] \end{bmatrix} \qquad [B] = \begin{bmatrix} -[M] & [0] \\ [0] & [K] \end{bmatrix} \qquad \{X\} = \begin{Bmatrix} \{0\} \\ \{Q\} \end{Bmatrix} \tag{2.140}$$

The free vibration solution of (2.139) is then of the form

$$\{x\} = \{y\}e^{\lambda t} \tag{2.141}$$

which, when substituted into (2.139), provides the first-order eigenvalue problem

$$[\lambda[A] + [B]]\{y\} = \{0\}_{2n\times 1} \tag{2.142}$$

and associated adjoint problem

$$\{z\}^T[\lambda[A] + [B]] = \{0\}^T \tag{2.143}$$

The solution of (2.142) and (2.143) yields the $2n$ values λ_j, the right state vectors $\{y_j\} = [\lambda_j\{u_j\}^T \ \{u_j\}^T]^T$, and the left state vectors $\{z_j\} = [\lambda_j\{v_j\}^T \ \{v_j\}^T]^T$. These right and left state vector sets satisfy the biorthogonality conditions

$$\begin{aligned} \{z_k\}^T[A]\{y_j\} &= N_j\delta_{jk} \\ \{z_k\}^T[B]\{y_j\} &= -\lambda_j N_j\delta_{jk} \end{aligned} \tag{2.144}$$

With the same normalization of the modes, (2.144) is equivalent to (2.138).

Critical Speeds and Modes. A critical speed of order ρ of a rotor system is defined as a *rotational speed* of one of the system's rotating assemblies when a multiple ρ of that speed coincides with one of the system's natural frequencies of whirl. An excitation frequency line $\omega_{\text{exc}} = \rho\Omega_{\text{ref}}$ is included in Fig. 2.29, and the intersection of this line with the natural whirl frequency curve ω_j defines the critical speed n_j. Thus, when Ω_{ref} equals n_j, the excitation frequency ρn_j creates a resonance condition. Excitation due to unbalance in the reference rotor occurs at frequency Ω_{ref}; that is, the excitation is synchronous with Ω_{ref} and $\rho = 1$. Excitation due to unbalance in another rotor rotating at a rate ρ times the rotational rate of the reference rotor occurs at frequency $\rho\Omega_{\text{ref}}$. Excitation of superharmonics occurs at integer multiples of these frequencies, and excitation of subharmonics occurs at integer submultiples of these frequencies. One approach for determining critical speeds is then to simply generate the whirl speed map, include all excitation frequency lines of interest, and graphically note the intersections to obtain the critical speeds associated with each excitation.

If the goal of an analysis is simply to obtain the critical speeds for a particular excitation, the above approach may require more computational effort than necessary. It is possible to evaluate the critical speeds directly without generating the whirl speed map. For conservative gyroscopic systems with isotropic bearing stiffness, the process is simplified since the whirl modes are planar and circular relative to a fixed reference frame. If the coordinate transformation (2.14), with $\theta = \omega t$, is applied to the fixed reference system equations of motion (2.129), the result is

$$[K]\{q'\} = \omega^2\left([M] + \frac{1}{\rho}[\hat{G}]\right)\{q'\} \tag{2.145}$$

where $[\hat{G}]$ is a symmetric array obtained from the transformation of the skew symmetric gyroscopic matrix $[G]$ (see Nelson and McVaugh, 1976). The $x'z'$ and $y'z'$ planes of motion in (2.145) are decoupled and may be considered separately. One one plane needs to be considered, so the eigenvalue problem is of order $n/2$. For a particular excitation frequency ratio $\rho = \omega/\Omega$, the eigenvalue problem (2.145) yields the related eigenvalues ω_i^2 and critical speeds $n_i = \omega_i/\rho$. The stiffness matrix $[K]$ is positive definite for adequately constrained rotors. The array $([M] + (1/\rho)[\hat{G}])$, however, may not be positive definite for some choices of the ratio ρ. In these cases negative values of ω_i^2 will result. They are extraneous roots and correspond to situations when the gyroscopic stiffening is such as to prevent intersection of the excitation line with one or more of the natural frequencies of whirl curves (Den Hartog, 1956, pp. 253–265).

For general symmetric undamped systems, Childs and Graviss (1982) present a procedure which allows for direct calculation of the system critical speeds associated with multiples or fractions of the reference spin speed. This is accomplished in fixed reference coordinates. Rajan et al. (1986) present a procedure based on sensitivity analysis for directly determining the system critical speeds for general linear systems including damping. This approach is based on an optimal search procedure rather than eigenvalue extraction.

Forced Vibration Analyses

The forced response of a rotor system may be determined directly in physical coordinates or indirectly in modal coordinates. Analytical procedures for determining steady unbalance, static, and general transient response of linear systems are presented below using physical coordinates. A separate section introduces modal analysis procedures using two different types of modes.

Unbalance Response. For a multishaft system, the unbalance force vector in (2.129) has the form

$$\{Q_u\} = \sum_{\sigma=1}^{n_\sigma} (\{Q_{uc}\}_\sigma \cos \Omega_\sigma t + \{Q_{us}\}_\sigma \sin \Omega_\sigma t) \tag{2.146}$$

where the index σ denotes the shaft number for a total of n_σ shafts. The cosine and sine individual shaft unbalance vectors are assembled as illustrated by (2.113). Since the system is linear, the n_σ solutions can be superposed, and the steady-state response is

$$\{q_u\} = \sum_{\sigma=1}^{n_\sigma} (\{q_{uc}\}_\sigma \cos \Omega_\sigma t + \{q_{us}\}_\sigma \sin \Omega_\sigma t) \tag{2.147}$$

where the cosine and sine component vectors are obtained from

$$\begin{Bmatrix} \{q_{uc}\} \\ \{q_{us}\} \end{Bmatrix}_\sigma = \begin{bmatrix} [K] - \Omega_\sigma^2[M] & \Omega_\sigma[\Delta] \\ -\Omega_\sigma[\Delta] & [K] - \Omega_\sigma^2[M] \end{bmatrix}^{-1} \begin{Bmatrix} \{Q_{uc}\} \\ \{Q_{us}\} \end{Bmatrix}_\sigma \tag{2.148}$$

The total translation (u, v) and rotation (α, β) response at each rotor station is then the superposition of n_σ elliptic vibrations with frequencies corresponding to the shaft rotation speeds. Each shaft unbalance excitation has the potential of producing a resonance condition at each system natural whirl speed.

Static Response. For a static loading such as from gravity, bearing misalignment, gear forces, or steady maneuvers, the system excitation vector $\{Q_s\}$ is constant and the associated static response, from (2.129), is

$$\{q_s\} = [K]^{-1}\{Q_s\} \tag{2.149}$$

Transient Response. The transient response of large-order linear systems, such as modeled by (2.129), can be obtained directly in terms of the physical coordinates or in terms of modal coordinates associated with a linear transformation using a set of system modes. Equations 2.129 with specified initial conditions may be directly integrated by using one of several different integration algorithms. These algorithms attempt to satisfy the equations of motion in a dynamic equilibrium sense at discrete time intervals Δt apart. In addition, assumptions are made as to how the displacement, velocity, and acceleration vary within each interval. Integration procedures, such as the central difference method, that utilize the dynamic equilibrium condition at time t to step to the solution at $t + \Delta t$ are called *explicit integration methods.* Other procedures (such as the Houbolt, Wilson, and Newmark methods) require dynamic equilibrium at $t + \Delta t$ in establishing the solution at $t + \Delta t$. They are referred to as *implicit integration methods.*

Linear transformation of the system equations by using a modal expansion is presented in the next section. The primary advantage here is that modal truncation can normally be used to reduce the order of the system equations of motion. In addition, the transformed equations are decoupled if an orthogonal set of eigenmodes is utilized. These resulting decoupled equations may then be solved by whatever procedure is appropriate for the form of excitation including numerical integration. Methods that utilize a transformation of coordinates are referred to as *indirect methods.* Details on various numerical integration procedures are included in Bathe (1982, p. 268) and Craig (1981, Chap. 18).

Modal Analysis

Several approaches can be taken to implement a modal analysis of rotor systems. Two approaches are presented here, and additional information is provided in a later section on substructure modal synthesis. The first approach makes use of real modes obtained from the undamped symmetric problem associated with (2.129). Any damping, gyroscopic, and circulation forces in this case are treated as pseudo applied forces for the system. The second approach makes use of the complex right and left eigenvector sets from the general nonsymmetric problem, using either the second-order form (2.129) or first-order form (2.139).

Real Modes. The symmetric undamped homogeneous equation from (2.129) is

$$[M]\{\ddot{q}\} + [S]\{q\} = \{0\} \tag{2.150}$$

with a solution of the form $\{q\} = \{\varphi\}e^{j\omega t}$ and associated eigenvalue problem

$$[S]\{\varphi\} = \omega^2[M]\{\varphi\} \tag{2.151}$$

This eigenvalue problem yields n real-valued nonnegative eigenvalues ω_i^2 and related vectors $\{\varphi_i\}$, $i = 1, 2, \ldots, n$, which satisfy the orthogonality conditions (Craig, 1981, p. 303).

$$\begin{aligned} \{\varphi_j\}^T[M]\{\varphi_i\} &= \mu_i\delta_{ij} \\ \{\varphi_j\}^T[S]\{\varphi_i\} &= \mu_i\omega_i^2\delta_{ij} = \kappa_i\delta_{ij} \end{aligned} \tag{2.152}$$

The μ_i and κ_i are defined as the modal mass and stiffness, respectively, of the ith mode. An assumed-modes approach, using the real vectors from (2.151) as approximate system eigenvectors, is taken by introducing the transformation of coordinates

$$\{q\}_{n\times 1} = \sum_{i=1}^{\hat{n}\le n} \{\varphi_i\}\eta_i(t) = [\hat{\varphi}]\{\eta\}_{\hat{n}\times 1} \tag{2.153}$$

The introduction of this transformation into (2.129) and the use of conditions (2.152) give the transformed set of system equations

$$\begin{bmatrix} \diagdown & & \\ & \hat{\mu} & \\ & & \diagdown \end{bmatrix}\{\ddot{\eta}\} + [\hat{\varphi}]^T([D] + [G])[\hat{\varphi}]\{\dot{\eta}\} + \left(\begin{bmatrix} \diagdown & & \\ & \hat{\kappa} & \\ & & \diagdown \end{bmatrix} + [\hat{\varphi}]^T[C][\hat{\varphi}]\right)\{\eta\} = [\hat{\varphi}]^T\{Q\} \tag{2.154}$$

Even though this set of equations is still coupled in the modal coordinates η_i, many analysts have successfully utilized the approach to obtain rotordynamic response. Since $\hat{n}$ is usually much less than n, the order of (2.154) is substantially decreased, thereby reducing subsequent computational effort with particular savings for transient analyses (Dennis et al., 1975).

Complex Modes. In the complex modal approach (Glasgow and Nelson, 1980; Nelson et al., 1983) it is convenient to use the state equations (2.139) and introduce the eigenvector expansion

$$\{x\} = \sum_{i=1}^{2\hat{n}\le 2n} \{y_i\}\zeta_i(t) \tag{2.155}$$

in terms of the system right state vectors from (2.142). The substitution of this transformation into (2.139) and use of the biorthogonality conditions (2.144) yield the following set of decoupled equations, in terms of the complex modal coordinates $\zeta_i(t)$:

$$N_i\dot{\zeta}_i - \lambda_i N_i\zeta_i = \{z_i\}^T\{X\} = Z_i \qquad i = 1, 2, \ldots, 2\hat{n} \le 2n \tag{2.156}$$

with $Z_i(t)$ representing the ith complex modal force. Equation 2.156 may be solved independently by whatever procedure is appropriate for the excitation. The physical coordinate response is then obtained by back transformation using (2.155). If a complex mode is retained in (2.155), then it is also necessary to retain its conjugate so that a real response is obtained upon back transformation.

Substructure Modal Synthesis

For some multishaft systems, a finite-element model of the entire system may contain so many degrees of freedom that it is not computationally efficient to directly analyze the system model. In these situations it often proves useful to divide the system into a set of substructures, perform some analysis on each substructure, and then develop a reduced-order mathematical model of the system by using the substructure equations. This approach is referred to as *substructure modal synthesis*. The displacement vector of each substructure is considered to be partitioned into a subset of boundary displacements, associated with interconnection with other system components, and a complement subset of interior displacements.

The equations of motion of a particular component have the first-order form of (2.139), and the displacement vector is reordered and partitioned into the boundary and interior subsets. The partitioned form of the substructure state vector is then $\{x\} = [\{x_B\}^T \{x_I\}^T]^T$ with $\{x_B\}^T = [\{\dot{q}_B\}^T \{q_B\}^T]$ and $\{x_I\}^T = [\{\dot{q}_I\}^T \{q_I\}^T]$. The corresponding substructure state equations are

$$\begin{bmatrix} [A_{BB}] & [A_{BI}] \\ [A_{IB}] & [A_{II}] \end{bmatrix} \begin{Bmatrix} \{\dot{x}_B\} \\ \{\dot{x}_I\} \end{Bmatrix} + \begin{bmatrix} [B_{BB}] & [B_{BI}] \\ [B_{IB}] & [B_{II}] \end{bmatrix} \begin{Bmatrix} \{x_B\} \\ \{x_I\} \end{Bmatrix} = \begin{Bmatrix} \{X_B\}_c \\ \{0\} \end{Bmatrix} + \begin{Bmatrix} \{X_B\} \\ \{X_I\} \end{Bmatrix}) \quad (2.157)$$

The components of $\{X\}^T = [\{X_B\}^T \{X_I\}^T]$ contain all applied forces such as gravity, unbalance, and maneuver loads. The boundary subset of interconnection forces $\{X_B\}_c$ can include contributions from both linear and nonlinear connection components. Section 2.3.3 discusses the nonlinear case in more detail, while linear situations are of interest here.

In the substructure modal synthesis approach presented here, the displacement of any point in the substructure is represented as the superposition of two types of modes:

1. Fixed interface modes with displacements relative to fixed substructure boundaries, that is, $\{x_B\} = \{0\}$
2. Static constraint modes with displacements associated with the motion of the boundary coordinates

The fixed interface modes are determined by setting the boundary state vector $\{x_B\}$ to zero and then utilizing the free vibration problem from (2.157) for the interior coordinates, i.e.,

$$[A_{II}]\{\dot{x}_I\} + [B_{II}]\{x_I\} = \{0\} \quad (2.158)$$

The eigenvalue problem and adjoint problem from (2.158), analogous to (2.142) and (2.143), gives the following set of eigenvalues, right state vectors, and left state vectors:

$$\lambda_i \qquad \{y_i\} = \begin{Bmatrix} \lambda_i\{u_i\} \\ \{u_i\} \end{Bmatrix} \qquad \{z_i\} = \begin{Bmatrix} \lambda_i\{v_i\} \\ \{v_i\} \end{Bmatrix} \qquad i = 1, 2, \ldots, 2n_I \quad (2.159)$$

The interior motion of the substructure relative to fixed boundary coordinates is then expressed by the eigenvector expansion

$$\{x_I\} = \sum_{i=1}^{2n_I} \{y_i\}\zeta_i = [Y]\{\zeta\} \tag{2.160}$$

where $[Y]$ is the matrix of right state vectors for the substructure. The related matrix of left state vectors is $[Z]$, and these two sets of vectors satisfy the biorthogonality conditions of (2.144).

The static constraint modes are determined from the homogeneous form of the substructure stiffness matrix, written in the partitioned form

$$\begin{bmatrix} [K_{BB}] & [K_{BI}] \\ [K_{IB}] & [K_{II}] \end{bmatrix} \begin{Bmatrix} \{q_B\} \\ \{q_I\} \end{Bmatrix} = \begin{Bmatrix} \{Q_B\} \\ \{0\} \end{Bmatrix} \tag{2.161}$$

As in Sec. 2.3.1, the following dependency relation is obtained between the interior and boundary displacements:

$$\{q_I\} = -[K_{II}]^{-1}[K_{IB}]\{q_B\} = [\chi]\{q_B\} \tag{2.162}$$

It is then assumed that the velocities obey the same relationship, so that the following transformation may be written:

$$\begin{Bmatrix} \{x_B\} \\ \{x_I\} \end{Bmatrix} = \begin{bmatrix} [I] & [0] \\ [\underline{\chi}] & [Y] \end{bmatrix} \begin{Bmatrix} \{x_B\} \\ \{\zeta\} \end{Bmatrix} \tag{2.163}$$

with

$$[\underline{\chi}] = \begin{bmatrix} [\chi] & [0] \\ [0] & [\chi] \end{bmatrix}$$

The first half of (2.163) is simply an identity involving the real boundary states, and the second half states that the total state of the interior coordinates is the superposition of static constraint motion due to boundary states plus relative motion represented by a complex right vector expansion of the fixed interface modes. If the substructure under consideration is a rotating assembly, the right vectors possess the properties of a conservative gyroscopic system, and the precessional modes are undamped and circular. If the substructure is a proportionally damped nonrotating structure such as a support structure, the right vectors are real as associated with a conservative nongyroscopic system.

The primary purpose of substructure modal synthesis is to reduce the total system degrees of freedom. Thus, the right vector expansion in (2.160) is usually truncated so as to exclude the higher-frequency fixed interface modes. The transformation (2.163) is then approximated by

$$\begin{Bmatrix} \{x_B\} \\ \{x_I\} \end{Bmatrix} = \begin{bmatrix} [I] & [0] \\ [\underline{\chi}] & [\hat{Y}]_{2n_I \times 2\hat{n}_I} \end{bmatrix} \begin{Bmatrix} \{x_B\} \\ \{\hat{\xi}\} \end{Bmatrix} \tag{2.164a}$$

or

$$\{x\} = [\beta]\{\underline{x}\} \tag{2.164b}$$

where $[\beta]$ is the truncated transformation matrix and

$$[\bar{\beta}] = \begin{bmatrix} [I] & [0] \\ [\underline{\chi}] & [\hat{\underline{Z}}] \end{bmatrix} \tag{2.165}$$

is the corresponding truncated left vector matrix. The number of retained vectors $2\hat{n}_I$ is generally much less than $2n_I$ for large-order substructures. The introduction of (2.164*a*) into (2.157), premultiplication by the transpose of (2.165), and use of the biorthogonality conditions (2.144) give

$$[\underline{A}]\{\dot{\underline{x}}\} + [\underline{B}]\{\underline{x}\} = \{\underline{X}\}_c + \{\underline{X}\} \tag{2.166}$$

where

$$\{\underline{X}\}_c = \begin{Bmatrix} \{X_B\}_c \\ \{0\} \end{Bmatrix} \qquad \{\underline{X}\} = \begin{Bmatrix} \{X_B\} + [\underline{\chi}]^T\{X_I\} \\ [\hat{\underline{Z}}]^T\{X_I\} \end{Bmatrix} \tag{2.167}$$

Equation 2.166 is the substructure state equation in terms of the boundary states plus a set of retained fixed interface modal coordinates. The order of (2.166) is equal to the sum of the number of boundary states (displacements and velocities) plus the number of retained complex modes.

With the reduced-order equation of motion (2.166) developed for each substructure, the next step in the process is to assemble them to form the reduced-order system state equations. To this end, designate $\{x_s\}$ as the system state vector composed of all boundary states, denoted by $\{x_s^B\}$, and the retained fixed interface modal coordinates of each substructure, typically denoted by $\{\zeta_i\}$ ($i = 1, 2, \ldots, N$). Thus,

$$\{x_s\}^T = [\{x_s^B\}^T\{\zeta_1\}^T\{\zeta_2\}^T \cdots \{\zeta_N\}^T] \tag{2.168}$$

Each substructure's reduced state vector $\{\underline{x}_i\}$ is then related to the system state (2.167) by a geometric constraint relation of the form

$$\{\underline{x}_i\} = \begin{Bmatrix} \{x_{Bi}\} \\ \{\zeta_i\} \end{Bmatrix} = [\gamma_i]\{x_s\} \qquad i = 1, 2, \ldots, N \tag{2.169}$$

For a typical substructure equation of motion given by (2.166), with index i,

$$[\underline{A}_i]\{\dot{\underline{x}}_i\} + [\underline{B}_i]\{\underline{x}_i\} = \{\underline{X}_i\}_c + \{\underline{X}_i\} \tag{2.170}$$

Substitute the transformation (2.169), premultiply by $[\gamma_i]^T$, and sum over all substructures to obtain

$$[A_s]\{\dot{x}_s\} + [B_s]\{x_s\} = \{X_s\} \tag{2.171}$$

where

$$\begin{aligned} [A_s] &= \sum_{i=1}^{N} [\gamma_i]^T[\underline{A}_i][\gamma_i] \\ [B_s] &= \sum_{i=1}^{N} [\gamma_i]^T[\underline{B}_i][\gamma_i] \\ \{X_s\} &= \sum_{i=1}^{N} [\gamma_i]^T\{\underline{X}_i\} \end{aligned} \tag{2.172}$$

Equation 2.171 is the reduced-order system equation of motion in state form.

For linear systems, (2.171) may be utilized to determine any of the previously discussed linear dynamic characteristics, including whirl speeds and modes, unbalance response, static response, and transient response. The primary advantage, however, is with transient response simulation that requires numerical integration. The prudent truncation of substructure modes can result in the order of (2.171) being substantially lower than that of the original system state equations (2.139).

The primary nonlinear mechanisms of a rotordynamic system are usually associated only with the boundary states $\{X_s^B\}$. The above procedure is particularly advantageous in these cases since the presence of nonlinear components only modifies the RHS of (2.171). An additional force vector which involves only the real boundary states $\{x_s^B\}$ is included. This situation is discussed and illustrated in more detail in Sec. 2.3.3.

Several other substructure modal synthesis approaches are available in addition to the static constraint mode/fixed interface mode approach presented above. Childs (1974) utilized a rotor-fixed modal simulation model for flexible rotating equipment. In his work he approximated the distributed parameter characteristics of the rotor as an assemblage of elastically connected rigid bodies. A modal transformation using free-free planar modes at zero spin speed was introduced. This approach has been successfully utilized and verified on several large-order systems and was generalized (Childs, 1976) for flexible asymmetric rotors. A detailed discussion of component mode synthesis of flexible rotor systems is included in Li and Gunter (1982). The formulation presented by Li and Gunter utilizes undamped normal modes which are associated with the symmetric part of the substructure model. Although they are not eigenmodes of the substructure, these assumed modes still provide a good Ritz approximation and have the advantage of being real rather than complex, thereby simplifying the computer code development.

2.3.2 Transfer Matrix Method

The other common approach for analyzing the dynamics of rotor bearing systems is called the state vector transfer matrix method or simply the transfer matrix method (TMM). This method is particularly well suited for "chainlike" structures such as large-order multishaft rotor systems. The first use of this method is usually atttributed to Holzer (1921) in the area of torsional vibrations. Subsequent contributors (Myklestad, 1944; and Prohl, 1945) adapted the procedure to lateral vibrations of beams and rotor systems. More recently Lund (1967, 1974a, 1974b) has presented procedures for the use of this method for almost all aspects of rotor dynamic analysis. Other refinements, to be discussed later, have made this a reliable and widely used analytical procedure for engine manufacturers and rotordynamic experts throughout the world.

The primary advantage of the TMM over the DSM is that it does not require the storage and manipulation of large system arrays. Rather, the method utilizes a marching procedure, starting with the boundary conditions at one side of the system, and successively marches along the structure to the other side of the system. The satisfaction of the boundary conditions at all boundary points is the basis of locating a solution. The "state" of the rotor system at a particular "point" is defined by means of a "state vector," and information is transferred between successive points by means of transfer matrices.

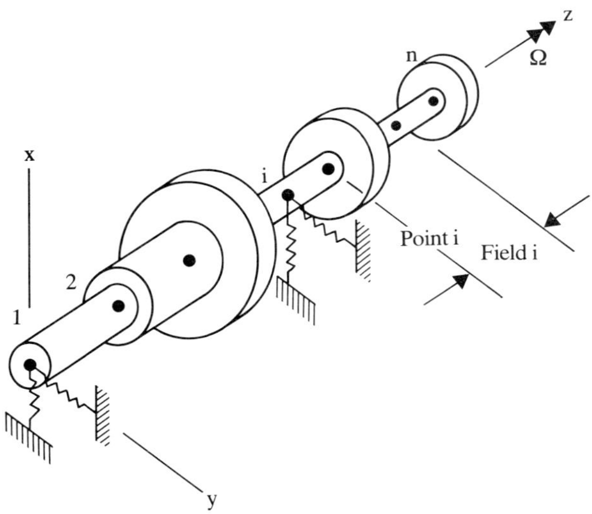

FIGURE 2.31 Transfer matrix method for rotating shaft model.

State Vector

A typical single-shaft rotor system is shown in Fig. 2.31. In the TMM, several points are identified along the axis of the shaft with the points numbered from 1 to n starting with the left end. The number n can be quite large without substantially increasing the computation time. The number should be held to a minimum, however, as a general rule of modeling. The state vector at a particular point is defined as a vector whose elements include the displacements and internal forces of the element for a shaft cross section located at the point. In the case of lateral shaft vibrations, the state vector is defined as

$$\{z\} = [u \;\; v \;\; \alpha \;\; \beta \;\; \tau_y \;\; \tau_x \;\; f_y \;\; f_x]^T \tag{2.173}$$

with the displacements and internal forces illustrated in Fig. 2.32 for a beam cross section with a positive outward normal. The sign convention used here is that positive internal forces act in positive (negative) coordinate directions on beam cross sections with a positive (negative) outward normal. The order of the elements of $\{z\}$ is arbitrary; however, the usual practice is to order the elements so that the jth and $(9-j)$th elements are conjugate in the sense that their product represents work done at the point. The state vector for the rotating $x'y'z'$ reference is

$$\{z'\} = [u' \;\; v' \;\; \alpha' \;\; \beta' \;\; \tau_{y'} \;\; \tau_{x'} \;\; f_{y'} \;\; f_{x'}]^T \tag{2.174}$$

and the state vector associated with deformation in the $x'z'$ plane is

$$\{z'\}_{x'z'} = [u' \;\; \beta' \;\; \tau_{y'} \;\; f_{x'}]^T \tag{2.175}$$

The "state" on the left-hand side of a structural point i is generally different from that on the right-hand side. Superscripts L and R are used to denote the left and right sides, respectively, of point i.

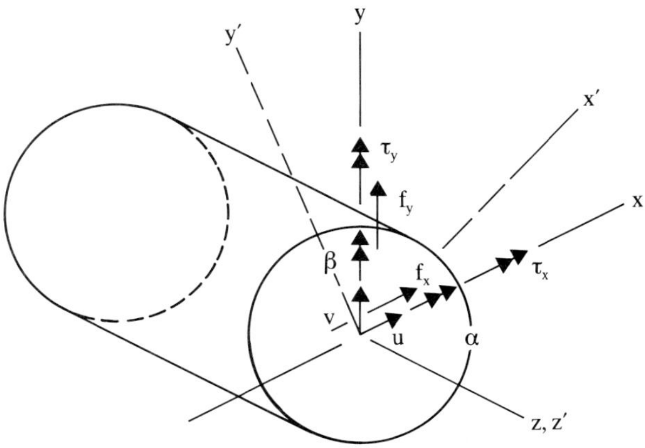

FIGURE 2.32 Beam state components $[u\ v\ \alpha\ \beta\ \tau_y\ \tau_x\ f_y\ f_x]$.

Transfer Matrices

The rotating shaft, as illustrated in Fig. 2.31, is considered to consist of rigid disks and/or linear bearings at the "points" which are separated by elastic beam segments called "fields." Here the fields are modeled as massless with inertia effects included by the rigid disks at the endpoints of the fields. A continuous mass field may be utilized as well and is presented by Lund and Orcutt (1967). For simplicity of presentation, however, a massless field is utilized in the development that follows. Shear deformation is included, but internal damping effects are omitted. Two different types of transfer matrices are used: a point transfer matrix and a field transfer matrix.

Point Transfer Matrix. A point transfer matrix relates the state vector at the right-hand side of a point to the state vector at the left-hand side of the same point. A typical point is illustrated in Fig. 2.33, and it includes a rigid disk with a linear translational bearing connected to a rigid support. The equations of motion for this rigid disk bearing point are

$$\begin{bmatrix} m & 0 & 0 & 0 \\ 0 & m & 0 & 0 \\ 0 & 0 & I_d & 0 \\ 0 & 0 & 0 & I_d \end{bmatrix}_i \begin{Bmatrix} \ddot{u}_i \\ \ddot{v}_i \\ \ddot{\alpha}_i \\ \ddot{\beta}_i \end{Bmatrix} + \begin{bmatrix} c_{xx} & c_{xy} & 0 & 0 \\ c_{yx} & c_{yy} & 0 & 0 \\ 0 & 0 & 0 & \Omega I_a \\ 0 & 0 & -\Omega I_a & 0 \end{bmatrix}_i \begin{Bmatrix} \dot{u}_i \\ \dot{v}_i \\ \dot{\alpha}_i \\ \dot{\beta}_i \end{Bmatrix}$$
$$+ \begin{bmatrix} k_{xx} & k_{xy} & 0 & 0 \\ k_{yx} & k_{yy} & 0 & 0 \\ 0 & 0 & 0 & 0 \\ 0 & 0 & 0 & 0 \end{bmatrix}_i \begin{Bmatrix} u_i \\ v_i \\ \alpha_i \\ \beta_i \end{Bmatrix} = \begin{Bmatrix} f_x^R - f_x^L \\ f_y^R - f_y^L \\ \tau_x^R - \tau_x^L \\ \tau_y^R - \tau_y^L \end{Bmatrix}_i + m\Omega^2 \operatorname{Re}\left(\begin{Bmatrix} a_{x'} + i a_{y'} \\ -i(a_{x'} + i a_{y'}) \\ 0 \\ 0 \end{Bmatrix}_i e^{i\Omega t} \right) \tag{2.176}$$

The disk/bearing point is considered to be of zero axial extent so that the displacements on the right- and left-hand sides of the point are considered to be identical, i.e.,

$$[u_i^L\ v_i^L\ \alpha_i^L\ \beta_i^L] = [u_i^R\ v_i^R\ \alpha_i^R\ \beta_i^R] \tag{2.177}$$

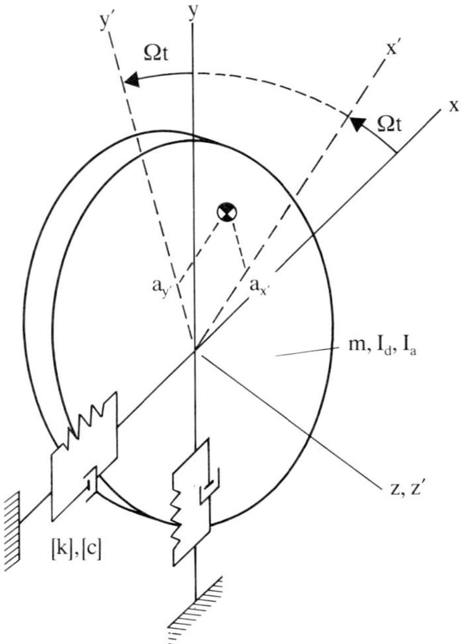

FIGURE 2.33 Typical rotor point, disk/bearing.

For free vibration, the solution of (2.176) is of the form $\{z_i\} = \{\bar{z}_i\}e^{\lambda t}$ or

$$\begin{Bmatrix} u \\ v \\ \alpha \\ \beta \end{Bmatrix}_i = \begin{Bmatrix} \bar{u} \\ \bar{v} \\ \bar{\alpha} \\ \bar{\beta} \end{Bmatrix}_i e^{\lambda t} \quad \text{with} \quad \begin{Bmatrix} f_x \\ f_y \\ \tau_x \\ \tau_y \end{Bmatrix}_i = \begin{Bmatrix} \bar{f}_x \\ \bar{f}_y \\ \bar{\tau}_x \\ \bar{\tau}_y \end{Bmatrix}_i e^{\lambda t} \tag{2.178}$$

where $\lambda = \sigma + i\omega$ represents an eigenvalue with ω equal to the damped natural frequency and σ equal to the damping coefficient.

The substitution of (2.178) into (2.176), utilization of the identity (2.177), rearrangement, and use of the definition (2.173) provide the following transfer relation:

$$\{\bar{z}_i\}^R = [P_i]\{\bar{z}_i\}^L \tag{2.179}$$

where

$$[P_i] = \begin{bmatrix} 1 & 0 & 0 & 0 & 0 & 0 & 0 & 0 \\ 0 & 1 & 0 & 0 & 0 & 0 & 0 & 0 \\ 0 & 0 & 1 & 0 & 0 & 0 & 0 & 0 \\ 0 & 0 & 0 & 1 & 0 & 0 & 0 & 0 \\ 0 & 0 & -\Omega I_a\lambda & I_d\lambda^2 & 1 & 0 & 0 & 0 \\ 0 & 0 & I_d\lambda^2 & \Omega I_a\lambda & 0 & 1 & 0 & 0 \\ b_{yx} & b_{yy} + m\lambda^2 & 0 & 0 & 0 & 0 & 1 & 0 \\ b_{xx} + m\lambda & b_{xy} & 0 & 0 & 0 & 0 & 0 & 1 \end{bmatrix}_i \tag{2.180}$$

($b_{yx} = c_{yx}\lambda + k_{yx}$, etc.) is the point transfer matrix for a rigid disk/linear bearing point. The premultiplication of the rotor state at the left-hand side of point i by this transfer matrix provides the rotor state at the right-hand side of the point.

Field Transfer Matrix. A typical uniform-cross-section beam segment is illustrated in Fig. 2.34, and it represents a flexible "field" connection between

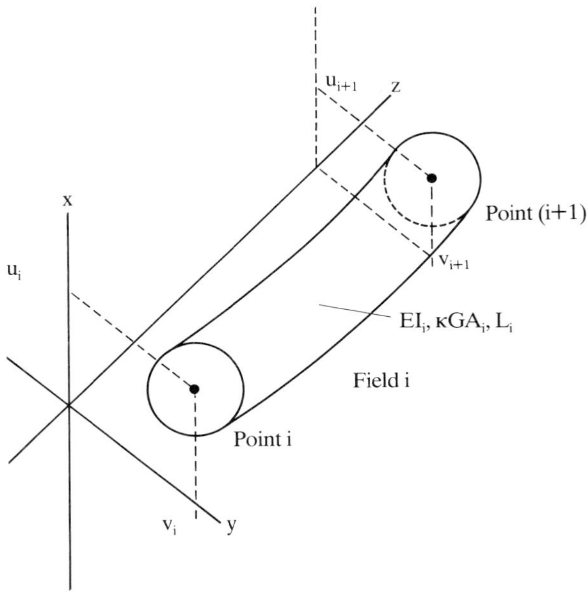

FIGURE 2.34 Typical rotor field.

point i at its left end and point $i+1$ at its right end. It is assumed here that the inertia effects have been included by rigid disks at the endpoints. Thus, the shaft segment is considered to be massless, and the equilibrium requirements for the xz plane are

$$(f_x^R)_i - (f_x^L)_{i+1} = 0 \qquad (\tau_y^L)_{i+1} - (\tau_y^R)_i + (f_x^L)_{i+1}L_i = 0 \tag{2.181}$$

and from Timoshenko beam theory, the deflection relations are

$$\begin{aligned} u_{i+1} &= u_i + \beta_i L_i + (\tau_y^L)_{i+1}\frac{L_i^2}{2EI_i} + (f_x^L)_{i+1}\frac{L_i^3}{3EI_i}(1 + 3\varepsilon_i) \\ \beta_{i+1} &= \beta_i + (\tau_y^L)_{i+1}\frac{L_i}{EI_i} + (f_x^L)_{i+1}\frac{L_i^2}{2EI_i} \end{aligned} \tag{2.182}$$

Similar relations are obtained for the yz plane with some polarity changes due to sign convention. Rearrangement of (2.182) and use of the equilibrium conditions (2.181) produce the following relation between states at the ends of the shaft segment:

$$\{\bar{z}_{i+1}\}^L = [F_i]\{\bar{z}_i\}^R \tag{2.183}$$

with

$$[F_i] = \begin{bmatrix} 1 & 0 & 0 & L & L^2/(2EI) & 0 & 0 & -L^3(1-6\varepsilon)/(6EI) \\ 0 & 1 & -L & 0 & 0 & -L^2/(2EI) & -L^3(1-6\varepsilon)/(6EI) & 0 \\ 0 & 0 & 1 & 0 & 0 & L/(EI) & L^2/(2EI) & 0 \\ 0 & 0 & 0 & 1 & L/(EI) & 0 & 0 & -L^2/(2EI) \\ 0 & 0 & 0 & 0 & 1 & 0 & 0 & -L \\ 0 & 0 & 0 & 0 & 0 & 1 & L & 0 \\ 0 & 0 & 0 & 0 & 0 & 0 & 1 & 0 \\ 0 & 0 & 0 & 0 & 0 & 0 & 0 & 1 \end{bmatrix}_i$$

Equation 2.183 is the field transfer relation for a massless cylindrical shaft segment. The premultiplication of the rotor state at the left end of the segment (RHS of point i) by the field transfer matrix provides the rotor state at the right end of the segment (LHS of point $i + 1$).

System Transfer Matrix All the various analyses for a rotor system, whether free or forced, require the development of a system transfer matrix. This matrix is used to transfer the rotor state associated with all the LHS system boundary points to the rotor state at the RHS system boundary points. One LHS boundary point and one RHS boundary point exist for each shaft of the system. The development of this system transfer matrix is presented below first for a single-shaft system and then for multishaft systems with interconnection elements between shafts.

Single-Shaft Systems. The system transfer matrix is obtained by utilizing the point and field transfer matrices to sequentially transfer the rotor state from the left boundary point 1 to the right boundary point n. This sequential application of (2.179) and (2.183) gives

$$\begin{aligned} \{\bar{z}_1\}^R &= [P_1]\{\bar{z}_1\}^L \\ \{\bar{z}_2\}^L &= [F_1]\{\bar{z}_1\}^R \\ &\vdots \\ \{\bar{z}_n\}^L &= [F_{n-1}]\{\bar{z}_{n-1}\}^R \\ \{\bar{z}_n\}^R &= [P_n]\{\bar{z}_n\}^L \end{aligned} \tag{2.184a}$$

or

$$\{\bar{z}_n\}^R = [U]\{\bar{z}_1\}^L \tag{2.184b}$$

where

$$[U] = [P_n] \prod_{i=n-1}^{1} [F_i][P_i] \tag{2.185}$$

is the system (overall) transfer matrix.

Multipleshaft Systems. It is assumed here, for illustration purposes, that the system consists of two shafts. The procedure can then be generalized in straightforward fashion to situations with more than two shafts. A two-shaft system is illustrated in Fig. 2.35 with the shafts denoted by A and B. A typical point on shaft (A, B) is denoted by the indices (i, j) each starting with 1 and terminating with (n, m). The point and field transfer relations, defined above for single-shaft systems, are generalized as follows (Li et al., 1981):

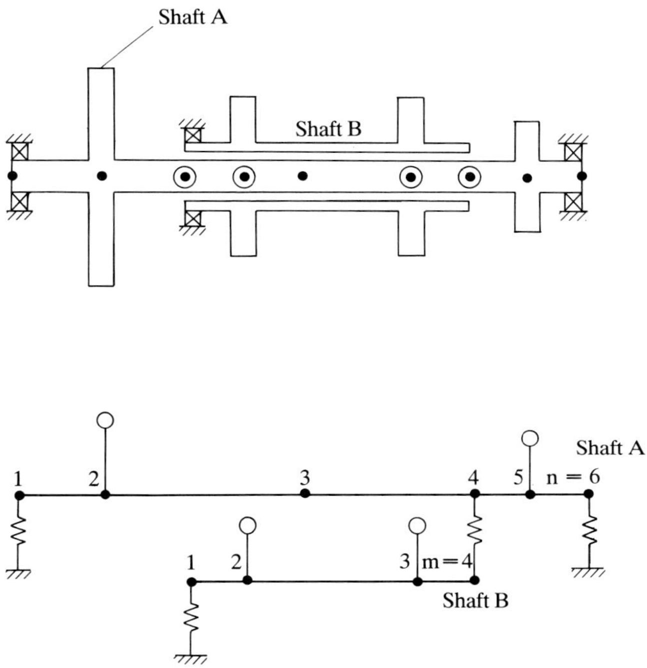

FIGURE 2.35 Dual-shaft TMM model and schematic.

Point transfer equation:

$$\begin{Bmatrix} {}^{A}\{\bar{z}_i\}^{R} \\ {}^{B}\{\bar{z}_j\}^{R} \end{Bmatrix} = \begin{bmatrix} \delta_A\ {}^{A}[P_i] & [0] \\ [0] & \delta_B\ {}^{B}[P_j] \end{bmatrix} \begin{Bmatrix} {}^{A}\{\bar{z}_i\}^{L} \\ {}^{B}\{\bar{z}_j\}^{L} \end{Bmatrix} \tag{2.186}$$

Field transfer equation:

$$\begin{Bmatrix} {}^{A}\{\bar{z}_{i+1}\}^{L} \\ {}^{B}\{\bar{z}_{j+1}\}^{L} \end{Bmatrix} = \begin{bmatrix} \delta_A\ {}^{A}[F_i] & [0] \\ [0] & \delta_b\ {}^{B}[F_j] \end{bmatrix} \begin{Bmatrix} {}^{A}\{\bar{z}_i\}^{R} \\ {}^{B}\{\bar{z}_j\}^{R} \end{Bmatrix} \tag{2.187}$$

with

$$i = 1, 2, \ldots, n$$
$$j = 1, 2, \ldots, m$$

where

$$\delta_A\ {}^{A}[P_i] = \begin{cases} [I] & \text{when transferring along shaft } B \text{ only} \\ {}^{A}[P_i] & \text{otherwise} \end{cases}$$
$$\delta_B\ {}^{B}[P_j] = \begin{cases} [I] & \text{when transferring along shaft } A \text{ only} \\ {}^{B}[P_j] & \text{otherwise} \end{cases} \tag{2.188}$$

with analogous definitions for the field transfer operators.

The development of the system transfer matrix proceeds as for a single-shaft system starting with point 1 on both shafts. The marching process continues until an interconnection (or linking) member between the two shafts is encountered and continues further, if necessary, until the position on both shafts corresponds to the interconnection location. A transfer across the interconnection, such as shown between point $i = 4$ and $j = 4$ on shafts A and B, respectively, requires an interconnection transfer matrix. This transfer matrix for a linear damped translational interconnection is shown below. Inertia properties are included as rigid disks at the interconnection points.
Interconnection transfer matrix:

$$\begin{Bmatrix} ^A\{\bar{z}_i\}^R \\ ^B\{\bar{z}_j\}^R \end{Bmatrix} = \begin{bmatrix} ^A[P_i] + [C_{ij}] & -[C_{ij}] \\ -[C_{ij}] & ^B[P_j] + [C_{ij}] \end{bmatrix} \begin{Bmatrix} ^A\{\bar{z}_i\}^L \\ ^B\{\bar{z}_j\}^L \end{Bmatrix} \tag{2.189}$$

where

$$[C_{ij}]_{8\times 8} = \begin{bmatrix} 0 & 0 & 0 & \cdots & 0 \\ \hline 0 & 0 & 0 & \cdots & 0 \\ b_{yx} & b_{yy} & 0 & \cdots & 0 \\ b_{xx} & b_{xy} & 0 & \cdots & 0 \end{bmatrix}_{ij} \tag{2.190}$$

The development of the system transfer matrix continues along each shaft, point to point, until the RHS boundary points are encountered for both shafts. The system transfer matrix is then of the form

$$\begin{Bmatrix} ^A\{\bar{z}_n\}^R \\ ^B\{\bar{z}_m\}^R \end{Bmatrix} = [U]_{16\times 16} \begin{Bmatrix} ^A\{\bar{z}_1\}^L \\ ^B\{\bar{z}_1\}^L \end{Bmatrix} \tag{2.191}$$

The above process is generalized to multishaft systems by altering the size of the transfer relations (2.186), (2.187), and (2.189) so that they are compatible with the system to be analyzed.

Solution Procedures
The system transfer matrix of (2.191) and some variations to it, as presented below, can be used for several different types of rotordynamic analyses. This includes the determination of damped whirl speeds and precessional modes, undamped critical speeds and modes, and synchronous unbalance response.

Whirl Speeds and Precessional Modes. For a specified set of shaft spin speeds, the system transfer matrix (2.191) is generated, and the boundary conditions are then imposed at the left and right boundary points of each shaft. Boundary conditions for typical end conditions for the xz plane are as follows: free: $f_x = 0$, $\tau_y = 0$; pinned: $u = 0$, $\tau_y = 0$; fixed: $u = 0$, $\beta = 0$. Analogous conditions exist for the yz plane. After the introduction of the appropriate boundary conditions, it is convenient to rearrange and partition (2.191) so that the zero conditions appear as a subset, such as the lower portion of the system boundary state, as shown in (2.192). Then from the lower portion of

$$\begin{Bmatrix} \{z^R\} \\ \{0\} \end{Bmatrix} = \begin{bmatrix} [U_{11}] & [U_{12}] \\ [U_{21}] & [U_{22}] \end{bmatrix} \begin{Bmatrix} \{z^L\} \\ \{0\} \end{Bmatrix} \tag{2.192}$$

the following homogeneous relation is obtained

$$[U_{21}(\lambda)]\{z^L\} = \{0\} \tag{2.193}$$

and a nontrivial solution for $\{z^L\}$ exists if and only if the coefficient determinant vanishes. The values of λ that render the coefficient determinant zero are the system eigenvalues analogous to (2.133).

The great advantage of the transfer matrix method is that the order of the matrices $[U_{ij}]$ in (2.192) remains the same no matter how many segments there are between the left and right boundaries. The associated drawback is the buildup of numerical error (Uhrig, 1966) as the $[U_{ij}]$ are constructed by repeated matrix multiplications of the individual segment point and field transfer operators. In general, the system transfer matrix becomes increasingly ill-conditioned as the whirl frequency is increased. The onset of the problem can be delayed by using double-precision arithmetic and by scaling all state variables to have magnitudes of the order unity.

Several procedures can be used to determine the characteristic values from (2.193). A Newton–Raphson approach, developed by Lund (1974a), has proved to be successful and reliable with convergence normally within a few iterations. Lund includes a numerical procedure for evaluation of the gradient of the determinant which is required in the Newton–Raphson procedure. Some caution in applying this procedure needs to be taken to ensure that all values of interest have been located. A second approach is to treat the determinant of (2.193) as a nonlinear function of λ and then utilize one of several different types of nonlinear root finders, which are available in commercial computer packages. Murphy and Vance (1983) have shown that by rearranging the calculations performed in the TMM is is possible to calculate the coefficients of the characteristic polynomial directly. The roots may then be extracted by standard polynomial root finders such as Muller's method or Bairstow's method. In addition, with the coefficients of the polynomial available, it is also possible to investigate the nature of the roots by using the Routh-Hurwitz criteria (Loewy and Piarulli, 1969, p. 39). A stability survey of the system is then possible without the necessity of calculating the system eigenvalues.

Critical Speeds and Modes. For an undamped rotor with isotropic supports, the orbits of the natural modes of whirl are circular relative to the fixed xyz reference frame. In a rotating $x'y'z'$ reference frame with a rotation rate equal to the whirl speed, the whirl mode appears as a fixed curve in a plane through the z' axis. Because of the axial symmetry, the order of the system equations can be cut in half by taking this plane to be one of the coordinate planes, say, the $x'z'$ plane. The point and field transfer relations for this special case are, respectively,

$$\{\bar{z}_i'\}^R = [P_i']\{\bar{z}_i'\}^L \tag{2.194}$$

and

$$\{\bar{z}_{i+1}'\}^L = [F_i']\{\bar{z}_i'\}^R \tag{2.195}$$

where

$$\{\bar{z}_i'\} = [\bar{u}' \;\; \bar{\beta}' \;\; \bar{\tau}_{y'} \;\; \bar{f}_{x'}]_i^T \tag{2.196}$$

and

$$[P_i'] = \begin{bmatrix} 1 & 0 & 0 & 0 \\ 0 & 1 & 0 & 0 \\ 0 & -\omega^2\left(I_d - \dfrac{1}{\rho} I_a\right) & 1 & 0 \\ k - \omega^2 m & 0 & 0 & 1 \end{bmatrix}_i$$

$$[F_i'] = \begin{bmatrix} 1 & L & L^2/(2EI) & -L^3(1-6\varepsilon)/(6EI) \\ 0 & 1 & L/(EI) & -L^2/(2EI) \\ 0 & 0 & 1 & -L \\ 0 & 0 & 0 & 1 \end{bmatrix}_i$$

For a specified whirl-to-spin ratio ρ, the method for determining the critical speeds of a single-shaft system proceeds as above with relations (2.194) and (2.195), taking the place of (2.179) and (2.183). The eigenvalues for this analysis are purely real ω_j^2 and the associated critical speed is $n_j = (1/\rho)\,|\omega_j|$. The mode shape associated with this critical speed is obtained by arbitrarily choosing the unspecified state at the left endpoint of the shaft and then successively evaluating the state at other rotor points, using the transfer relations (2.194) and (2.195). The modes can then be normalized as desired.

For multishaft systems in a planar undamped whirl mode, the interconnection transfer relation, taking the place of (2.189), is

$$\begin{Bmatrix} {}^A\{\bar{z}_i'\}^R \\ {}^B\{\bar{z}_j'\}^R \end{Bmatrix} = \begin{bmatrix} {}^A[P_i'] + [C_{ij}'] & -[C_{ij}'] \\ -[C_{ij}'] & {}^B[P_j'] + [C_{ij}'] \end{bmatrix} \begin{Bmatrix} {}^A\{\bar{z}_i'\}^L \\ {}^B\{\bar{z}_j'\}^L \end{Bmatrix} \tag{2.197}$$

where

$$[C_{ij}'] = \begin{bmatrix} 0 & 0 & 0 & 0 \\ 0 & 0 & 0 & 0 \\ 0 & 0 & 0 & 0 \\ k & 0 & 0 & 0 \end{bmatrix} \tag{2.198}$$

The undamped critical speed analysis of multishaft systems requires the whirl-to-spin ratio to be specified for each shaft. If $\rho_{\text{ref}} = \omega/\Omega_{\text{ref}}$ is the whirl-to-spin ratio for the reference shaft, then

$$\rho_k = \frac{\omega}{\Omega_k} = \sigma_k \rho_{\text{ref}} \tag{2.199}$$

is the whirl-to-spin ratio for the kth typical shaft with $\sigma_k = \Omega_{\text{ref}}/\Omega_k$. The eigenvalues ω_j are then evaluated in a manner equivalent to whirl speed evaluation, and the critical speeds, related to the reference shaft, are $(n_{\text{ref}})_j = (1/\rho_{\text{ref}})\,|\omega_j|$.

Unbalance Response. For systems with general linear damped bearings and interconnections, the steady-state unbalance response is determined by utilizing "extended" state vectors and associated transfer matrices. The extended state is defined as the state of (2.173) with a unit element adjoined to form

$$\{\underline{\bar{z}}\}_{9\times 1} = [\bar{u}\ \ \bar{v}\ \ \bar{\alpha}\ \ \bar{\beta}\ \ \bar{\tau}_y\ \ \bar{\tau}_x\ \ \bar{f}_y\ \ \bar{f}_x\ \ 1]^T \tag{2.200}$$

The overbar denotes the complex amplitude of the response with $u = \mathrm{Re}\,(\bar{u}e^{i\Omega t})$ etc. The underbar denotes the extended state. The associated extended point and field transfer relations, are, respectively,

$$\{\underline{\bar{z}}_i\}^R = [\underline{P}_i]\{\underline{\bar{z}}_i\}^L \tag{2.201}$$

and

$$\{\underline{\bar{z}}_{i+1}\}^L = [\underline{F}_i]\{\underline{\bar{z}}_i\}^R \tag{2.202}$$

with

$$[\underline{P}_i]_{9\times 9} = \left[\begin{array}{c|c} [P_i] & \begin{matrix} 0 \\ \vdots \\ 0 \\ im_i\Omega^2(a_{x'} + ia_{y'}) \\ -m_i\Omega^2(a_{x'} + ia_{y'}) \end{matrix} \\ \hline [0 \cdots 0] & 1 \end{array}\right] \tag{2.203}$$

and

$$[\underline{F}_i]_{9\times 9} = \begin{bmatrix} [F_i]_{8\times 8} & \{0\} \\ \{0\}^T & 1 \end{bmatrix} \tag{2.204}$$

The extended interconnection relation between shafts A and B, replacing (2.189) is

$$\begin{Bmatrix} {}^A\{\underline{\bar{z}}_i\}^R \\ {}^B\{\underline{\bar{z}}_j\}^R \end{Bmatrix} = \begin{bmatrix} {}^A[\underline{P}_i] + [\underline{C}_{ij}] & -[\underline{C}_{ij}] \\ -[\underline{C}_{ij}] & {}^B[\underline{P}_j] + [\underline{C}_{ij}] \end{bmatrix} \begin{Bmatrix} {}^A\{\underline{\bar{z}}_i\}^L \\ {}^B\{\underline{\bar{z}}_j\}^L \end{Bmatrix} \tag{2.205}$$

with

$$[\underline{C}_{ij}] = \begin{bmatrix} [C_{ij}] & \{0\} \\ \{0\}^T & 1 \end{bmatrix} \tag{2.206}$$

For a multishaft system, the system unbalance response is determined by successively utilizing the unbalance excitation from each shaft. The total response is then the superposition of the separate shaft responses.

For a specified spin speed of the shaft under consideration, the system transfer matrix is determined as before except that the "extended state" relations (2.201), (2.202), and (2.205) are utilized. The result is a set of equations, as illustrated below for a dual-shaft system, which relate the system boundary states.

$$\begin{Bmatrix} {}^A\{\underline{\bar{z}}_n\}^R \\ {}^B\{\underline{\bar{z}}_m\}^R \end{Bmatrix} = [\underline{U}] \begin{Bmatrix} {}^A\{\underline{\bar{z}}_1\}^L \\ {}^B\{\underline{\bar{z}}_1\}^L \end{Bmatrix} \tag{2.207}$$

After the introduction of the specified boundary conditions, it is convenient to rearrange and partition (2.207) as follows:

$$\begin{Bmatrix} \{\bar{z}\}_u^R \\ \{0\} \\ 1 \end{Bmatrix} = \begin{bmatrix} [\underline{U}_{11}] & [\underline{U}_{12}] & \{\underline{U}_{13}\} \\ [\underline{U}_{21}] & [\underline{U}_{22}] & \{\underline{U}_{23}\} \\ [\underline{U}_{31}] & [\underline{U}_{32}] & \underline{U}_{33} \end{bmatrix} \begin{Bmatrix} \{\bar{z}\}_u^L \\ \{0\} \\ 1 \end{Bmatrix} \tag{2.208}$$

where the u subscript refers to the unspecified states. For a well-posed problem, the unspecified state at the left boundary points is determined from the second row of (2.208), i.e.,

$$\{\underline{\bar{z}}\}_u^L = -[\underline{U}_{21}]^{-1}\{\underline{U}_{23}\} \tag{2.209}$$

With the unspecified states known at the left boundary points from (2.209), it is then possible to determine the response at all other rotor points by marching across the system, using the appropriate transfer relations (2.201), (2.202), and (2.205). The process is then repeated as required at different spin speeds to complete an unbalance response survey for the excitation shaft under consideration. The response due to other shaft unbalances is obtained in the same manner, and the results are superposed to produce the total system unbalance response.

Support Structure Dynamics. The inclusion of the support structure dynamics, in utilizing the TMM of analysis, is not as straightforward as for the DSM. An exception exists, however, if the coupling through the support structure between support points is negligible. In this situation, a single pedestal mass with damping and stiffness to ground may be included at each support point. These masses may each be treated as additional single-point shafts, and the analysis may proceed as discussed above for multishaft systems. The pedestal mass damper stiffness may also be replaced by a dynamic stiffness element where the dynamic stiffness is determined from analysis or experiment. For more detailed support structure models, the reader is referred to Wang and Lund (1984) and Li et al. (1981).

2.3.3 Nonlinear Systems

In the previous sections of this chapter, the emphasis is on rotordynamic systems whose dynamic characteristics can be expressed by a set of linear differential equations. Large-order systems with considerable complexity were modeled by using either the direct stiffness method or the transfer matrix method to form a set of coupled ordinary differential equations. When these equations are linear, many standard procedures are available for obtaining a solution. The application of these procedures may be computationally quite intensive and on occasion may be numerically sensitive; however, a solution is always possible, at least in theory. Linear systems possess the property of superposition, their solutions are unique, and the frequency content of the response is identical with that of the excitation. When the set of differential equations is nonlinear, analytical solutions are generally not possible. Those special cases with known closed-form solutions are usually weakly nonlinear and of small order. For large-order nonlinear problems, such as multishaft flexible rotor systems, only a few options are available to the analyst for use in obtaining solutions. Exact solutions are not possible, except in very special cases, and approximate solutions can only be obtained numerically. Simulation of the dynamic behavior by numerical integration of the differential equations of motion is the primary approach used by practicing engineers. Several options for implementing such a simulation by numerical integration are discussed below. For rotordynamic systems, the unbalance excitation of the rotating assemblies is quite often the dominant form of excitation. There are several procedures, other than numerical integration, that have been utilized for searching for the possible response of large-order systems to harmonic excitation, and these are also briefly discussed and referenced below.

Note that nonlinear systems generally possess characteristics which are totally different from those of linear systems. In particular, the method of superposition is not valid; thus all excitation sources of interest must be considered simultaneously. Multiple solutions may exist for certain areas of the parameter space, and the stability nature of each solution may be different. The possibility always exists for jump phenomena or bifurcation of the solutions to occur as the system parameters, e.g., spin speed, are varied. Chaotic behavior is another characteristic of nonlinear systems which, although not presently well documented for rotordynamic systems, should be kept in mind by the analyst.

Most of the components that comprise a rotordynamic system can be quite accurately modeled as linear, as discussed in previous sections of this chapter. The strongly nonlinear components of the system, if present, are usually localized in the sense that they are directly coupled to only a small number of the system coordinates. Such components include fluid-film bearings, squeeze-film dampers, piecewise linear supports (e.g., deadbands, rubs), working fluid interaction, etc. The number of nonlinear components in most applications is usually small. This fact often makes it advantageous to partition the system displacement vector $\{q\}$ into two subvectors; $\{q_2\}$ which is usually of small order and contains the system displacements associated with the nonlinear components and $\{q_1\}$ which is the complement to $\{q_2\}$. The opportunity then exists to utilize condensation procedures which relate the motion of these two subvectors. When this is done, there is often a substantial reduction in computational effort required for most simulations.

Numerical Integration

The direct stiffness method and the transfer matrix method were discussed earlier as two separate approaches for studying rotordynamic systems, and several analytical procedures were presented for these methods. The emphasis in this section is on the use of the direct stiffness method for simulation of nonlinear rotordynamic systems. The usual strategy is to generalize the system equation (2.129) by including a nonlinear force vector $\{Q_N\} = \{Q_N(\{q_2\}, \{\dot{q}_2\})\}$ on the right-hand side of the equation. Thus, the nonlinear system equations are of the form

$$[M]\{\ddot{q}\} + [\Delta]\{\dot{q}\} + [K]\{q\} = \{Q\} + \{Q_N\} \tag{2.210}$$

and $\{Q_N\}$ is treated as a pseudo force vector.

As an illustration of (2.210), consider the dual-shaft system of Fig. 2.25 with the addition of a nonlinear intershaft bearing between station 4 on shaft 1 and station 10 on shaft 2. For this example, consider the bearing to be undamped with the radial bearing force modeled by

$$f = k_1\delta + k_3\delta^3 \tag{2.211}$$

with $\delta = +[(u_{10} - u_4)^2 + (v_{10} - v_4)^2]^{1/2}$ representing the radial deformation of the bearing. The x and y force components are then

$$\begin{aligned} Q_{u_4} &= \frac{u_{10} - u_4}{\delta} f = -Q_{u_{10}} \\ Q_{v_4} &= \frac{v_{10} - v_4}{\delta} f = -Q_{v_{10}} \end{aligned} \tag{2.212}$$

and the 40×1 nonlinear force vector for this case is

$$\{Q_N\} = \begin{Bmatrix} \{0\}_{12\times 1} \\ \begin{Bmatrix} 0 \\ 0 \\ Q_{u_4} \\ Q_{v_4} \end{Bmatrix}_{4\times 1} \text{------}\leftarrow \text{station 4} \\ \{0\}_{20\times 1} \\ \begin{Bmatrix} 0 \\ 0 \\ Q_{u_{10}} \\ Q_{v_{10}} \end{Bmatrix}_{4\times 1} \text{------}\leftarrow \text{station 10} \end{Bmatrix}_{40\times 1} \tag{2.213}$$

With the initial conditions $\{q(0)\}$ and $\{\dot{q}(0)\}$ and force vector $\{Q\}$ specified, the system physical coordinate response can be obtained by numerical integration of (2.210) by using step-by-step procedures such as the Wilson θ method or the Newmark β method (see Craig, 1981, pp. 452–463).

Equation 2.210 is in second-order form in terms of the physical coordinates. An alternative approach is to convert the set of second-order equations (2.210) to first-order form as typified by (2.139). With the addition of a pseudo force vector, (2.139) becomes

$$[A]\{\dot{x}\} + [B]\{x\} = \{X\} + \{X_N(\{x_2\})\} \tag{2.214}$$

with

$$\{X_N(\{x_2\})\} = \begin{Bmatrix} \{0\} \\ \{Q_N\} \end{Bmatrix} \tag{2.215}$$

The vector $\{x_2\} = \lfloor \{\dot{q}_2\}^T \; \{q_2\}^T \rfloor^T$ contains the state variables associated with the nonlinear components of the system. Equation 2.214 may then be numerically integrated by using step-by-step procedures for systems of first-order ordinary differential equations, e.g., the fourth-order Runge–Kutta method.

A primary disadvantage of direct integration in terms of the physical coordinates such as with the second-order form of (2.210) or the first-order form of (2.214) is that the computational effort required for a simulation is often unreasonably large. For the illustration associated with the 40-degree-of-freedom system shown in Fig. 2.25, (2.210) is of order 40 and (2.214) is of order 80.

One approach for reducing the number of nonlinear ordinary differential equations which must be integrated is to include the nonlinear pseudo force vector in a set of transformed system equations in terms of modal coordinates. Equation 2.154, which utilizes real modes, would then be extended to

$$\begin{bmatrix} \diagdown & & \\ & \hat{\mu} & \\ & & \diagdown \end{bmatrix} \{\ddot{\eta}\} + [\hat{\varphi}]^T([D] + [G])[\hat{\varphi}]\{\dot{\eta}\} + \left(\begin{bmatrix} \diagdown & & \\ & \hat{\kappa} & \\ & & \diagdown \end{bmatrix} + [\hat{\varphi}]^T[C][\hat{\varphi}] \right)\{\eta\}$$
$$= [\hat{\varphi}]^T\{Q\} + [\hat{\varphi}]^T\{Q_N\} \tag{2.216}$$

By means of modal transformation (2.153), the nonlinear force $\{Q_N\}$ can be expressed in terms of the modal coordinates η_i, $i = 1, 2, \ldots, \hat{n}$. From the modal transformation (2.153) and the orthogonality condition (2.152), the initial conditions for the modal coordinates may be written as

$$\begin{aligned} \eta_i(0) &= \frac{1}{\mu_i}\{\varphi_i\}^T[M]\{q(0)\} \\ \dot{\eta}_i(0) &= \frac{1}{\mu_i}\{\varphi_i\}^T[M]\{\dot{q}(0)\} \end{aligned} \tag{2.217}$$

With the initial coordinates $\{q(0)\}$ and $\{\dot{q}(0)\}$ and the force vector $\{Q\}$ specified, the modal coordinate response $\{\eta(t)\}$ is then obtained by numerical integration of the set of second-order equations (2.216), and the physical coordinate response is obtained by back transformation using (2.153). When the number of retained real modes $\hat{n}$ is much less than the system order n, the computational effort required for a dynamic simulation is substantially reduced. Considerable caution should be utilized in the choice of modes to be included in the transformation of (2.153). The frequency content of the excitation vector $\{Q\}$ and the location of the nonlinear components are two factors which might strongly influence the choice.

Another option for reducing the order of the set of nonlinear equations is to utilize the modal transformation in terms of complex modes as given by (2.155). The inclusion of the nonlinear pseudo force vector $\{Q_N\}$ in modal equations (2.156) then yields

$$N_i\dot{\zeta}_i - \lambda_i N_i \zeta_i = \mathrm{Z}_i + \mathrm{Z}_{N_i}(\{\zeta\}) \qquad i = 1, 2, \ldots, 2\hat{n} \le 2n \tag{2.218}$$

where

$$\mathrm{Z}_{N_i} = \{v_i\}^T\{Q_N\} \tag{2.219}$$

is a pseudo-nonlinear modal force. The force vector $\{Q_N\}$ is a function of the physical states which are directly coupled to the nonlinear components. From the transformation (2.155), the Z_{N_i} are then clearly nonlinear functions of the complex modal coordinates. The initial values of the modal coordinates from (2.155), using the biorthogonality relation (2.144), are

$$\zeta_i(0) = \frac{1}{N_i}\{z_i\}^T[A]\{x(0)\} \qquad i = 1, 2, \ldots, 2\hat{n} \tag{2.20}$$

Given the initial system state $\{x(0)\} = \lfloor\{\dot{q}(0)\}^T\ \{q(0)^T\rfloor^T$ and the physical coordinate excitation $\{Q\}$, Eq. 2.218 may be integrated numerically to determine the complex modal coordinate responses $\zeta_i(t)$, $i = 1, 2, \ldots, 2n$. The physical coordinate response is then obtained by back transformation using (2.155). It is important here, as for any modal approach, to use care in selecting the modes used in the transformation.

Nonlinear components can also be included in systems which are analyzed by various substructure modal synthesis strategies. The system equations (2.171) are based on a substructuring approach which utilizes the superposition of static constraint modes and fixed interface normal modes to represent the absolute motion of the substructure. This approach is particularly convenient for including localized nonlinear components at the boundary coordinates of the substructure. The motion of each nonlinear component is expressed in terms of the physical boundary coordinate motion and is not directly dependent upon the fixed interface modes of the substructures. Details of this approach may be found in Nelson et al. (1983).

Steady Periodic Response

The dominant loads in most rotordynamic systems are usually constant side loads and steady rotating unbalance loads at the shaft spin speeds. The response of the nonlinear system due to this type of loading is normally a steady periodic response or responses, if multiple solutions are present. This general situation is quite difficult to analyze, and most analysts utilize numerical integration procedures, such as discussed above, for simulating the system dynamics.

A special situation of steady motion exists if the stiffness and damping characteristics of all system components are isotropic and if the excitation is due only to rotating unbalance at a single frequency. In this case, the whirling response consists of circular orbits centered about the undeformed bearing centerline and whirls in circular orbits with a frequency which is generally synchronous with the rotating unbalance. Such an analysis is often called a *synchronous centered circular orbit analysis* of the rotordynamic system, and it is useful for designing components such as squeeze-film dampers.

For simplicity of presentation, the development that follows is restricted to systems where the nonlinear components, e.g., squeeze-film dampers, interconnect a station on a shaft and a rigid support. The generalization is straightforward to allow for intershaft connections and connections to compliant support structures. The system equations of motion, excluding the nonlinear components, are of the form of (2.129). The excitation is represented by (2.146) with only one frequency, Ω, present. The introduction of the coordinate transformation (2.14) into the fixed reference equations of motion (2.129), with $\theta = \Omega t$ and consideration restricted to a constant response in the rotating reference (circular in the fixed reference), gives

$$[H(\Omega)]\{q'\} = \{Q'\} \tag{2.221}$$

where

$$[H(\Omega)] = [K] - \Omega^2([M] + [\hat{G}]) \tag{2.222}$$

is the symmetric dynamic stiffness with $\Omega[\hat{G}] = [T]^T[G][\dot{T}]$. The unbalance force $\{Q'\}$ in the rotating $x'y'$ reference is constant.

It is convenient to rearrange and partition the physical coordinates vector $\{q'\}$ into a subvector $\{q_2'\}$ which contains the physical coordinates associated with the nonlinear components and $\{q_1'\}$ which is the complement to $\{q_2'\}$. The partitioned form of (2.221), including the nonlinear forces, is

$$\begin{bmatrix} [H_{11}] & [H_{12}] \\ [H_{21}] & [H_{22}] \end{bmatrix} \begin{Bmatrix} \{q_1'\} \\ \{q_2'\} \end{Bmatrix} = \begin{Bmatrix} \{Q_1'\} \\ \{Q_2'\} + \{P_2'\} \end{Bmatrix} \tag{2.223}$$

The vector $\{P_2'(\{q_2'\})\}$ contains the nonlinear force components which act on the rotor at the locations of the nonlinear components (see McLean and Hahn, 1983). From the first row of (2.223),

$$\{q_1'\} = [H_{11}]^{-1}(\{Q_1'\} - [H_{12}]\{q_2'\}) \tag{2.224}$$

which, when substituted into the second row, gives

$$[H_r]\{q_2'\} + \{Q_r\} - \{P_2'(\{q_2'\})\} = \{0\} \tag{2.225}$$

where

$$[H_r] = [H_{22}] - [H_{21}][H_{11}]^{-1}[H_{12}] \tag{2.226a}$$

$$\{Q_r\} = [H_{21}][H_{11}]^{-1}\{Q_1'\} - \{Q_2'\} \tag{2.226b}$$

represent a reduced dynamic stiffness matrix and unbalance force vector, respectively. Equation 2.225 is a set of nonlinear algebraic equations in terms of the elements of $\{q_2'\}$. For many practical systems the order of $\{q_2'\}$ is usually small, so that (2.225) can be solved quite efficiently by using nonlinear equation solvers available in most computer software libraries.

Once a solution is located for $\{q_2'\}$, the complementary set of unknowns $\{q_1'\}$ can be obtained from (2.224). In some regions of the system parameter space, multiple solutions may exist. Thus, caution must be exercised in solving (2.225) to be certain that all solutions of interest have been found. The choice of different starting values for the unknowns may cause the search process to converge to different solutions. The stability characteristics of each solution are generally different, and these should be examined as part of the analysis. The usual approach is to investigate the stability of motion in the small by studying the characteristics of perturbed motions in the neighborhood of a steady solution (see Meirovitch, 1980, 223–226).

2.4 APPLICATION AND INTERPRETATION OF RESULTS

The analytical techniques presented above play an important role in virtually all phases of preliminary design, detailed design, development, production (i.e., balancing), and operation (i.e., fault monitoring, detection, and diagnosis) of rotating machinery.

2.4.1 Preliminary Design

As implied in Sec. 1.3, a key design focus in the preliminary design stage is the placement of critical speeds with respect to the operating speed range for the machine. Since unbalance is usually the most important excitation source in a high-speed machine, the forward synchronous critical speeds, $\rho = +1$ in (2.145), are the speeds of primary interest in a preliminary design. For these purposes a whirl speed map, such as illustrated in Fig. 2.29, is usually generated for a reference spin speed range from zero to a speed which is above, e.g., 25 percent, the maximum operating speed of the unit. By this process the system's natural frequencies and associated mode shapes are identified. In a preliminary design, the system damping is not usually well established, and the general practice is to consider undamped system models. The effects of damping are then considered in the detailed design stage.

In the preliminary design, the system parameters need to be adjusted so as to change the system natural frequencies of whirl and thereby locate the critical speeds as required. The parameters which have the most influence in this regard are the support stiffnesses (bearings and support structure), support locations, and the mass and stiffness distribution of the rotating assemblies. It is often useful to plot the system critical speeds versus a particular design parameter to obtain a

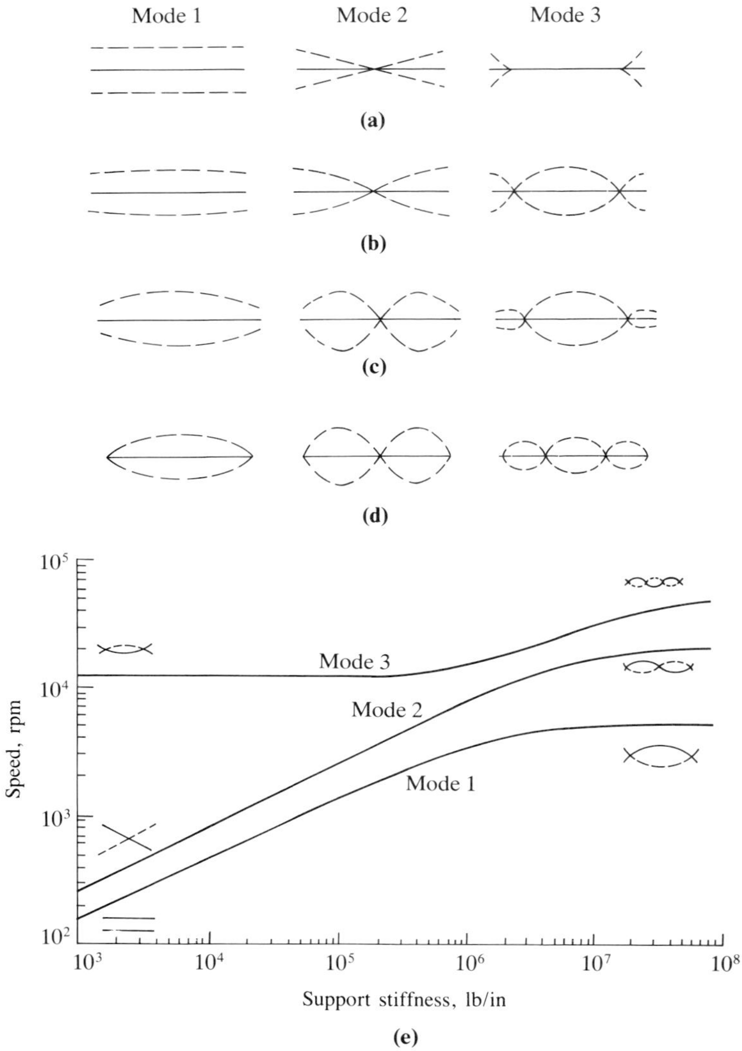

FIGURE 2.36 Critical speed map. (*a*) Rigid rotor in flexible bearings; (*b*) stiff rotor in flexible bearings (minor bending in first two modes); (*c*) flexible rotor in flexible bearings (substantial bending in first two modes); (*d*) flexible rotor in rigid bearings (no rigid-body modes); (*e*) influence of bearing stiffness on flexible rotor modes.

graphical description of the sensitivity of the critical speeds to this parameter. Such a plot is called a *critical speed map,* and one is illustrated in Fig. 2.36 with an isotropic support stiffness k representing a design variable. Several of these types of plots can be generated, and they often prove useful in the design process. They also may be utilized in attempting to improve mathematical models based on experimental results from an actual rotor system.

It is also quite useful to generate energy maps of the system which display the kinetic energy and/or potential (strain) energy distribution associated with a natural mode of whirl. For this purpose the system damping is neglected, and the supports are considered as isotropic. The shaft whirl modes are then circular, and the energy distribution remains constant throughout the orbit. Since the amplitude of a free whirl mode is arbitrary, the energy distribution is usually displayed as a percentage of the total energy of the system for the mode of interest. A typical energy map is displayed in Fig. 2.37. A whirl mode with a large

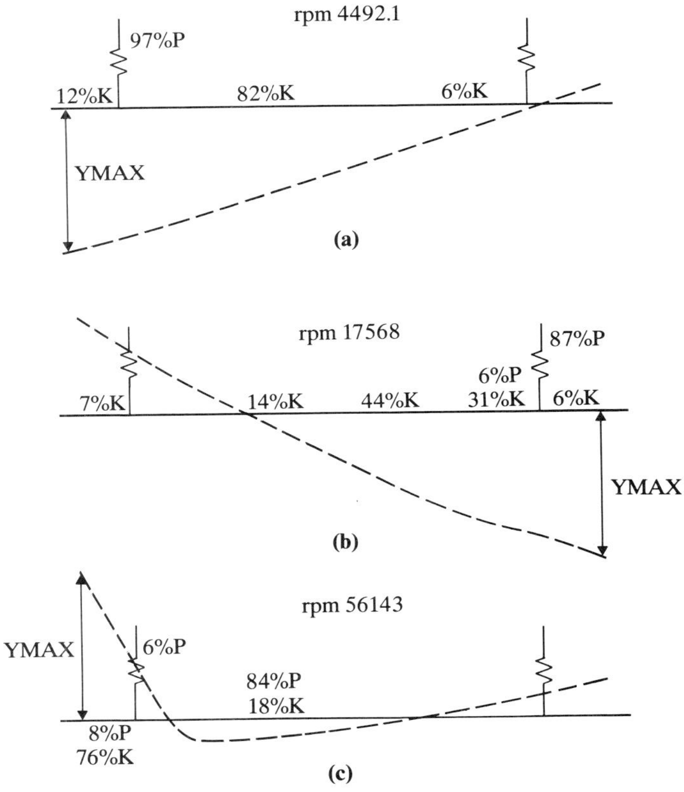

FIGURE 2.37 Energy distribution map: P = potential or strain energy; K = kinetic energy.

percentage of the total strain energy in the supports, e.g., >70 percent, is characterized as a rigid rotor mode, while a mode with a large percentage of strain energy in the rotating assembly is characterized as a flexible rotor mode. System modes with a large percentage of the strain energy in the static support structure are usually referred to as *support structure modes,* and these modes are generally not strongly excited by normal unbalance loads in the rotating assemblies.

The energy maps are useful in refining a design so as to move critical speeds out of the operating speed range or to raise the first critical speed to reduce the susceptibility of instability of the mode. The whirl speeds are generally quite sensitive to changes in bending stiffness in regions of high curvature which are usually associated with regions of high potential energy density. If the amplitude of motion at a support point is large for a particular mode, the associated whirl frequency will be quite sensitive to changes in support stiffness at that location. Decreasing (increasing) mass locally to increase (decrease) critical speeds is most effective in regions of large kinetic energy density. These regions normally correspond to large-amplitude motion. The critical speeds are most sensitive to "gyroscopic stiffening" [see (2.78)] in regions of large angular motion.

The above concepts can be automated to a large extent by utilizing parameter sensitivity coefficients (Lund, 1979; Rajan et al., 1986) in conjunction with an optimization routine. It is possible to utilize these ideas for optimal placement of the critical speeds (Rajan et al., 1987) and for other analyses which require the adjustment or determination of system dynamic characteristics due to changes in the design parameters.

2.4.2 Detailed System Design

For the initial phases of a detailed design, response maps such as shown in Fig. 2.38 become useful. These response maps generally are related to the displacement (absolute or relative) at important locations of the system, bearing loads, and/or internal forces at crucial cross sections of the rotating assemblies. The system damping should be estimated as soon as possible in the design, and the response maps associated with realistic unbalance distributions should be generated. These results can then be compared with allowable deflections, clearance

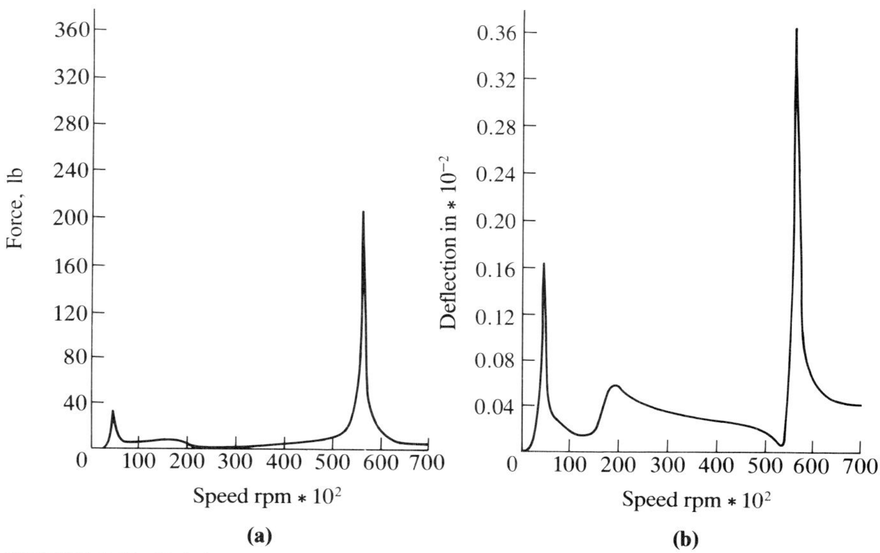

FIGURE 2.38 Unbalance response plots (5 g · in unbalance at midspan): (*a*) force on left bearing; (*b*) displacement at midshaft.

closures, bearing loads, and internal forces as required by the design standards of the industry. The response maps are clearly a function of the unbalance distribution utilized in the simulation. The usual practice is to determine limits on the expected production mass center eccentricities and then establish worst-case unbalance distributions which will strongly excite the modes associated with the forward synchronous critical speeds located in the operating speed range. Unbalance distributions which are of the same general shape as a mode tend to produce a large modal unbalance force for that mode. Thus, the worst-case unbalance distribution for one mode may be quite different from that of another mode.

Particular attention should be paid to regions of the system that must run with close rotor-stator clearance. These regions should be screened to avoid large clearance closure, particularly at speeds in the neighborhood of the critical speeds that must be traversed in the operation of the machine. If dampers, e.g., squeeze-film dampers, are to be included, they should be located in regions where there is significant relative rotor-stator motion, and their effectivity must be quantitatively assessed and optimized. The effectiveness of isolation systems must be assessed and optimized to prevent transmission of vibration from the machine to the environment or in isolating the machine from excitation by environmental stimuli. Guidance must be provided in identifying locations on the static structure which have low-amplitude response where auxiliary equipment can be safely mounted. Locations with large-amplitude motion, particularly for all the modes encountered, must be identified as appropriate for mounting fault monitoring and detection transducers and, if the rotor is to be balanced in situ, for locating sensors to be used in the balancing operation. Machines that will be mounted in a variety of configurations must be evaluated at all the extremes of installation to ensure smooth, safe, and reliable operation for all circumstances.

As soon as it is possible to identify and quantify the major sources of damping, including cross-coupling terms such as discussed in the bearings subsection of Sec. 2.2.3, a stability survey of the system should be conducted. The damping coefficients as well as other coefficients of the system are a function of spin speed, thus a whirl speed map such as displayed in Fig. 2.29 should be developed. The modal damping coefficients σ_i then provide information regarding the level of damping of each mode and the possibility of a particular mode becoming unstable, that is, $\sigma_i > 0$. Since damping coefficients are usually difficult to estimate to high accuracy, the sensitivity of the system dynamic characteristics should be examined so as to provide appropriate stability margins.

In the final stages of design, it is useful to conduct a complete review of the system dynamic characteristics to ensure that no major areas of concern have been overlooked. The most refined model, with the best estimates of the system parameters, should be used in this step. If the parameter values are known within upper and lower bounds, the limiting cases should be included so as to evaluate the variation in the system behavior over the range of design parameters. This review should include prediction of the system natural frequencies and modes of whirl and the stability of those modes. The critical speeds should be identified with particular emphasis given to the forward synchronous speeds which would be strongly excited by rotating unbalance. A complete study of the unbalance response of the system due to worst-case unbalance distributions should be made including displacements, support loads, and internal forces at the critical regions of the system. Additional studies may be required such as associated with the transient response of the system due to a blade loss event or to an aircraft maneuver or hard landing.

2.5 NOMENCLATURE

A = amplitude, area
$\{A\}$ = typical vector
$[A]$, $[B]$ = state matrices
a = mass center component
a, b = semimajor, semiminor axes
b = $c + \lambda k$
$[B]$ = connectivity matrix
c = damping coefficient
$[c]$, $[C]$ = damping matrix
$[C]$ = circulation matrix
$[D]$ = dissipation matrix
E = Young's modulus
e = Naperian base
f = force
$\bar{f}$ = complex force
$\{f\}$ = force vector
$[F]$ = field transfer matrix
g = acceleration of gravity
$[g]$ = gyroscopic matrix
G = shear modulus
$[G]$ = gyroscopic matrix
$[\hat{G}]$ = transformed gyroscopic matrix
H = angular momentum, dynamic stiffness
$\{H\}$ = angular momentum vector
$[H]$ = dynamic stiffness matrix
i = $\sqrt{-1}$
I = mass, area moment of inertia
$[I]$ = inertia, identity matrix
Im = imaginary part
k = stiffness coefficient
$[k]$ = stiffness matrix
$[K]$ = stiffness matrix
L = length, lowest common multiple
m = mass coefficient
$[m]$ = mass matrix
M = moment
$[M]$ = mass matrix
n = critical speed, subharmonic order
n_σ = number of shafts
N = norm
$\{P\}$ = linear momentum vector
$[P]$ = point transfer matrix
$\{q\}$ = displacement vector for x, y, z
$\{q'\}$ = displacement vector for x', y', z'
Q = shear force
$\{Q\}$ = force vector for x, y, z
r, R = radii
Re = real part
$\{r\}$, $\{R\}$ = position vectors
$[S]$ = stiffness matrix
t = time
T^* = kinetic coenergy
$[T]$ = rotation transformation matrix
u, v = x, y translations
$\{u\}$ = unit vector
$\{u\}$, $\{v\}$ = right, left displacement vectors
$[U]$ = overall transfer matrix
V = potential energy, velocity
w = z displacement
$\bar{w}$ = complex lateral displacement
W = work
$\{x\}$ = state vector
$\{X\}$ = state force vector
$\{y\}$, $\{z\}$ = right, left state vectors
$[Y]$ = right vector matrix
x, y, z = stationary reference
x', y', z' = rotating reference

$\{z\}$ = state vector
Z = generalized complex force
$[Z]$ = left vector matrix

α = coefficient
α, β = y, ξ rotations
$\bar{\gamma}$ = complex rotation angle
$[\gamma]$ = constraint transformation matrix
δ = variational operator, kronecker delta, displacement
$[\Delta]$ = (dissipation + gyroscopic) matrix
Δt = time increment
ε = $EI/(\kappa GAL)$
ζ = complex modal coordinate
$\{\zeta\}$ = complex modal coordinate vector
η = modal coordinate
$\{\eta\}$ = modal coordinate vector
θ = rotation angle
κ = shape factor, modal stiffness
λ = eigenvalue $(\sigma + j\omega)$
μ = modal mass, coefficient
ν = Poisson's ratio, coefficient
ξ, η, ζ = body reference
ξ = generalized coordinate
Ξ = generalized force
π = pi
ρ = mass density, ω/Ω
σ = damping exponent, normal stress
τ = torque
$\bar{\tau}$ = complex torque
φ = phase angle, rotation angle, shape function
$\{\varphi\}$ = normal mode vector
$[\chi]$ = coordinate transformation matrix
ψ = attitude angle
$[\Psi]$ = coordinate transformation matrix
ω = whirl speed
$\{\omega\}$ = angular velocity vector
Ω = spin speed

Subscripts

a = axial, active
A = amplitude
b = backward
B = reference point, boundary
c = cosine, centering, interconnection
d = disk, diametral, dependent
exc = excitation
f = forward
g = gravity
i, j, k = indices
I = interior
L = left
N = nonlinear
o = left end, lowest
p = prescribed
r = radial
r, s = indices
R = rotational, right
ref = reference
s = sine, static
T = translational
u = unbalance
u, v = x, y components
xyz = relative to xyz reference
x, y, z = x, y, z components
x', y', z' = x', y', z' components
1, 2, 3 = principal axes
$\xi\eta\zeta$ = relative to ξ, η, ζ reference
ξ, η, ζ = ξ η ζ components
σ = shaft index
θ = hoop

Superscripts

A, B = shaft indices
b = bearing
c = coupling
i = station number
R, L = right, left
T = transpose
$\hat{}$ = truncated
$\dot{}$ = d/dt
$\bar{}$ = complex quantity
$'$ = x', y', z' components

2.6 REFERENCES

Adams, M., 1980, "Nonlinear Dynamics of Multibearing Flexible Rotors," *Journal of Sound and Vibration,* 71(1): 129–144.

———, and J. Padovan, 1981, "Insights into Linearized Rotor Dynamics," *Journal of Sound and Vibration,* 76: 129–142.

Bathe, K. J., 1982, *Finite Element Procedures in Engineering Analysis,* Prentice-Hall, Englewood Cliffs, N.J.

Childs, D. W., 1974, "A Rotor-Fixed Modal Simulation Model for Flexible Rotating Equipment," *Journal of Engineering for Industry, Transactions, ASME,* 96(2): 659–669.

———, 1976, "A Modal Transient Rotordynamic Model for Dual-Rotor Jet Engine Systems," *Journal of Engineering for Industry, Transactions, ASME* 98(3): 876–882.

———, and K. Graviss, 1982, "A Note on Critical-Speed Solutions for Finite-Element Based Rotor Models," *Journal of Mechanical Design, Transactions ASME,* 104(2): 412–416.

Craig, R. R., Jr., 1981, *Structural Dynamics,* Wiley, New York.

Crandall, S. H., 1962, "Rotating and Reciprocating Machines," *Handbook of Engineering Mechanics* (W. Flugge, ed.), McGraw-Hill, New York, Chap. 58.

———, D. C. Karnopp, E. F. Kurtz, Jr., and D. C. Pridmore-Brown, 1982, *Dynamics of Mechanical and Electromechanical Systems,* Krieger Publishing Co., Malabar, Fla.

Den Hartog, J. P., 1956, *Mechanical Vibrations,* McGraw-Hill, New York.

Dennis, A. J., R. H. Erikson, and L. H. Seitelman, 1975, "Transient Response Analysis of Damped Rotor Systems by the Normal Mode Method," ASME Paper 75-GT-58, Gas Turbine Conference, Houston.

Dimaragonas, A. D., and S. A. Paipetis, 1983, *Analytical Methods in Rotor Dynamics,* Applied Science Publishers, London.

Dimentberg, F. M., 1961, *Flexural Vibrations of Rotating Shafts,* Butterworths, London.

Fawzy, I., and R. E. D. Bishop, 1976, "On the Dynamics of Linear Non-Conservative Systems," *Proceedings of the Royal Society of London,* A352: 25–40.

Genta, G., and A. Gugliotta, 1988, "A Conical Element for Finite Element Rotor Dynamics," *Journal of Sound and Vibration,* 120(2): 175–182.

Glasgow, D. A., and H. D. Nelson, 1980, "Stability Analysis of Rotor-Bearing Systems Using Component Mode Synthesis," *Journal of Mechanical Design, Transactions ASME,* 102(2): 352–359.

Greenhill, L. M., W. B. Bickford, and H. D. Nelson, 1985 "A Conical Beam Finite Element for Rotor Dynamics Analysis," *Journal of Vibration, Acoustics, Stress, and Reliability in Design, Transactions ASME,* 107(4): 421–430.

Gunter, E. J., Jr., 1966, "Dynamic Stability of Rotor-Bearing Systems," NASA SP-113, 29, Office of Technical Utilization, US Government Printing Office, Washington, D.C.

Guyan, R. J., 1965, "Reduction of Stiffness and Mass Matrices," *AIAA Journal,* 3(2): 380.

Holzer, H., 1921, "Die Berechnung de Drehschwingungen," *Julius Springer,* Berlin.

Irons, B. M., 1965, "Structural Eigenvalue Problems: Elimination of Unwanted Variables," *AIAA Journal,* 3(5): 961–962.

Jeffcott, H. H., 1919, "The Lateral Vibration of Loaded Shafts in the Neighborhood of a Whirling Speed—The Effect of Want of Balance," *Philosophical Magazine,* 6(37): 304–314.

Kirk, R. G., and E. J. Gunter, 1974, "Transient Response of Rotor-Bearing Systems," *Journal of Engineering for Industry,* B96(2): 682–693.

Leung, A. Y. T., 1988, "Inverse Iteration for the Quadratic Eigenvalue Problem," *Journal of Sound and Vibration,* 124(2): 249–267.

Li, D. F., and E. J. Gunter, 1982, "Component Mode Synthesis of Large Rotor System," *J. of Engineering for Industry, Transactions ASME,* 104(3): 552–560.

———, ———, L. E. Barrett, and P. E. Allaire, 1981, "The Dynamic Analysis of Multi-Level Flexible Rotor Systems Using Transfer Matrices and Component Mode Synthesis," Final Report UVA/628145/MAE81/102, University of Virginia, Charlottesville.

Loewy, R. G., and V. J. Piarulli, 1969, *Dynamics of Rotating Shafts,* SVM-4, Naval Publication and Printing Service Office, Washington.

Lund, J. W., 1974a, "Stability and Damped Critical Speeds of a Flexible Rotor in Fluid-Film Bearings," *Journal of Engineering for Industry, Transactions ASME,* 96(2): 509–517.

———, 1974b, "Modal Response of a Flexible Rotor in Fluid Film Bearings," *Journal of Engineering for Industry, Transactions ASME,* 96(2): 525–553.

———, 1979, "Sensitivity of the Critical Speeds of a Rotor to Changes in the Design," *Journal of Mechanical Design, Transactions ASME,* 102: 115–121.

——— and F. K. Orcutt, 1967, "Calculations and Experiments on the Unbalance Response of a Flexible Rotor," *Journal of Engineering for Industry, Transactions ASME,* 89(4): 785–796.

McLean, L. J., and E. J. Hahn, 1983, "Unbalance Behavior of Squeeze Film Damped Multi-Mass Flexible Rotor Bearing Systems," *Journal of Lubrication Technology, Transactions ASME,* 105(1): 22–28.

Meirovitch, L., 1980, *Computational Methods in Structural Dynamics,* Sijthoff and Noordhoff, Alphen aan den Rijn, Netherlands.

Murphy, B. T., and J. M. Vance, 1983, "An Improved Method for Calculating Critical Speeds and Rotordynamic Stability of Turbomachinery," *Journal of Engineering for Power, Transactions ASME,* 105(3): 591–595.

Myklestad, N. O., 1944, "A New Method of Calculating Natural Modes of Uncoupled Bending Vibration of Airplane Wings and Other Types of Beams," *Journal of Areonautical Science,* 11(2): 153–162.

Nelson, H. D., 1980, "A Finite Rotating Shaft Element Using Timoshenko Beam Theory," *Journal of Mechanical Design, Transactions ASME,* 102: 793–803.

——— and J. M. McVaugh, 1976, "The Dynamics of Rotor-Bearing Systems Using Finite Elements," *Journal of Engineering for Industry, Transactions ASME,* 98(2): 593–600.

———, W. L. Meacham, D. P. Fleming, and A. Kascak, 1983, "Nonlinear Analysis of Rotor-Bearing Systems Using Component Mode Synthesis," *Journal of Engineering for Power, Transactions ASME,* 105(3): 606–614.

Oden, J. T., 1967, *Mechanics of Elastic Structures,* McGraw-Hill, New York.

Padovan, J., M. Adams, D. Fertis, I. Zeid, and P. Lam, 1984, "Nonlinear Transient Finite Element Analysis of Rotor Bearing Systems," *Computers and Structures,* 18(4): 629–634.

Prohl, M. A., 1945, "A General Method for Calculating Critical Speeds of Flexible Rotors," *Journal of Applied Mechanics, Transactions ASME,* pp. A142–A148.

Rajan, M., H. D. Nelson, and W. J. Chen, 1986, "Parameter Sensitivity in the Dynamics of Rotor-Bearing Systems," *Journal of Vibration, Acoustics, Stress, and Reliability in Design, Transactions ASME,* 108: 197–206.

———, S. D. Rajan, H. D. Nelson, and W. J. Chen, 1987, "Optimal Placement of Critical Speeds in Rotor-Bearing Systems," *Journal of Vibration, Acoustics, Stress, and Reliability in Design, Transactions ASME,* 109: 152–157.

Rao, J. S., 1983, *Rotor Dynamics,* Wiley, New York.

Rieger, N. F., 1982, *Vibrations of Rotating Machinery, Part I: Rotor-Bearing Dynamics,* The Vibration Institute, Clarendon Hills, Ill.

———, 1986, *Balancing of Rigid and Flexible Rotors,* SVN 12, The Shock and Vibration Center.

Roark, R. J., 1965, *Formulas for Stress and Strain,* 4th ed., McGraw-Hill, New York.

Rouch, K. E., and J. S. Kao, 1979, "A Tapered Beam Finite Element for Rotor Dynamics Analysis," *Journal of Sound and Vibration,* 66(1): 119–140.

To, C. W. S., 1981, "A Linearly Tapered Beam Finite Element Incorporating Shear Deformation and Rotary Inertia for Vibration Analysis," *Journal of Sound and Vibration,* 78(4): 475–484.

Tondl, A., 1965, *Some Problems of Rotor Dynamics,* Chapman and Hall, London.

———, 1973, "Notes on the Identification of Subharmonic Resonances of Rotors," *Journal of Sound and Vibration,* 31(1): 119–127.

Uhrig, R., 1966, "The Transfer Matrix Method as Seen as One Method of Structural Analysis among Others," *Journal of Sound and Vibration,* 4: 136–148.

Vance, J. M., 1988, *Rotordynamics of Turbomachinery,* Wiley, New York.

Wang, Z., and J. W. Lund, 1984, "Calculations of Long Rotors with Many Bearings on a Flexible Foundation," C291/84, *Proceedings of Third International Conference on Vibration in Rotating Machinery, Institute of Mechanical Engineers Conference* Publication, York, England, 1984-10: 13–16.

CHAPTER 3
BALANCING OF RIGID AND FLEXIBLE ROTORS

Dr. Edgar J. Gunter
Professor of Mechanical and Aerospace Engineering
University of Virginia

Charles Jackson, P.E.
Turbomachinery Consultant

3.1 BACKGROUND AND INTRODUCTION TO BALANCING

3.1.1 Introduction

Rotor balancing is a fundamental requirement for the smooth operation of turbomachinery. Ideally, in the operation of all rotating machinery, the inertia axis of the rotor lies along the rotor spin axis. In actuality, this does not occur, and centrifugal forces and moments are generated which can result in high forces transmitted to the bearings and the supporting structure. Excessive rotor unbalance may lead to large amplitudes of motion or even failure of the shaft, bearings, and the foundation. Unbalance and shaft misalignment are recognized as two of the major factors that can lead to machinery malfunction and even catastrophic failure.

The unbalance in a machine may result from its initial manufacturing process or may occur as a result of various operating factors such as machine erosion, thermal effects, and unbalance buildup of process material on impellers and surfaces of the rotor. A rotor will experience some residual unbalance during manufacturing because of machine tolerances and material inhomogeneity. Multistage turborotors such as compressors, pumps, and turbines are susceptible to unbalance due to the assembly of multiple components. In the initial manufacturing of most rotors ranging from simple motors to elaborate pumps, compressors, and aircraft engines, the manufacturers generally employ elaborate procedures to ensure the initial balance of their machinery. The balancing may involve individual balancing of all the components and a final balance of the assembly. These balancing procedures usually involve commercial balancing machines based on either the soft bearing or the hard bearing support concept. These rotors are balanced in two planes to extremely high accuracies. For very flexible rotors, the manufacturer may even employ a high-speed spin pit facility to apply final trim balance.

Once the rotor is placed in service, however, unbalance may occur in the system due to any number of situations. Table 3.1 by Rieger (1986) shows

TABLE 3.1 Possible Causes and Signs of Rotor Unbalance (Rieger, 1986)

Cause of unbalance	Observable signs*
Disk or component eccentric on shaft	Detectable runout on slow rotation (center of gravity runs to bottom on knife-edges)
Dimensional inaccuracies	Measurable lack of symmetry
Eccentric machining or forming inaccuracies	Detectable runout
Oblique-angled component	Detectable angular runout; measured with dial gauge on knife-edges
Bent shaft; distorted assembly; stress relaxation with time	Detectable runout on slow rotation, often heavy vibration during rotation
Section of blade or vane broken off	Visually observable; bearing vibration during operation; possible process pulsations
Eccentric accumulation of process dirt on blade	Bearing vibration
Differential thermal expansion	Shaft bends and throws out center of gravity; source of heavy vibration
Nonhomogeneous component structure; subsurface voids in casting	Rotor machined concentric, bearing vibration during operation; center of gravity runs to bottom on knife-edges
Nonuniform process erosion	Bearing vibration
Loose bolt or component slip	Vibration reappears after balancing because of component angular movement; possible vibration magnitude and phase changes
Trapped fluid inside rotor, possible condensing or vaporizing with process cycle	Vibration reappears after balancing; apparent angular movement of center of gravity occurs; possible vibration magnitude and phase changes
Ball-bearing wear	Bearing vibration; eccentric orbit with possible multiloops; frequency of vibration is 1, 2, or more per revolution

* Unless otherwise indicated, the frequency of vibration is once per revolution.

possible causes and signs of rotor unbalance. After a rotor is placed in service, the procedure for correction balance may be considerably different from the initial process employed by the manufacturer. It may be very expensive and time-consuming to remove the rotor to return to the manufacturer for reworking and balancing.

In many cases, it is necessary to apply an in situ or field balance on the machine in place. The object of the field balancing is to add correction weights in one or more axial planes along the rotor to maintain permissible levels of vibration. Some of these various methods of field balancing are discussed in detail in later sections.

The subject of balancing has been of interest to users of rotating machinery for over 100 years. The literature on the field is extensive, and thousands of references concerning rigid and flexible rotor balancing have been written as well as balancing standards developed by various organizations. Rieger's (1986) authoritative work will be of interest to those who wish to further study the theory and practice of balancing and standards in extensive detail.

3.1.2 Classification of Rotors

The International Standards Organization (ISO) (1973) has issued documentation on rotor classification and balance quality of rotating rigid bodies which are also discussed by Rieger (1986).

Rotors may be classified as either rigid or flexible systems according to their dynamic behavior at operating speeds. The classification of a rotor may be readily determined by performing a critical speed analysis on the system. If the strain or potential energy in the bearings is over 80 percent of the system's total strain energy, then the rotor may be generally classified as rigid. A rigid-body rotor is one which may be balanced in two arbitrary planes. The rotor will appear to maintain balance throughout its operating speed range.

On the other hand, if the strain energy of the shaft begins to exceed 20 percent of the system strain energy and the rotor is operating through one or more critical speeds, then it may be considered a flexible or quasi-flexible rotor. Under these circumstances, a two-plane rigid-body balance may not be adequate, and additional trim weights may have to be placed along the shaft to minimize the vibration amplitude at speed.

The ISO has classified rotors to describe the type and quality of balance needed for each particular instance. Table 3.2 presents the ISO classification of rotors as presented by Rieger (1986). Five basic classifications are presented by the ISO:

Class 1: Rigid rotors: These rotors may be balanced in any two arbitrary axial planes and will remain in balance throughout the operating speed range.

Class 2: Quasi-flexible rotors: These rotors are not perfectly rigid but may be adequately balanced in a low-speed balancing machine and will maintain smooth operation throughout the speed range.

Class 3: Flexible rotors: These rotors cannot be balanced in a low-speed balancing machine and require one or more high-speed trim plane corrections.

Class 4: Flexible-attachment rotors: These rotors can be categorized as class 1, class 2, or class 3 rotors but have components within themselves or flexibly attached.

Class 5: Single-speed flexible rotors: These rotors could be classified as class 3 flexible rotors but are balanced for operation at one speed only.

TABLE 3.2 ISO Classification of Rotors (ISO, 1973)

Class	Description	Example
1	Rigid rotor. Unbalance can be corrected in any two (arbitrarily selected) planes, and after that correction, unbalance does not significantly change at any speed up to maximum service speed.	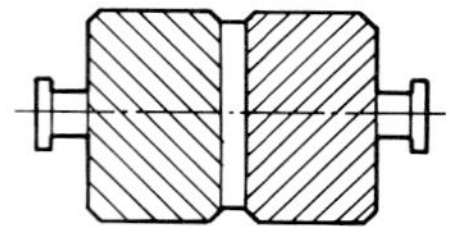 (a) Gear wheel
2	Quasi-flexible rotors: rotors that cannot be considered rigid but can be balanced in a low-speed balancing machine	
2A*	A rotor with a single transverse plane of unbalance (e.g., single mass on a light shaft whose unbalance can be neglected)	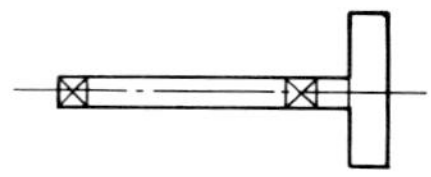 (b) Shaft with grinding wheel
2B*	A rotor with two axial planes of unbalance (e.g., two masses on a light shaft whose unbalance can be neglected)	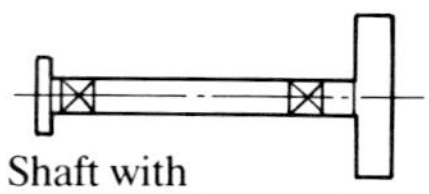 (c) Shaft with grinding wheel and pulley
2C*	A rotor with more than two transverse planes of unbalance	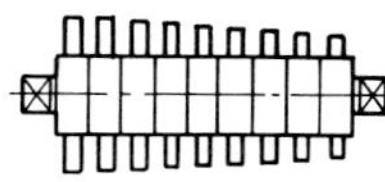 (d) Jet-engine compressor rotor
2D*	A rotor with uniformly distributed unbalance	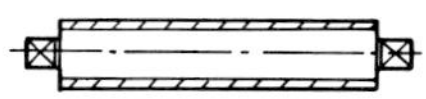 (e) Printing-press roller
2E*	A rotor consisting of a rigid mass of significant axial length supported by a flexible shaft whose unbalance can be neglected	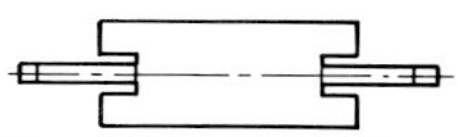 (f) Computer memory drum
2F†	A symmetric rotor, with two end correction planes, whose maximum speed does not significantly approach second critical speed, whose service speed range does not contain first critical speed, and with a controlled initial unbalance	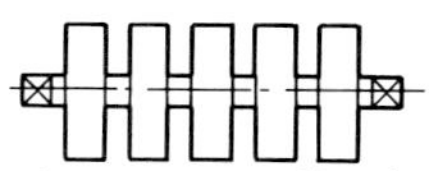 (g) Five-stage centrifugal pump
2G†	A symmetric rotor with two end correction planes and a central correction plane whose maximum speed does not significantly approach second critical speed and with a controlled initial unbalance	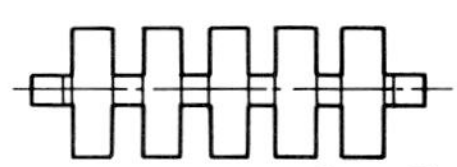 (h) Multistage pump impeller

TABLE 3.2 (*Continued.*)

Class	Description	Example
2H†	An asymmetric rotor with controlled initial unbalance treated in a similar manner as class 2F rotors	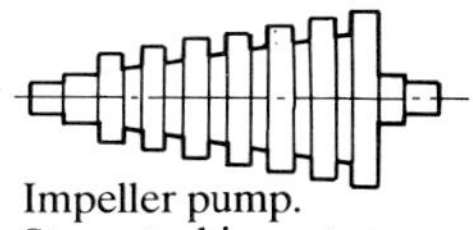 (i) Impeller pump. Steam turbine rotor
3	Flexible rotors: rotors that cannot be balanced in a low-speed balancing machine and require high-speed balancing	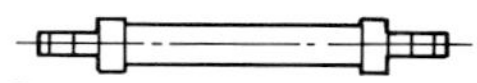 (j) Generator rotor
4	Special flexible rotors: rotors that could fall into classes 1, 2, or 3 but have in addition one or more components that are themselves flexible or are flexibly attached	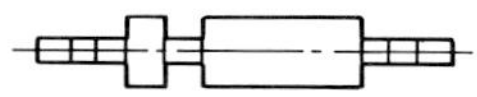 (k) Rotor with centrifugal switch
5	Single-speed flexible rotors: rotors that could fall into class 3 but for some reason (e.g., economy) are balanced only for a single service speed	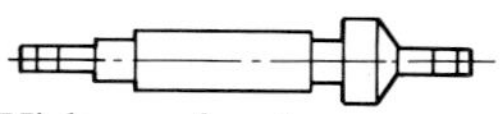 (l) High-speed motor

* Rotors where the axial distribution of unbalance is known.
† Rotors where the axial distribution of unbalance is not known.

Single- and two-plane constant-speed balancing is usually adequate for class 1 and class 2 rotors. For class 3 and 4 flexible rotors, a least-squared-error influence coefficient or combined modal technique is preferred.

3.1.3 Types of Balancing

Balancing of rotating machinery may be generally classified in two broad areas. The first general area is balancing in the factory done on either hard or soft bearing balancing machines. This procedure is done at the manufacturer, and components may be individually balanced as well as the final assembly balanced as a rigid rotor. If the components are individually balanced before final assembly as well as at final assembly, then even extremely long flexible rotors may operate very successfully without further adjustments in the field. For certain classes of rotors such as generators in particular, it may be necessary to further balance the rotor in a spin pit facility. This is usually only done with class 3 rotors that must operate through a number of critical speeds. This balancing procedure is very expensive and time-consuming. The facilities available to do this type of balancing are very limited.

The second broad area of balancing is field or in situ balancing in which the rotor is balanced in place. In field balancing, the turborotor is balanced in its own bearings, and suitable instrumentation is placed on the shaft, bearing housing, or foundation to monitor motion. We now give a brief summary of the type of field balancing procedures that may be employed.

Single-Plane Balancing by the Influence Coefficient Method

The single-plane single-speed influence coefficient method of balancing is the simplest of all procedures. Phase-angle conventions are different between displacement probe measurements and phase angles obtained directly with strobe lights. Also some balancing instruments will have different conventions as to the initiation of phase measurement. In this method, a trial or calibration weight is placed on the shaft. The new vibration response of the rotor is recorded after the placement of the trial weight. The change in vibration divided by the trial weight represents an influence coefficient. The influence coefficient represents the response of the rotor due to the placement of a unit trial weight at a particular speed. It is important to realize that this influence coefficient is a vector quantity which represents an amplitude and a phase angle. The phase angle represents the relative phase relationship between the forcing function and the vibration measurement.

In field balancing by the influence coefficient method, it is imperative that one fully understand the phase measurements employed. The phase measurements may be made with direct-reading noncontact or displacement probes, velocity pickups, or accelerometers. Phase may be determined by an electronic key phasor or by means of a strobe light to measure the phase angle on the shaft. One should be fully acquainted with the phase measurement procedure to be employed before attempting single- or multiplane balancing.

In single-plane balancing by the influence coefficient method, the system is assumed to be linear and a single correction weight is placed on the shaft based on the computed influence coefficient. This procedure is referred to as *trim balancing* and is the method generally used in 90 percent of the balancing situations encountered in the field. The application of the single-plane balance correction may theoretically reduce the rotor amplitude to zero at that particular speed. The major problem with the single-plane single-speed influence coefficient method of balancing is that even though the rotor amplitude may be reduced to small vibrations at that particular location and speed, other points along the rotor may exhibit higher vibrations. At other speeds, the rotor may appear not be in balance.

The advantage of the single-plane method is that it may be performed rapidly in the field, and the influence coefficient may be calculated by a graphical procedure. Once the influence coefficient is computed for a particular rotor, the machine may be rebalanced without the necessity of adding a trial weight. This method of balancing is often referred to as the "one-shot" method of balancing and is based on a predetermined value of the influence coefficient.

Two-Plane Balancing by the Influence Coefficient Method

There are many rotors in which a single phase of unbalance will not suffice. The application of a balance correction at one plane may cause an excessive amount of vibration at another location. Under these circumstances, a simultaneous two-plane influence coefficient balancing procedure must be employed.

The father of modern field balancing is Thearle (1934), who presented the general two-plane influence coefficient balancing procedure in 1934. The procedure developed by Thearle was a semigraphical method to determine the shaft influence coefficients and the two-plane balance corrections. This procedure is very popular and is still currently used by industry. The procedure is very simple to program on a hand calculator.

In the two-plane balancing procedure, a trial weight is placed on the first or near plane, and the responses at the near and far planes are recorded. From these measurements, two influence coefficients are computed. Next the weight is removed and placed at the second or far plane. The procedure is repeated to generate two more influence coefficients. This then forms a 2×2 matrix of complex influence coefficients which must be inverted to solve for the balancing. The 2×2 matrix of complex influence coefficients is equivalent to a 4×4 matrix of real numbers. Thus, two-plane balancing is equivalent to inverting a 4×4 matrix in order to determine four quantities which represent the respective rotor balancing magnitudes and relative phase-angle locations. This method is tedious to compute by hand, but fairly simple to perform on a calculator or desktop computer.

Single-Plane Balancing Using Static and Dynamic Components

This method is a variation of the single-plane influence coefficient procedure, except that vibration measurements are recorded at both ends of the shaft. The two vibration recordings are vectorially added and subtracted to determine the static and/or dynamic components. If the average static component appears to be highest, then either one or two weights are placed on the rotor in phase. If the dynamic or out-of-phase components of vibration appear to be highest, then two weights are placed 180° out of phase. This method is, in essence, a simplified modal procedure using the influence coefficient method.

Single-Plane Balancing by the Influence Coefficient Method Using Linear Regression

In the single-plane influence coefficient balancing method at one speed, it is possible to overbalance the rotor at the one speed and have extremely high vibrations at other speeds. The linear regression method of balancing is the simplest of the least-squared error methods of rotor balancing. Instead of measuring the rotor response at one speed, the vibration characteristics are measured over a speed range. The linear regression procedure produces a balancing magnitude which will not cause a large excitation at the other speeds. Balancing magnitudes computed by this procedure are always smaller than the magnitude computed from the single-plane influence coefficient method. This procedure ensures that one does not overcorrect by the application of too large a balance weight. This method is useful for balancing rotors with shaft bows and large initial runouts.

Generalized Influence Coefficient Method Using Pseudoinversion

The generalized influence coefficient method of balancing using two or more planes and multiple speed measurements is referred to as the *generalized influence coefficient method.* This method is also referred to as the *pseudoinverse method* or *least-squared-error method* of balancing. Unlike the two-plane method which is referred to as the *exact-speed point method* of balancing, more speed measurements may be used than balancing planes. This method leads to a best fit of the balancing data and is the preferred method for balancing large turbine-generators. The generalized least-squared-error method of balancing was initially presented by Goodman (1964) and has been improved upon by Lund and Tonnesen (1972), Badgley (1974), Thomson (1965), and Palazzola and Gunter (1977). The advantage of the generalized least-squared-error method of balancing is that it requires no prior rotordynamic knowledge of the system. It has been successfully used in multiplane balancing of turbine-generators through five critical speeds. Several variations of this procedure have been developed such as the weighted least-squared-error method which places emphasis on a weighting function on particular vibrations to minimize or ignore.

If the balance planes are improperly chosen, then it is possible that the least-squared-error method will generate extremely large balance magnitudes in adjacent planes, 180° out of phase, which are impossible to incorporate into the rotor. The least-squared-error method cannot limit the magnitude of balance weights predicted for a particular plane.

Multiplane Balancing Using Linear Programming Techniques

Multiplane balancing procedure using linear programming theory proceeds in a fashion similar to the least-squared-error method of balancing in obtaining the influence coefficients; however, in linear programming, a constraint or upward bound may be placed on the magnitude of the computed balancing weights at a particular station. This method was reported on by Little (1971), and a practical linear programming approach has been developed for the microcomputer by Foiles and Gunter (1982). In most well-behaved systems, however, the results from linear programming and the generalized least-squared-error balance theory are very similar.

Modal Balancing

Flexible rotors operating through multiple critical speeds such as generators are best balanced by a modal method. In the modal method, the critical speed mode shapes of the rotor must be known from either theoretical or experimental measurements. The weights placed on the rotor are proportional to the rotor mode shape. Each modal distribution is used to individually balance one mode at a time. The object of modal balancing is to balance rotors with high amplification factors without upsetting the balance at other modes. The modal balance corrections to apply on a rotor are best computed by using the influence coefficient method to measure the rotor response and predict the modal influence balancing coefficients. M. Darlow (1989) has presented considerable material on the combined modal and influence coefficient method referred to as the *unified method of rotor balancing*. The accuracy of the method depended on the knowledge of the rotor mode shapes, which may become quite complex for modes higher than the second critical speed. This method has been applied with great success to generators and long, flexible centrifuges.

Three-Trial-Weight Method of Balancing

There are many instances in which it is not possible to obtain an accurate phase measurement. Blake (1967) pioneered a method referred to as the *three-trial-weight or four-run method*. By applying a trial weight at three different locations, a locus of balancing points may be generated to determine the magnitude of the unbalance and the balance location. This method has been very successfully applied to fans in which the speed is unsteady or beating is encountered. It is extremely accurate and compensates for nonlinearities. This method of computation is graphical and does not require the use of a computer to perform the calculations.

Multiplane Balancing without Phase

The three-trial-weight method has been extended to balance flexible rotors through multiple critical speeds by Foiles and Gunter (1982). In this extension, a modal distribution of weights is placed on the rotor. The amplitude of the rotor is measured at the critical speed by using a fast Fourier transform (FFT) analyzer or using the peak hold data acquisition system to record the maximum amplitude at a particular critical speed. Each mode then may be separately balanced without exciting the previous critical speed modes.

This method has the advantage of balancing flexible rotors which have extremely high amplification factors and are difficult or impossible to balance by either the least-squared error or the linear programming method. It also has the advantage that it will accurately balance shafts with substantial shaft bow.

The disadvantage of this method is that the rotor mode shapes must be known and that three runs must be performed for each critical speed to be balanced. After this method has been performed, one may initiate a two-run method using the rotor modal sensitivity derived in the initial exercise.

Rotor Balancing without Trial Weights

Rotor balancing without trial weights is referred to as a *one-shot method* of balancing. Once the influence coefficients for a particular rotor are determined, a balance may be predicted by using the current vibration measurements with the influence coefficients.

A modal method of balancing without trial weights was developed by Palazzolo and Gunter (1982). In this procedure, the Nyquist plot of rotor amplitude is used to determine the rotor amplification factor at a particular mode. By computing the rotor amplification factor, the modal unbalance eccentricity for a particular critical speed may be determined. From the polar or Nyquist plot, the phase locations of the modal weight distribution may be determined to balance a particular critical speed.

The disadvantage of this method is that the rotor modal mass and critical speed mode shape must be known. The advantage of this method is that the balance predictions may be used as the initial trial run for further refinement by the influence coefficient method of balancing.

Coupling Trim Balancing

There are many instances in which two machines are perfectly balanced but, when interconnected with a coupling, appear to have high vibrations. Since the two rotors are initially well balanced, it is undesirable to attempt to rebalance the units. In the coupling method of balancing developed by Winkler (1983), the coupling is rotated 180°. From the new resulting vibration measurements, the optimum position of the coupling may be computed for trim balancing.

3.1.4 Instrumentation and Vibration Measurement Techniques for Balancing

When a rotor is balanced at low speed as a component, the procedure will generally take place in an integrated general-purpose balancing machine with built-in transducers, signal conditioning, and often a highly automated data reduction and presentation facility. On the other hand, when rotor balancing is performed in situ at operating speed in the factory or the field, the instrumentation and data systems, whether removable or permanently installed, must be tailored to the specific design and to its operating environment and circumstances. It is therefore appropriate and most efficient that any vibration measurement systems included for such other functions as condition monitoring, performance verification, fault diagnosis, and/or parameter identification be specified to include the data acquisition requirements of the selected balancing procedure—

typically, the amplitude and phase of that portion of the signal which is synchronous with the rotor speed at the specified number of locations and at the specified rotor speeds. A general discussion of such general-purpose vibration instrumentation can be found in Sec. 4.2.

3.1.5 Nomenclature

A = amplification factor, dim

A_c = amplification factor at critical speed

$\mathbf{a}$ = influence coefficient

a = acceleration vector, in/s^2 (m/s^2)

a_t = tangential acceleration, in/s^2 (m/s^2)

a_n = normal acceleration, in/s^2 (m/s^2)

C_c = critical damping, lb · s/in^2 (N · s/m)

D = shaft diameter, in (m)

$\mathbf{e}_u$ = unbalance eccentricity vector, in (m)

$\mathbf{e}_r$ = unit vector in radial direction, rotating with angular velocity ω

$\mathbf{e}_t$ = unit vector in tangential direction, rotating with angular ω

F_b = bearing reaction, lb (N)

F_u = rotating unbalance force, lb (N)

f = frequency ratio, dim

g = gravity, in/s^2 (m/s^2)

I = shaft second moment of area, in^4 (m^4)

I_{ij} = moment-of-inertia matrix

I_p = polar moment of inertia, lb · in · s^2 (kg · m^2)

I_t = transverse moment of inertia, lb · in · s^2 (kg · m^2)

$[\mathbf{I}]$ = moment-of-inertia matrix

K = bearing or shaft spring rate, lb/in (N/m)

L = length, in (m)

L_b = bearing span, in (m)

M = rotor mass, lb · s^2/in (kg)

m_b = small rotor balancing mass, lb · s^2/in (kg)

m_i = unbalance masses, lb · s^2/in (kg)

m_u = effective unbalance mass, lb · s^2/in (kg)

$\mathbf{M}_u$ = unbalance moment, lb · in (N · m)

n = number of unbalance weights

N = rotor speed, rpm

N = number of balance planes

$\mathbf{P}$ = position vector

R = shaft radius or radius of motion

t = disk thickness, in (m)

$\mathbf{U}$ = unbalance vector, g · in, oz · in, lb · s^2 (kg · m, g · mm)

U_b = rotor radial balance correction, g · in, oz · in, lb · s^2 (kg · m, g · mm)

$\mathbf{V}$ = velocity vector, in/s (m/s)

V = velocity of motion, in/s (m/s)

W = total rotor weight, lb (kg)

w_u = unbalance weight, lb (kg)

X, Y = absolute cartesian coordinates of rotor motion

x_i, y_i = relative cartesian locations of unbalance weight W_i

x, y, z = cartesian coordinate system fixed in shaft with z along spin axis Z

X', Y', Z' = principal coordinates fixed in disk or cylinder

Z = rotor spin axis or axial distance, in (m)

$\mathbf{Z}$ = complex displacement

α = angular acceleration, rad/s^2

α_{ij} = influence coefficient matrix

β = relative mass-displacement phase angle, deg

δ = shaft deflection, in (m)

ϕ = phase angle, deg

ρ = material density, lb · s^2/in^4 (kg/m)

θ = cylindrical coordinate, deg

τ = disk skew, deg or rad

ω = angular velocity, rad/s

ω_{cr} = rotor critical speed on rigid supports, rad/s

3.2 RIGID ROTOR BALANCING

3.2.1 Introduction to Rigid-Body Balancing

The concept of rigid-body balancing is fundamental to the balancing of all rotating machinery. In reality, no rotor or equipment is truly rigid. The term *rigid body* may be applied to specific rotors which are operating substantially below their flexible critical speeds. An ideal rigid-body rotor may be accurately balanced in any two arbitrary planes.

A rigid body may be described by its mass, location of the center of gravity, and inertia properties. For the case of an idealized rotor represented only by a point mass, the rotor may be balanced by the application of a single balancing correction weight. In a more generalized rigid body such as an electric motor or a gyroscope, the body must be described by an inertia matrix as well as by the rotor mass and location of the center of gravity. The body has three principal orthogonal axes about which there are no products of inertia. An arbitrary free-spinning body will rotate in a stable condition about either its minimum or its maximum principal moment of inertia axis.

For the case of rotating machinery, such as an idealized rigid-body motor, the rotor or motor is constrained to spin about a fixed axis because of the action of the bearings. Balancing is required when the spin axis of the rotor does not correspond to either the maximum or the minimum principal inertia axis. If the mass center of the rigid body is displaced from the spin axis, then the rotor is said to have *static unbalance*. If the mass center lies along the spin axis of the shaft, but the minimum or maximum inertia axis does not coincide with the spin axis, then the rotor is said to have *dynamic unbalance*.

In the first case, a single correction plane is required for static balancing. In the second case, two weights of equal magnitude but 180° out of phase are required for the dynamic correction. An example is the old procedure of balancing automobile tires by means of a bubble balance. The wheel may be accurately balanced for static unbalance by a single-plane correction, but may be severely dynamically unbalanced due to the action of unbalance couples. After the wheel is statically balanced, two additional weights are required on opposite sides of the wheel and 180° out of phase to each other to provide dynamic balance.

In general, a combination of static and dynamic unbalance will exist simultaneously in a rotor. Thus rigid-body balancing, in general, will require two distinct balancing planes situated normal to the spin axis of the rotor. The object of rigid-body balancing, therefore, is to make the maximum or minimum rigid-body inertia axis correspond identically with the rotor spin axis. When this occurs, the rotor is said to have rigid-body balance.

The theoretical conditions for rigid-body balancing are independent of machine speed provided that the rotor speed is well below the rotor flexible critical speed. Various types of low-speed rigid-body balancing machines have been developed to utilize the concept of rigid-body balancing in which the inertia axis corresponds with the rotor spin axis. In one set of balancing machines, called *soft bearing balancing* machines, the rotor is supported by bearings with very low spring rates. This causes the rotor to spin about its inertia axis. The displacement, magnitude, and phase are recorded at each end of the rotor. Weights are placed on the rotor until the runouts (displacements) are corrected to a minimal value. By this means, the rotor principal inertia axis is made to correspond to the spin axis.

In another class of balancing machine, the rotor is supported by very stiff bearings (Schenck Jrebel, 1980). The bearing supports are calibrated, and the bearing forces and phase angles are recorded at a given speed. This type of balancing machine is referred to as a *hard bearing balancing* machine. From the magnitude and phase of the recorded forces, the unbalance corrections may be determined for the rotor regardless of the rotor size.

A rotor is said to be a rigid rotor if it may be perfectly balanced by the application of suitable correction balance weights in any two arbitrary planes. The assumption of a rigid-body rotor is reasonable for electric motors but not for large turbine-generators due to shaft flexibility effects. A rotor, therefore, may be perfectly balanced as a rigid rotor in a balancing machine but may exhibit high unbalance characteristics at high speeds due to rotor flexibility effects when operating near a bending critical speed.

The process of rigid-body balancing is the application of suitable balance weights at two separate balancing planes along the rotor. The rotor is in static balance when the rotor mass center lies on the spin axis of the rotor. The rotor is in dynamic balance when the principal inertia axis coincides with the rotor spin axis. Dynamic balancing is equivalent to the condition that the product of inertia terms corresponding to the rotor spin axis and any other orthogonal axis acting through the rotor mass center be zero.

3.2.2 Single-Plane Balancing

Single-plane rotor balancing represents the most fundamental procedure in rotor balancing. With single-plane balancing, the rotor mass center is assumed to be offset from the spin axis of the shaft. The amount of radial offset is often referred to as the *rotor unbalance eccentricity* e_u or unbalance eccentricity vector $\mathbf{e}_u$.

In single-plane balancing, the suitable small balance correction weights are placed on the rotor to shift the mass center of the system to lie along the rotor spin axis. This process is static balancing.

Forces Caused by Single-Plane Rotating Unbalance

Consider the rotating mass M which is displaced from the origin point 0 by a fixed radius R, as shown in Fig. 3.1. The mass is rotating about the Z axis and is

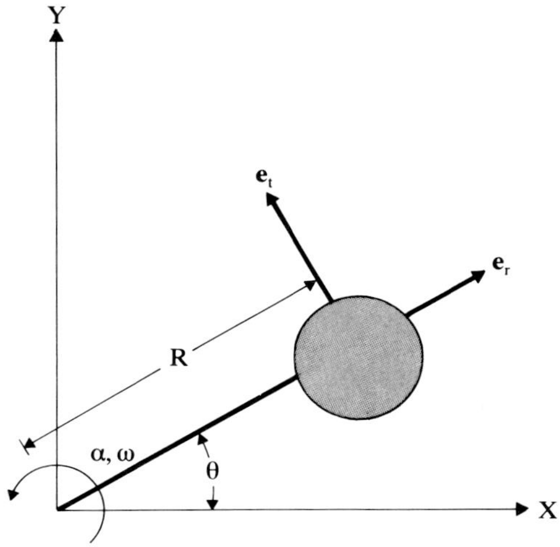

FIGURE 3.1 Schematic diagram of rigid-body point mass moving in XY plane.

moving in a circle of radius R in the XY plane. A position vector to the mass M from the origin 0 is given by

$$\mathbf{R} = R\mathbf{e}_r \tag{3.1}$$

The velocity of the mass is given by

$$\mathbf{V} = \dot{\mathbf{R}} = \dot{R}\mathbf{e}_r + R\dot{\mathbf{e}}_r = R\omega\mathbf{e}_t \qquad \text{[since } R \neq R(t),\ |R| \text{ is constant]}$$

The acceleration of the mass center is given by

$$\mathbf{a} = R\alpha\mathbf{e}_t - R\omega^2\mathbf{e}_r = a_t\mathbf{e}_t + a_n\mathbf{e}_r \tag{3.2}$$

where

$$a_t = R\alpha = \text{tangential acceleration}$$

$$a_n = -R\omega^2 = \text{centripetal acceleration}$$

The forces generated in the bar are given by Newton's second law of motion:

$$\mathbf{F} = M\mathbf{a} = M(R\alpha\mathbf{e}_t - R\omega^2\mathbf{e}_r) = F_r\mathbf{e}_r + F_t\mathbf{e}_t \tag{3.3}$$

The reaction forces acting on the bar at point 0 are equal and opposite to the forces generated in the bar and are given by

$$\begin{aligned} F_r &= MR\omega^2 \\ F_t &= -MR\alpha \end{aligned} \tag{3.4}$$

The radial reaction force F_r is often referred to as the *centrifugal force* caused by the centripetal acceleration $-R\omega^2$.

The total reaction is given by

$$F_{\text{reaction}} = MR\omega^2\left(1 + \frac{\alpha^2}{\omega^2}\right)^{1/2} \tag{3.5}$$

For acceleration rates normally encountered with rotating machinery, the rigid rotor bearing reaction force is approximated by

$$F_{\text{reaction}} = MR\omega^2 \tag{3.6}$$

The product of the mass M and radial distance R may be expressed as **U**, which is the *radial unbalance*. The generalized reaction force may be expressed in terms of the radial unbalance vector **U** and angular velocity ω as follows (considering negligible system acceleration):

$$\mathbf{F} = \mathbf{U}\omega^2 \tag{3.7}$$

Example 3.1: Calculation of Centrifugal Force. A rotor of 45.36 kg (100 lb) has its mass center displaced by 25.4 μm (0.001 in, or 1 mil). The centrifugal force generated at 3600 rpm is given by

$$\omega = 3600 \text{ rpm} \times \frac{2\pi}{60} = 376.91 \text{ rad/s}$$

$$F_{3600} = (45.36 \text{ kg})(25.4 \times 10^{-6})(376.91)^2 = 163.68 \text{ N}$$

or

$$F_{3600} = \left(\frac{100 \text{ lb}}{386 \text{ in/s}^2}\right)(0.001)(376.91)^2 = 36.8 \text{ lb}$$

At 10,000 rpm the rotating force is

$$F_{10,000} = 36.8\left(\frac{10,000}{3600}\right)^2 = 284 \text{ lb}$$

The rotating weight of 100 lb (0.259 lb · s/in) displaced from the center of rotation by 1 mil (0.001 in) produces a rotating force of 37 lb at 3600 rpm. At 10,000 rpm, this rotating load has increased almost eightfold to 284 lb. Therefore, small displacement of the mass center of a rotor can lead to large rotating loads at high speeds.

The product of mass and radial distance, or eccentricity, may be expressed in metric units, U.S. Customary System (USCS) units, or as a product of the two systems. Typical units of unbalance **U** may be gram-micrometers, gram-inches, or ounce-inches.

Example 3.2. Express the equivalent unbalance **U** of Example 3.1 in terms of gram-inches and ounce-inches.

$$\mathbf{U}_{\text{oz}\cdot\text{in}} = (100 \text{ lb} \times 16 \text{ oz/lb})(0.001 \text{ in}) = 1.6 \text{ oz}\cdot\text{in}$$

or

$$\mathbf{U}_{\text{g}\cdot\text{in}} = \frac{100 \text{ lb} \times 1000 \text{ g/kg}}{2.2046 \text{ lb/kg}} \times 0.001 \text{ in} = 45.4 \text{ g}\cdot\text{in}$$

The rotating force generated by unbalance U expressed in terms of ounce-inches and gram-inches, respectively, is given by

$$F_{\text{lb}} = 1.775 \times 10^{-6} U_{\text{oz}\cdot\text{in}} \times (N \text{ rpm})^2 \tag{3.8}$$

$$F_{\text{lb}} = 6.26 \times 10^{-8} U_{\text{g}\cdot\text{in}} \times (N \text{ rpm})^2 \tag{3.9}$$

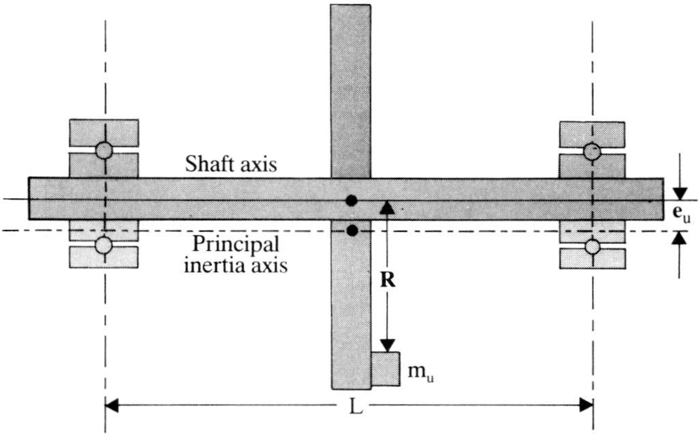

FIGURE 3.2 Single-mass disk with radial unbalance.

***Relationship between Rotor Unbalance Eccentricity* $\mathbf{e}_u$ *and Unbalance* U**

Figure 3.2 represents a single-mass disk of mass M with a small unbalance mass of m_u placed at a radial distance **R** from the disk shaft axis. The effective mass center of the system is shifted radially by a small magnitude of $\mathbf{e}_u$ from the original axis of rotation. The amount of the shift $\mathbf{e}_u$ is given by

$$\mathbf{e}_u = \frac{\mathbf{R}m_u}{M + m_u} = \frac{\mathbf{U}}{M + m_u} \approx \frac{\mathbf{U}}{M} \tag{3.10}$$

The small radial unbalance mass m_u at radius **R** causes a translation of the principal inertia axis from the shaft neutral axis by the vector magnitude $\mathbf{e}_u$.

The unbalance weight is assumed to be much smaller than the total rotor weight, and the radius at which the weight is placed is much larger than the rotor unbalance eccentricity. If these conditions are not met, then the balancing and the rotordynamics may not be described by linear equations of motion. The unbalance eccentricity vector $\mathbf{e}_u$ thus lies in the same direction as the rotor unbalance vector **U**.

In Eq. 3.10, the rigid-body radial unbalance magnitude remains constant for all speeds. Such is not the case when flexible shaft effects are taken into consideration.

Single-Plane Balancing of Rigid-Body Rotor

In Fig. 3.2, let a small balance weight m_b be placed at a radius R on the rim of the wheel 180° out of phase to the eccentricity vector $\mathbf{e}_u$. The resulting centrifugal or inertia loading is given by

$$\mathbf{F}_u = M\mathbf{e}_u\omega^2 - m_b\mathbf{R}\omega^2 \tag{3.11}$$

The value of m_b may be chosen such that the net centrifugal loading is zero. In this case, the wheel is said to be statically balanced, and the mass center lies along the spin axis of the shaft. In general, we will show that for a rigid body to be in balance, the principal inertia axis must lie along the spin axis.

The vector magnitude of small mass m_b at radius R, out of phase to the mass center eccentricity vector, may be referred to as the *rotor radial balancing* vector and is given by

$$\mathbf{U}_b = m_b\mathbf{R} \tag{3.12}$$

The condition for rigid-body balancing the single-plane disk is given by

$$\mathbf{U} + \mathbf{U}_b = 0 \tag{3.13}$$

It is important to note that both $\mathbf{U}$ and $\mathbf{U}_b$ are vectors, not scalars. This implies that the balancing correction $\mathbf{U}_b$ can be represented by two scalar quantities. The scalar quantities may be either the magnitude and phase (or orientation of the balance weight) or the local cartesian XY components of the balancing correction with respect to an arbitrary reference frame fixed mark in the disk.

In rigid rotor balancing, it is assumed that only small correction weights must be placed on the disk or rotor to balance it.

Example 3.3: Calculation of Balance Magnitude. Compute the balancing magnitude $\mathbf{U}_b$ for Example 3.1 in terms of ounce-inches and gram-inches for a disk with a radius of 3 in.

$$\begin{aligned}\mathbf{U}_b = -M\mathbf{e}_u &= -100\text{ lb} \times 0.001\text{ in} = -0.1\text{ lb}\cdot\text{in}\\ &= -0.1\text{ lb}\cdot\text{in} \times 16\text{ oz/lb} = -1.6\text{ oz}\cdot\text{in}\\ &= -0.1\text{ lb}\cdot\text{in} \times \frac{1000\text{ g/kg}}{2.2\text{ lb/kg}} = -45.46\text{ g}\cdot\text{in}\end{aligned}$$

The correction weight that should be placed on the wheel at the 3-in radius is 15 g placed out of phase to the $\mathbf{e}_u$ vector.

Representation of Multiple Unbalances by a Single Unbalance Vector

Figure 3.3 represents a disk of mass M and radius $\mathbf{R}$ with multiple of unbalances m_i, $i = 1$ to n, at a radius of R_i as measured from the center. It is assumed that

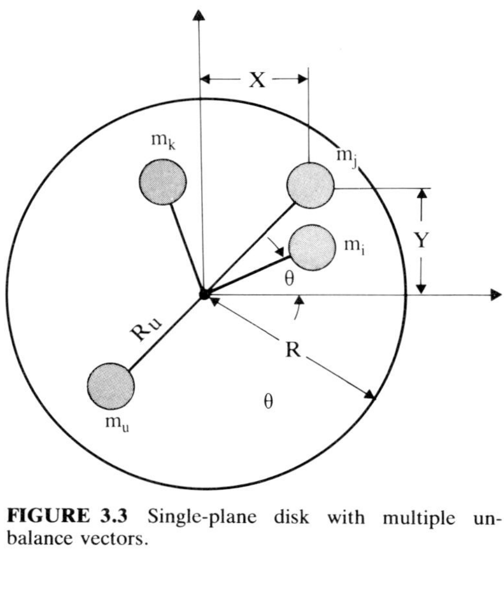

FIGURE 3.3 Single-plane disk with multiple unbalance vectors.

the sum of the unbalance weights is an order of magnitude smaller than the disk weight M. That is,

$$\sum_{i=1}^{n} m_i \ll M \tag{3.14}$$

The number of unbalances n produces a rotating load given by

$$\mathbf{F}_u = \sum_{i=1}^{n} m_i \mathbf{R}_i \omega^2 \tag{3.15}$$

The rotating unbalance load vectors may be replaced by a single mass m_u acting at a vector radius $\mathbf{R}_u$ as follows:

$$\mathbf{F}_u = m_u \mathbf{R}_u \omega^2 \tag{3.16}$$

Equating the two expressions for rotating unbalance force, we have

$$m_u \mathbf{R}_u = \sum m_i R_i = \mathbf{U} \tag{3.17}$$

For the case of a perfectly rigid shaft rotating about the shaft axis, the rotating force generated is

$$\mathbf{F}_u = m_u \mathbf{R} \omega^2 = \mathbf{U} \omega^2 \approx M \mathbf{e}_u \omega^2 \tag{3.18}$$

For a perfectly rigid shaft with the disk located in the center of the bearing span, the bearing reaction $\mathbf{F}_b$ is

$$\mathbf{F}_b = \frac{-M \mathbf{e}_u \omega^2}{2} = \frac{-\mathbf{F}_u}{2} \tag{3.19}$$

Such a statement will not be true for the flexible rotor where the bearing forces transmitted may be more than the rigid-body interia loads (critical and subcritical speed operation) or less than the rigid-body inertia loads (supercritial speed operation). For rigid-body operation with radial unbalance **U**, the bearing forces increase as the square of the speed.

Note that in a ball-bearing- or roller-element-supported rotor, the bearing life varies approximately inversely to the cube of the bearing loads. Therefore, a doubling of the bearing loads due to unbalance will cause a reduction of the bearing life by a factor of 8, suggesting that ball-bearing-supported rotors must be very carefully balanced.

Equation 3.17 for the total unbalance magnitude is a vector equation and may also be resolved into scalar equations. An arbitrary relative xy axis may be inscribed on the disk with the x axis passing through a reference mark scribed on the disk. The radius vectors and the unbalance components may be expressed in terms of the loacl cartesian reference system fixed in the disk. The placement of the reference (or timing) mark on the disk is arbitrary. Let

$$\mathbf{R}_u = x_u \mathbf{i}' + y_u \mathbf{j}'$$

$$\mathbf{R}_i = x_i \mathbf{i}' + y_i \mathbf{j}'$$

The scalar equations for the cartesian components of unbalance are given by

$$U_x = m_u x_u = \sum m_i x_i = \sum U_{ix} \tag{3.20}$$

$$U_y = m_u y_u = \sum m_i y_i = \sum U_{iy} \tag{3.21}$$

where

$$x_i = R_i \cos \theta_i$$

$$y_i = R_i \sin \theta_i$$

The angle θ from the x axis at which the effective unbalance mass m_u is located is given by

$$\theta = \tan^{-1} \frac{y_u}{x_u}$$

$$= \tan^{-1} \frac{\sum m_i y_i}{\sum m_i x_i} \tag{3.22}$$

The effective unbalance mass m_u is

$$m_u = \frac{[(\sum m_i x_i)^2 + (\sum m_i y_i)^2]^{1/2}}{R_u} \tag{3.23}$$

The total or effective vector unbalance expressed in terms of the local xy cartesian coordinate system fixed in the disk is given by

$$\mathbf{U} = \sum_{i=1}^{n} U_{ix}\mathbf{i}' + \sum_{i=1}^{n} U_{ix}\mathbf{j}' = M\mathbf{e}_u$$

Therefore, a number of unbalance vectors which lie in a plane may be resolved into one single unbalance vector similar to Eq. 3.10.

Example 3.4: Resolution of Multiple Balance Weights into a Single Balance. We are given a 50-lb wheel with 36 holes equally spaced at a 6-in radius, as shown in Fig. 3.4. The holes are labeled in a counterclockwise

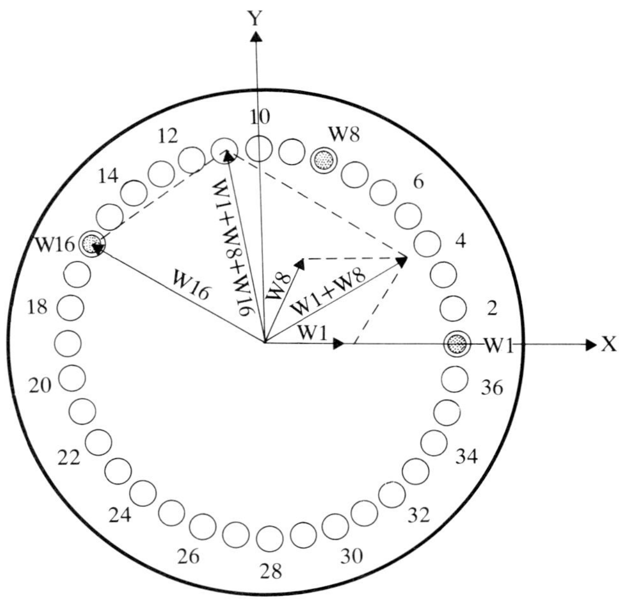

FIGURE 3.4 Vector addition of balance weights.

direction. There are three weights placed in various holes. Replace the three weights by a single equivalent weight in the nearest hole. The balance weights are

$$w_1 = 1 \text{ oz at hole } 1$$

$$w_8 = 1 \text{ oz at hole } 8$$

$$w_{16} = 2 \text{ oz at hole } 16$$

Solution. A relative reference axis is drawn on the disk passing through the disk center and hole 1. This line establishes the relative x axis. The holes are spaced at intervals of 10°. The angular location and moments generated by the balance weights are given in Table 3.3. The equivalent balance weight W_b is given

TABLE 3.3 Resolution of Multiple Balance Weights

Number	Weight	Hole	Angle	x_i	y_1	$w_i x_i$	$w_i y_i$
1	1	1	0°	6	0	6	0
2	1	8	70°	2.05	5.64	2.05	5.64
3	2	16	150°	−5.20	3.00	−10.39	6.00
Total	2	11	101°	−1.15	5.89	−2.34	11.64

by

$$w_b = \frac{[(-2.34)^2 + (11.64)^2]^{1/2}}{6 \text{ in}}$$

$$= \frac{11.87 \text{ oz} \cdot \text{in}}{6 \text{ in}} = 1.98 \text{ oz}$$

$$\theta = \tan^{-1}\left(\frac{11.64}{-2.34}\right) = 101.37°$$

An approximate equivalent balance resultant is obtained by placing a 2-oz weight at hole 11. It is seen in the above example that any number of balance weights acting in a single plane may be combined into one single balance vector of weight w_b placed at a radius vector **R**.

The approach used to sum the unbalance magnitudes is called *vector addition.* If $\mathbf{U}_{bi}$ are the various individual balance magnitudes acting at a given xy plane, then the effective balance $\mathbf{U}_b$ acing at the plane is given by the following vector sum:

$$\mathbf{U}_b = \mathbf{U}_{b1} + \mathbf{U}_{b2} + \cdots + \mathbf{U}_{bi} + \cdots + \mathbf{U}_{bn} \qquad i = 1 \text{ to } n \tag{3.24}$$

Example 3.4 may also be solved graphically. Figure 3.4 represents a polar plot of the locations of the balance weights. If plotted on polar paper, then one major division represents 0.5 oz, and full scale is 2.5 oz. As a first step, the three balance vectors are drawn to scale to represent the weights placed in holes 1, 8, and 16. Next the balancing weights W_1 and W_8 are vectorially added (by using a compass to construct the parallelogram). The resultant vector $\mathbf{W}_1 + \mathbf{W}_8$ is now added to $\mathbf{W}_{16}$ to generate the final resultant balance vector $\mathbf{W}_b$. From the construction, we see that the resultant balance is approximately 2 oz at hole 11.

In a similar fashion, a single balancing weight may be resolved into two or more arbitrary vectors. This process is very important, in balancing because it is often encountered in field balancing where the balance hole specified for weight placement already is filled with balance weights. In this case, the equivalent balance may be achieved by placing the proper weights in holes on either side of the specified balancing hole.

So a number of unbalance or balancing weights placed on a disk may always be resolved into a single balancing (or unbalance) weight positioned with respect to a relative *xy* reference frame fixed in the disk.

Single-Plane Balancing by the Influence Coefficient Method

The rotating unbalance in the rigid body rotor creates a force in the bearings that is proportional to the unbalance $\mathbf{U}$. By measuring the forces transmitted through the bearing pedestal via a calibrated strain gauge or force transducer, one may balance the rotor using the influence coefficient method.

In the application of this method, the relative phase angle of the force (or pedestal displacement) with respect to a timing mark on the shaft must be measured along with the magnitude of the force. The force or displacement measured at the pedestal is given by

$$\mathbf{Z} = \mathbf{aU} \tag{3.25}$$

The measured force is a function of the system influence coefficient $\mathbf{a}$ and the unknown rotor unbalance $\mathbf{U}$.

A trial or calibration weight is next placed on the rotor. The new response $\mathbf{Z}_t$ is measured after the application of the calibration unbalance:

$$\mathbf{Z}_t = \mathbf{a}(\mathbf{U} + \mathbf{U}_t) \tag{3.26}$$

By vector subtraction of the calibration run and the initial run, the system influence coefficient $\mathbf{a}$ is

$$\mathbf{a} = \frac{\mathbf{Z}_t - \mathbf{Z}}{\mathbf{U}_t} = \frac{\Delta \mathbf{Z}}{\mathbf{U}_t} \tag{3.27}$$

The influence coefficient $\mathbf{a}$ represents the response of this particular rigid-body rotor to a unit unbalance. All rotors of this same class will have the same influence coefficient response. Note that this value represents a vector quantity since it involves both amplitude and phase information.

The inverse of the influence coefficient $\mathbf{a}$ is the rotordynamic *stiffness* or *impedance coefficient* $\mathbf{d}$. Once the rotordynamic impedance has been determined, the balancing correction may be computed directly from the value of the dynamic impedance $\mathbf{d}$ and the initial vibration $\mathbf{Z}$ as follows:

$$\mathbf{U}_b = -\mathbf{U} = -\mathbf{a}^{-1} \times \mathbf{Z} = -\mathbf{dZ} \tag{3.28}$$

The value of the rotordynamic impedance $\mathbf{d}$ will be the same for all rotors of a similar class when operated at the particular speed at which the initial measurements were made. Therefore, a new rotor with an unknown value of unbalance could be directly balanced from the value of the vibration reading by using the previously determined rotordynamic impedance value **d**. The process of balancing a rotor without the addition of trial weights is referred to as *one-shot balancing*.

The single-plane balancing procedure is well suited to a graphical solution since the vibration values and influence coefficients are two-dimensional vectors. Figure 3.19 is an example of the single-plane balancing graphical procedure. In the graphical solution, the $\Delta\mathbf{Z}$ vector is constructed by vector subtraction of the trial response minus the initial response, and the resultant magnitude and phase angle are measured on polar paper:

$$\Delta\mathbf{Z} = \mathbf{Z}_t - \mathbf{Z} = |\Delta\mathbf{Z}| \angle \Delta\theta^\circ$$

Equation 3.27 may be rewritten for the graphical solution in the form

$$\mathbf{U}_b = -\mathbf{U}_t \frac{\mathbf{Z}}{\Delta\mathbf{Z}} = -U_t \angle \theta_t^\circ \left|\frac{\mathbf{Z}}{\Delta\mathbf{Z}}\right| \angle(\theta^\circ - \Delta\theta^\circ)$$

$$= U_t \left|\frac{\mathbf{Z}}{\Delta\mathbf{Z}}\right| \angle(\theta_t^\circ + \theta^\circ - \Delta\theta^\circ - 180^\circ) = U_b \angle \theta_b^\circ \qquad (3.29)$$

In the graphical solution, the magnitude of the trial unbalance $\mathbf{U}_t$ is multiplied by the scalar value of $|\mathbf{Z}/\Delta\mathbf{Z}|$ to produce the balance magnitude of $\mathbf{U}_b$. The angle at which the balance weight should be placed is rotated from the position of the trial balance by the angle of $\Theta^\circ - \Delta\Theta^\circ - 180^\circ$.

3.2.3 Two-Plane Rigid Rotor Balancing

Figure 3.5 represents a two-mass rotor mounted on rigid shaft in rigid bearings. The disks are assumed to have masses M_1 and M_2, respectively. A small

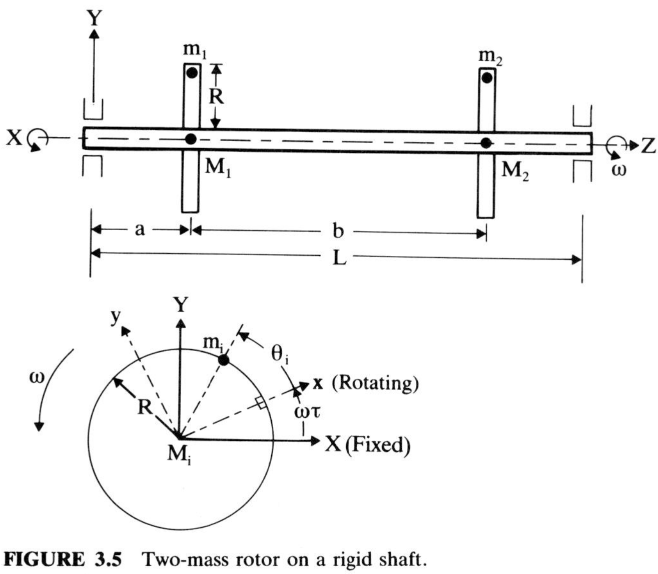

FIGURE 3.5 Two-mass rotor on a rigid shaft.

unbalance weight is attached to each disk, m_1 to M_1 and m_2 to M_2, at angles of θ_1 and θ_2, respectively, as measured from a reference mark on the shaft.

The vector forces generated by the two rotating masses are given by

$$\begin{aligned} \mathbf{F}_{u1} &= m_1\mathbf{R}_1\omega^2 = \mathbf{U}_1\omega^2 \\ \mathbf{F}_{u2} &= m_2\mathbf{R}_2\omega^2 = \mathbf{U}_2\omega^2 \end{aligned} \tag{3.30}$$

The unbalances $\mathbf{U}_1$ and $\mathbf{U}_2$ may be resolved in terms of xy components with respect to a local xy coordinate system fixed in the rotor;

$$U_{ix} = m_iR_i \cos \theta_i \qquad U_{iy} = m_iR_i \sin \theta_i \qquad i = 1, 2 \tag{3.31}$$

The vector forces may be resolved into cartesian components in the fixed x and y newtonian reference frame.

$$\begin{aligned} F_{uX} &= \omega^2 \sum_{i=1}^{2} m_iR_i \cos(\omega t + \theta_i) = \omega^2 \sum_{i=1}^{2} U_i \cos(\omega t + \theta_i) \\ &= \omega^2 \sum_{i=1}^{2} (U_{ix} \cos \omega t - U_{iy} \sin \omega t) = F_{uX_1} + F_{uX_2} \\ F_{uY} &= \omega^2 \sum_{i=1}^{2} m_iR_i \sin(\omega t + \theta_i) = \omega^2 \sum_{i=1}^{2} U_i \sin(\omega t + \theta_i) \\ &= \omega^2 \sum_{i=1}^{2} (U_{ix} \sin \omega t + U_{iy} \cos \omega t) = F_{uY_1} + F_{uY_2} \end{aligned} \tag{3.32}$$

At the instant the reference mark on the shaft corresponds to the horizontal axis, the inertial forces generated in the fixed xy reference frame are given by

$$\begin{aligned} F_{uX} &= \omega^2 \sum U_{ix} \\ F_{uY} &= \omega^2 \sum U_{iy} \end{aligned} \tag{3.33}$$

For a perfectly rigid shaft in rigid bearings, the mass or disk moments of inertia do not contribute to the inertia loading gravitational weight. It is useful to refer to *D'Alembert's principle,* which is simply a rearrangement of Newton's second law of motion and is stated as follows for a system of particles:

$$-\sum M\mathbf{a} + \sum \mathbf{F}_{\text{external}} = 0 \tag{3.34a}$$

where

$$\sum \mathbf{F}_{\text{external}} = \sum_{i=1}^{n} \mathbf{F}_{\text{unbalance}} + \sum_{i=1}^{2} \mathbf{F}_{\text{bearing}} \tag{3.34b}$$

From D'Alembert's principle, the sum of the inertia loads and the external forces (bearing reactions and unbalance forces) must form a zero system.

It is of interest that the mass of the disks does not enter into the expression for the dynamic bearing forces $\mathbf{F}_b$ transmitted due to the two planes of rotor unbalance. In a hard bearing support, low-speed balancing machine, the supports are designed to be sufficiently stiff that the rotor is operating as a rigid rotor. Hence, the bearing forces measured are a direct function of the rotor unbalance, and the size of the rotor is theoretically not important. Thus, the sensitivity of a hard bearing support balancing machine should be relatively independent of size.

In Fig. 3.5, the unbalance forces acting on the two disks may be resolved into an equivalent force $\mathbf{F}_u$ and the moment or couple $\mathbf{M}_u$ acting at the mass center of the system as follows:

$$\mathbf{F}_u = \omega^2(\mathbf{U}_1 + \mathbf{U}_2) \tag{3.35a}$$

$$\mathbf{M}_u = \omega^2(\mathbf{L}_1 \times \mathbf{U}_1 + \mathbf{L}_2 \times \mathbf{U}_2) \tag{3.35b}$$

In the above formulations, where L_1 and L_2 are the axial distances from the rotor center of gravity to the planes of unbalance, the unbalance magnitudes $\mathbf{U}_1$ and $\mathbf{U}_2$ must be expressed in mass-displacement units.

Vector Resolution of Arbitrary Unbalance into Two Planes

Figure 3.6 represents a rigid-body rotor with an arbitrary unbalance of $\mathbf{U}(m_u\mathbf{R})$ situated at an axial distance of l_u from the rotor mass center. When the x reference axis fixed in the shaft corresponds with the stationary X axis, the equivalent rotating force and moment acting about the mass center are given by

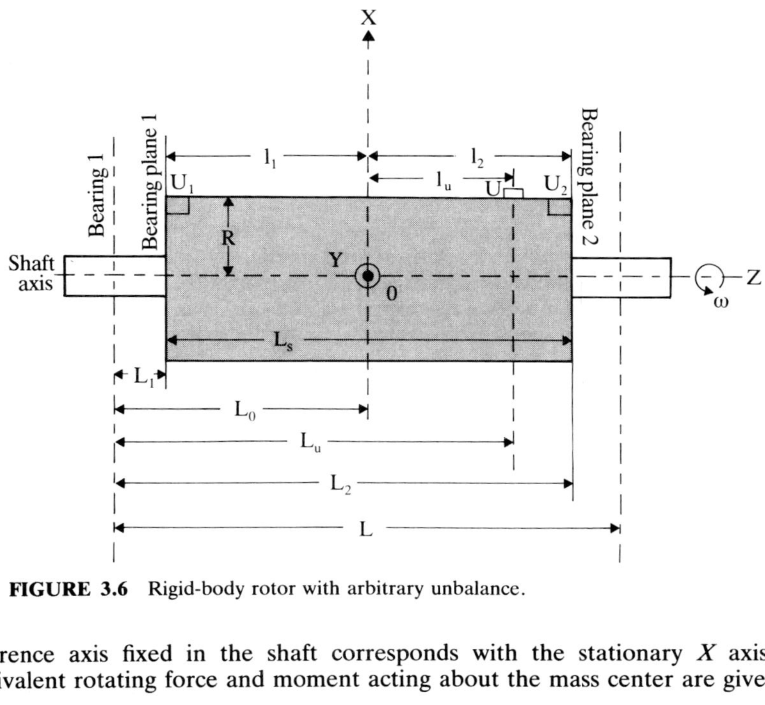

FIGURE 3.6 Rigid-body rotor with arbitrary unbalance.

$$\mathbf{F}_u = \mathbf{F}_{ux} = \omega^2 U\mathbf{i} \tag{3.36a}$$

$$\mathbf{M}_u = \mathbf{M}_{uy} = \omega^2 L_u U\mathbf{j} \tag{3.36b}$$

The arbitrary unbalance $\mathbf{U}$ for a rigid body may be specified in terms of the equivalent unbalance magnitudes of $\mathbf{U}_1$ and $\mathbf{U}_2$ located at two specified planes in the rotor.

The unbalance magnitudes $\mathbf{U}_1$ and $\mathbf{U}_2$ will generate an equivalent force and couple about the rotor mass center as the original arbitrary unbalance $\mathbf{U}$. Assuming that the unbalances $\mathbf{U}_1$ and $\mathbf{U}_2$ lie along relative x axis attached to the rotor,

$$F_{ux} = \omega^2(U_1 + U_2) = \omega^2 U \tag{3.37a}$$

$$M_{uy} = \omega^2(l_2 U_2 - l_1 U_1) = L_u \omega^2 U \tag{3.37b}$$

Solving for U_1 and U_2 gives

$$U_1 = U\frac{l_2 - l_u}{l_1 + l_2} \qquad U_2 = U\frac{l_1 + l_u}{l_1 + l_2} \tag{3.38}$$

And U_1 and U_2 may also be expressed in terms of measurements from the bearing 1 centerline as

$$U_1 = U\frac{L_2 - L_u}{l_s} \qquad U_2 = U\frac{L_u - L_1}{l_s} \tag{3.39}$$

Example 3.5. A 40-in rotor has an unbalance of 10 g · in located 30 in from the end. Calculate the equivalent two-plane unbalance for (*a*) balance planes at the ends of the rotor and (*b*) balance planes at $L_1 = 35$ and $L_2 = 40$.
(*a*) $L_2 = 40$ in, $L_1 = 0$ in, $L_2 = 40$ in, $L_u = 30$ in, and $l_s = L_2 - L_1 = 40$ in.

$$U_1 = 10\left(\frac{40 - 30}{40}\right) = 2.5 \text{ g} \cdot \text{in} \qquad \text{balance plane 1}$$

$$U_2 = 10\left(\frac{30 - 0}{40}\right) = 7.5 \text{ g} \cdot \text{in} \qquad \text{balance plane 2 (in phase to plane 1)}$$

(*b*) $L_2 = 40$ in, $L_1 = 35$ in, $L_s = 5$ in, $L_u = 30$ in, and $l_s = L_2 - L_1 = 40 - 35 = 5$ in.

$$U_1 = 10\left(\frac{40 - 30}{5}\right) = 20 \text{ g} \cdot \text{in} \qquad \text{balance plane 1}$$

$$U_2 = 10\left(\frac{30 - 35}{5}\right) = -10 \text{ g} \cdot \text{in} \qquad \text{balance plane 2} - 180° \text{ to plane 1}$$

In Example 3.5*b*, if the two arbitrary balance planes selected fall outside the original plane of unbalance, then large values of U_1 and U_2 will be generated. Such a situation can occur when one is attempting two-plane balancing on an overhung rotor. A small unbalance weight at the center bull gear can translate into large balance couples on an overhung wheel.

Resolution of Two-Plane Unbalance into Static and Dynamic Components
A single plane or two arbitrary planes of unbalance may always be resolved into equivalent static and dynamic components $\mathbf{U}_s$ and $\mathbf{U}_d$, respectively. Figure 3.5, e.g., represents a two-mass system with arbitrary unbalances $\mathbf{U}_1$ and $\mathbf{U}_2$ on the wheels. Figure 3.7 represents the unbalances resolved into static and dynamic components. The vector relationship is given by

$$\mathbf{U}_1 = \mathbf{U}_s - \mathbf{U}_d \tag{3.40a}$$

$$\mathbf{U}_2 = \mathbf{U}_s + \mathbf{U}_d \tag{3.40b}$$

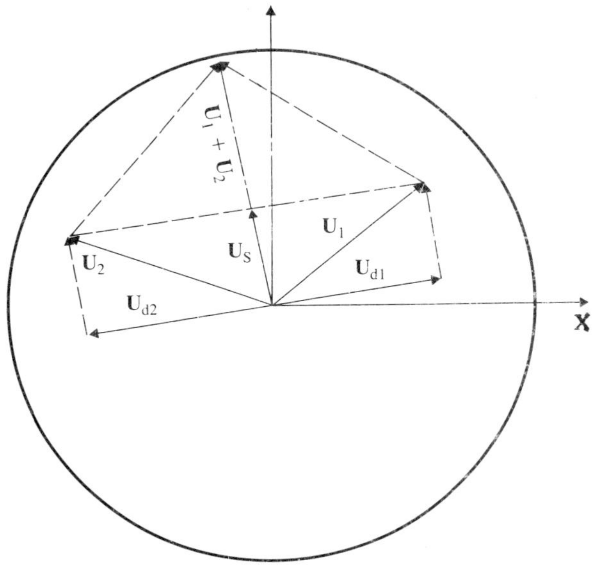

FIGURE 3.7 Resolution of two-plane unbalance into static and dynamic components.

Adding the above two equations, we obtain the effective static unbalance on each wheel as follows:

$$\mathbf{U}_s = \frac{\mathbf{U}_1 + \mathbf{U}_2}{2} \tag{3.41a}$$

The dynamic component or couple is given by

$$\mathbf{U}_d = \frac{\mathbf{U}_2 - \mathbf{U}_1}{2} \tag{3.41b}$$

In two-plane balancing, it is often advantageous to resolve the unbalance or displacements into static and dynamic components. This resolution represents a modal decomposition into static ϕ_s (in-phase) components and dynamic ϕ_d (out-of-phase) components as follows:

$$\phi_s = \begin{Bmatrix} 1 \\ 1 \end{Bmatrix} \quad \text{static or in-phase mode}$$
$$\phi_d = \begin{Bmatrix} -1 \\ 1 \end{Bmatrix} \quad \text{dynamic or out-of-phase mode} \tag{3.42}$$

The two-plane unbalance vectors $\mathbf{U}_1$ and $\mathbf{U}_2$ are represented in terms of modal components as follows:

$$\begin{Bmatrix} \mathbf{U}_1 \\ \mathbf{U}_2 \end{Bmatrix} = \{\mathbf{U}\} = \mathbf{U}_s\phi_s + \mathbf{U}_d\phi_d$$
$$\{\mathbf{U}\} = \mathbf{U}_s\begin{Bmatrix} 1 \\ 1 \end{Bmatrix} + \mathbf{U}_d\begin{Bmatrix} -1 \\ 1 \end{Bmatrix} \tag{3.43}$$

Therefore, any two arbitrary planes of unbalance may be vectorially resolved into a static component $\mathbf{U}_s$ and a dynamic component $\mathbf{U}_d$. The $\mathbf{U}_s$ component at each plane is acting in the same radial direction and generates a centrifugal force $\mathbf{F}_u$. It causes a shift in the mass eccentricity vector $\mathbf{e}_u$. The dynamic components $\mathbf{U}_d$ at each plane are acting 180° out of phase to each other. They do not cause a shift of the mass unbalance eccentricity vector but instead create a dynamic moment $\mathbf{M}_u$.

If $\mathbf{U}_s \gg \mathbf{U}_d$, then the system is said to have static unbalance and only one balance plane is required. If $\mathbf{U}_d \gg \mathbf{U}_s$, then the system is said to have dynamic unbalance and two balance planes are required. For dynamic balancing, two balance weights of equal magnitude, placed 180° out of phase to each other, are required at the two balance planes.

3.2.4 Dynamic Balancing of Arbitrary Rigid-Body Rotors

A rigid-body rotor has static balance when the shaft axis of rotation passes through the mass center of the rigid body. The mass eccentricity vector $\mathbf{e}_u$ is zero, hence no net radial centrifugal unbalance forces are generated. For the case of static unbalance, only one balance correction plane is required to achieve balance. If two planes are used, then equal weights may be placed at each plane to achieve static balance. If the rotor is placed on knife-edge supports or a bubble balance (e.g., a tire-balancing machine), then no apparent unbalance appears to exist in the rotor.

The most general case of rotor unbalance is any arbitrary unbalance distribution. This arbitrary unbalance distribution can be resolved into static and dynamic components. Upon spinning a rotor that has been statically balanced, however, large bearing forces may be encountered. This is caused by the existence of dynamic unbalance in the rotor, creating a dynamic moment about the rotor mass center.

Dynamic unbalance exists whenever the spin axis of the shaft does not correspond to the rotor principal inertia axis. The rotor may be in static balance because the spin axis passes through the rotor center of gravity, but may be dynamically unbalanced if the spin axis or axis of shaft rotation does not correspond to the principal inertia axis. Dynamic unbalance normally cannot be corrected by a single-plane balancing correction. In general, two balancing planes are required for dynamic rigid-body rotor unbalance.

Rigid-Body Constrained Motion about Z Axis

Assume that the rotor of Fig. 3.6 is constrained to rotate about the Z axis. This condition is similar to the situation of a hard bearing balancing machine.

The dynamic moments created about the x, y, z axes fixed in the rotor are given by

$$\begin{aligned} M_x &= \alpha I_{zz} - \omega^2 I_{yz} \\ M_y &= \alpha I_{yz} + \omega^2 I_{xz} \\ M_z &= \alpha I_{zz} \end{aligned} \tag{3.44}$$

For constant angular speed of rotation ω ($\alpha = 0$), the constrained Euler equations reduce to

$$\begin{aligned} M_x &= -\omega^2 I_{yz} \\ M_y &= +\omega^2 I_{xz} \end{aligned} \tag{3.45}$$

The product-of-inertia terms for a rigid body are defined by

$$I_{xz} = -\int_v \rho xz\, dv \qquad I_{yz} = -\int_v \rho yz\, dv$$

For a rigid-body rotor to be dynamically balanced, the product-of-inertia terms I_{yz} and I_{xz} must be zero.

Example 3.6: Dynamic Moment Caused by Unbalance. Calculate the product of inertia caused by a 2-g weight on mass 2 and the dynamic moment generated at 3600 rpm in Fig. 3.6. Assume $R = 5$ in and $L_2 = 10$ in. Then

$$I_{yz} = -\sum m_i y_i z_i = 0$$

$$I_{xz} = -\sum m_i x_i z_i = \frac{-10\text{ g} \times 2.2\text{ lb/kg} \times 5\text{ in} \times 10\text{ in}}{1000\text{ g/kg} \times 386\text{ in/s}^2}$$

$$I_{xz} = -2.85 \times 10^{-3}\text{ lb} \cdot \text{in} \cdot \text{s}^2$$

At 3600 rpm, the dynamic moment about the rotating **y** axis generated by the unbalance weight is given by

$$M_y = \omega^2 I_{xz} = \left(\frac{3600}{60} \times 2\pi\right)^2 I_{xz} = 405\text{ in} \cdot \text{lb}$$

Euler's Equations for Principal Axes

A set of coordinate axes may always be determined in which the product-of-inertia terms I_{xz} and I_{yz} vanish. A rigid body spinning freely will tend to rotate about either the maximum or the minimum principal inertia axis. This is the basic principle of a soft bearing balancing machine. The soft bearing spring rate allows the rotor to spin about a principal inertia axis.

Figure 3.8 represents a disk of mass M, radius R, and thickness T which is attached to a rigid shaft at an inclined angle of τ degrees from the XY plane fixed

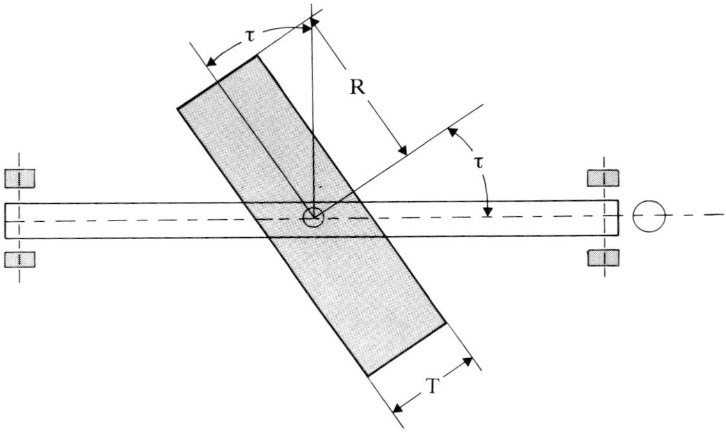

FIGURE 3.8 Skewed circular disk on massless shaft.

in the shaft. The x, y, z axes are a set of principal axes fixed in the disk. The shaft is assumed to rotate about the Z axis (directed along the bearing centerline) with an angular velocity of ω.

The principal axes of the disk are the x, y, and z coordinates fixed in the disk. If the bearing constraints were removed, then the disk would tend to rotate about the local $\mathbf{z}$ axis, which is normal to the disk. By constraining the skewed disk to rotate about the fixed Z axis, a dynamic moment or couple is created. This dynamic moment may be expressed either in terms of products of inertia by using the X, Y, Z axes set fixed in the shaft or in terms of principal moments of inertia and skew angle by using the $\mathbf{x}$, $\mathbf{y}$, $\mathbf{z}$ principal axes fixed in the disk.

Euler's equations for principal axes reduce to

$$
\begin{aligned}
M_x &= \dot{\omega}_x I_{xx} + \omega_y \omega_z (I_{zz} - I_{yy}) \\
M_y &= \dot{\omega}_y I_{yy} + \omega_x \omega_z (I_{xx} - I_{zz}) \\
M_z &= \dot{\omega}_z I_{zz} + \omega_x \omega_y (I_{yy} - I_{xx})
\end{aligned}
\tag{3.46}
$$

The angular velocity vector $\boldsymbol{\omega}$ is given in terms of the $\mathbf{x}$, $\mathbf{y}$, $\mathbf{z}$ axes fixed in the disk by

$$
\boldsymbol{\omega} = \omega \mathbf{K} = \omega_x \mathbf{i} + \omega_y \mathbf{j} + \omega_z \mathbf{k} \tag{3.47}
$$

$$
\begin{gathered}
\boldsymbol{\omega} = -\omega \sin \tau \mathbf{i} + \omega \cos \tau \mathbf{k} \\
\omega_x = -\omega \sin \tau \qquad \omega_z = \omega \cos \tau
\end{gathered}
\tag{3.48}
$$

For constant angular velocity ω, the moment expressions are given by

$$
\begin{aligned}
M_x &= 0 \\
M_y &= -(\omega^2 \sin \tau \cos \tau)(I_{xx} - I_{zz}) \\
M_z &= 0
\end{aligned}
\tag{3.49}
$$

For a uniform disk or a cylinder of radius R and length L, the principal moment of inertia I_{zz} is referred to as the *polar moment of inertia* I_p, and the principal moments of inertia I_{yy} and I_{zz} are called the *transverse moments of inertia* I_t. These values are given by

$$
I_{zz} = I_p = M \frac{R^2}{2} \tag{3.50}
$$

$$
I_{xx} = I_{yy} = I_t = M\left(\frac{R^2}{4} + \frac{L^2}{12}\right) \tag{3.51}
$$

The dynamic moment M_y generated by the rotating skewed disk or cylinder is given by

$$
M_y = \frac{\mathbf{I}_p - \mathbf{I}_t}{2} \omega^2 \sin 2\tau \tag{3.52}
$$

Since the y and Y axes coincide, $M_y = M_Y$, we conclude that the product-of-inertia term as measured in shaft fixed axes is related to the disk polar and transverse moments of inertia and skew angle as follows:

$$
I_{xz} = \frac{I_p - I_t}{2} \sin 2\tau \tag{3.53}
$$

$$
\approx \tau(I_p - I_t) \qquad \text{for small skew angles} \tag{3.54}
$$

Hence in a soft bearing balancing machine, the runout angle τ is given by

$$\tau = \frac{I_{xy}}{I_p - I_t} \tag{3.55}$$

Figure 3.9*a* represents a balanced disk of radius R and thickness $L = T$. Two unbalance weights of mass m are placed at a radius of R on either side of the

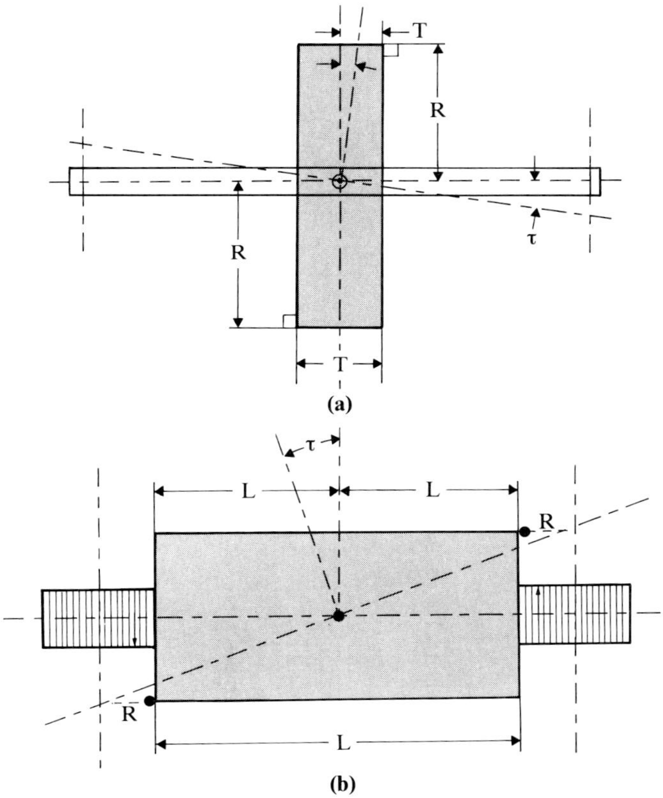

FIGURE 3.9 (*a*) Disk with two-plane couple unbalance; (*b*) cylinder with two-plane couple unbalance.

faces of the disk 180° out of phase to each other. If the disk is rotated about the Z axis with an angular velocity of ω rad/s, a dynamic couple or moment acting along the relative or rotating y axis is

$$\begin{aligned} \mathbf{M}_y &= \mathbf{R}_1 \times \mathbf{F}_1 + \mathbf{R}_2 \times \mathbf{F}_2 \\ &= \frac{T}{2}\mathbf{k} \times \omega^2 R m_1 \mathbf{i} + \left(-\frac{T}{2}\mathbf{k}\right) \times (-\omega^2 R m_2 \mathbf{i}) \\ &= \omega^2 m R T \mathbf{j} = \omega^2 U T \mathbf{j} \\ &= \omega^2 U_c \mathbf{j} \end{aligned} \tag{3.56}$$

where

$$U_c = \text{unbalance couple} = UT$$

Since the unbalance weights lie in the xz plane attached in the disk, the product-of-inertia term I_{xz} corresponding to the x, z axes is given by

$$I_{xz} = -\sum_{i=1}^{2} m_i x_i z_i$$

$$= -\left[mR\left(\frac{T}{2}\right) + m(-R)\left(\frac{-T}{2}\right)\right] = -mRT = -U_c \tag{3.57}$$

The product-of-inertia term P_{xz} in various engineering texts is given as

$$P_{xz} = -I_{xz}$$

therefore

$$P_{xz} = mRT = U_c = -I_{xz} \tag{3.58}$$

Hence, we see that the placement of two weights of equal magnitude but 180° out of phase at a plane separation distance of T, will generate an unbalance moment or couple $\mathbf{U}_c$. The product-of-inertia value P_{xz} is equal to the unbalance couple $\mathbf{U}_c$.

The unbalance couple will cause a small rotation of the principal axes about the **y** axis by angle τ as follows:

$$\tau = \frac{I_{xz}}{I_p - I_t} = \frac{-\mathbf{U}_c}{I_p - I_t} \tag{3.59}$$

For a disk, the polar and transverse moments of inertia are

$$I_p = \frac{MR^2}{2} = \text{polar moment of inertia} = I_{zz}$$

$$= M\left(\frac{R^2}{4} + \frac{T^2}{12}\right) = \text{transverse moment of inertia} = I_{xx} = I_{yy}$$

If $T/R \ll 1$, then the disk is considered to be thin and

$$I_t = \frac{I_p}{2}$$

For a thin disk, the skew angle τ is given by

$$\tau_{\text{disk}} = \frac{-U_c}{I_p/2} = \frac{-mRT}{MR^2/4} = -\frac{8T}{D}\frac{m}{M} \quad \text{rad} \qquad \frac{m}{M} \ll 1$$

(The unbalance mass m is assumed to be small in comparison to the disk mass.) The dynamic or inertia moment generated about the rotating **y** axis is given by

$$M_{y,\,\text{inertia}} = -M_y = -\omega^2 I_{xz} = \omega^2 P_{xz} = \omega^2 U_c$$

If bearings of very low spring rate were placed at the bearing locations, the disk would rotate about the **z** principal inertia axis as shown in Fig. 3.9*a*.

Figure 3.9*b* is similar to Fig. 3.9*a* except that the unbalance weights have been placed on the ends of a cylinder of length L. A cylinder is differentiated from a disk in that the transverse moment of inertia I_t is greater than the polar moment of inertia:

$$I_t > I_p \qquad \text{for a cylinder}$$

For a uniform circular cylinder

$$I_t = M\left(\frac{R^2}{4} + \frac{L^2}{12}\right) \qquad I_p = \frac{MR^2}{2}$$

Hence, an object may be considered to be a cylinder if

$$\frac{L}{R} > 0.866 \tag{3.60}$$

The angle τ for the principal inertia axes for a cylinder is given by

$$\tau_{\text{cylinder}} = \frac{-U_c}{MR^2/4 - ML^2/12} = +\frac{R\,m}{L\,M}\frac{12}{(1-3(R/L)^2)} \tag{3.61}$$

Thus the inclusion of a dynamic unbalance couple on a rigid body will cause a shift in the principal inertia axes by a small amount τ. The rotational angle τ is negative for a disk and positive for a cylinder.

3.2.5 Two-Plane Rigid-Body Balancing by the Influence Coefficient Method

In general, two planes are required to balance a rigid-body rotor with arbitrary unbalance. Although the selection of the two balance planes is arbitrary, there are numerous practical considerations for proper selection. On long cylindrical rotors, the balance planes may be at the ends of the rotor. On a turbocharger shaft assembly, the planes may be the opposite disk faces rather than the extreme ends of the rotor.

The method of two-plane balancing was first presented by Thearle in 1934. Before the advent of the hand calculator and computer-generated solutions of the complex influence coefficients, the two-plane balancing method as originally presented by Thearle was a semigraphical procedure. The concepts presented by Thearle form the basis for the current influence coefficient method of balancing.

In balancing by the influence coefficient method, the values of the rotor mass or bearing properties do not have to be known. Measurements must be taken at two planes of motion with sufficient separation to yield two distinct sets of readings. The balancing planes do not have to correspond to the measurement planes.

The measurements may be of the bearing forces or supports (as in a hard bearing balancing machine), or of displacements of the rotor, or even velocity or acceleration of the casing or foundation. A third transducer, called the *key phasor,* is required to establish a timing reference mark on the shaft. The timing reference mark is compared to the peak amplitude (or zero crossing) of the vibration transducer to produce a phase reference angle of the vibration with respect to the timing mark.

The basic assumption of the influence coefficient method is that the vibration measured at a particular location at a fixed speed is the product of a linear combination of unbalances $\mathbf{U}_i$ and rotor influence coefficients $\mathbf{a}_{ij}$. The influence coefficients, therefore, are not a function of unbalance or loading, but vary only with speed. This condition is identical to the assumption that the rotordynamic equations of motion are linear. For the normal situation in which the rotor unbalance weights are very small in comparison to the total rotor weight, this condition is closely approximated. For the case of a rotating gyroscope with large unbalance, the equations of motion are highly nonlinear, and an iterative balancing procedure must be used.

In employing the influence coefficient method of balancing, the rotor is spun to a fixed rotational speed. The synchronous amplitude and phase at the two vibration planes are recorded. These vibrations must remain constant in amplitude and phase with time, or else other factors such as thermal warping, bowing, or loose parts must be investigated.

The initial vibration readings at the two planes for the fixed speed ω are given by

$$\mathbf{Z}_1(\omega) = Z_1\angle\phi_1 \qquad \mathbf{Z}_2(\omega) = Z_2\angle\phi_2$$

where Z_i = amplitude of ith transducer and ϕ_i = relative phase angle in degrees.

The vibrations $\mathbf{Z}_1$ and $\mathbf{Z}_2$ are assumed to be linear combinations of the unknown unbalances $\mathbf{U}_1$ and $\mathbf{U}_2$ as follows:

$$\begin{aligned} \mathbf{Z}_1 &= \mathbf{a}_{11}(\omega)\mathbf{U}_1 + \mathbf{a}_{12}(\omega)\mathbf{U}_2 \\ \mathbf{Z}_2 &= \mathbf{a}_{21}(\omega)\mathbf{U}_1 + \mathbf{a}_{22}(\omega)\mathbf{U}_2 \end{aligned} \tag{3.62}$$

To determine the speed-dependent influence coefficients, a trial weight $\mathbf{U}_t$ is placed at each balance plane, and the new vibrations are recorded.

After the placement of the trial balance weight $\mathbf{U}_{t1}$ at plane 1, these vibrations are recorded:

$$\begin{aligned} \mathbf{Z}_{11} &= \mathbf{a}_{11}(\mathbf{U}_1 + \mathbf{U}_{t1}) + \mathbf{a}_{12}\mathbf{U}_2 \\ \mathbf{Z}_{21} &= \mathbf{a}_{21}(\mathbf{U}_1 + \mathbf{U}_{t1}) + \mathbf{a}_{22}\mathbf{U}_2 \end{aligned} \tag{3.63}$$

where $\mathbf{Z}_{11}$ and $\mathbf{Z}_{21}$ are the new vibrations recorded at planes 1 and 2 due to the trial weight placed in plane 1.

The influence coefficients $\mathbf{a}_{11}$ and $\mathbf{a}_{21}$ may now be computed as follows:

$$\mathbf{a}_{11} = \frac{\mathbf{Z}_{11} - \mathbf{Z}_1}{\mathbf{U}_{t1}} \qquad \mathbf{a}_{21} = \frac{\mathbf{Z}_{21} - \mathbf{Z}_2}{\mathbf{U}_{t1}} \tag{3.64a}$$

In the normal influence coefficient method, the first trial weight is removed and a second trial weight $\mathbf{U}_{t2}$ is placed in the second plane (it is not necessary that the trial weights be identical). The resulting vibrations are

$$\mathbf{Z}_{12} = \mathbf{a}_{11}\mathbf{U}_1 + \mathbf{a}_{12}(\mathbf{U}_1 + \mathbf{U}_{t2}) \tag{3.64b}$$

$$\mathbf{Z}_{22} = \mathbf{a}_{21}\mathbf{U}_1 + \mathbf{a}_{22}(\mathbf{U}_2 + \mathbf{U}_{t2}) \tag{3.64c}$$

The influence coefficients $\mathbf{a}_{12}$ and $\mathbf{a}_{22}$ are given by

$$\mathbf{a}_{12} = \frac{\mathbf{Z}_{12} - \mathbf{Z}_1}{\mathbf{U}_{t2}} \qquad \mathbf{a}_{22} = \frac{\mathbf{Z}_{22} - \mathbf{Z}_2}{\mathbf{U}_{t2}} \tag{3.64d}$$

In matrix form,

$$\begin{Bmatrix} \mathbf{Z}_1 \\ \mathbf{Z}_2 \end{Bmatrix} = \begin{bmatrix} \mathbf{a}_{11} & \mathbf{a}_{12} \\ \mathbf{a}_{21} & \mathbf{a}_{22} \end{bmatrix} \begin{Bmatrix} \mathbf{U}_1 \\ \mathbf{U}_2 \end{Bmatrix} \tag{3.64e}$$

The balance correction weights are

$$\begin{Bmatrix} \mathbf{U}_{b1} \\ \mathbf{U}_{b2} \end{Bmatrix} = -\begin{Bmatrix} \mathbf{U}_1 \\ \mathbf{U}_2 \end{Bmatrix} = -\begin{bmatrix} \mathbf{a}_{11} & \mathbf{a}_{12} \\ \mathbf{a}_{21} & \mathbf{a}_{22} \end{bmatrix}^{-1} \begin{Bmatrix} \mathbf{Z}_1 \\ \mathbf{Z}_2 \end{Bmatrix} \tag{3.65}$$

The balance correction $\mathbf{U}_{b1}$ for plane 1 is given by

$$\mathbf{U}_{b1} = -\frac{\begin{vmatrix} \mathbf{Z}_1 & \mathbf{a}_{12} \\ \mathbf{Z}_2 & \mathbf{a}_{22} \end{vmatrix}}{\begin{vmatrix} \mathbf{a}_{11} & \mathbf{a}_{12} \\ \mathbf{a}_{21} & \mathbf{a}_{22} \end{vmatrix}} = -\frac{\mathbf{Z}_1\mathbf{a}_{22} - \mathbf{Z}_2\mathbf{a}_{12}}{\mathbf{a}_{11}\mathbf{a}_{22} - \mathbf{a}_{12}\mathbf{a}_{21}} \tag{3.66}$$

The balance correction $\mathbf{U}_{b2}$ for plane 2 is given by

$$\mathbf{U}_{b2} = -\frac{\begin{vmatrix} \mathbf{a}_{11} & \mathbf{Z}_1 \\ \mathbf{a}_{21} & \mathbf{Z}_2 \end{vmatrix}}{\begin{vmatrix} \mathbf{a}_{11} & \mathbf{a}_{12} \\ \mathbf{a}_{21} & \mathbf{a}_{22} \end{vmatrix}} = -\frac{\mathbf{Z}_2\mathbf{a}_{11} - \mathbf{Z}_1\mathbf{a}_{21}}{\mathbf{a}_{11}\mathbf{a}_{22} - \mathbf{a}_{12}\mathbf{a}_{21}} \tag{3.67}$$

Example 3.7: Two-Plane Rigid Rotor Balancing by the Influence Coefficient Method. The initial vibration readings on a rigid-body rotor are

$$\mathbf{Z}_1 = 8.6 \text{ mils } \angle 63^\circ \qquad \mathbf{Z}_2 = 6.5 \text{ mils } \angle 206^\circ$$

A trial balance weight of $U_{t1} = 10$ g is placed at a relative phase angle of 270°:

$$\mathbf{U}_{t1} = 10 \text{ g } \angle 270^\circ$$

The resulting vibrations at planes 1 and 2 due to the placement of the trial weight at plane 1 are

$$\mathbf{Z}_{11} = 5.9 \text{ mils } \angle 123^\circ \qquad \mathbf{Z}_{21} = 4.5 \text{ mils } \angle 228^\circ$$

The influence coefficients $\mathbf{a}_{11}$ and $\mathbf{a}_{21}$ may be calculated as follows:

$$\mathbf{a}_{11} = \frac{\mathbf{Z}_{11} - \mathbf{Z}_1}{\mathbf{U}_{t1}} = \frac{5.9\angle 123^\circ - 8.6\angle 63^\circ}{10\angle 270^\circ} = \frac{7.61\angle 200.9^\circ}{10\angle 270^\circ}$$

$$= 0.7612\angle -69.13^\circ = 0.7612\angle 290.9^\circ$$

$$\mathbf{a}_{21} = \frac{\mathbf{Z}_{21} - \mathbf{Z}_2}{\mathbf{U}_{t1}} = \frac{4.5\angle 228^\circ - 6.5\angle 206^\circ}{10\angle 270^\circ} = 0.287\angle 80^\circ$$

The first balance trial weight is removed, and a trial weight of $\mathbf{U}_{t2} = 12$ g at 180° is placed at plane 2. The resulting vibration readings are

$$\mathbf{Z}_{12} = 6.2 \text{ mils } \angle 36° \qquad \mathbf{Z}_{22} = 10.4 \text{ mils } \angle 162°$$

The influence coefficients at planes 1 and 2 caused by the trial weight placed at plane 2 are given by

$$\mathbf{a}_{12} = \frac{\mathbf{Z}_{12} - \mathbf{Z}_1}{\mathbf{U}_{t2}} = \frac{6.2\angle 36° - 8.6\angle 63°}{12\angle 180°} = 0.347\angle 105°$$

$$\mathbf{a}_{22} = \frac{\mathbf{Z}_{22} - \mathbf{Z}_2}{\mathbf{U}_{t2}} = \frac{10.4\angle 162° - 6.5\angle 206°}{12\angle 180°} = 0.6076\angle 303°$$

Let

$$\begin{aligned} \Delta &= \mathbf{a}_{11}\mathbf{a}_{22} - \mathbf{a}_{12}\mathbf{a}_{21} \\ &= (0.762\angle 291°)(0.608\angle 303°) - (0.347\angle 105°)(0.287\angle 80°) \\ &= 0.398\angle 245° \end{aligned}$$

Balance correction $\mathbf{U}_{b1}$ is given by

$$\begin{aligned} \mathbf{U}_{b1} &= -\frac{(8.6\angle 63°)(0.60\angle 303°) - (6.5\angle 206°)(0.384\angle 105.5°)}{0.398\angle 245°} \\ &= -10.75 \text{ g } \angle 149° = +10.75 \text{ g } \angle 329° \end{aligned}$$

and

$$\begin{aligned} \mathbf{U}_{b2} &= -\frac{\mathbf{a}_{11}\mathbf{Z}_2 - \mathbf{a}_{21}\mathbf{Z}_1}{\Delta} = -\frac{4.95\angle 136.8 - 2.47\angle 143°}{0.398\angle 295°} \\ &= -6.27 \text{ g } \angle 245° = +6.27 \text{ g } \angle 65° \end{aligned}$$

Figure 3.10 represents the relative positions on balance planes 1 and 2 of the trial balance weights and the final balance corrections. On each balance plane, 16 balance holes have been assumed. The balance holes have been labeled against rotation. In Fig. 3.10*a*, the first trial weight $\mathbf{U}_{t1}$ is shown placed at 270°, which corresponds to balance hole 13. The final balance correction $\mathbf{U}_{b1}$ is a total of 10.76 g, and this should be placed between holes 15 and 16. A vector split of the 10.76-g balance weight is given as 5.31 g at hole 15 and 5.66 g at hole 16.

The balance weights satisfy the vector equation

$$\mathbf{U}_{b1} = \mathbf{U}_{15} + \mathbf{U}_{16} \tag{3.68}$$

Figure 3.10*b* represents the placement of the trial balance weights and final balance correction for the second balance plane. The trial weight of 12 g at 180° corresponds to a weight placement at balance hole 9. The final balance correction $\mathbf{U}_{b2}$ is shown as 6.2 g to be placed in balance hole 4. If one chooses not to remove the initial trial weight from hole 9 for balance plane 2, then a trim balance correction of 15.63 g may be placed in balance hole 2. In cases where a trial weight is welded on, it may be desirable to add an additional trim weight rather than to remove the original trial weight.

The trim, trial, and balance weights in plane 2 satisfy the vector relationship

$$\mathbf{U}_{b2} = \mathbf{U}_{t2} + \mathbf{U}_{\text{trim }2} \tag{3.69}$$

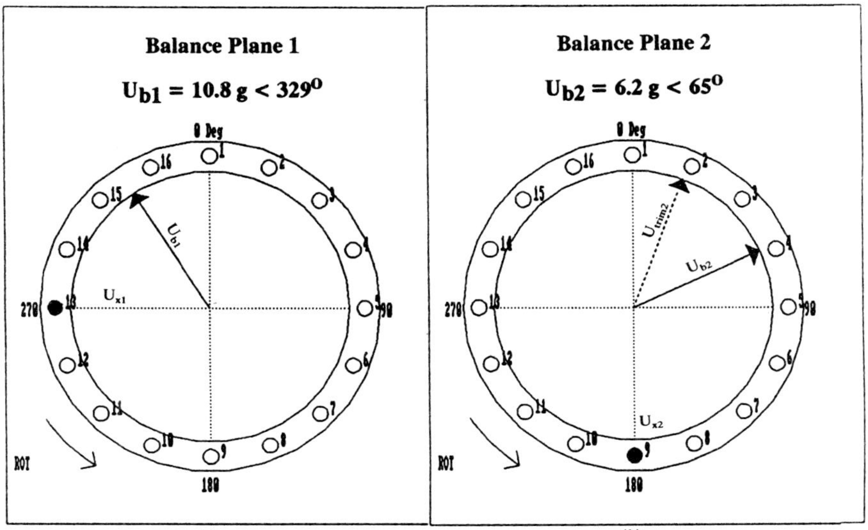

(a) (b)

FIGURE 3.10 Radial and couple unbalance caused by an offset mass on a rigid body.

In cases where the balance correction is greater than 120° from the trial balance weight, it is preferable to remove the original trial weight. In this case, with the trial weight of 12 g left in hole 9, it would require a large trim weight of almost 16 g.

3.3 THEORY OF FLEXIBLE ROTOR BALANCING

3.3.1 Single-Mass Jeffcott Rotor

Description of Rotor

The idealized single-mass flexible rotor is shown in Fig. 3.11*a*. It represents a single-mass disk situated on a massless elastic shaft of stiffness K. The mass center of the disk is displaced from the center of the elastic shaft by distance $\mathbf{e}_u$.

The analysis of the idealized single-mass rotor was first performed by H. H. Jeffcott (1919), to examine the effect of unbalance on the whirl amplitude and forces transmitted to the bearings.

Figure 3.11*b* represents a cross section of the Jeffcott rotor illustrating the mass-displacement relationships for the single-mass rotor. In the Jeffcott rotor, the mass is assumed to be located at the disk, and the shaft is assumed to be massless. The disk is assumed to be rotating with constant speed ω.

The mass center M is displaced from the shaft elastic centerline C by the distance $\mathbf{e}_u$. The distance $\mathbf{e}_u$ is referred to as the *disk unbalance eccentricity*. The total rotor effective unbalance $\mathbf{U}$ is eual to the product of the mass of the disk or rotor and the unbalance eccentricity $\mathbf{e}_u$.

The displacement of the disk mass M from the shaft elastic axis C causes a centrifugal unbalance to be exerted on the shaft at point C. This rotating

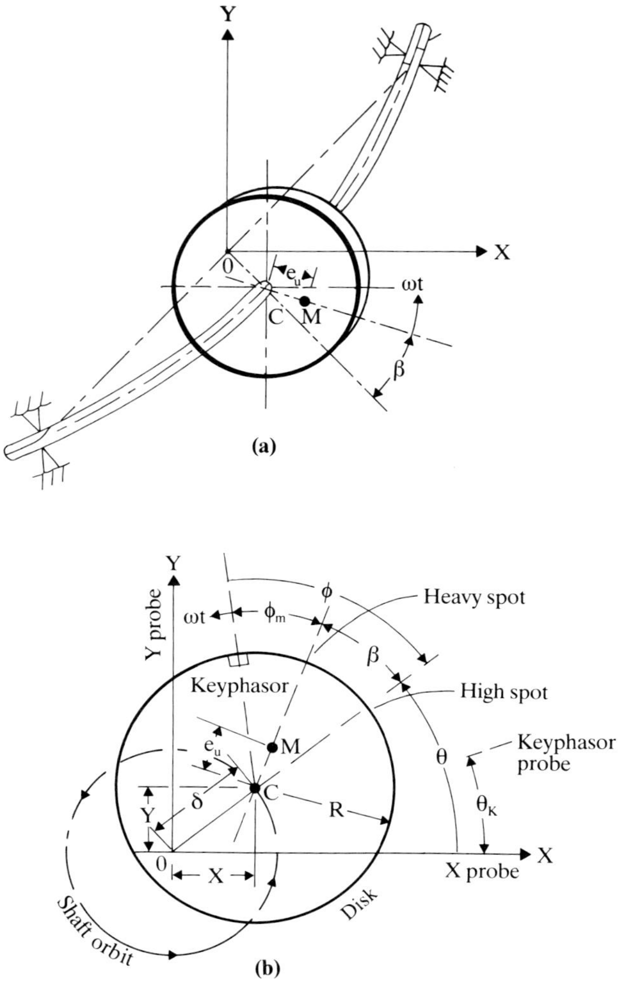

FIGURE 3.11 (*a*) Schematic view of deflected single-mass Jeffcott rotor; (*b*) cross section of Jeffcott rotor illustrating the mass displacement–phase relationship.

unbalance force causes the shaft centerline to precess or orbit about the idealized bearing centerline, point *O*, with a radius of δ at an angular velocity of ω rad/s. This motion is referred to as *synchronous circular precession*.

A reference mark called the *key phasor* is prescribed on the disk. It is used to determine a relative phase between the timing reference mark and the displacement vector $\boldsymbol{\delta}$. The angular location ϕ_m of the mass center or eccentricity vector $\mathbf{e}_u$ is measured from the keyphasor to the line of *CM* extended in a direction opposite to the direction of rotation.

The total inertial loading on the bearings is a function of the vector sum of the displacement $\boldsymbol{\delta}$ and the mass unbalance eccentricity vector $\mathbf{e}_u$. At low speeds (below the rotor critical speed or natural resonance frequency), the unbalance eccentricity vector $\mathbf{e}_u$ and the shaft displacement vector $\boldsymbol{\delta}$ are in phase.

The line of OC extended to the disk radius R marks a point on the shaft referred to as the *high spot of motion.* (For example, if one were to touch the orbiting shaft with chalk or a marking pen, the resulting mark would indicate the high spot of motion.)

The line of CM extended represents the line of action of the rotating unbalance load. The intersection of this line with the disk radius R is referred to as the disk *heavy spot.* At any given speed ω, the relative angular position of the triad of points O, C, and M is fixed with respect to each other. Points C and M orbit about the origin O with respect to a relative phase angle β, which is speed-dependent.

Flexible Rotor Equations of Motion

The forces acting on the flexible disk are the shaft elastic restoring force, rotor damping, rotor unbalance, and inertia force due to the mass of the disk. These forces may be combined by D'Alembert's principle to yield the rotor equation of motion

$$\mathbf{F}_{\text{elastic}} = -K\boldsymbol{\delta} \tag{3.70}$$

where K = shaft elastic stiffness coefficient and $\boldsymbol{\delta}$ = vector deflection of the shaft centerline C from the bearing line of centers O. The angular speed ω rad/s of the rotor is assumed to be constant and is represented by the time derivative of the angle formed by the line CM fixed in the disk with respect to the fixed reference axis OX. The rotation of the disk causes a rotating unbalance force with magnitude

$$\mathbf{F}_u = M\mathbf{e}_u\omega^2 = \mathbf{U}\omega^2 \tag{3.71}$$

The damping force opposing the motion is assumed to be proportional to the velocity and is given by

$$\mathbf{F}_d = -C\mathbf{V} = -C[\dot{X}\mathbf{i} + \dot{Y}\mathbf{j}] \tag{3.72}$$

The equations of motion for the single mass with radial unbalance is given in fixed cartesian coordinates by

$$\begin{aligned} M\ddot{X} + C\dot{X} + KX &= Me_u\omega^2 \cos(\omega t - \phi_m) \\ M\ddot{Y} + C\dot{Y} + kY &= Me_u\omega^2 \sin(\omega t - \phi_m) \end{aligned} \tag{3.73}$$

The equations of motion for the single-mass Jeffcott rotor may be combined into one complex vector equation by means of the complex variable transformation $\mathbf{Z} = X + \mathbf{i}Y$ as follows:

$$M\ddot{\mathbf{Z}} + C\dot{\mathbf{Z}} + K\mathbf{Z} = Me_u\omega^2 e^{i(\omega t - \phi_m)} \tag{3.74}$$

The complex variable $\mathbf{Z}$ represents the motion of the shaft centerline C.

Synchronous Unbalance Response of the Jeffcott Rotor

The vibration of the rotor as governed by Eq. 3.74 consists of an initial transient motion and a steady-state whirling or synchronous response caused by the rotating unbalance force:

$$\mathbf{Z} = \mathbf{Z}(t)_{\text{transient}} + \mathbf{Z}(\omega)_{\text{steady state}}$$

The steady-state complex rotor motion due to unbalance is given by

$$\mathbf{Z}_{\text{steady state}} = \frac{Me_u\omega^2}{K - M\omega^2 + iC\omega} e^{i(\omega t - \phi_m)} \tag{3.75}$$

The complex rotor motion may also be expressed in terms of amplitude of the whirl orbit radius δ and the phase lag angle ϕ by

$$\mathbf{Z}(\omega) = \delta(\omega) e^{i(\omega t - \phi)} \tag{3.76}$$

where δ = whirl orbit radius
$= Me_u\omega^2[(K - M\omega^2)^2 + C\omega^2]^{-1/2}$
ϕ = phase lag = $\phi_m + \beta$ (3.77)
ϕ_m = angular location of mass center with respect to keyphasor
β = relative phase lag angle = $\tan^{-1} \dfrac{C\omega}{K - M\omega^2}$ (3.78)

The amplitude and phase may be further expressed in dimensionless form by using the following variables:

$\omega_{\text{cr}} = \sqrt{\dfrac{K}{M}}$ = natural frequency of shaft on rigid supports, rad/s

$C_c = 2M\omega_c$ = critical damping, lb · s/in

$\dfrac{C}{C_c} = \xi$ = damping ratio

$\dfrac{\omega}{\omega_{\text{cr}}} = f$ = frequency ratio

$\dfrac{C\omega}{K} = 2\xi f$

$A = \dfrac{\delta}{e_u}$ = amplification factor

The dimensionless shaft amplitude is given by

$$A = \frac{f^2}{\sqrt{(1 - f^2)^2 + (2\xi f)^2}} \tag{3.79}$$

The relative phase-angle lag of the deflection vector from the rotating unbalance load is given by

$$\beta = \tan^{-1} \frac{2\xi f}{1 - f^2} \tag{3.80}$$

There are three speed ranges of operation of the Jeffcott rotor that are of interest. These speed ranges are defined in terms of the frequency ratio f of the rotor angular velocity ω to the system critical speed ω_{cr}. The rotor amplitude and phase relationships vary considerably between the three speed regions.

From the standpoint of balancing, it is important to know in what speed range the rotor is operating, in order to properly place correction weights on the rotor for the determination of influence coefficients for balancing.

The three speed ranges of interest are as follows:

1. Subcritical speed operation: $f < 1,\ \beta < 90°$

$$A \approx f^2$$

(Unbalance eccentricity vector $\mathbf{e}_u$ and deflection vector $\boldsymbol{\delta}$ are in phase.)

2. Critical speed operation: $f = 1,\ \beta = 90°$

$$A = \frac{1}{2\xi} = A_{cr} \qquad \text{critical speed amplification factor}$$

(Unbalance eccentricity vector $\mathbf{e}_u$ is leading the deflection vector by $\approx 90°$.)

3. Supercritical speed operation: $f \gg 1,\ \beta > 90°$

$$A = 1$$

Unbalance eccentricity vector $\mathbf{e}_u$ is out of phase to the deflection vector $\boldsymbol{\delta}$.

Figure 3.12 represents the dimensionless rotor synchronous response versus the speed ratio f for various values of the damping parameter $D = 2\xi$. Below $f = 1$ and for the curve $D = 0$ (no rotor damping), the rotor unbalance response increases as the square of the speed. Figure 3.13a represents the relative phase-angle change β, between the unbalance inertia load and the shaft centerline displacement vector $\boldsymbol{\delta}$, as given by Eq. 3.80.

For low values of D and for speeds below the critical speed ($f < 1$), the

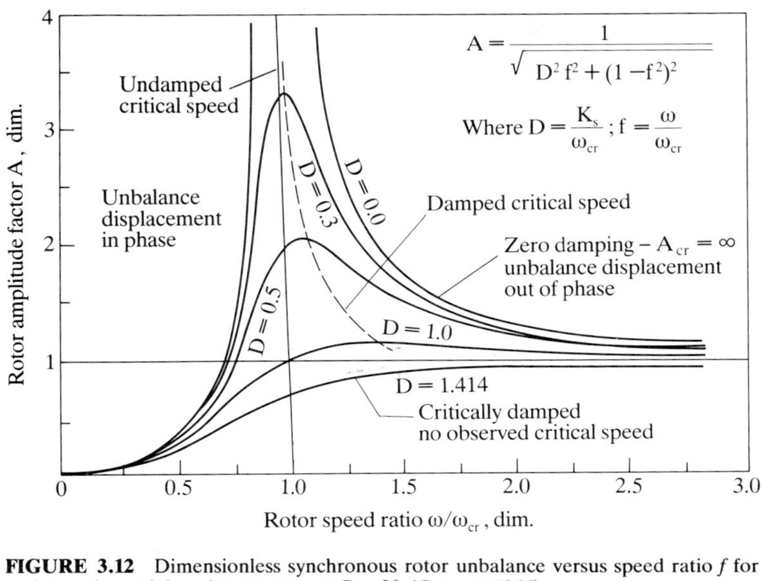

FIGURE 3.12 Dimensionless synchronous rotor unbalance versus speed ratio f for various values of damping parameter $D = 2\xi$ (Gunter, 1966).

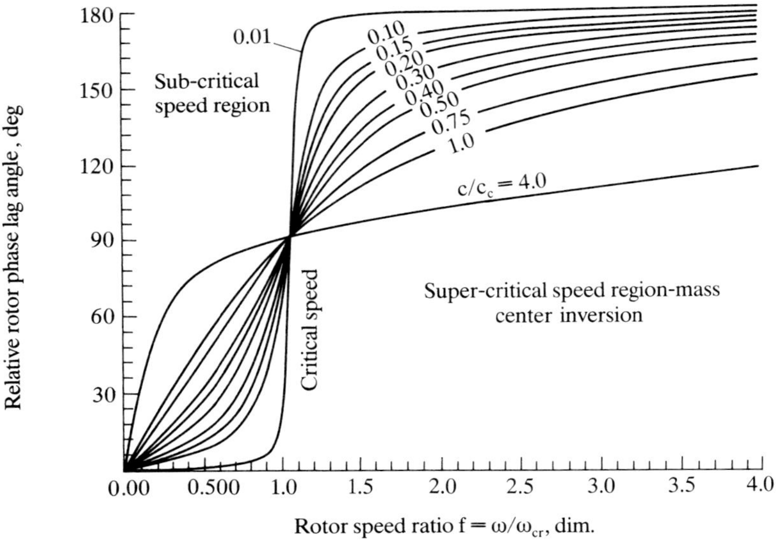

(a)

FIGURE 3.13 (*a*) Rotor relative unbalance force-displacement phase lag angle for various values of dimensionless damping ξ; (*b*) rotor displacement-mass phase relationship below, at, and above the critical speed; (*c*) definition of high spot with noncontacting proximity probe (Bently-Nevada Company, 1984a, b).

deflection vector $\boldsymbol{\delta}$ lies along the same line of action as the rotating unbalance force.

At speeds below the rotor critical speed ($N < N_{cr}$), the rotor mass unbalance eccentricity vector $\mathbf{e}_u$ and the shaft elastic displacement vectors are in phase. Under these circumstances, the relative mass-displacement phase angle β is small, and the shaft high spot and heavy spot are said to coincide.

In Fig. 3.13*a*, e.g., the value of $\xi = 0.01$ represents a lightly damped rotor ($A_{cr} = 50$). For speeds up to 80 percent of the critical speed, there are only a few degrees of phase lag of the displacement vector from the forcing function (rotor unbalance). As the rotor damping increases, the phase change below the critical speed becomes more pronounced.

In Fig. 3.11*b*, the motion of the flexible rotor may be observed by the use of two noncontacting proximity probes. These probes are illustrated as the *X* probe and *Y* probe and are shown in the figure. By combining the sinusoidal signals as observed by the probes, the orbit of the shaft centerline may be observed. This is represented by the orbit formed by the line *OC*. For the ideal Jeffcott rotor with no bearing asymmetry, the orbit is circular at all speeds.

The line of *OC* extended to the radius of the disk represents the high spot of motion. The high spot is the part of the shaft that would first contact a closely fitting labyrinth seal, e.g., and would leave a mark on a shaft at this location. The extension line *CM* represents the line of action of the rotating unbalance load. The intersection of line *CM* on the disk radius represents the heavy spot, point *U*. The extension of line *CM* to point *B* represents the location on the disk wheel where the balance correction weight should be applied to counteract the unbalance eccentricity of the shaft.

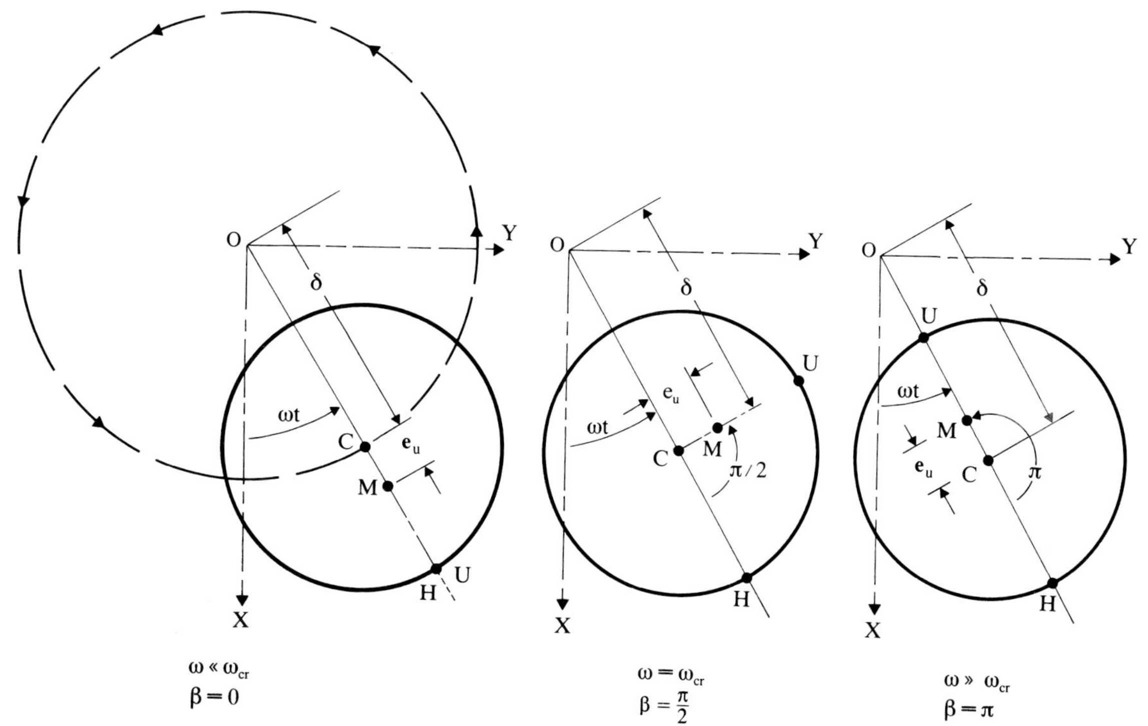

(b)

FIGURE 3.13 (*Continued.*)

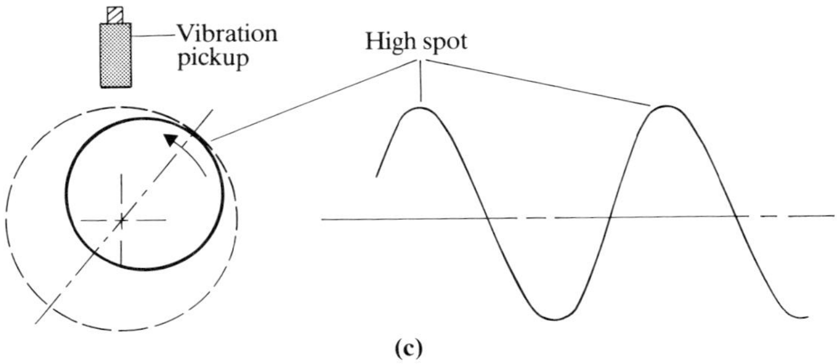

FIGURE 3.13 (*Continued*.)

For speed values up to 80 percent of the critical speed and light damping, Fig. 3.13*a* shows that the phase-angle lag β increases only a few degrees. This means that point *U*, the heavy spot, and point *H*, the high spot, coincide. Figure 3.13*b* represents a schematic sketch of the relationship between the high spot and heavy spot below, at, and above the critical speed. Diagram *A* of Fig. 3.13*b* for $f \ll f_{cr}$ shows that the relative phase lag angle β is approximately 0 and the shaft heavy and high spots *U* and *H* correspond.

When the rotor speed *N* is equal to undamped critical speed N_{cr} ($f = 1$), the mass center is leading the deflection vector by 90° regardless of the value of damping in the rotor, as given by Eq. 3.80. At this speed, the amplitude of the rotor is determined by the amount of damping in the rotor system. For $f = 1$, the rotor dimensionless amplitude is given by

$$A_{ct} = \frac{1}{2\xi} \tag{3.81}$$

This amplitude is called the *critical speed amplification factor*. For the case of zero damping, the amplification factor at the rotor center would be theoretically inflnite. The rotor unbalance response would have unacceptably high amplitudes of vibration. After the rotor has passed through the critical speed region, the response reduces.

At the critical speed, the mass center is leading the displacement vector by 90°. If, e.g. a shaft were to rub while passing through the critical speed, the high-spot mark on shaft point *H* would not correspond to the heavy spot. The high spot *H* is lagging the heavy spot *U* by 90° as shown in diagram *B* of Fig. 3.13*b*. A balance correction weight, therefore, should be placed approximately 90° from the rub mark opposite to the direction of rotation. (This angle is reduced from 90° depending on the system damping.)

For the case of an undamped or lightly damped rotor, Fig. 3.13*a* shows that there is a rapid phase shift while passing through the rotor critical speed. This phase shift changes abruptly from 0° to almost 180° depending on the level of damping. Therefore, balancing a lightly damped rotor in the critical speed region is extremely difficult because of the rapid phase change. Ball-bearing-supported rotors have inherently low damping and hence have high amplification responses when operated in the vicinity of a critical speed. In addition, rapid phase changes occur over a small speed range. Hence, this class of rotor is extremely difficult to balance if data are taken near the critical speed region.

After the rotor passes through the critical speed ($f > 1$), the relative phase angle β increases beyond 90° and approaches 180° for speeds well above the critical speed ($f \gg 1$). This speed of operation is referred to as the supercritical speed range.

When the rotor is operating in the supercritical speed range, the eccentricity mass vector $\mathbf{e}_u$ and the deflection vector $\boldsymbol{\delta}$ are out of phase. The mass center M approaches point O, the spin axis center, in the limit as the rotor speed becomes very high.

Under these conditions, the rotor is said to be *self-balanced*. The heavy spot U approaches point O and is 180° out of phase to the rotor high spot, point H, as shown in diagram C of Fig. 3.13*b*.

Figure 3.12 shows that the rotor dimensionless amplitude A approaches unity at speeds well above the critical speed. This implies that the rotor whirl radius δ approaches the unbalance eccentricity $\mathbf{e}_u$ in magnitude, but 180° out of phase. If a chalk mark were placed on the rotor to indicate the high spot during supercritical speed operation, it would be 180° out of phase to the mass eccentricity vector. Therefore, an initial trial balancing weight should be placed at the location of the chalk mark or indicated high spot.

The high spot of shaft motion may be observed experimentally by means of a noncontact displacement probe mounted near the shaft. Figure 3.13*c* represents a shaft moving in a circular synchronous orbit. The vibration pickup records the sinusoidal motion as observed in that plane.

The harmonic waveform may be displayed on an oscilloscope. The high spot is represented by the peak amplitude on the waveform. As the shaft passes under the vibration pickup, the gap between the pickup and shaft reduces until the maximum amplitude of motion occurs when the shaft is directly under the probe.

To find the high spot on the shaft, a timing mark must be provided on the shaft as a reference point to measure the phase between the moment when the timing mark passes under the probe and the maximum amplitude.

Figure 3.14 represents the transient motion of a rotor with a suddenly applied unbalance while operating at the critical speed. The situation simulates the loss of a turbine blade on an engine. Figure 3.14*a* represents the motion of the rotor at the critical speed with no damping in the system. The motion shown in this figure is similar to the motion that would occur with a ball bearing rotor attempting to operate in the critical speed region. This figure shows that the rotor amplitude increases with time in a linear fashion. The unbalance orbit will become unbounded unless the action of nonlinear shaft forces creates a limit cycle orbit.

If the rotor, however, is accelerated and the speed is increased above the critical speed, then the undamped rotor amplitude will decrease. Therefore, with sufficient acceleration, an undamped or lightly damped rotor may be forced to traverse the critical speed region without encountering unbounded amplitudes of response.

If the undamped rotor is slowly decelerated, when it reaches the critical speed region, the amplitude of motion will again start to increase linearly with time, unless there is a sufficiently rapid deceleration rate to remove it from the critical speed region.

On these orbits, timing marks for each cycle of motion have been placed to represent the effect of a keyphasor probe. These timing marks represent a 90° phase shift opposite to the direction of rotation from the location of the timing marks at speeds well below the critical speed. The observation of the location of the timing mark on the orbit is useful in determining the location of initial trial weights for balancing.

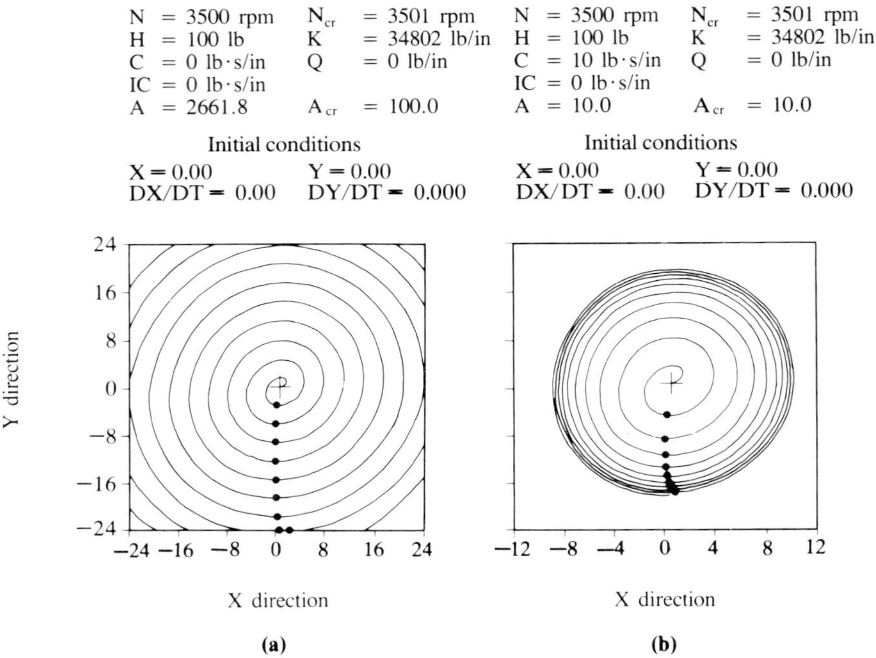

FIGURE 3.14 Transient motion at critical speed with suddenly applied blade loss (Kirk and Gunter, 1973): (*a*) no rotor damping, (*b*) $A_c = 10$.

The keyphasor probe fires when the shaft notch passes under the keyphasor probe. This signal may be inserted into the *Z* axis of an oscilloscope to form a timing mark on the orbit. For the case of Fig. 3.14*b* it is known that the shaft orbit corresponds to the rotor critical speed. In this case, the mass center is leading the displacement vector by 90°.

In Fig. 3.14*b*, the rotor has finite damping and an amplification factor of 10. This implies that the maximum response at the critical speed will be 10 times the unbalance eccentricity vector. Hence, the lower the amplification factor, the lower the response will be at the critical speed. Figure 3.14*b* shows that the orbit initially spirals outward in a fashion similar to the case of the undamped rotor. However, due to the action of the damping, the initial transient motion is suppressed, and a limit cycle synchronous whirl orbit of 10 times the unbalance eccentricity is achieved.

Polar Representation of Rotor Synchronous Amplitude and Phase Response

The synchronous Jeffcott rotor response and phase-angle change with speed as given by Eqs. 3.77 and 3.78 may be combined in a polar form, as shown in Fig. 3.15. In Fig. 3.15, the individual amplitude and phase plots, commonly referred to as *Bode plots,* are shown on the right-hand side (Bently-Nevada, 1984a, b). The amplitude and phase may be combined into a polar or Nyquist plot similar to the theory of transfer functions as employed in electric circuits. In this figure, the

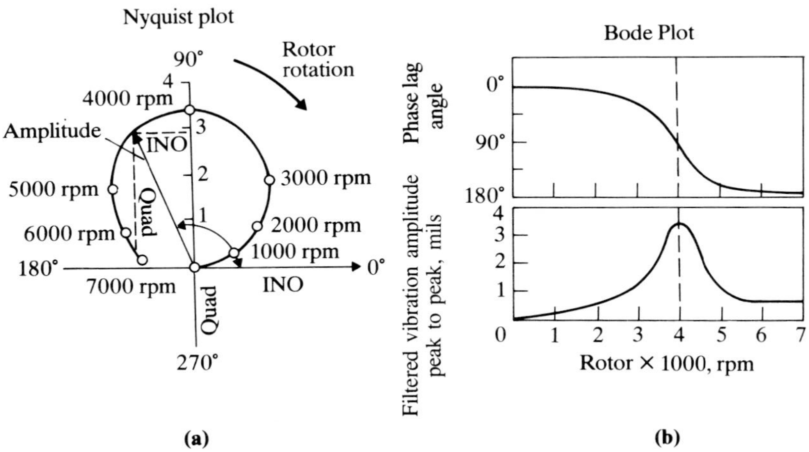

FIGURE 3.15 Typical Nyquist (polar) and Bode plots (Bently-Nevada Company, 1984a, b).

mass center is assumed to be located along the line of the timing mark or the keyphasor.

The rotor phase angle lag is measured from the X axis which is the 0° reference mark. As the rotor speed increases, the amplitude and phase increase, forming a polar plot. The phase-angle increase for the Jeffcott rotor is plotted in a direction opposite to that of shaft rotation.

The component of amplitude measured along the **X** axis is often referred to as the *co-component,* and the amplitude of motion measured along the **Y** axis is referred to as the *quad component.* For the Jeffcott rotor, the dimensionless co-component and quad component are given as

$$\text{Co} = \frac{f^2(1-f^2)}{(1-f)^2 + (2\xi f)^2} \tag{3.82a}$$

$$\text{Quad} = \frac{-2\xi f^3}{(1-f^2)^2 + (2\xi f)^2} \tag{3.82b}$$

The zero crossing of the in-phase or co-component indicates the location of the critical speed. The maximum value of the quad component, which occurs at the zero crossing of the in-phase curve, is equal to the rotor amplification factor. For example, in the Nyquist plot of Fig. 3.15 at a rotor speed of 4000 rpm, the rotor phase angle has shifted by 90°. The 90° phase shift occurs at the rotor undamped critical speed.

As the rotor speed increases, the rotor amplitude reduces and the phase angle becomes greater than 90°. At speeds above approximately twice the rotor speed, and the quad or out-of-phase component of the amplitude diminishes until a 180° phase shift is obtained. For the sample rotor plotted in Fig. 3.15, the phase inversion occurs approximately at speeds above 7000 rpm. Under these circumstances, the rotor displacement is 180° out of phase with the rotor mass center. The distance from the origin to the 7000 rpm speed point on the polar plot represents twice the rotor unbalance eccentricity value, assuming that the plot of rotor amplitude is peak to peak. The rotor amplification factor may be quickly estimated by measuring the maximum quad amplitude (90° phase shift) divided by the amplitude after the 180° phase shift.

In an actual rotor, the mass center does not generally lie along the line of the timing mark or keyphasor. The mass center is displaced from the keyphasor axis by the phase angle ϕ_m. Figure 3.16, e.g., represents a typical polar plot of a single-mass rotor (Bently-Nevada, 1984a, b).

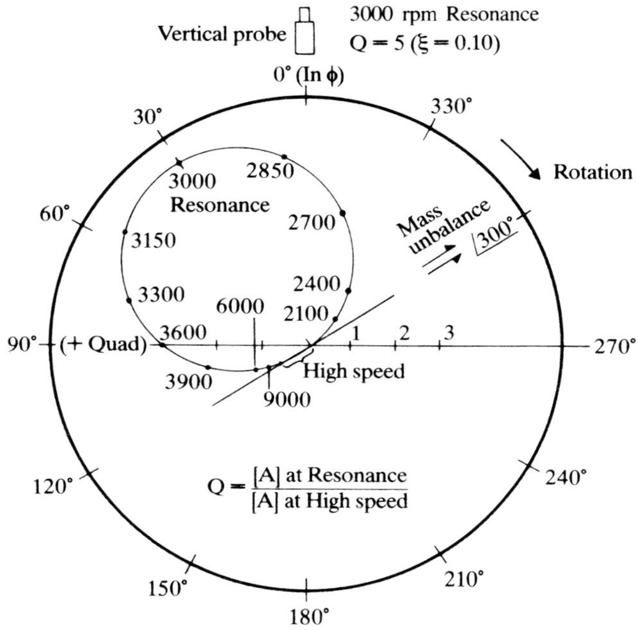

FIGURE 3.16 Polar plot for vertical probe (Bently-Nevada Company, 1984a, b).

In the displacement phase convention adopted by Bently-Nevada, the 0° reference mark is placed at the location of the monitoring probe. In this particular case, the probe is located at the vertical axis. Probe orientation is not restricted to either the horizontal or the vertical axis, but may, in general, be in any arbitrary orientation.

In the polar plot shown in Fig. 3.16, the direction of rotation is clockwise. The phase-angle change for the Jeffcott Rotor (in the absence of shaft bow or rub effects) is always against rotation. In this example, the phase lag angle is plotted in the clockwise direction.

The rotor has an amplification factor of 5. The resonance frequency occurs at 3000 rpm. At 9000 rpm which is 3 times the critical speed, the phase increases to 180°. This represents the inversion speed. A tangent drawn from this point to the origin represents the line of action of the mass unbalance.

For the rotor system illustrated, the mass unbalance lies along the 300° mark with respect to the vertical probe. Balancing weights, therefore, should be applied at the 120° location. Figure 3.16 represents the dimensionless polar (Nyquist) plot for Q or the amplification factor for $A = 5$ (or $Q = 5$) for various values of the speed ratio f (Bently-Nevada, 1984a, b). In this idealized polar plot of the dimensionless rotor amplitude, the angular location of the rotor mass center ϕ_m from the timing mark is 300° against rotation. The peak amplitude is shown to occur at 3000 rpm with a corresponding phase lag angle of 30°. This represents a 90° phase shift from the low-speed value of 300°.

Phase Lag Angle at Maximum Rotor Unbalance Response Speed
It is important to understand that the maximum rotor response does not occur at the critical speed corresponding to the 90° phase shift. The maximum rotor response occurs at slightly higher speed and is given by

$$\omega_u = \text{unbalance response speed} = \frac{\omega_{cr}}{(1 - 2\xi^2)^{1/2}} \tag{3.83}$$

For $A = 5$ or $\xi = 0.10$, the speed ratio of maximum response is $f_{max} = 1.0102$. At this speed ratio the relative phase lag angle of the deflection vector to the rotor eccentricity vector $\mathbf{e}_u$ is 95.8°.

The speed at which maximum unbalance response is achieved, the *resonance response speed,* is higher than the rotor undamped critical speed depending on the system damping value, given by Eq. 3.83. For low to moderate values of ξ, the difference between the undamped critical speed and the maximum response speed is only a few percent.

There are three resonance frequencies or eigenvalues associated with the Jeffcott rotor. These frequencies are the system damped natural frequency ω_d, the system undamped natural frequency or critical speed ω_{cr}, and the speed at which maximum unbalance response amplitude is obtained ω_u. The relationship between these frequencies is

$$\omega_u > \omega_{cr} > \omega_d \tag{3.84}$$

where ω_d = damped natural frequency = $\omega_{cr}(1 - \xi^2)^{1/2}$
ω_u = maximum unbalance response frequency = $\omega_{cr}/(1 - 2\xi^2)^{1/2}$

The damped natural frequency (or damped system eigenvalue) is the frequency observed by striking or impacting the rotor. This frequency is always lower than the undamped rotor critical speed for the Jeffcott rotor. When the rotor speed reaches the value of the theoretical system natural frequency or critical speed, the mass center is leading the deflection vector by 90°. However, the theoretical critical speed (based on the undamped system eigenvalue) does not precisely represent the speed at which the maximum unbalance response will be obtained.

The speed of maximum unbalance response ω_u will always be slightly higher than the theoretical critical-speed value. Hence, the phase-angle relationship between the rotor mass center and the rotor deflection vector at maximum rotor response will be larger than 90°. The amount of phase increase beyond 90° is dependent upon the system damping.

Table 3.4 presents the relative phase angle β between the mass center and rotor deflection at the maximum unbalance response speed for various values of dimensionless damping ξ. For rotors with amplification factors above 10, the mass center is leading the deflection vector at the maximum response speed by approximately 90°. As the damping in the rotor system increases, the relative phase lag between the mass center and the displacement vector increases. For the

TABLE 3.4 Phase Angle and Frequency Ratio at Maximum Unbalance Response for Various Values of Damping ξ or Rotor Amplification Factor A_{cr}

ξ	A_{cr}	A_{max}	f_{max}	β_{max}	$\phi_{balance}$
0.05	10	10.0125	1.0025	92.9	87.1
0.10	5	5.025	1.0102	95.8	84.2
0.125	4	4.03	1.0160	97.2	82.8
0.15	3.33	3.371	1.0233	98.7	81.3
0.1667	3.0	3.042	1.0290	99.7	80.3
0.20	2.5	2.552	1.0426	101.8	78.2
0.25	2.0	2.06	1.0690	105.0	75.0
0.30	1.667	1.747	1.1043	108.3	71.7
0.40	1.250	1.364	1.2127	115.9	64.1
0.50	1.00	1.15	1.4142	125.0	55.0

relatively large amplification factor of 10, there is approximately a 3° increase in the phase angle beyond 90° at the maximum response speed.

As the system damping increases and the rotor amplification reduces, the speed at which the maximum unbalance response occurs increases. For example, for an amplification vector of 5, there is only a 1 percent increase in the forced response frequency above the natural frequency. However, the relative mass-displacement phase angle β is 96° at peak amplitude.

For highly damped rotors, the phase shift angle β at the maximum unbalance response speed may be substantial. For example, with a well-damped rotor with an amplification factor of only 2, the phase angle between unbalance and displacement at the forced response resonance frequency is 105°.

The value of $\xi = 0.5$ may be taken as the value of critical damping required for the Jeffcott rotor. For $\xi = 0.5$, the rotor critical speed amplification factor A_{cr} is 1.0. However, the maximum unbalance response occurs at 41 percent of the rotor critical speed. The maximum amplification factor at this speed is 1.15. The phase-angle shift observed at this speed is 125°. This represents a 35° shift beyond the theoretical value at the undamped critical speed.

3.3.2 Balancing the Single-Mass Jeffcott Rotor

There are various methods of balancing the flexible single-mass Jeffcott rotor. The first is sometimes referred to as factory balancing and consists of simply balancing the rotor and assembly as a rigid body on either a hard bearing or soft bearing balancing machine.

The other methods are often referred to as field balancing and involve operating the rotor through various speeds below, at, or above the critical speed and recording the appropriate amplitude and phase of motion. These various methods of field balancing are referred to as the influence coefficient method, the one-shot balance method, the modal method of balancing without trial weights, and the three-trial-run procedure without phase measurements. Each of these procedures has its distinct advantages and disadvantages.

Balancing by the Influence Coefficient Method—Ideal Single-Plane Balancing Analytical Procedure with Noncontact Probes

The influence coefficient method of balancing is one of the standard and widely used procedures today for field balancing of rotating equipment. The advantage of the influence coefficient method is that it requires very little knowledge of the rotating system (although this fact can also be a serious disadvantage). The procedure involves measuring the rotor synchronous amplitude and phase at one or more speeds. The rotor is then stopped, and a known trial or calibration weight is placed on the rotor at a specific radius and angular location as measured from a reference mark on the disk. Although the placement of the trial or calibration weight is arbitrary, it is preferable to place it so as to reduce the rotor response. Proper placement of the calibration weight may often be quickly determined by viewing the polar plot of the rotor response, if available.

For example, to place a calibration weight on the sample rotor as shown in Fig. 3.16, the timing mark on the disk would be lined up with the vertical probe. A calibration weight would be placed at 120° from the timing mark opposite the direction of rotation. This would place a component of correction unbalance at approximately 180° out of phase to the mass unbalance in the shaft. The magnitude of the calibration weight may be predicted or computed from the polar plot or computed by using the basic guideline that the calibration weight should create a rotating unbalance load of approximately 10 percent of the rotor static weight. This will ensure that an excessive amount of trial weight is not placed on the rotor, which could result in permanent shaft bowing or distortion, particularly with a built-up rotor configuration.

Equation 3.75 represents the complex vector response of the system with an unbalance. The rotor response $\mathbf{Z}$ may best be represented by a complex influence coefficient $\mathbf{a}$, multiplied by the system unbalance $\mathbf{U}_u$. When the influence coefficient method is used, it is assumed that the influence coefficient $\mathbf{a}$ is a function of rotor speed only. This implies that if a small correction weight is placed on the rotor, the influence coefficient at a particular speed will not change.

In an actual situation, the influence coefficient, for a given speed, does change with respect to the level of unbalance. This is due to various nonlinear effects in the system. Therefore, if a flexible rotor is badly out of balance, several balancing runs may be required to obtain a low vibration level due to the change in the influence coefficients with unbalance loading.

The rotor synchronous amplitude response $\mathbf{Z}$ may be expressed in terms of a complex influence coefficient $\mathbf{a}$ and the vector unbalance $\mathbf{U}_u$ as follows:

$$\mathbf{Z} = \mathbf{a}(\omega)\mathbf{U}_u \tag{3.85}$$

where

$$\mathbf{a}(\omega) = \frac{f^2}{M(1 - f^2 + i2\xi f)}$$

and

$$\mathbf{U}_u = Me_u e^{-i\phi_m} = Ue^{-i\phi_m}$$

The idealized influence coefficient is linear and is dependent on the speed. Therefore, it is not a function of the magnitude of the balance weight. In actual practice, the influence coefficient is affected by large balance weights due to various nonlinear effects in the system such as fluid-film bearings, seals, and nonlinear shaft effects under large deformation.

The vector unbalance $\mathbf{U}_u$ is expressed in terms of the product of the rotor mass M and unbalance eccentricity $\mathbf{e}_u(\mathbf{U})$ and the phase lag angle ϕ_m. This angle represents the location where the unbalance $\mathbf{U}_u$ is located from the timing reference mark on the shaft, as measured in a direction opposite to rotation.

When the idealized single-mass Jeffcott rotor is balanced by using the influence coefficient method, it is assumed that the influence coefficients are repeatable at any given speed (no thermal bowing) and that the shaft is straight with no appreciable amount of runout. The object of the single-plane balancing by the influence coefficient method is to be able to determine the amount and angular location of the balance correction weight to be placed on the shaft. Thus, in single plane balancing, two unknown quantities must be solved for.

To employ the influence coefficient method of balancing, a trial or calibration weight is placed on the shaft at a given radius R and at a known angular location from the shaft timing or reference mark. The location of the trial weight should be such that the overall amplitude of motion is reduced upon application of the weight. The rotor response is then measured at the same speed as the initial vibration reading. The new amplitude of vibration may be expressed in terms of the influence coefficient $\mathbf{a}$ and trial unbalance $\mathbf{U}_t$ as

$$\mathbf{Z}_t = \mathbf{a} \times (\mathbf{U}_u + \mathbf{U}_t) \tag{3.86}$$

By subtracting the initial amplitude, the rotor complex influence coefficient may be obtained as follows:

$$\mathbf{a} = \frac{\mathbf{Z}_t - \mathbf{Z}}{\mathbf{U}_t} \tag{3.87}$$

The amount of unbalance in the system is given by

$$\mathbf{U}_u = \frac{\mathbf{Z}}{\mathbf{a}} = \mathbf{Z} \times \frac{\mathbf{U}_t}{\mathbf{Z}_t - \mathbf{Z}} \tag{3.88a}$$

The balance correction weight $\mathbf{U}_b$ is considered to be equal and opposite to the rotor unbalance vector $\mathbf{U}$. Hence the final balancing correction weight is

$$\mathbf{U}_b = -\mathbf{U}_u = \mathbf{Z} \times \frac{\mathbf{U}_t}{\mathbf{Z} - \mathbf{Z}_t} \tag{3.88b}$$

Example 3.8: Single-Plane Balancing Via a Mathematical Solution—Noncontact Probe. A rotor weighing 1000 lb has a recorded amplitude of 3.0 mils (peak to peak) at a phase lag angle of 270° using a standard noncontact displacement probe. A trial weight of 0.174 oz is placed at a radius of 9 in at a position of 30° from the timing mark against rotation. The equivalent trial unbalance is 1.57 oz · in. The rotating force caused by this magnitude of unbalance at 6000 rpm is

$$F_u = \frac{1.57 \text{ oz} \cdot \text{in}}{16 \text{ oz/lb} \times 386}\left(\frac{6000}{60} \times 2\pi\right)^2$$
$$= 100 \text{ lb}$$

The trial balance $\mathbf{U}_t$ thus exerts a force of 10 percent of the rotor static weight. This represents a reasonable magnitude of balance weight to apply to the rotor. One must be careful not to apply an excessive amount of trial weight that would cause unbalance forces equal to the magnitude of the rotor weight, as this would lead to nonlinear effects.

After the trial weight is applied, a new rotor response is recorded with a magnitude of 2 mils at 170°. The influence coefficient is given by

$$\mathbf{a} = \frac{\mathbf{Z}_t - \mathbf{Z}}{\mathbf{U}_t}$$

$$= \frac{2e^{-i170} - 3e^{-i270}}{1.57e^{-i30}}$$

$$= \frac{2[\cos(-170) + i\sin(-170)] - 3[\cos(-270) + i\sin(-270)]}{U_t}$$

$$= -2.469e^{i89.5} - 2.469e^{-i90.5}$$

The balance correction is given by

$$\mathbf{U}_b = -\frac{\mathbf{Z}}{\mathbf{a}} = \frac{-3.0e^{-270}}{2.469e^{-i90.5}}$$

$$= -1.21e^{-i179.5}$$

$$= 1.21e^{-i359.5}$$

Therefore, a correction balance of 1.21 oz · in (0.134 oz at 9-in radius) should be placed at the 0° reference mark.

Trim Balance Weight Correction

If the balance weight is left in the rotor, since the overall rotor amplitude is reduced, the resulting balance correction weight is referred to as a *trim balance correction*. The trim balance correction is based on Eq. 3.86, the rotor response after the initial trial weight has been added to the shaft. For the case where a weight is welded onto the rotor, it may not be convenient to remove the initial trial weight. If, however, the rotor response is much larger than the original rotor amplitude, not only should the trial weight be removed, but also the trial run $\mathbf{Z}_t$ should be repeated with either a reduced trial weight or a change in the angular location of the weight. The trim balance correction vector is given by

$$\mathbf{U}_{\text{trim}} = -(\mathbf{U}_u + \mathbf{U}_t) = -\frac{\mathbf{Z}_t}{\mathbf{a}}$$

$$= \frac{2.0e^{-i170}}{2.469e^{-90.5}} = -0.810e^{-i79.5} \qquad (\text{since } -1 = e^{-i180})$$

$$= 0.810e^{-i259.5} \tag{3.89}$$

Therefore, the trim magnitude is 0.810 oz · in placed at an angle of 295.5° against rotation.

Figure 3.17 represents the balance plane in which the trial weight is added. In this example the balance plane is shown with 12 balance holes spaced evenly around the disk. In this case, the first hole is placed at the location of the keyphasor or timing reference mark on the shaft. The holes are labeled in the opposite direction to rotation. This is the preferred direction in which to label the balance holes. The holes will then appear in increasing sequence when the shaft is rotated or rolled over.

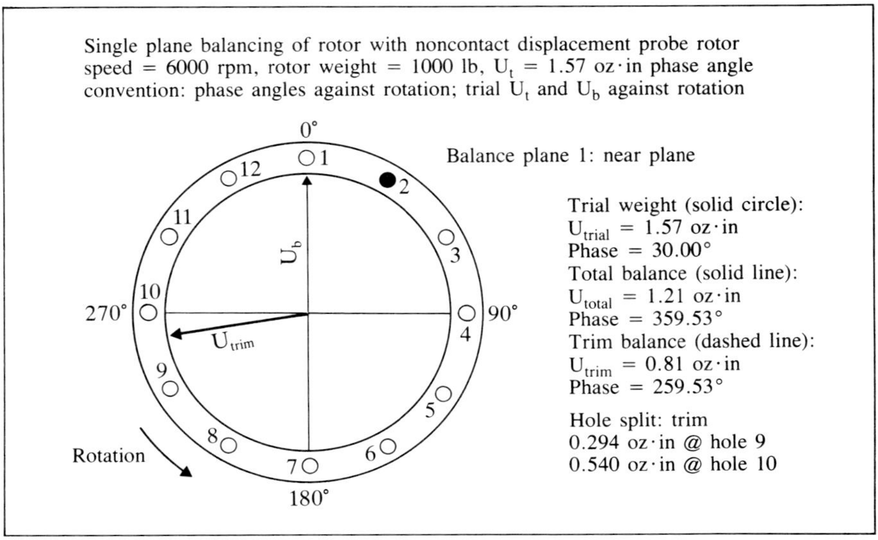

FIGURE 3.17 Single-plane balancing with noncontacting vertical displacement probe.

For an impeller or fan, the fan blades should also be labeled opposite to the direction of rotation.

In this example, the placement of the trial balance component of $\mathbf{U}_t = 1.57e^{-i30}$ oz · in implies the placement of 0.174 oz (5 g) at balance hole 2, which is at a radius of 9 in. The trial balance phase angle of −30° implies placing a weight in an angular position 30° against rotation, from the timing shaft reference mark.

The dark dot on Fig. 3.17 represents the placement of the trial balance weight in the second balance hole. If the trial weight is totally removed, then the calculated balance weight is 1.21 oz · in/9 in = 0.134 oz (or 3.831 g) to be placed in hole 1. Thus, the correct balance weight is 77 percent of the original trial unbalance, and the corrected angle is 30° clockwise from the placement of the original trial weight. If the correction angle is more than 90° from the location of the original trial weight, then an additional trim balance run may be needed for further improvement.

It is possible to leave the original trial weight in place and add a trim weight. Figure 3.17 shows that the trim weight should be 0.81 oz · in/9 in = 0.09 oz or 2.55 g between balance holes 9 and 10. Since the trim weight does not lie on top of a balancing hole, the trim vector may be split into two components for holes 9 and 10 to produce an equivalent effect. The hole split for holes 9 and 10, respectively, is 0.294 and 0.540 oz · in or 0.926 and 1.7 g to be placed in holes 9 and 10.

Thus the final balance correction for the wheel could consist of removing the original trial weight at hole 2 and placing 3.81 g in hole 1, or leaving in the original trial weight and placing approximately 1 g in hole 9 and $1\frac{3}{4}$ g in hole 10. Note that the hole-splitting computations could easily be performed with holes 8 and 11 if holes 9 and 10 already contain balance weights.

In the calculations shown, a right-handed coordinate system of units has been adapted to correspond to Fig. 3.11*b* with the rotation vector in the +**Z** direction (counterclockwise) and positive angles measured from the X axis in the counterclockwise direction. Hence, in this convention, positive angles in an orthogonal right-handed coordinate system are measured in the direction of rotation (lead), and lag angles are measured against rotation. Since the same phase conventions for the balance weights and rotor response have been chosen (lag), identical results may be obtained by reversing the signs of the phase angles in the calculations.

Table 3.5 presents a summary of the balance calculations for the simple-plane correction using a noncontact displacement probe. The table shows the value of the initial amplitude and phase and the amplitude and phase after the trial or calibration weight has been placed on the shaft. The resulting influence coefficient and total balance and trim balance connections are shown.

The standard phase-angle convention used with displacement probes is a lag phase-angle convention. This phase convention was initiated by Bently (1982) and is illustrated in Fig. 3.18 for shaft rotation in clockwise and counterclockwise directions.

The upper figure illustrates the phase convention for shaft clockwise rotation. When the timing reference pickup (or keyphasor probe) lines up with the timing mark on the shaft, a timing signal is generated. The timing reference signal is shown as a dot on the waveform, as illustrated on the right side of the shaft configuration.

The rotation or orbiting of the shaft generates a sinusoidal waveform, as shown in Fig. 3.18. In many cases, a pure harmonic waveform is not obtained. Under these circumstances, the output of the vibration sensor to a synchronous tracking filter will produce the required harmonic waveform. The phase angle is measured from the timing mark on the waveform to the peak amplitude. Figure 3.18*a* depicts a phase lag angle of 135° and clockwise rotation. From the measured phase angle, the high spot on the shaft may be determined. The shaft high spot is obtained by first lining up the keyphasor pickup to the timing mark on the shaft. The high spot on the shaft is determined by measuring from the probe (not the keyphasor!) in a direction opposite to rotation.

Figure 3.18*b* represents the phase convention with counterclockwise rotation. Note that, regardless of shaft rotation, the phase lag angle is measured on the waveform, in the direction of increasing time, to the first high spot on the waveform.

When synchronous tracking filters are used to obtain amplitude and phase, it is important to determine what phase convention the device is following. For example, one could measure from the timing mark to the zero crossover point, or one could measure from the peak amplitude to the timing mark. If this procedure is adopted, then one has a lead phase-angle convention.

In Table 3.5, to be consistent with the phase lag convention, the balance weight angle locations are measured from the timing mark against rotation. Therefore, it is critical in balancing to record the phase convention adopted for the balancing calculation and the type of instruments used in obtaining the experimental values of the amplitude and phase.

In Fig. 3.17, hole 1 has been assigned as the 0° reference position. The 0° position is usually placed at the location of the shaft reference mark. The angular coordinate system is labeled against rotation.

In summary, the phase angle measured with a noncontact probe is taken as the phase angle recorded from the point where the timing mark is triggered to the point where the peak amplitude of response is encountered. This procedure is

TABLE 3.5 Single-Plane Balancing with Noncontacting Displacement Probe—Lag Phase Convention

Rotor balancing computed results

Single-plane balancing of rotor with noncontact displacement probe: rotor speed, 6000 rpm; rotor weight, 1000 lb; $U_t = 1.57$ oz · in. Phase-angle convention: phase angles against rotation; trial U_t and U_b against rotation

System influence coefficients: units of mils or oz · in. All probe readings are in the lag convention.

Balance shot									Influence coefficient		
No.	Amp	Phase	Speed rpm	Prb no.	Amp 1, mils	Phase 1, deg	Amp 2, mils	Phase 2, ded	(Mag.)	Deg	Relative lag*
1	1.57	30	6	1	3.000	270	2.000	170	2.474	90	300

Initial and current response and predicted residual response

		Initial		Current		Predicted		Change in amp	
Speed no.	Probe no.	Amp mils	Phase, deg	Amp mils	Phase deg	Amp mils	Phase deg	Predicted initial	Predicted current
6	1	3.000	270	2.000	170	0.000	204	−3.000	−2.000

Balance correction weights

		Total balance				Trim balance†	
No.	Balance plane location	Mag., oz · in	Phase (lag)	**X**	**Y**	Mag., oz · in	Phase (lag)
1	Near plane	1.21	360	1.2	−0.0	0.81	260

* Relative lag = amount the influence coefficient lags the trial weight.

† Note: Trim balance = (total balance) − (trial weight left in rotor).

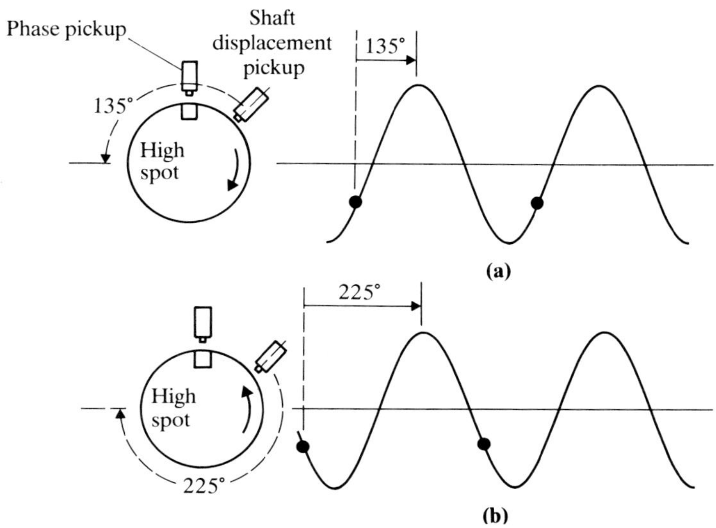

FIGURE 3.18 Phase lag measurements with shaft noncontacting probe for (*a*) clockwise and (*b*) counterclockwise rotation (Bently, 1982).

independent of the direction of shaft rotation. The high point on the shaft is obtained by measuring the required angle backward (against rotation) from the probe, after the timing mark has been lined up with the keyphasor probe. The identical result is also obtained by rolling the shaft forward (in the direction of rotation) by the required angle. The indicated high spot will lie under the probe. It is also preferable to use a consistent coordinate system for the placement of the balance weight. The balance weight locations are also measured from the keyphasor opposite to the direction of shaft rotation.

A lead phase convention could also be adopted for the balancing procedure when noncontact probes are used. There are currently on the market several designs of synchronous tracking filters that either give the phase angle as a lead angle or may be manually switched from a lag to lead phase convention. The lead phase convention is simply the complement of the lag angle; it is the angle measured from the peak response to the timing mark. If one adopts a lead phase convention, then it is desirable to measure the balance weight angles also in a lead convention. The lead convention for the angular placement of weights is taken as positive for angles measured in the direction of rotation.

Graphical Solution of Balancing with Noncontacting Probes

The balancing equation given by Eq. 3.89 may be represented by a graphical solution. It is only in the last decade that computer solutions to balancing have become standard with the introduction of small inexpensive handheld calculators.

Before the advent of the scientific handheld calculators (particularly with the ability to perform complex arithmetic), the standard procedure for single-plane balancing was to use a graphical solution. Graphical solutions are still highly desirable as checks on computer calculations.

Figure 3.19 represents the graphical balancing solution for the single-plane

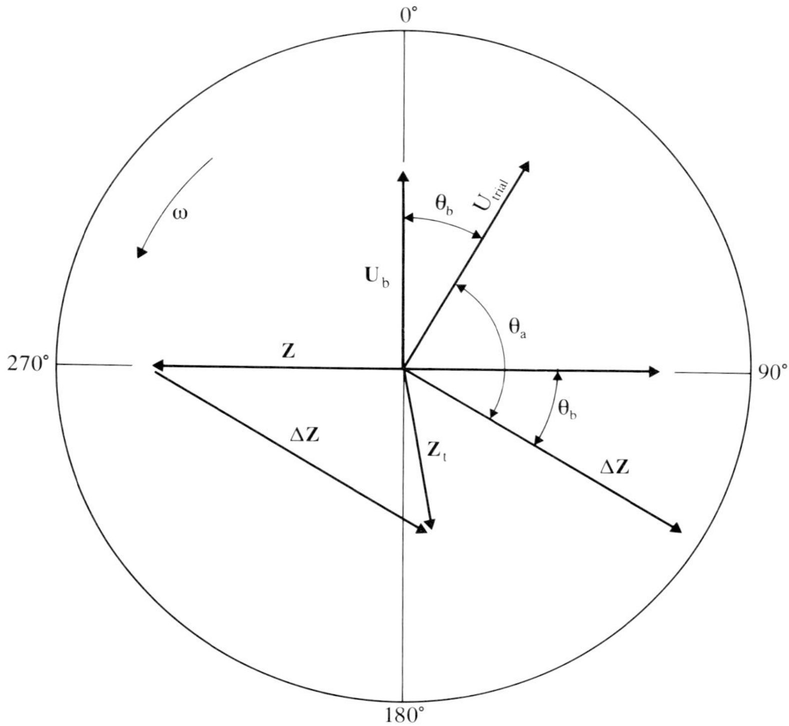

FIGURE 3.19 Single-plane balancing graphical solution with noncontacting displacement probes.

rotor using noncontact displacement probes. In the figure, the direction of rotation is shown as counterclockwise. The angles for phase and balancing are labeled in the clockwise direction, opposite to rotation.

The initial vector $\mathbf{Z} = 3.0\angle 270°$ is drawn on the figure. The vector response $\mathbf{Z}_t = 2.0\angle 170°$, obtained after the placement of the trial weight, is drawn next. The difference vector $\Delta\mathbf{Z}$ is now constructed by drawing a vector from $\mathbf{Z}$ to the end of vector $\mathbf{Z}_t$. The value of $\Delta\mathbf{Z}$ is given by

$$\Delta\mathbf{Z} = \mathbf{Z}_t - \mathbf{Z} = 2.0\angle 170° - 3.0\angle 270° = 3.88\angle 120.5°$$

A parallelogram is now constructed by using vectors $\mathbf{Z}$ and $\Delta\mathbf{Z}$. Thus, $\Delta\mathbf{Z} = 3.88$ at 120° is graphically constructed.

Next the mirror image of $\mathbf{Z}$ is constructed. The angle between $\Delta\mathbf{Z}$ drawn through the origin and the mirror of $\mathbf{Z}$ is measured. This angle is approximately 30°. It is desired that the placement of the new weight create a $\Delta\mathbf{Z}$ vector which is equal to the original $\mathbf{Z}$ vector but acts in the opposite direction or along the mirror of $\mathbf{Z}$. To have $\Delta\mathbf{Z}$ act along the inverse $\mathbf{Z}$ direction, the balance weight must be shifted by 30° in the clockwise direction, as shown in Fig. 3.19.

The original trial balance had a magnitude of 1.57 oz · in. This trial balance caused a net rotor response of 3.88 mils. The magnitude of the influence coefficient is

$$|\mathbf{a}| = \frac{|\Delta\mathbf{Z}|}{|\mathbf{U}_t|} = \frac{3.88}{1.57} = 2.47 \text{ mils/(oz} \cdot \text{in)}$$

Since the magnitude of the original vector $\mathbf{Z}$ is only 3 mils, the final correction weight is smaller than the trial weight and is given by

$$\mathbf{U}_b = \frac{|\mathbf{Z}|}{|\boldsymbol{a}|} = \frac{3.0}{2.47} = 1.215 \text{ oz} \cdot \text{in}$$

Thus the final balance is achieved by removing the trial weight and attaching a new balance magnitude of 1.21 oz · in from the original trial correction by approximately 30° in the counterclockwise direction.

The complete influence coefficient for this case may now be determined from the vector diagram. The trial vector is acting at 30° while the vector response is acting at approximately 120°. Thus the trial unbalance vector is leading the response vector by 90° in the direction of rotation. For the Jeffcott rotor, the 90° phase angle for the influence coefficient implies that the balancing calculation is being performed on top of the critical speed. One must be extremely careful when performing balancing calculations by the influence coefficient method at the critical speed since small changes in speed cause considerable changes in the influence coefficient. (We will show that the three-run method may be used to produce accurate balancing by recording the rotor amplitude at the critical speed.)

If the influence coefficient phase angle θ_a is greater than 90° but less than 180°, then the rotor is operating in the supercritical speed region and the mass center has inverted and is approaching the axis of rotation. It is preferable to take balancing readings in either the subcritical or supercritical speed region.

If, however, the influence coefficient phase angle θ_a, is greater than 180°, then the system is not a true single-mass Jeffcott rotor. The fact that the value of the influence coefficient phase angle is greater than 180° implies the influence of shaft bow, disk skew, or the approach of a second critical speed. Therefore, additional balance planes may have to be considered to achieve a low level of vibration response.

Single-Plane Balancing with Velocity Pickup and Strobe Light

For the case of flexible elastic rotor with little casing motion, it is preferable to balance by using noncontacting probes to monitor the shaft motion. For fairly stiff rotors mounted in rolling-element bearings, such as various types of motors, pumps, and fans, the relative shaft-bearing displacement may be small. There are a wide class of machines in which substantial vibration may be imparted to the bearing housing. Under these circumstances, the vibration may be monitored with a velocity pickup.

Another method to measure phase uses a strobe light which is triggered to flash based on the signal received from the vibration analyzer. The strobe light is made to flash on the shaft, freezing it in place. The strobe light flashes as the signal goes from negative to positive or positive to negative depending on the analyzer design (see Jackson, 1979).

The phase angle between the sensor's position and the distance that the high spot has traveled at the time the strobe light freezes the reference mark, placed on the rotor shaft end or a balancing ring, is very important in balancing. With a tunable analyzer, the phase can vary with the filter adjustment and hence is correct only at the center band of the filter. There is also a phase shift with decreasing speed due to the frequency response of the velocity pickup, if it is a seismic sensor.

When a stobe light is used for a phase reference indication in balancing, the determination of the rotor high spot is not as simple or straightforward as when a noncontact probe and a keyphasor reference mark are used. Figure 3.20

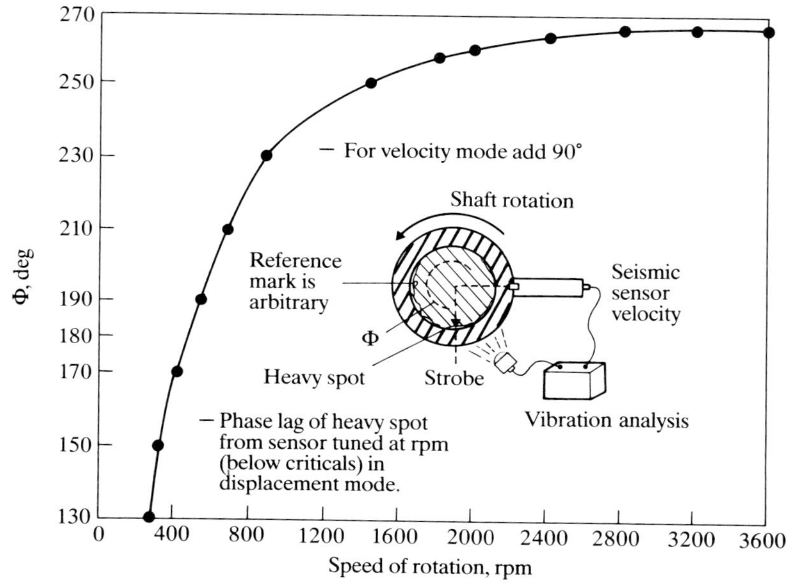

FIGURE 3.20 Phase lag from seismic velocity sensor to the rotor heavy spot.

represents the phase angle for a typical seismic vibration analyzer in the displacement mode. The seismic pickup measures the velocity of the bearing housing or foundation. Integrating the velocity signal to produce the displacement causes a 90° shift. In addition to the 90° phase shift, when one is integrating to obtain displacement, there is an inherent electronic phase shift of the instrument as well as a speed-dependent rolloff phase shift due to the inertia effects of the velocity sensor at low frequencies.

Figure 3.20 was developed by Jackson (1979) to help find the rotor heavy spot for guidance in placement of the initial trial weight for influence coefficient balancing. For example, if balancing of a subcritical speed rotor is to be accomplished at 1200 rpm, Fig. 3.20 shows that the phase angle of the high spot is located 244° from the sensor in the direction of rotation.

The rotor then is positioned where the reference mark is shown at 1200 rpm. (If an analyzer with a tuned filter is used, the analyzer must be carefully tuned to 1200 cycles/min, or a significant error will result.) If the rotor is assumed to be subcritical, then the high spot will coincide. A trial weight should then be placed 180° out of phase to the high spot for the first trial run to find the system influence coefficient.

Figure 3.21*a* represents the phase angle obtained with a perfectly balanced rotor after a 2-g trial weight has been placed on the rotor (Mechanalysis). The

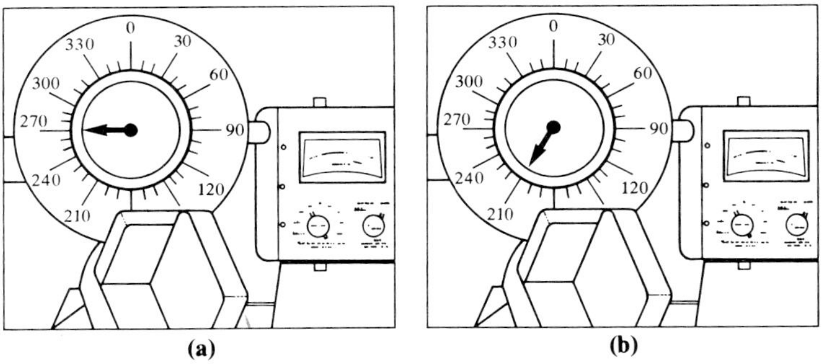

FIGURE 3.21 Phase reading with 2-g unbalance on balanced rotor: (*a*) placed on balanced rotor; (*b*) shifted 60° clockwise. (Mechanalysis).

phase angle here is shown drawn in a clockwise direction. In Fig. 3.21*b*, the weight is moved clockwise by 60°. The new phase angle recorded here is 210°. Hence, when a strobe is used to measure phase angles, movement of the rotor weight in the clockwise direction causes the phase angle to move in the counterclockwise direction. This effect is the same regardless of the direction of rotation.

When one is seeking a vector balancing solution with a strobe light reference system, the balancing procedure is modified. If the addition of the rotor trial weight has shifted the new vector $\mathbf{Z}_t$ in the clockwise direction, then the balance correction weight is moved from the relative angular location of the trial weight by the angle θ_a in the counterclockwise direction. Thus, the relative weight move is always opposite to the direction of shift of the trial vector $\mathbf{Z}_t$ regardless of the direction of rotation.

Figure 3.22 represents the phase convention for the vibration signals and the

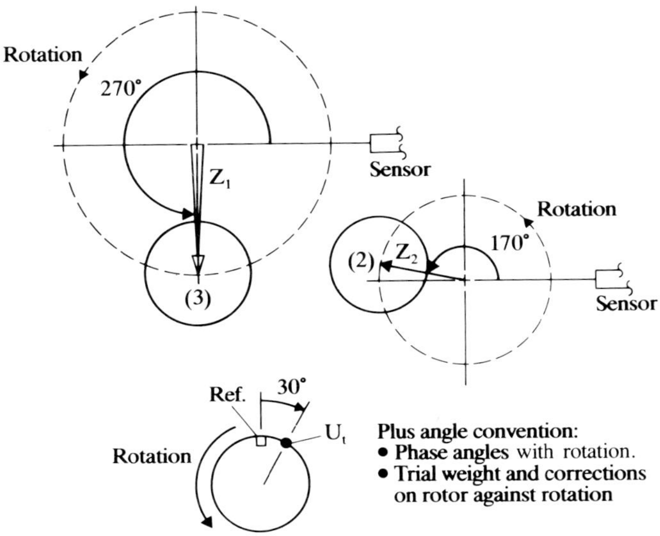

FIGURE 3.22 Single-plane balancing solution with strobe phase measurements.

measurement of the angular location of the rotor trial weight for strobe phase measurements and a velocity seismic pickup. The direction of rotation for this example is counterclockwise. The direction of positive phase angles is in the direction of rotation. The direction of the positive angular placement of the trial weight and corrections is against rotation. The resulting influence coefficients and angular correction are similar to those shown in Fig. 3.17.

Figure 3.23 (Jackson, 1979) shows the graphical solution for single-plane

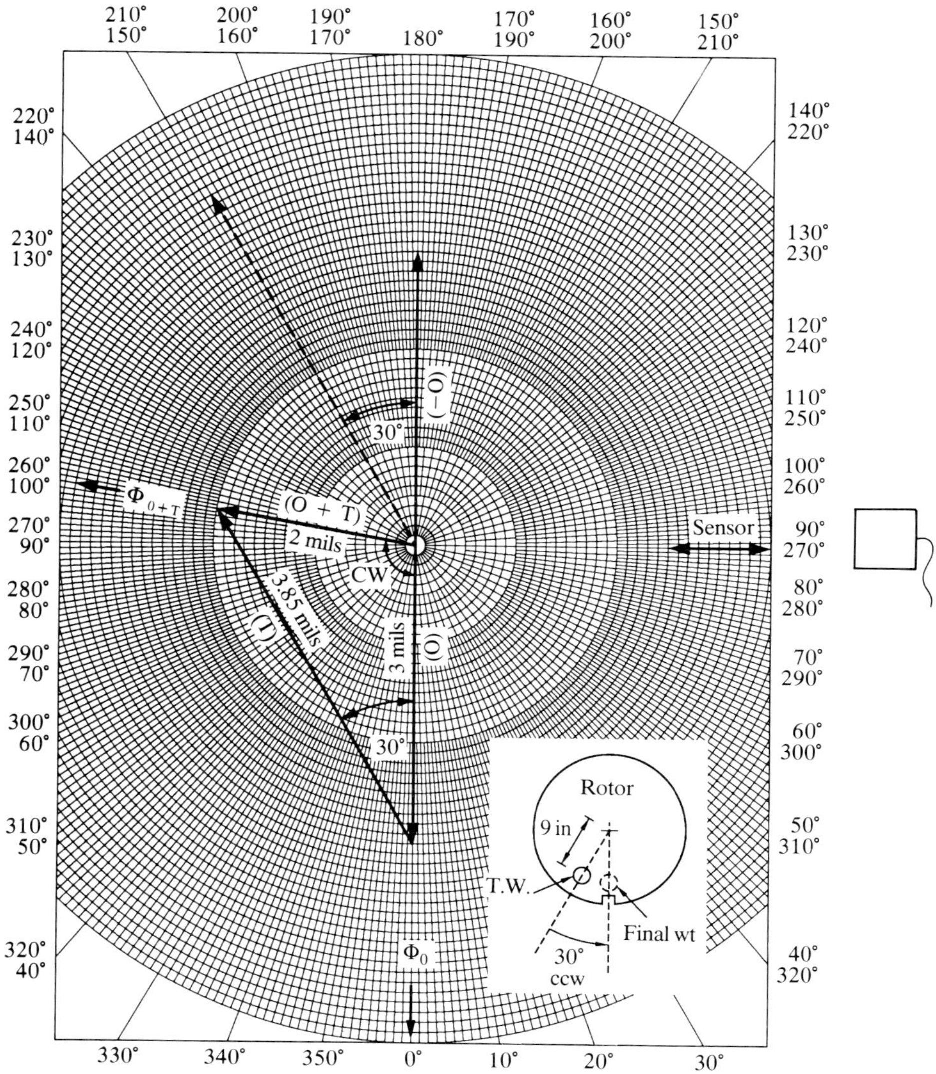

FIGURE 3.23 Graphical solution for single-plane balancing with strobe phase measurements (Jackson, 1979).

balancing with a strobe phase reference using the IRD convention of weight movement. In this figure the initial vector **O** is drawn from the seismic sensor in the direction of shaft rotation (counterclockwise) by 270°. Afte a trial weight is placed on the shaft, the vector **O** + **T** results. This vector is 170° counterclockwise from the sensor. It has moved in the clockwise direction in relationship to **O**. The trial vector **T** forms an angle of 30° with respect to the negative initial vector (−**O**). Using the balance phase convention for strobe phase measurements, the final balance weight is moved 30° counterclockwise from the position of the initial trial weight since the vector **O** + **T** (or $\mathbf{Z}_t$) moved clockwise.

3.3.3 Jeffcott Rotor with Shaft Bow

The dynamic unbalance response and balancing of the Jeffcott rotor with shaft bow were extensively investigated by Nicholas et al. (1976). Figure 3.24 represents the end and side views of the Jeffcott rotor with shaft bow. The amount of shaft bow is given by δ_r which represents the radial displacement or warp of the shaft from the theoretical axis of rotation. If the shaft were placed on knife-edges and slowly rotated, the disk centerline would describe an orbit of radius δ_r.

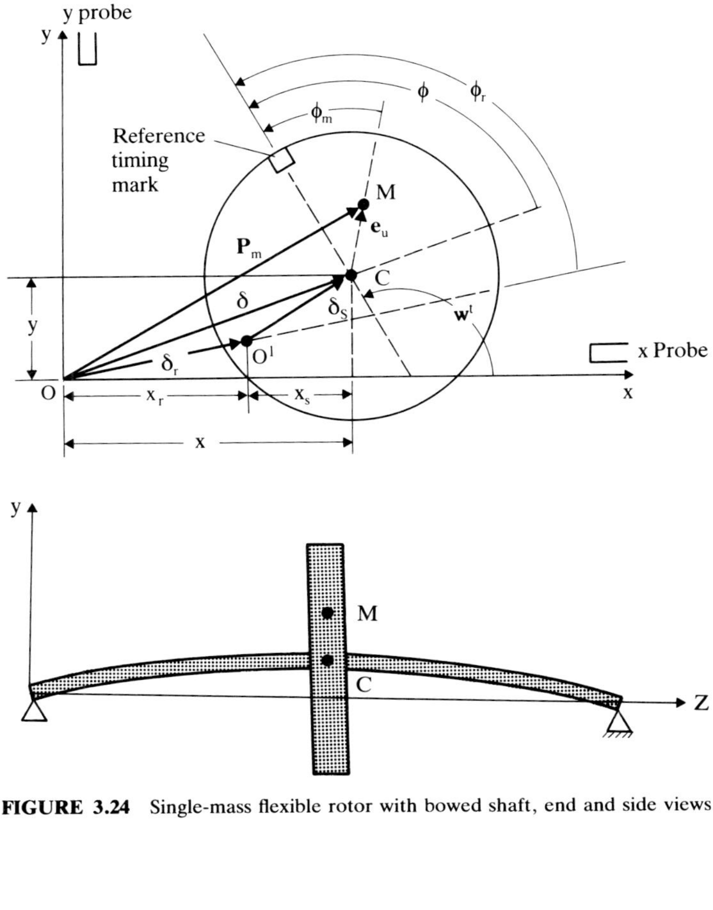

FIGURE 3.24 Single-mass flexible rotor with bowed shaft, end and side views.

The shaft bow is a vector and requires two quantities to describe it. The shaft bow is specified in terms of radial bow displacement δ_r and the phase lag angle ϕ_r from the shaft timing reference mark to the line of action of the bow. The shaft bowing produces an equivalent unbalance of magnitude $\mathbf{U}_r = M\delta_r$. Therefore, only a few mils of shaft bow may result in a large equivalent unbalance.

In fact, this effective unbalance may be so large that it may be physically impossible to place enough correction weight on the shaft to compensate for the bow. Under these circumstances, the shaft bow must be reduced. This has often been accomplished in the field by peening the shaft on the high side to draw it.

In Fig. 3.24, the shaft X and Y probes observe the motion of the shaft centerline, point C. The complex vector equation of motion to describe the rotor response of the shaft centerline C is given by

$$M\ddot{\mathbf{Z}} + C\dot{\mathbf{Z}} + K\mathbf{Z} = Me_u\omega^2 e^{i(\omega t - \phi_m)} + K\delta_r e^{i(\omega t - \phi_r)} \tag{3.90}$$

The steady-state synchronous response is given by

$$\mathbf{Z}(\omega) = \mathbf{a}_u\mathbf{e}_u + \mathbf{a}_r\boldsymbol{\delta}_r \tag{3.91}$$

where

$$\mathbf{a}_u = \frac{M\omega^2}{K - M\omega^2 + ic\omega} = \frac{f^2}{1 - f^2 + i2\xi f}$$

$$\mathbf{a}_r = \frac{K}{K - M\omega^2 + ic\omega} = \frac{f^1}{1 - f^2 + i2\xi f}$$

The rotor response is determined by the magnitude of unbalance and shaft bow times their respective influence coefficients $\mathbf{a}_u$ and $\mathbf{a}_r$. For speeds below the critical speed ($f < 1$), the influence coefficient for shaft bow has a greater effect than the influence coefficient for unbalance. At the undamped critical speed where $f = 1$, the two influence coefficients have identical values. As the speed becomes much larger than the critical speed, the influence coefficient for shaft bow becomes much smaller in comparison to the effect of shaft bow.

The relative phase-angle relationship of the shaft bow ϕ_r with respect to unbalance ϕ_m is important. If ϕ_r and ϕ_m are in phase, then the combined effects of shaft bow and unbalance will act to increase the overall amplitude of vibration. If the unbalance and shaft bow vectors are out of phase, then the two effects will help to reduce the overall vibration. If the vectors are exactly 180° out of phase, then there will exist a speed at which zero response is obtained. The dimensionless speed at which this occurs is

$$f = \left(\frac{\delta_r}{e_u}\right)^{1/2} \tag{3.92}$$

This speed is referred to as the self-balancing speed. The rotor amplitude of motion is zero only at this speed. Figure 3.25 represents the rotor response in which the shaft bow vector is one-half the value of the unbalance eccentricity vector ($\delta_r = 0.5$). The dimensionless rotor amplitude is plotted as a function of the frequency ratio f. At a speed of 70 percent of the critical speed, the effects of shaft bow and unbalance compensate each other.

Although the rotor is well balanced at 70 percent of the critical speed, clearly its unbalance characteristics are not satisfactory at other speeds. This is typical of the balance that would be achieved by the conventional single-plane method of balancing if performed at this speed. Conventional single-plane balancing would

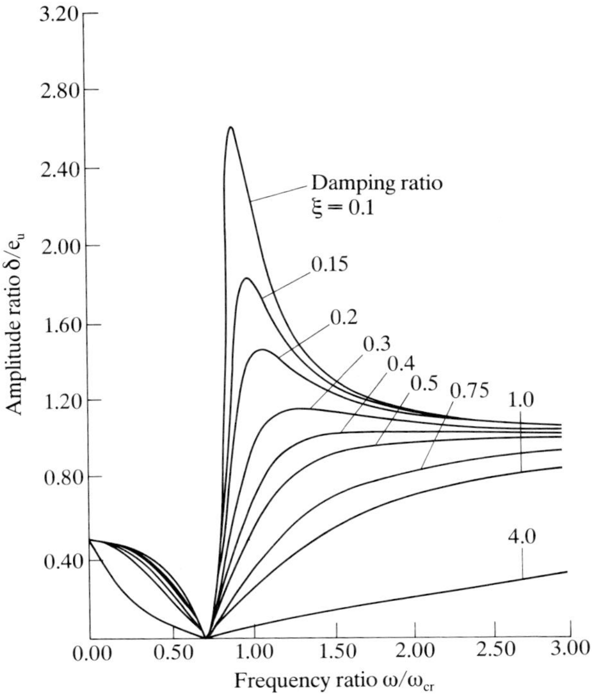

FIGURE 3.25 Response of a warped shaft with small shaft bow, out of phase to unbalance ($\delta_r = 0.5$, $\phi_r = 180°$) (Nicholas et al., 1976).

reduce the rotor response to zero, but the rotor would still be badly out of balance at the critical speed.

If the rotor amplitude starts to diminish before the critical speed, this then may be an indication that shaft bow is present in the shaft and out of phase with the unbalance. The relative magnitude of the shaft bow with respect to the rotor unbalance eccentricity may be obtained by observing the low-speed amplitude (the bow vector $\boldsymbol{\delta}_r$) in comparison to the high-speed response above the critical speed (the $\mathbf{e}_u$ vector).

In Fig. 3.25, the low-speed or slow-roll amplitude is only one-half the magnitude of the response above $f = 2$. This, then, is an indication that the effective rotor unbalance is twice the shaft bow. In this case, balancing will result in adding additional weight to the rotor in phase with the current bow vector, to reduce the overall response.

Figure 3.26 represents the rotor phase change for the shaft bow vector that is one-half the magnitude of the unbalance eccentricity vector and out of phase to it. The low-speed phase angle starts at 180° (since the bow vector is dominant). The phase angle increases until 70 percent of the critical speed is achieved. At this speed the rotor approaches zero amplitude. There is a 180° abrupt phase change after passing through this region. If the shaft bow vector is not exactly 180° out of phase, then a phase reversal will occur.

If the shaft bow is larger than and out of phase with the unbalance eccentricity,

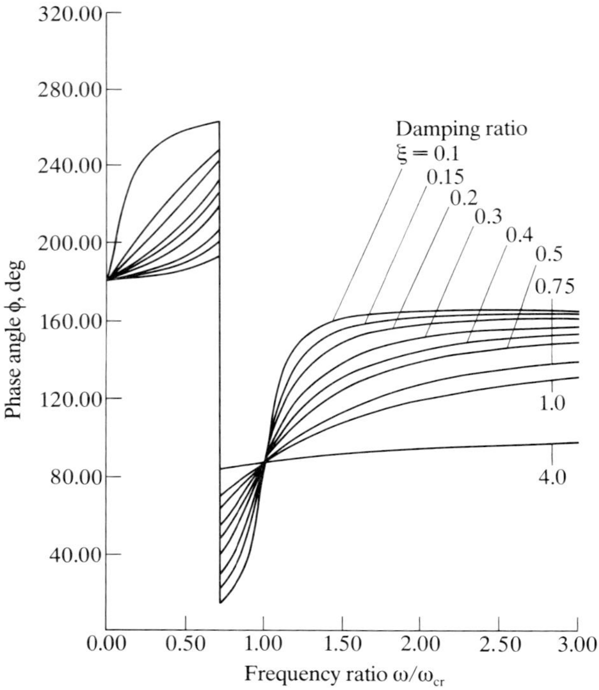

FIGURE 3.26 Phase angle of warped shaft ($\delta_r = 0.5$, $\phi_r = 180°$). Phase-angle curves with $\delta_r = 0.5$, $\phi_r = 180°$ for various damping ratios.

then the speed at which zero rotor response is observed will be above the critical speed. Figure 3.27 represents the rotor response in which the shaft bow vector is twice the unbalance eccentricity vector. In this case, the shaft self-balancing speed is 1.41 times the rotor critical speed.

If the shaft bow is identically equal to, but out of phase with, the rotor unbalance eccentricity vector, then the rotor will be self-balanced at the critical speed. Figure 3.28 shows the case where the shaft bow and unbalance eccentricity are equal but out of phase. At low speed the shaft runout is observed. As the rotor approaches the critical speed, the effects of shaft bow and unbalance cancel and the rotor motion reduces to zero. Upon a further increase in speed, this rotor amplitude increases and approaches $\mathbf{e}_u$ in the limit. To have the unbalance be exactly equal to and out of phase with the shaft bow is quite unusual in practice. However, such a situation has been observed by Jackson (1979) on acceptance tests for a large steam turbine.

The shaft bow vector is usually at an angle with respect to the rotor unbalance. Figure 3.29 represents the unbalance response of a bowed rotor in which the shaft bow is one-half the rotor unbalance eccentricity and lagging by 90°. By comparing the low-speed rotor amplitude to the response of the rotor above the rotor critical speed, one can determine the relative magnitude of shaft bow in comparison to

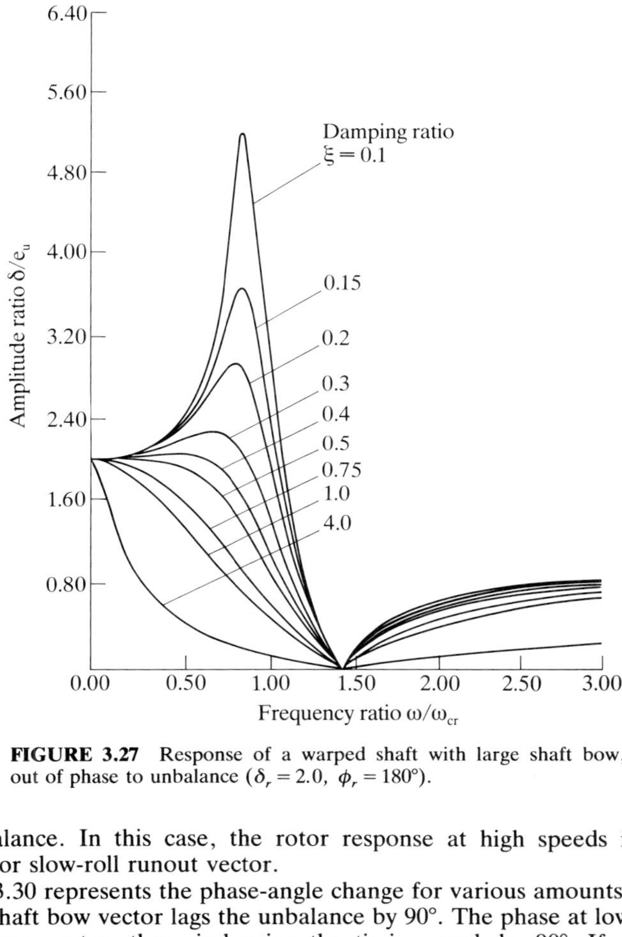

FIGURE 3.27 Response of a warped shaft with large shaft bow, out of phase to unbalance ($\delta_r = 2.0$, $\phi_r = 180°$).

rotor unbalance. In this case, the rotor response at high speeds is twice the low-speed or slow-roll runout vector.

Figure 3.30 represents the phase-angle change for various amounts of damping when the shaft bow vector lags the unbalance by 90°. The phase at low speed is at 90°. The bow vector, then, is lagging the timing mark by 90°. If one were to attempt to straighten out the shaft mechanically, one would first line up the shaft timing mark with the keyphasor probe. Then one would roll the shaft in the direction of rotation by 90°. The corresponding point on the shaft under the displacement probe represents the position of the shaft bow. This bow could be reduced by either mechanical loading or shaft peening.

One must be careful not to confuse a constant-runout vector with true shaft bow. If one is, e.g., monitoring the motion of a rotor at the impeller hub, rather than the shaft, one usually obtains a large runout vector. This runout vector is a constant and is superimposed on the actual shaft motion. The constant vector is due to the fact that it is impossible to machine a perfectly round and concentric hub at a large diameter. One always picks up a considerable runout vector. With a shaft bow, the vector response is not constant but varies with speed in magnitude and phase according to Eq. 3.91.

In Fig. 3.30, the initial phase is at 90°. As the speed is increased, the phase

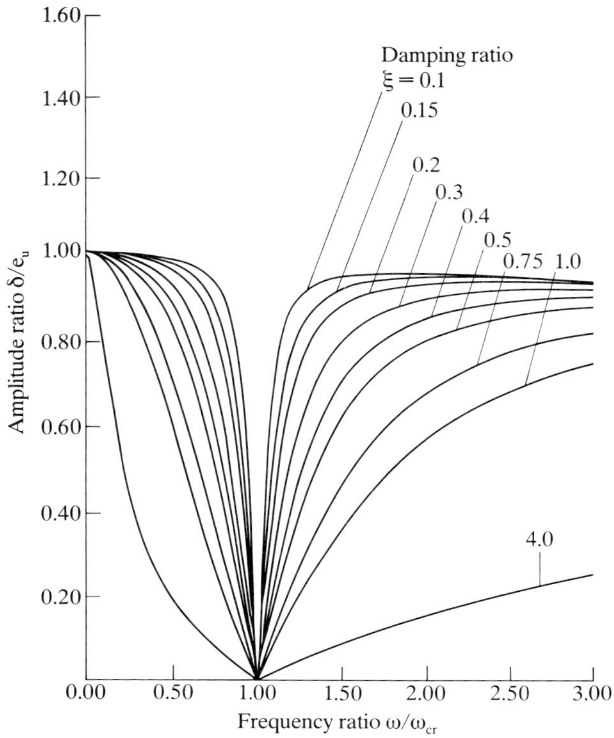

FIGURE 3.28 Response of a warped shaft with shaft bow equal to unbalance, out of phase ($\delta_r = 1.0,\ \phi_r = 180°$).

angle reduces as the unbalance eccentricity vector begins to dominate. The minimum phase is observed at 70 percent of the critical speed. After this speed is reached, the total phase angle begins to increase. Since the phase reversal occurs at 70 percent of the critical speed, the shaft bow is less than the unbalance eccentricity. The ratio of shaft bow to unbalance is given as the square root of the speed at which phase reversal occurs. Hence δ_r is one-half of the unbalance eccentricity.

The rotor phase angle begins to increase until it reaches its maximum value at 180°. From this value, it is apparent that the unbalance eccentricity lies at the 0° position.

The total phase change of the rotor with shaft bow is only 180°. If the total phase change is over 180°, then other effects are occurring such as a second mode, disk skew, or foundation effects.

One can often identify the presence of shaft bow in a rotor by observing the timing mark on the rotor orbit. (This is similar to recording the phase.) Figure 3.31 is a schematic showing the movement of the timing mark on an oscilloscope screen for various values of shaft bow δ_r and angular location ϕ_r for an underdamped rotor.

In Fig. 3.31*a* we see the location of the shaft bow in relationship to the

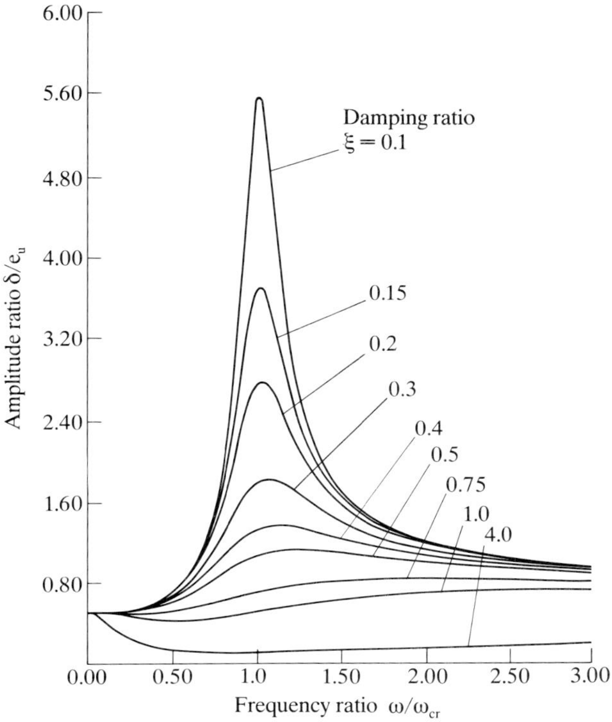

FIGURE 3.29 Response of a warped rotor with small shaft warp, 90° phase lag to unbalance ($\delta_r = 0.5$, $\phi_r = 90°$).

unbalance eccentricity vector, which is situated at the zero reference mark. Figure 3.31*b* shows the path of the timing mark for the case of no shaft bow. At low speed, $f = 0$, the timing reference mark is at 0°. As speed increases, the timing mark moves opposite to the direction of rotation until it is 180° out of phase at high speed ($f \gg 1$).

Figure 3.31*c* represents a rotor with a relative shaft bow of $\boldsymbol{\delta}_r = 0.5$ at 90° lag from the unbalance. At low speed the initial phase is recorded at 90° (which is the influence of the bow vector only). As the speed increases to 60 percent of the critical speed, the phase angle reduces to 60°. Upon a further increase in speed, the phase angle increases, and at the critical speed the angle is 120° rather than 90°, as for pure shaft unbalance. Upon a further increase in speed, the phase angle approaches 180° in the limit as the rotor speed greatly exceeds the critical speed. In all cases shown, the timing mark will eventually reach 180° since at high speeds the mass unbalance eccentricity vector predominates and the shaft bow straightens out.

In Fig. 3.31*d*, the shaft dimensionless bow, $\boldsymbol{\delta}_r$ is twice the unbalance eccentricity vector $\mathbf{e}_u$ and is at 135°. Since the shaft bow is larger than the unbalance eccentricity, the shaft bow motion predominates until after the rotor has passed through the critical speed. At 110 percent of the critical speed, the maximum phase angle of 245° is observed. As the speed further increases, the phase angle reverses and approaches 180° in the limit.

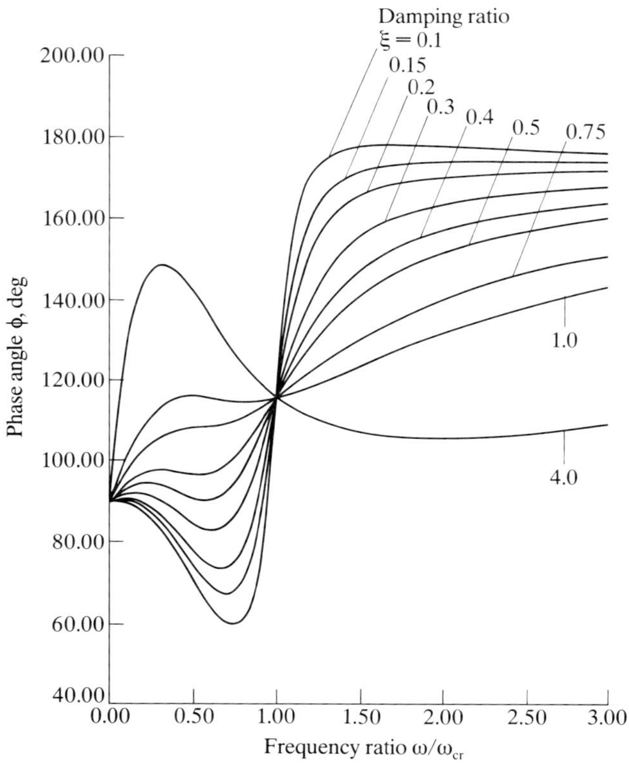

FIGURE 3.30 Phase angle with small shaft warp, 90° phase lag to unbalance ($\delta_r = 0.5$, $\phi_r = 90°$).

Hence, we see from these diagrams that the position of the inital timing mark on the orbit shows the orientation of the shaft bow. If phase reversal occurs below the critical speed, then the shaft bow is smaller than the unbalance eccentricity vector. If, however, the phase reversal occurs above the critical speed, then the shaft bow is larger than the unbalance eccentricity. It may not be desirable to attempt to balance a rotor with a large shaft bow since the required weights may be excessive and the bow vector may change with time.

Balancing the Single-Mass Jeffcott Rotor with Shaft Bow

There are several methods to balance a rotor with shaft bow. These methods may be best illustrated by the following example.

Example 3.9: Balancing a Bowed Rotor. We are given a rotor of 3.8 lb (1.73 kg) with a stiffness $K = 467$ lb/in (317.8 N/cm) and a damping $C = 0.22$ lb · s/in (0.385 N · s/cm). The unbalance eccentricity $e_u = 0.001$ in (1 mil) and is at a phase location of $\phi_m = 0°$.

The unbalance is in line with the timing mark. The shaft bow vector $\boldsymbol{\delta}_r$ is 0.001 in and lags the unbalance vector by 90°. A schematic diagram of the unbalance and shaft bow is shown in Fig. 3.32 (Nicholas et al., 1976). The

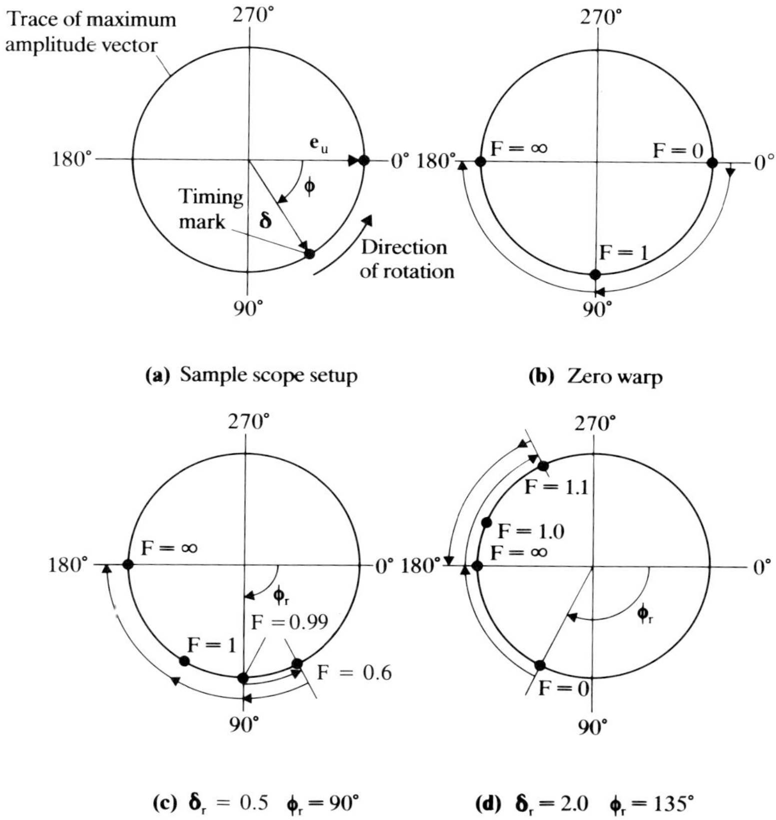

FIGURE 3.31 Schematic showing the movement of the timing mark on an oscilloscope screen as speed increases for various δ_r and ϕ_r values (Nicholas et al., 1976).

equivalent mechanical unbalance in the rotor is

$$\mathbf{U}_u = 3.8 \text{ lb} \times 0.001 \text{ in} = 0.0038 \text{ lb} \cdot \text{in} = 1.72 \text{ g} \cdot \text{in}$$

$$\omega_{cr} = \sqrt{\frac{k}{m}} = \sqrt{\frac{467 \times 386}{3.8}} = 217.8 \text{ rad/s}$$

$$N_{cr} = 2080 \text{ rpm} = \text{critical speed}$$

$$C_c = 2M\omega_{cr} = 2 \times 3.8 \times \frac{217.8}{386} = 4.23 \text{ lb} \cdot \text{s/in}$$

$$\xi = \frac{C}{C_c} = \frac{0.22}{4.28} = 0.0514$$

$$A_{cr} = \frac{1}{2\xi} = 9.73$$

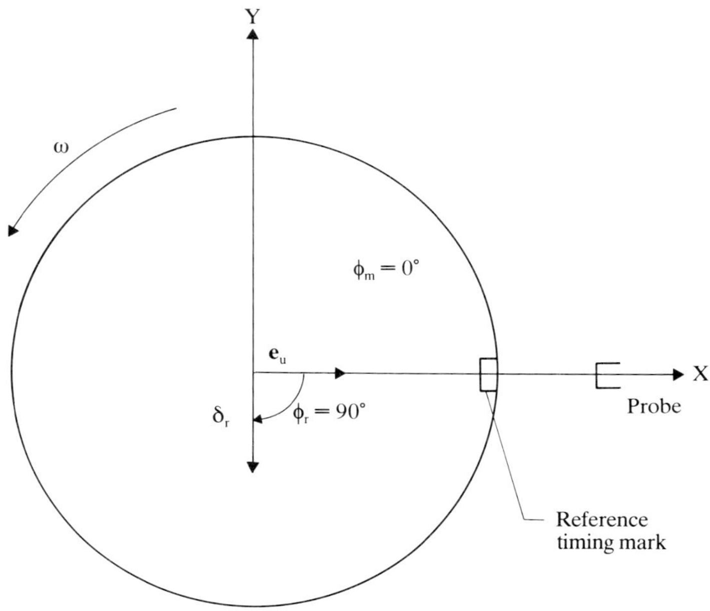

FIGURE 3.32 Rotor schematic for shaft with unbalance and shaft bow.

The shaft unbalance and bow vectors are given in complex notation:

$$\mathbf{e}_u = 1.0e^{i360}\text{ mils} = 0.025e^{i360}\text{ mm}$$

$$\boldsymbol{\delta}_r = 1.0e^{-i90}\text{ mils} = 0.025e^{-i90}\text{ mm}$$

Balance Method I: Standard Influence Coefficient Method. In balance method I, the standard influence coefficient method is used to reduce the rotor motion to zero amplitude. The balancing speed is taken at

$$N_b = 1800\text{ rpm } (217.9\text{ rad/s})$$

which is at a speed ratio of 86 percent of the rotor critical speed. The influence coefficients for unbalance and shaft bow in complex polar notation are given as follows (where the complex angular notation $e^{-i\phi}$ is represented by $\angle -\phi$):

$$\mathbf{a}_u = \frac{f^2 e^{-i\phi}}{[(1-f^2)^2 + (2\xi f)^2]^{1/2}} = 2.79\angle -19.4° \tag{3.93}$$

$$\mathbf{a}_r = \frac{e^{-i\phi}}{[(1-f^2)^2 + (2\xi f)^2]^{1/2}} = 3.77\angle -19.4° \tag{3.94}$$

where

$$\phi = \tan^{-1}\frac{2\xi f}{1-f^2} = 19.4°$$

Note that the influence coefficient phase lag angle for the effect of shaft bow is identical to that for the unbalance response. Also note that since the balancing is performed below the critical speed, the influence coefficient due to shaft bow is larger than the influence coefficient for unbalance by the relationship

$$\frac{\mathbf{a}_u}{\mathbf{a}_r} = \frac{1}{f^2} \tag{3.95}$$

The total initial response at the balancing speed is given by

$$\begin{aligned} \mathbf{Z}_1 &= \mathbf{a}_u \mathbf{e}_u + \mathbf{a}_r \boldsymbol{\delta}_r \\ &= 2.79\angle -19.4° + 3.77\angle -109° \\ &= 4.69\angle -72.9° \text{ mils } (11.9 \times 10^{-3} \angle -72.9° \text{ cm}) \end{aligned}$$

If a trial weight is placed on the shaft, the resulting amplitude $\mathbf{Z}_t$ is given by

$$\mathbf{Z}_t = \mathbf{a}_u(\mathbf{e}_u + \mathbf{e}_t) + \mathbf{a}_r \boldsymbol{\delta}_r \tag{3.96}$$

The influence coefficient for the effect of rotor mechanical unbalance is obtained by subtracting the initial response and dividing by the trial unbalance eccentricity $\mathbf{e}_t$:

$$\mathbf{a}_u = \frac{\mathbf{Z}_t - \mathbf{Z}_1}{\mathbf{e}_t} = 2.79\angle -72.9° \tag{3.97}$$

(Note that the units of $\mathbf{a}_u$ may vary depending on how one wishes to express the trial calibration component; mils eccentricity, gram-inches, etc.)

In the standard influence coefficient method of balancing it is assumed that the rotor response is dependent on only the mechanical unbalance. The balance eccentricity is given by

$$\begin{aligned} \mathbf{Z}_{\text{balance}} &= \mathbf{a}_u(\mathbf{e}_u + \mathbf{e}_b) = \mathbf{Z}_1 + \mathbf{a}_u \mathbf{e}_b = 0 \\ \mathbf{e}_b &= -\frac{\mathbf{Z}_1}{\mathbf{a}_u} = \frac{-4.69\angle -72.9°}{2.79\angle -19.4°} \\ &= 1.681\angle -233.5 \text{ mils } (0.0427 \text{ mm}) \end{aligned} \tag{3.98}$$

Assuming a balance radius of 1.5 in, the balance correction weight in grams is

$$\begin{aligned} \mathbf{U}_b &= w\mathbf{e}_b = w_b \mathbf{R} = 3.8 \text{ lb} \times 0.00168 \text{ in} \\ &= 0.00639 \text{ lb} \cdot \text{in} = 2.90 \text{ g} \cdot \text{in} \end{aligned}$$

$$w_b = \frac{2.90 \text{ g} \cdot \text{in}}{1.5 \text{ in}} = 1.93 \text{ g}$$

The balance weight to produce zero amplitude is 1.93 g, and it should be placed at an angular location of 233.5° from the timing mark opposite to the direction of rotation.

Although the rotor is apparently perfectly balanced at 1800 rpm, the rotor will have a substantial response at the critical speed.

The influence coefficient for both unbalance and shaft bow at the critical speed is given by

$$\mathbf{a}_u = \mathbf{a}_r = A_{\text{cr}}\angle -90° = 9.74\angle -90°$$

The amplitude at the critical speed is given by

$$\begin{aligned}\mathbf{Z}_{cr} &= (\mathbf{e}_u + \boldsymbol{\delta}_r) \times \mathbf{A}_{cr} \\ &= (1.0\angle 0° + 1.0\angle -90°) \times 9.74\angle -90° = 13.774\angle -135°\end{aligned} \tag{3.99a}$$

The rotor response at the critical speed is given by

$$\begin{aligned}\mathbf{Z}_{cr,b} &= \mathbf{a}_u(\mathbf{e}_u + \mathbf{e}_b) + \mathbf{a}_r\boldsymbol{\delta}_r \\ &= 3.41 \text{ mils } \angle 0° \ (0.0866 \text{ mm})\end{aligned} \tag{3.99b}$$

Figures 3.33 and 3.34 show the rotor amplitude and phase before and after balancing by method I (the conventional influence coefficient method). The rotor

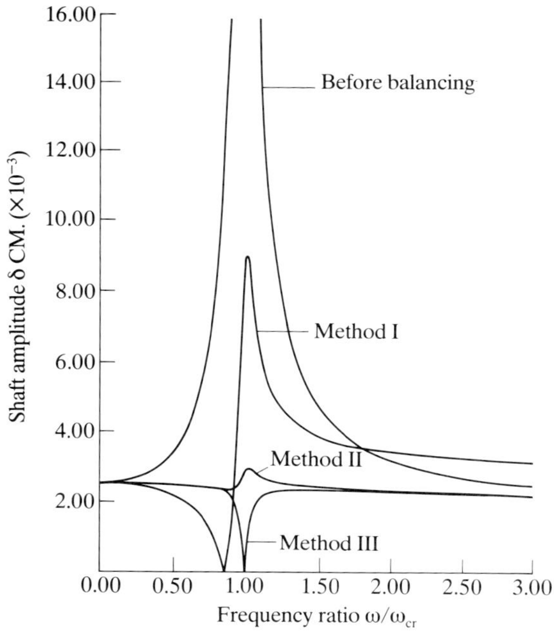

FIGURE 3.33 Bowed rotor response curves before and after balancing by methods I, II, and III.

amplitude, after balancing at 1800 rpm results in zero response at this speed. As the speed is increased, the shaft amplitude increases and reaches 3.41 mils, or 8.66×10^{-3} cm. The total unbalance in the rotor is $\mathbf{e}_u + \mathbf{e}_b = 1.35\angle -270°$, which is out of phase to the bow vector.

The ratio of the total unbalance in the rotor to the bow vector is

$$\frac{e_{\text{total}}}{\delta_r} = \frac{1}{f_b^2} = 1.35 \tag{3.100}$$

The rotor is balanced to zero amplitude at 1800 rpm by method I. This causes a

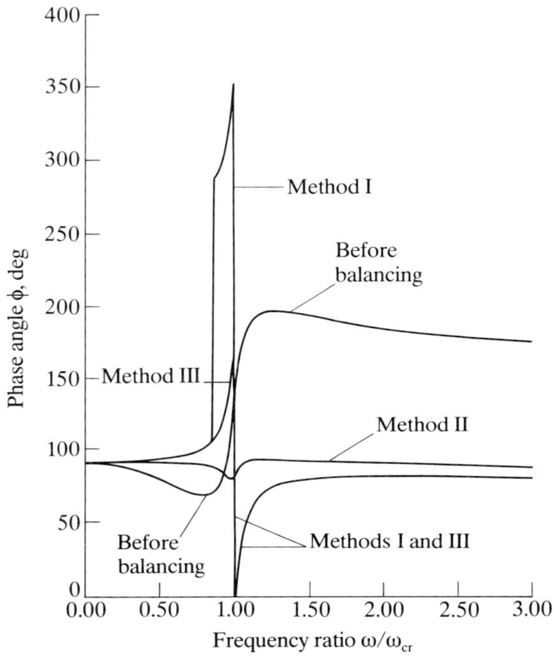

FIGURE 3.34 Phase-angle curves before and after balancing a bowed shaft by methods I, II, and III.

180° phase reversal above 1800 rpm and a large amplitude response is encountered at the critical speed. Balancing by method I is clearly unsatisfactory.

Balance Method II: Runout Subtractions. The procedure of placing a trial weight on the shaft will determine only the influence coefficient due to mechanical unbalance. The bow influence coefficient cannot be directly measured, e.g., by applying a trial bow vector.

The rotor response is assumed to be represented by the approximate equation

$$\mathbf{Z} = \mathbf{a}_u \mathbf{e}_u + \mathbf{a}_r \boldsymbol{\delta}_r \approx \mathbf{a}_u \mathbf{e}_u + \boldsymbol{\delta}_r \tag{3.101}$$

Solving for the balancing eccentricity vector gives

$$\mathbf{e}_b = -\mathbf{e} = -\frac{\mathbf{Z} - \boldsymbol{\delta}_r}{\mathbf{a}_u}$$

$$= 1.31 \text{ mils } \angle -228.8° \tag{3.102}$$

This procedure is equivalent to minimizing the elastic shaft deflection at the balance speed.

The balance correction by method II is

$$\mathbf{U}_b = w\mathbf{e}_b = 2.267 \text{ g} \cdot \text{in } \angle -228.8°$$

$$w_b = \frac{2.267 \text{ g} \cdot \text{in}}{1.5} = 1.5 \text{ g}$$

The amplitude at the critical speed after balancing by method II is

$$\mathbf{Z}_{cr} = 1.18 \text{ mils } \angle -84.2° \ (0.0299 \text{ mm})$$

Method III: Bow Vector Compensation. In this method, the influence coefficient for the bow vector is computed indirectly from Eq. 3.95:

$$\mathbf{a}_r = \frac{\boldsymbol{a}_u}{f_b^2} \tag{3.103}$$

where

$$f_b = \frac{N_b}{N_{cr}}$$

This procedure requires that the rotor critical speed be accurately known. If the balancing procedure is carried out at the critical speed, then the amplitude at the critical speed with the balancing correction added is given by

$$\begin{aligned} \mathbf{Z}_{cr,b} &= \mathbf{a}_u(\mathbf{e}_u + \mathbf{e}_b) + \mathbf{a}_r\boldsymbol{\delta}_r \\ &= \mathbf{a}_{cr}(\mathbf{e}_u + \mathbf{e}_b + \boldsymbol{\delta}_r) = 0 \end{aligned} \tag{3.104}$$

Since $f_b = 1$ at the critical speed, the influence coefficients due to shaft bow and unbalance become identical.

The balancing criterion in this method is that

$$\mathbf{e}_b = -(\mathbf{e}_u + \boldsymbol{\delta}_r) \tag{3.105}$$

This is identical to the condition that the mechanical unbalance eccentricity vector be equal but out of phase to the shaft bow vector.

If the balancing speed does not correspond to the critical speed, then Eq. 3.104 is given by

$$\mathbf{Z}_b = \mathbf{a}_u\left(\mathbf{e}_b + \mathbf{e}_u + \frac{\boldsymbol{\delta}_r}{f_b^2}\right) \neq 0 \tag{3.106}$$

The balance eccentricity vector $\mathbf{e}_b$ is given by

$$\mathbf{e}_b = \frac{\mathbf{Z}_1}{\mathbf{a}_u} + \boldsymbol{\delta}_r\left(1 - \frac{1}{f_b^2}\right) \tag{3.107}$$

With the value of shaft bow δ_r measured at low speed and the ratio f_b of the balance speed to the critical speed, the value of the balancing eccentricity is computed

$$\mathbf{e}_b = 1.41 \text{ mils } \angle -225° \ (0.0359 \text{ mm})$$

The corresponding balance correction weight for a radius of 1.5 in is 1.62 g. Figure 3.33 shows that the rotor amplitude at the critical speed will reduce to zero by balancing method III. Similar experimental and theoretical response curves for balancing a uniform flexible shaft with shaft bow are shown by Bishop and Parkinson (1965a, b; 1972).

Table 3.6 shows that balancing a bowed rotor by the standard influence coefficient method usually will result in overbalancing the rotor by 20 to 30 percent. From a practical standpoint, it is impossible to distinguish between shaft bow and conventional unbalance. In general, the vibration observed on the shaft

TABLE 3.6 Balance Corrections and Angular Location for a Bowed Rotor by Various Methods

Method	Balance corrections, g	Balance location, deg lag
I. Standard influence coefficient	1.93	233.5
II. Influence coefficient assuming constant runout vector	1.50	228.8
III. Bow vector compensation	1.62	225.0

may be a function of mechanical unbalance, shaft bow, and a constant runout vector $\boldsymbol{\delta}_0$ as follows:

$$\mathbf{Z} = \mathbf{a}_u \mathbf{e}_u + \mathbf{a}_r \boldsymbol{\delta}_r = \boldsymbol{\delta}_0 \tag{3.108}$$

At low speeds, where the effect of the unbalance is minimal (that is, $\mathbf{a}_u \ll 1$), the low-speed or residual runout vector $\mathbf{Z}_0$ is given by

$$\mathbf{Z}_0 = \boldsymbol{\delta}_r + \boldsymbol{\delta}_0 \tag{3.109}$$

It is not, in general, possible to distinguish between a shaft constant runout vector $\boldsymbol{\delta}_0$ and true mechanical shaft bow $\boldsymbol{\delta}_r$.

In comparing methods II and III, we see that method II generates an extremely good value of balance for the bowed rotor by treating the bow as a constant vector. A modified influence coefficient method of balancing is used in which the slow-roll vector $\mathbf{Z}_0$ is recorded.

The modified balancing procedure is similar to the standard single-plane balancing in Eq. 3.90 except that the procedure is modified to subtract the slow-roll runout vector $\mathbf{Z}_0$ as follows:

$$\mathbf{U}_b = (\mathbf{Z} - \mathbf{Z}_0) \times \frac{\mathbf{U}_t}{\mathbf{Z} - \mathbf{Z}_t} \tag{3.110}$$

Figure 3.35 represents a single-mass rotor that was balanced at 1500 rpm by the standard influence method without the shaft bow considered in the balancing calculations. The rotor was balanced at 1800 rpm to zero amplitude. However, at the critical speed the rotor amplitude is 6 mils. In Fig. 3.36 the residual shaft bow vector was subtracted from the rotor response, and the rotor was balanced to the residual vector rather than to zero amplitude. The rotor amplitude goes to zero response at the critical speed. Balancing the rotor with slow-roll vector subtraction included results in a reduction of the balancing weight from 584 to 560 g and a shift in this phase angle from 261° to 241°.

There are currently several designs of synchronous digital vector tracking filters on the market with the capability of low-speed vector runout subtraction built in.

If possible, one should always measure the low-speed runout vector and subtract this vector from the rotor response obtained at speed.

Balancing the Bowed Rotor without Trial Weights

With the current generation of hand calculators and portable computers, it is possible to invert complex matrices of order 3 or 4 with little difficulty. It is thus

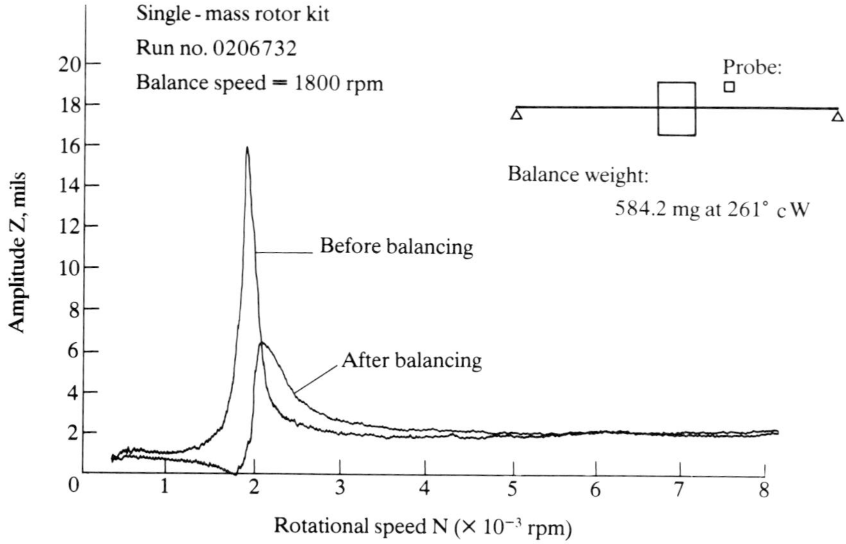

FIGURE 3.35 Unbalance response of a bowed rotor before and after balancing with residual bow neglected.

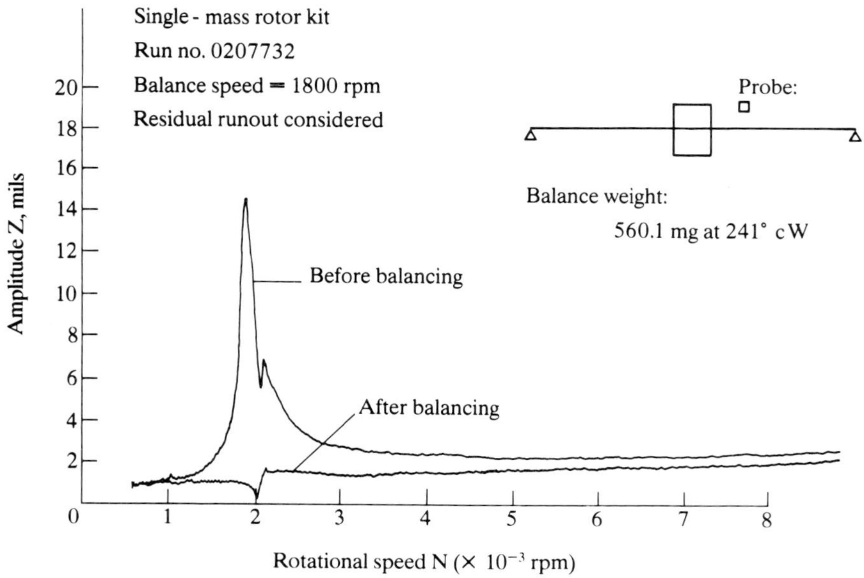

FIGURE 3.36 Unbalance response of a bowed rotor before and after balancing with residual bow considered.

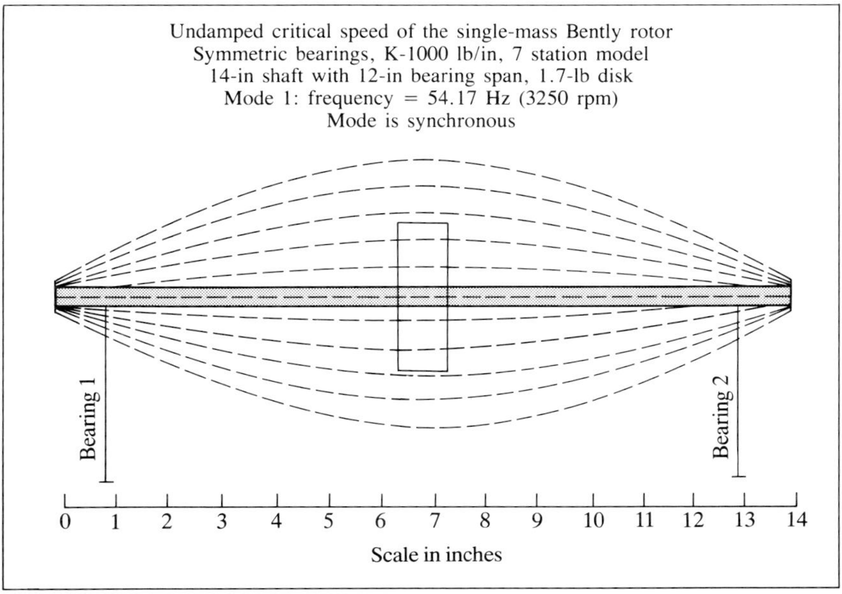

FIGURE 3.37 Single-mass rotor—animated mode shape of first critical speed.

possible to compute the initial balance of the rotor without the use of a trial weight. To do this, it is necessary to know the rotor critical speed, the rotor modal weight, and the axial location of the shaft probe along the shaft.

Figure 3.37 represents the rotor first critical speed mode shape. Normally the shaft probes are not located at the shaft center. The probes are located some distance z in from the end. At this distance z, the rotor dimensionless mode shape may be measured. The mode shape at the shaft center $\phi(L/2)$ is normalized to unity. For example, with the theoretical Jeffcott rotor with no bearing displacements, the mode shape $\phi(z)$ is given by

$$\phi(z) = \sin \frac{\pi z}{L} \tag{3.111}$$

The measurements at distance z along the shaft are corrected by the mode shape to arrive at the amplitude at the rotor center

$$\mathbf{Z}_{\text{center}} = \mathbf{Z} \times \frac{\phi(L/2)}{\phi(z)} \tag{3.112}$$

The rotor amplitude and phase are measured at three speeds. Since the critical speed is assumed known, the speed ratios $f_i = N_i/N_{\text{cr}}$ may be computed.

The rotor response at each speed is given by

$$M\ddot{\mathbf{Z}}_i + K(\mathbf{Z}_i - \boldsymbol{\delta}_r) + C\dot{\mathbf{Z}}_i = Me_u\omega^2 e^{i\omega t} \tag{3.113}$$

Assuming synchronous motion and normalizing the equation of motion by $M\omega_{\text{cr}}^2$;

we have

$$-f_i^2\mathbf{Z}_i + \mathbf{Z}_i - \boldsymbol{\delta}_r + i2\xi f_i\mathbf{Z}_i = \mathbf{e}_u f_i^2 \qquad i = 1, 2, 3 \tag{3.114}$$

Rearranging the equations of motion at the three different speeds, we obtain

$$\begin{aligned} f_1^2\mathbf{e}_u + \boldsymbol{\delta}_r - i2f_1\mathbf{Z}_1\xi &= (1 - f_1^2)\mathbf{Z}_1 \\ f_2^2\mathbf{e}_u + \boldsymbol{\delta}_r - i2f_2\mathbf{Z}_2\xi &= (1 - f_2^2)\mathbf{Z}_2 \\ f_3^2\mathbf{e}_u + \boldsymbol{\delta}_r - i2f_3\mathbf{Z}_3\xi &= (1 - f_3^2)\mathbf{Z}_3 \end{aligned} \tag{3.115}$$

Solving for rotor mechanical unbalance $\mathbf{e}_u$, shaft bow δ_r, and modal damping ξ as the system unknowns, we obtain

$$\begin{Bmatrix} \mathbf{e}_u \\ \boldsymbol{\delta}_r \\ \xi \end{Bmatrix} = \begin{bmatrix} f_1^2 & 1 & -2if_1\mathbf{Z}_1 \\ f_2^2 & 1 & -2if_2\mathbf{Z}_2 \\ f_3^2 & 1 & -2if_3\mathbf{Z}_3 \end{bmatrix}^{-1} \begin{Bmatrix} (1 - f_1^2)\mathbf{Z}_1 \\ (1 - f_2^2)\mathbf{Z}_2 \\ (1 - f_3^2)\mathbf{Z}_3 \end{Bmatrix} \tag{3.116}$$

Note that $\mathbf{e}_u$ and $\boldsymbol{\delta}_r$ are complex (or vector) quantities while the damping ξ is real. Thus five unknown quantities are to be computed.

The balance condition is given by

$$\mathbf{e}_b + \mathbf{e}_u + \boldsymbol{\delta}_r = 0$$

and the balance correction weight is

$$\mathbf{W}_b = \frac{\mathbf{U}_b}{\mathbf{R}} = -\frac{W_{\text{modal}}(\mathbf{e}_u + \boldsymbol{\delta}_r)}{\mathbf{R}} \tag{3.117}$$

This procedure may be extended to multimass rotors and is referred to as *first mode balancing without trial weights*. The problem of balancing the first mode of a multimass rotor may often be reduced to the balancing of an equivalent signle-mass Jeffcott rotor in which the rotor modal weight is given by

$$W_{\text{modal}} = \int_0^L \rho\phi_1^2(x)\,dx = \sum W_i\phi_{i1}^2(x)$$

3.3.4 Balancing the Jeffcott Rotor by the Least-Squared-Error Method

Since it is difficult to distinguish between true shaft bow and a constant shaft residual runout vector, one may use a multispeed balancing method based on minimization of the rotor response over a speed range.

In the procedure, the rotor low-speed runout vector $\mathbf{Z}_0$ is measured. The rotor response $\mathbf{Z}_i$ is then measured at N_i speeds ($i = 1$ to n values). The rotor amplitude is assumed to be of the form

$$\mathbf{Z}_i = \mathbf{a}_i\mathbf{U} + \mathbf{Z}_0 \qquad i = 1, \ldots, n \tag{3.118}$$

A trial or calibration weight is next attached to the rotor, and the new response is measured at the same speeds.

$$\mathbf{Z}_{it} = \mathbf{a}_i(\mathbf{U} + \mathbf{U}_t) + \mathbf{Z}_0 \tag{3.119}$$

The influence coefficients $\mathbf{a}_i$ are then computed:

$$\frac{\mathbf{Z}_{it} - \mathbf{Z}_i}{\mathbf{U}_t} = \mathbf{a}_i \tag{3.120}$$

Let $\mathbf{Z}_{ic} = \mathbf{Z}_i - \mathbf{Z}_0$ represent the compensated rotor vibration with slow-roll runout subtraction. The compensated vibration readings at the various speeds may be expressed as

$$\mathbf{Z}_{ic} = \mathbf{a}_i\mathbf{U} + \epsilon_i \qquad i = 1, \ldots, n \tag{3.121}$$

The error ϵ_i is given by

$$\epsilon_i = \mathbf{Z}_{ic} - \mathbf{a}_i\mathbf{U} \tag{3.122}$$

The error function ϵ_i is complex. Let ϵ_i^* be the complex conjugate error function.

A positive definite error function E_i may be constructed by multiplying the error function ϵ_i by its complex conjugate value ϵ_i^*:

$$E_i = \epsilon_i\epsilon_i^* = (\mathbf{Z}_{ic} - \mathbf{a}_i\mathbf{U})(\mathbf{Z}_{ic}^* - \mathbf{a}_i^*\mathbf{U}^*) \tag{3.123}$$

The total error function E is given by

$$E = \sum E_i = \sum (\mathbf{Z}_{ic} - \mathbf{a}_i\mathbf{U})(\mathbf{Z}_{ic}^* - \mathbf{a}_i^*\mathbf{U}^*) \tag{3.124}$$

Minimizing the total error function E with respect to U, we obtain the balancing conditions:

$$\mathbf{U}_b = -\mathbf{U} = \frac{-\sum \mathbf{a}_i^*\mathbf{Z}_{ic}}{\sum \mathbf{a}_i\mathbf{a}_i^*} \tag{3.125}$$

By taking vibration measurements at speeds both above and below the critical speed, excellent single-plane balancing may be achieved with a bowed rotor. This procedure also has the advantage that good balancing may be achieved by using the uncompensated vibration values $\mathbf{Z}_i$ corresponding to multiple speeds.

Example 3.10: Balancing a Bowed Rotor by the Least-Squared-Error Method. The rotor of Example 3.9 is to be balanced by measuring the vibration amplitude below and above the rotor critical speed, at 1800 and 2300 rpm.

The amplitude and phase at the three speeds are given by

$$\mathbf{Z}_{1800} = 4.69 \text{ mils } \angle -73° = \mathbf{Z}_1$$

$$\mathbf{Z}_{2080} = 13.77 \text{ mils } \angle -135° = \mathbf{Z}_{cr}$$

$$\mathbf{Z}_{2300} = 6.32 \text{ mils } \angle -192° = \mathbf{Z}_2$$

The compensated amplitude is given by subtracting the slow-roll vector $\mathbf{Z}_0$:

$$\mathbf{Z}_{1c} = 4.69\angle -73° - 1.0\angle -90° = 3.75\angle -68°$$

$$\mathbf{Z}_{2c} = 6.32\angle -192° - 1.0\angle -90° = 6.6\angle -201°$$

A trial weight of 0.576 g is placed at a radius of 1.5 in at 180° from the timing mark

$$\mathbf{e}_t = 0.5 \text{ mil}$$

The new vibration readings after the application of the trial weight are

$$\mathbf{Z}_{1t} = 4.02\angle -89°$$

$$\mathbf{Z}_{cr,t} = 10.9\angle -153°$$

$$\mathbf{Z}_{2t} = 4.0\angle -177°$$

The influence coefficients for 1800 and 2300 rpm are

$$\mathbf{a}_1 = 2.41 \text{ mils/g} \angle -20° \qquad \mathbf{a}_2 = 4.325 \text{ mils/g} \angle -152°$$

The complex conjugates of the influence coefficients $\mathbf{a}_1$ and $\mathbf{a}_2$ are

$$\mathbf{a}_1^* = 2.41\angle 20° \qquad \mathbf{a}_2^* = 4.325\angle 152°$$

The single-plane least-squared-error balancing procedure for the two speeds is

$$\begin{aligned}\mathbf{U}_b &= -\frac{\mathbf{Z}_{1c}\mathbf{a}_1^* + \mathbf{Z}_{2c}\mathbf{a}_2^*}{\mathbf{a}_1\mathbf{a}_1^* + \mathbf{a}_2\mathbf{a}_2^*} \\ &= \frac{-37.58\angle -48.7°}{24.5} = 1.53\angle -229° \text{ g}\end{aligned} \tag{3.126}$$

The balance correction computed with these two measurements is 1.53 g at a lag angle of 229° from the shaft reference mark.

Table 3.7 represents the Jeffcott rotor conditions before and after balancing using the vibration data at 1800 and 2300 rpm. In this table, the amplitude at the critical speed at 2080 rpm is shown as 13.8 mils at 135° lag.

In the bottom portion of the table, the initial readings, current vibration readings with trial weight, and predicted readings are given. The data indicates that if the computed balance correction of 1.53 g is placed on the rotor at 229°, then the response at the critical speed will drop from 13.8 to only 1.39 mils. The amplitude at the other low speeds is predicted to be only 1.0 mil at 80° lag. This is approximately the value of the low-speed runout vector. This situation represents an outstanding balanced condition.

Figure 3.38 represents the locations of the trial, trim, and total balance corrections. The mechanical unbalance of 1.15 g at 1.5 in is located at 0° which is at hole 1. The shaft bow of 1 mil radial is in the direction of hole 5 (90° lag angle). The trial weight of 0.58 g is shown at hole 9.

To balance the rotor, the trial weight of 0.58 g could be removed and 1.53 g could be placed in hole 11. If one did not wish to remove the trial weight (which is often the case), a trim weight of 1.23 g could be placed at hole 12. The trim and trial weight together would produce the same effect as the total balance correction of 1.53 g at 229°.

3.3.5 Single-Plane Balancing by the Three-Trial-Weight Method

One does not normally balance a lightly damped rotor at the critical speed by using the influence coefficient method because slight changes in speed cause

TABLE 3.7 Single-Plane Least-Squared-Error Balancing of Jeffcott Rotor with Shaft Bow Including Runout Compensation

Single-plane balancing of Jeffcott rotor with shaft bow, eu = 1.0 mil at 0°, shaft bow = 1.0 mil at 90° lag, two-speed balancing data taken below and above critical speed—slow roll compensation included—critical speed data not included.

System influence coefficients: units = mils (g). All probe readings are in the lag convention.

Balance shot			Speed	Predicted	Amplitude 1,	Phase 1,	Amplitude 2,	Phase 2,	Influence coefficient		Relative
No.	Amplitude	Phase	rpm	no.	mils	deg	mils	deg	Mag.	Deg	lag*
1	0.58	180	1800	1	4.700	73	4.020	89	2.410	20	160
1	0.58	180	2080	1	13.800	135	10.900	153	8.350	91	89
1	0.58	180	2300	1	6.320	192	4.700	212	4.325	152	28

Initial and current responses and predicted residual response†

		Initial		Current		Predicted		Change in amplitude	
Speed rpm	Probe no.	Amplitude mils	Phase deg	Amplitude mils	Phase deg	Amplitude mils	Phase deg	Predicted initial	Predicted current
1800	1	4.700	73	4.020	89	1.052	88	−3.648	−2.968
2080	1	13.800	135	10.900	153	1.392	93	−12.408	−9.508
2300	1	6.320	192	4.700	212	1.007	88	−5.313	−3.693

Balance correction weights

		Total balance				Trim balance‡	
No.	Balance plane location	Mag, g	Phase lag	**X**	**Y**	Mag, g	Phase lag
1	Center plane	1.53	229	−1.0	−1.2	1.23	249

* Relative lag = amount the influence coefficient lags the trial weight.
† Sum of squared residuals = 4.060; rms residual = 1.163; max residual = 1.392.
‡ Note: Trim balance = (total balance) − (trial weight left in rotor).

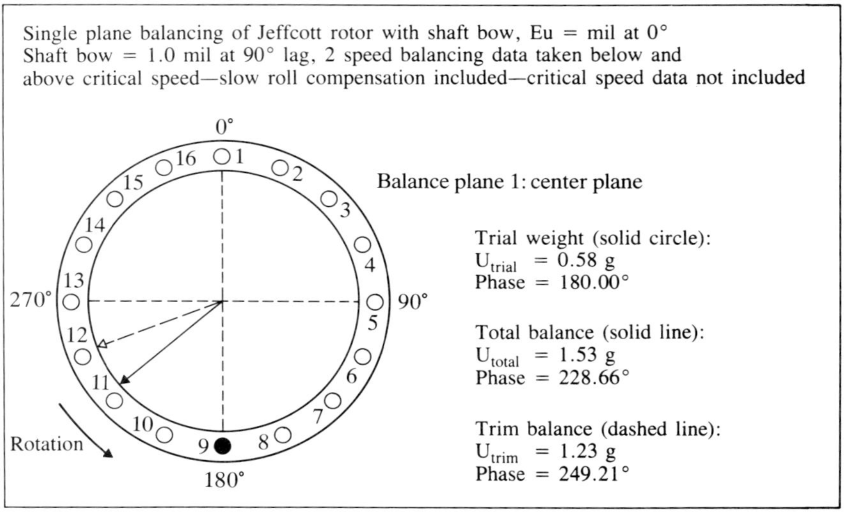

FIGURE 3.38 Single-plane balancing of Jeffcott rotor with shaft bow showing final and trim balance corrections.

considerable shifts in amplitude and phase. There is, however, a method that is ideally suited for balancing a lightly damped bowed rotor at the critical speed. This method is referred to as the *three-trial-weight or four-run* method, and it was first described by Blake (1967).

This method was first employed for balancing large fans under conditions where it was difficult to obtain a phase measurement. The procedure has the advantage that vibration can be recorded by a spectrum analyzer or fast Fourier transform (FFT) analyzer. The phase of the motion is not required. The basic procedure has been extended by Gunter, Springer, and Humphris (1982) to balance a multimass rotor through three critical speeds using modal balancing weight distributions.

The method has a fundamental drawback in that four runs of the rotor must be performed with trial weights moved in three different positions. It should not be used when thermal bowing is suspected.

As the first step in the three-trial-weight method, the initial rotor amplitude is recorded. Figure 3.39 represents the initial, trial, and final balance runs for the three-trial-weight method (Nicholas et al., 1976). The response Z_0 represents the original motion of a single-mass rotor with shaft bow. The amplitude of motion was plotted on an *XY* plotter to record the motion going through the critical speed. The peak motion passing through the critical speed Z_0 is recorded. A circle of radius 6.625 mils is drawn as shown in Fig. 3.40 to represent Z_0.

A trial weight of 0.34 g was selected and placed in hole 3 on the rotor disk. The rotor was rerun and the new amplitude of motion Z_1 was recorded. On the circumference of the circle Z_0, holes are labeled; Z_1 is drawn with a radius of 6.125 mils at hole 3 marked on circle Z_0.

The trial weight was then removed and placed in hole 8. The resulting vibration Z_2 was then recorded. A circle of diameter Z_2 was drawn with its origin at hole 8 marked on circle Z_0. The intersections of circles Z_1 and Z_2 represent possible balancing solutions.

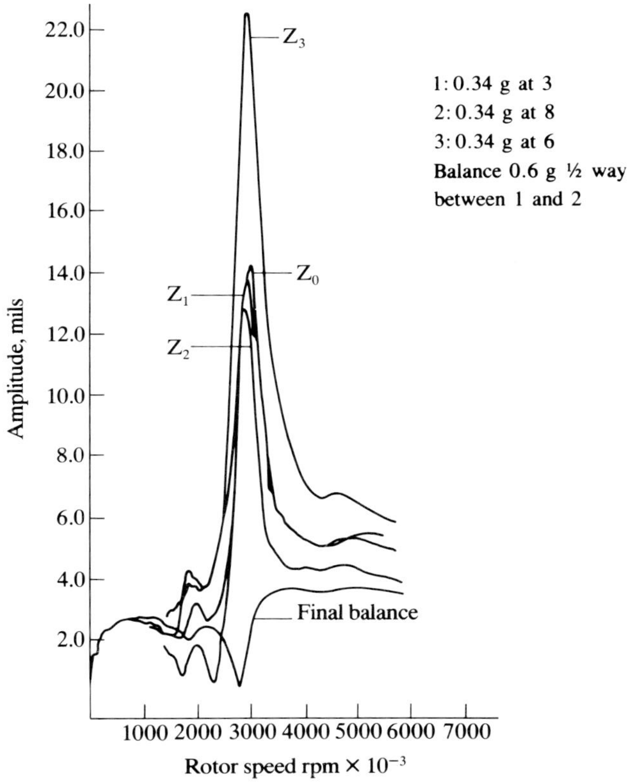

FIGURE 3.39 Initial, trial, and final balancing runs using the three-trial-weight method (Nicholas et al., 1976).

A third run is necessary to identify the final balance location and magnitude. The trial weight was next removed from hole 8 and placed in hole 6. The resulting amplitude Z_3 was recorded, and a circle was drawn with its origin at hole 6 on the circumference of Z_0.

Circles Z_1, Z_2, and Z_3 with origins lying on circle Z_0 represent the locus of points for various orientations of the unbalance vector $\mathbf{U}$ and the trial balance vector $\mathbf{U}_t$. The point, or vicinity, where all three circles intersect is the location of the balance correction weight, as shown in Fig. 3.40. A vector $\mathbf{e}_b$ is drawn from origin of Z_0 to the point of intersection of all three circles (or vicinity in case the circles do not exactly cross at a point). The line of action of $\mathbf{e}_b$ represents the balancing plane. The balance magnitude is given by

$$\mathbf{U}_b = \mathbf{U}_t \times \frac{\mathbf{Z}_0}{\mathbf{e}_b} \tag{3.127}$$

From Fig. 3.40, a balance correction of 0.6 g between holes 1 and 2 was computed. Figure 3.39 shows the final amplitude of motion after the required balance correction weight was placed on the shaft. Note the reduction in

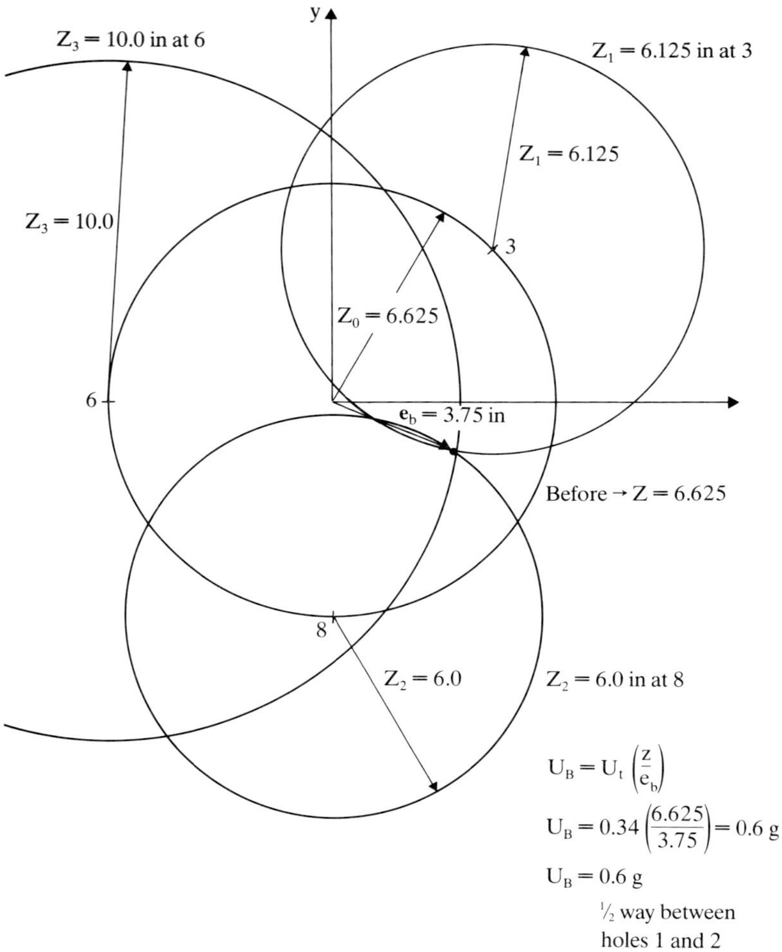

FIGURE 3.40 Graphical balancing solution using the three-trial-weight method (Nicholas et al., 1976).

amplitude near the critical speed. This is an indication that the rotor shaft bow and unbalance are of similar magnitudes and out of phase.

The three-trial-weight procedure is a very accurate method of single-plane balancing provided that one can easily attach and remove weights. The speed of the rotor does not have to be carefully controlled as in the influence coefficient method. One may use the peak hold capability of an analyzer and capture the maximum amplitude of motion while passing through the rotor critical speed. Since the influence coefficients, due to shaft bow and unbalance, are identical at the critical speed, a highly accurate balance is obtained. Therefore, it is not necessary to compensate for low-speed runout when this procedure is used.

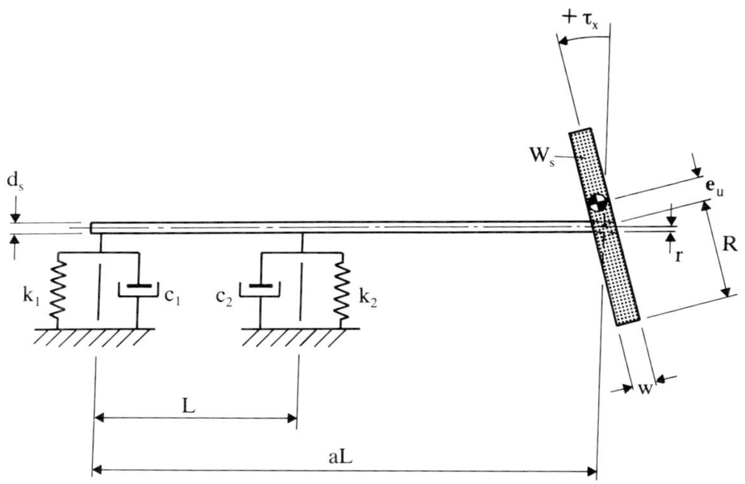

FIGURE 3.41 Overhung rotor with disk skew and unbalance (Benson, 1974).

3.3.6 Balancing the Overhung Rotor with Disk Skew

Figure 3.41 represents an overhung rotor with radial unbalance $\mathbf{e}_u$ and disk skew τ (Salamone and Gunter, 1978). The plane of the disk is inclined from the vertical by the small angle τ. The equations of motion of the overhung rotor with disk skew were extensively studied by Benson (1974). Balancing the overhung rotor with disk skew is considerably more complex than the conventional Jeffcott rotor which is symmetrically situated between the bearings.

The skew of the disk generates an unbalance moment with magnitude

$$M_{\text{gyro}} = \omega^2 \tau (I_p - I_t) \tag{3.128}$$

The sign of the moment is dependent on whether the polar moment of inertia I_p is larger than the transverse moment of inertia I_t. For $I_p > I_t$, which is the situation for a thin disk, there is normally only one synchronous critical speed due to the gyroscopic moment effects.

Figure 3.42 represents the response of an overhung disk as monitored at the near bearing for various combinations of rotor unbalance **U** and disk skew τ (Salamone and Gunter, 1978). If the disk skew is out of phase to the unbalance (case 4), the effects combine to increase the amplitude of motion at the critical speed. If, however, the disk skew is in phase to the disk unbalance, then the moment generated by the skewed disk will help to offset the radial unbalance force and the overall vibrational amplitude will be reduced.

If the disk skew is positive, as in case 5, there is a self-balancing speed at which zero amplitude of motion is achieved at the near bearing (the far bearing may have a very high amplitude of motion, however). It therefore can be readily demonstrated that for any given disk skew, one may choose a balance correction to reduce the rotor amplitude to zero for a particular speed. At any other speed, however, large amplitudes of motion may result.

Disk skew causes the rotor amplitude to increase with an increase in speed above the critical speed. This effect may often be confused with the approach of a

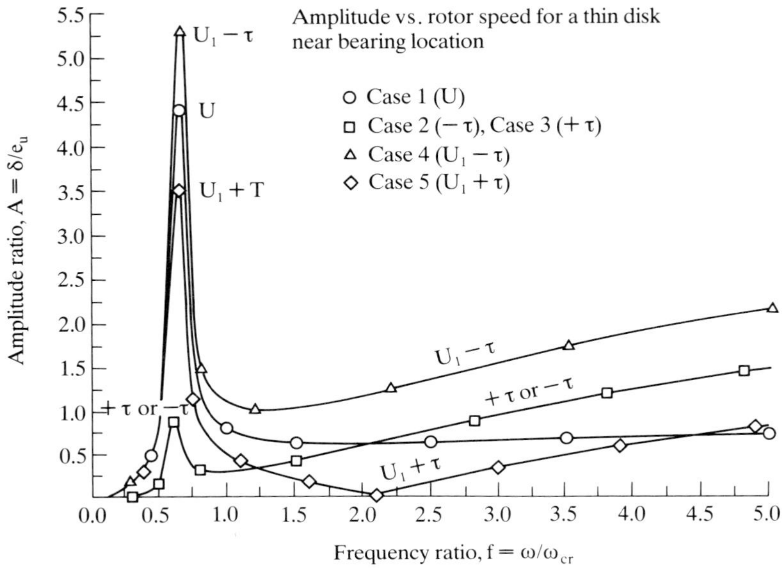

FIGURE 3.42 Amplitude versus rotor speed for a thin disk at the near-balancing location for various combinations of unbalance and disk skew τ (Salamone and Gunter, 1978).

second critical speed. It is often difficult to detect the presence of large disk skew by monitoring the motion at the near bearing.

When an overhung rotor is balanced, it is important to monitor the amplitude and phase at the far bearing as well as the near bearing. Figure 3.43 represents the phase-angle change observed at the far end bearing for an overhung rotor with various combinations of unbalance and disk skew. By observing the relative phase change at the far end bearing, it is very easy to distinguish between disk radial unbalance and disk skew effects. For the case of pure radial disk unbalance, a conventional 180° phase shift is observed while passing through the critical speed.

If, however, there is a substantial disk skew present in the rotor, then a phase reversal will be observed upon passing through the critical speed. If no phase reversal is observed at the far end bearing, then a single-plane balance correction on the disk will be sufficient to balance the rotor at all speeds. If the phase reversal is observed, then the single-plane correction cannot balance the rotor at all speeds. Attempts to precisely balance the near end bearing may result in large amplitudes at the far end bearing. In this case a compromise single-plane balancing may be achieved by using amplitude and phase readings obtained from both the near and far end bearings.

Figure 3.44 represents the near-bearing amplitude for an overhung rotor before and after various balancing procedures (Salamone and Gunter, 1978). In balance procedure 1, the rotor is balanced by a single-plane correction near the critical speed. Above the critical speed, the rotor amplitude increases rapidly. In balance run 3, the near plane is balanced to zero amplitude at $f = 3$, but large amplitudes result at the far bearing.

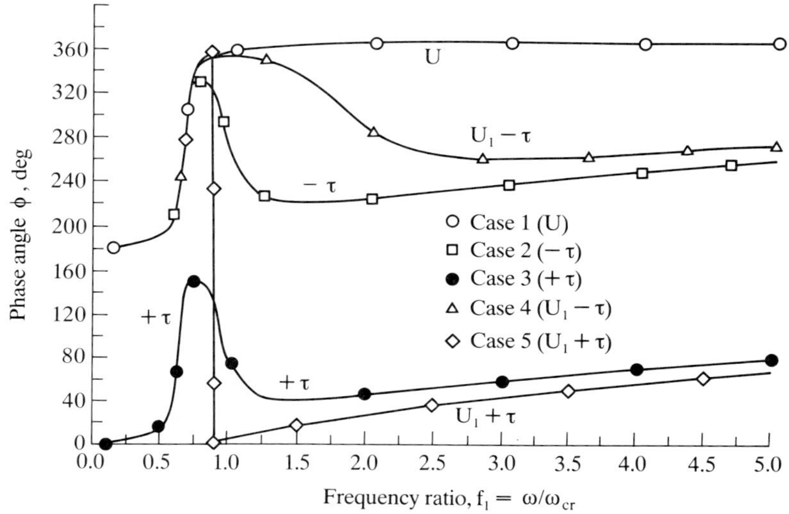

FIGURE 3.43 Phase angle versus rotor speed at the fan bearing for an overhung rotor for various combinations of rotor unbalance and disk skew (Salamone and Gunter, 1978).

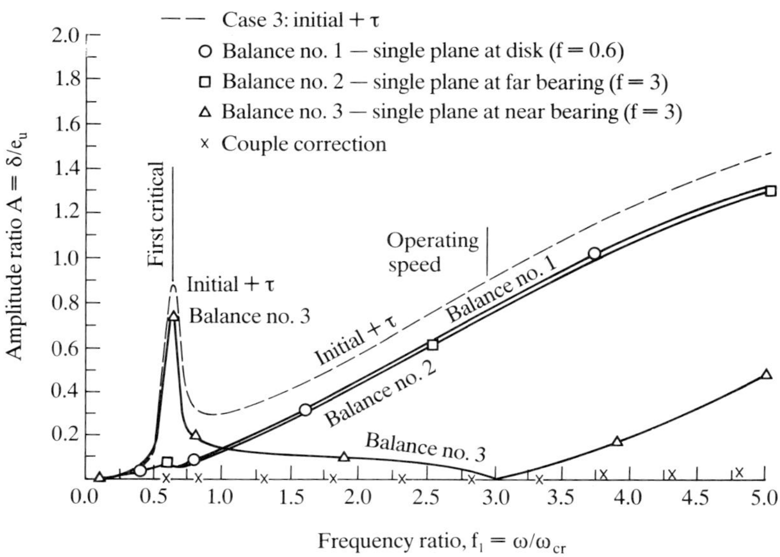

FIGURE 3.44 Near-bearing amplitude for an overhung rotor before and after various balancing procedures (Salamone and Gunter, 1978).

The recommended procedure for balancing an overhung rotor is to first attempt single-plane balancing of the disk by using the influence coefficients and vibrations obtained from monitoring both the near and far bearings. The balance correction is given by

$$\mathbf{U}_b = -\frac{\mathbf{Z}_n \mathbf{a}_n^* + \mathbf{Z}_f \mathbf{a}_f^*}{\mathbf{a}_n \mathbf{a}_n^* + \mathbf{a}_f \mathbf{a}_f^*} \tag{3.129}$$

This will produce a compromise balance that will not allow one bearing to have an excessive amount of vibration. If the resulting vibration is not sufficiently low, then a trial couple U_{tc} should be placed on the wheel and the influence coefficients for the couple obtained and used in the balancing calculations. The trial balance couple is generated by placing two weights 180° out of phase to each other on the disk with a plane separation distance of T.

The couple balance correction is given by

$$\mathbf{U}_{cb} = -\frac{\mathbf{Z}_{cn} \mathbf{a}_{cn}^* + \mathbf{Z}_{cg} \mathbf{a}_{cf}^*}{\mathbf{a}_{cn} \mathbf{a}_{cn}^* + \mathbf{a}_{cf} \mathbf{a}_{cf}^*} \tag{3.130}$$

This procedure results in a two-plane balancing of the disk. A similar balancing can be achieved by the two-plane balancing procedure in which trial weights are placed on the two planes of the disks and the vibrations for the near and far bearings recorded for the two trial runs. The solution of these balancing problems requires a computer to perform the necessary computations. The procedure presented here can be calculated by hand.

The required balance correction W_c (two weights 180° out of phase) for a given disk skew is given by

$$W_c = \frac{U_c}{TR} = \frac{\tau K (I_p - I_t)}{TR} \tag{3.131}$$

where τ = skew angle, rad
K = constant = 16 oz/lb (102 g/in)
I_p = polar moment of inertia, lb · in^2 (N · cm^2)
I_t = transverse moment of inertia, lb · in^2 (N · cm^2)
T = plane of axial separation, in (cm)
R = balance radial distance, in (cm)

Note that if the disk skew τ is excessive, then it may be physically impossible to place a sufficient couple on the disk to correct for the skew effect. The wheel must be removed from the rotor and realigned. It is often advisable to monitor the axial runout of the overhung wheel to check for disk skew.

3.3.7 General Two-Plane Balancing

The problem of balancing an overhung disk skew, as just discussed, is an example of a general two-plane balancing problem. The original procedure described by Thearle (1934) was a semigraphical method by which the two-plane balance corrections were computed. This semigraphical method was used into the late 1970s when it was replaced by the two-plane balancing solution on the handheld programmable calculators (see Jackson, 1979).

The vibration monitored at the near and far planes Z_n and Z_f is assumed to be a function of the two unknown values of unbalance $\mathbf{U}_1$ and $\mathbf{U}_2$.

$$\begin{aligned}\mathbf{Z}_n &= \mathbf{a}_{n1}\mathbf{U}_1 + \mathbf{a}_{n2}\mathbf{U}_2 \\ \mathbf{Z}_f &= \mathbf{a}_{f1}\mathbf{U}_1 + \mathbf{a}_{f2}\mathbf{U}_2\end{aligned} \tag{3.132}$$

To compute the values of the two planes of unbalance, a total of four influence coefficients must be determined. To determine these influence coefficients, a trial balance is placed in each plane, and the new resulting amplitudes of motion are measured. For example, the placement of a trial weight in the first plane results in

$$\begin{aligned}\mathbf{Z}_{n1} &= \mathbf{a}_{n1}(\mathbf{U}_1 + \mathbf{U}_{1t}) + \mathbf{a}_{n2}\mathbf{U}_2 \\ \mathbf{Z}_{f1} &= \mathbf{a}_{f1}(\mathbf{U}_1 + \mathbf{U}_{1t}) + \mathbf{a}_{f2}\mathbf{U}_2\end{aligned} \tag{3.133}$$

The influence coefficients $\mathbf{a}_{n1}$ and $\mathbf{a}_{f1}$ may now be determined as follows:

$$\begin{aligned}\mathbf{a}_{n1} &= \frac{\mathbf{Z}_{n1} - \mathbf{Z}_n}{\mathbf{U}_{1t}} \\ \mathbf{a}_{f1} &= \frac{\mathbf{Z}_{f1} - \mathbf{Z}_f}{\mathbf{U}_{1t}}\end{aligned} \tag{3.134}$$

The trial balance weight $\mathbf{U}_{1t}$ is now removed, and a trial balance weight $\mathbf{U}_{2t}$ is placed in the second balance plane. Note that it is not necessary to remove the first trial balancing weight. If the first balancing weight has caused a substantial reduction in the rotor response, then it is desirable to keep the trial weight in place. For example, removing a trial weight from the hot section of a gas turbine may require an extra day for the unit to cool down. Also complications may arise when an attempt is made to remove the balancing weight from the hot section, such as breaking the extraction tool or dropping the weight into the turbine passageway. When this occurs, the unit must be disassembled to remove the weight to prevent blade damage.

If the first trial weight $\mathbf{U}_{1t}$ is assumed to be left in place, and a second trial balance weight $\mathbf{U}_{2t}$ is placed in the second balance plane, the resulting vibration is given by

$$\begin{aligned}\mathbf{Z}_{n2} &= \mathbf{a}_{n1}(\mathbf{U}_1 + \mathbf{U}_{1t}) + \boldsymbol{a}_{n2}(\mathbf{U}_2 + \mathbf{U}_{2t}) \\ \mathbf{Z}_{f2} &= \mathbf{a}_{f1}(\mathbf{U}_1 + \mathbf{U}_{1t}) + \mathbf{a}_{f2}(\mathbf{U}_2 + \mathbf{U}_{2t})\end{aligned} \tag{3.135}$$

The influence coefficients $\mathbf{a}_{n2}$ and $\mathbf{a}_{f2}$ are obtained by subtracting the vibration readings after the application of the first trial weight as follows (if $\mathbf{U}_{1t}$ is left in):

$$\begin{aligned}\mathbf{a}_{n2} &= \frac{\mathbf{Z}_{n2} - \mathbf{Z}_{n1}}{\mathbf{U}_{2t}} \\ \mathbf{a}_{f2} &= \frac{\mathbf{Z}_{f2} - \mathbf{Z}_{f1}}{\mathbf{U}_{2t}}\end{aligned} \tag{3.136}$$

If the first trial weight $\mathbf{U}_{1t}$ is removed, then replace $\mathbf{Z}_{n1}$ and $\mathbf{Z}_{f1}$ by the original vibration readings $\mathbf{Z}_n$ and $\mathbf{Z}_f$.

The balance corrections $\mathbf{U}_{b1}$ and $\mathbf{U}_{b2}$ are given by

$$\begin{Bmatrix}\mathbf{U}_{b1} \\ \mathbf{U}_{b2}\end{Bmatrix} = -\begin{vmatrix}\mathbf{U}_1 + \mathbf{U}_{1t} \\ \mathbf{U}_2 + \mathbf{U}_{2t}\end{vmatrix} = -\begin{vmatrix}\mathbf{a}_{n1} & \mathbf{a}_{n2} \\ \mathbf{a}_{f1} & \mathbf{a}_{f2}\end{vmatrix}^{-1} \begin{Bmatrix}\mathbf{Z}_{n2} \\ \mathbf{Z}_{f2}\end{Bmatrix} \tag{3.137}$$

Solving for the final balance correction values gives

$$\mathbf{U}_{b1} = \frac{\mathbf{a}_{n2}\mathbf{Z}_{f2} - \mathbf{a}_{f2}\mathbf{Z}_{n2}}{\mathbf{a}_{n1}\mathbf{a}_{f2} - \mathbf{a}_{f1}\mathbf{a}_{n2}}$$
$$\mathbf{U}_{b2} = \frac{\mathbf{a}_{f1}\mathbf{Z}_{n2} - \mathbf{a}_{n1}\mathbf{Z}_{f2}}{\mathbf{a}_{n1}\mathbf{a}_{f2} - \mathbf{a}_{f1}\mathbf{a}_{n2}} \tag{3.138}$$

The values of $\mathbf{U}_{b1}$ and $\mathbf{U}_{b2}$ represent the values of trim balance correction required if both trial balance weights are left in place. If the balance computations are performed using the original vibration readings for the near and far planes $\mathbf{Z}_n$ and $\mathbf{Z}_f$, then the computed balances will correspond to the total original balance correction required on the rotor.

Example 3.11: Two-Plane Balancing Using the Influence Coefficient Method Including Runout Compensation, Rotor 99DM-1. Table 3.8 represents the initial and final vibration readings after trial weights were placed on a rotor at the near and far planes (Jackson, 1979). Shown on the table are the computed near and far plane balance corrections. In this calculation, a positive sign convention represents lag angles for both vibration and balance weights measures against rotation from the vibration data probe.

The compensated near and far amplitudes are first determined by subtracting the slow-roll vectors measured at the two ends:

$$\mathbf{Z}_{nc} = 1.8\angle 148° - 0.5\angle 272° = 2.12\angle 137°$$

$$\mathbf{Z}_{fc} = 3.6\angle 115° - 0.4\angle 123° = 3.2\angle 114°$$

The four influence coefficients are

$$\mathbf{a}_{n1} = 0.206\angle 175° \qquad \mathbf{a}_{f1} = 0.364\angle 194°$$

$$\mathbf{a}_{n2} = 0.341\angle 182° \qquad \mathbf{a}_{f2} = 0.170\angle 165°$$

Let

$$\Delta = \mathbf{a}_{n1}\mathbf{a}_{f2} - \mathbf{a}_{f1}\mathbf{a}_{n2} = 0.0986\angle 208°$$

Let $[R] = -[a]^{-1}$ be the negative of the inverse of the influence coefficient matrix.

$$\mathbf{R}_{11} = -\frac{\mathbf{a}_{f2}}{\Delta} = -\frac{0.170\angle 165°}{0.0986\angle 208°} = 1.724\angle 137°$$

$$\mathbf{R}_{12} = \frac{\mathbf{a}_{n2}}{\Delta} = 3.458\angle 334°$$

$$\mathbf{R}_{21} = \frac{\mathbf{a}_{f1}}{\Delta} = 3.69\angle 346°$$

$$\mathbf{R}_{22} = -\frac{\mathbf{a}_{n1}}{\Delta} = 2.09\angle 147°$$

The balancing solution for $\mathbf{U}_{b1}$ and $\mathbf{U}_{b2}$ is given in terms of the R matrix and compensated vibration readings as follows:

$$\mathbf{U}_{b1} = \mathbf{R}_{11}\mathbf{Z}_{nc} + \mathbf{R}_{12}\mathbf{Z}_{fc}$$
$$\mathbf{U}_{b2} = \mathbf{R}_{21}\mathbf{Z}_{nc} + \mathbf{R}_{22}\mathbf{Z}_{fc} \tag{3.139}$$

TABLE 3.8 Two-Plane Balance and Trim correction of Rotor Including Runout Compensations

Rotation counter-clockwise		Rotation clockwise

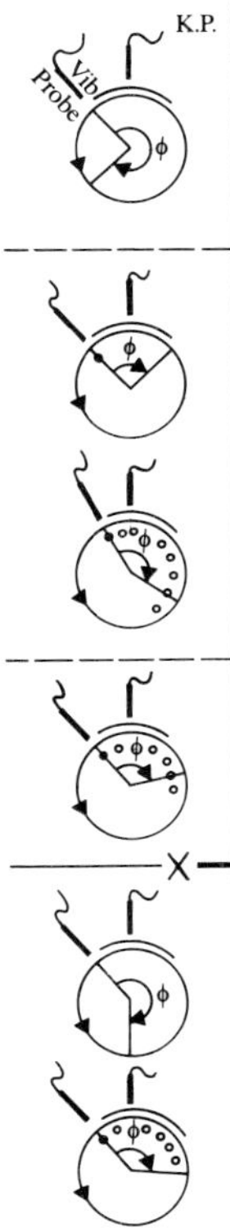

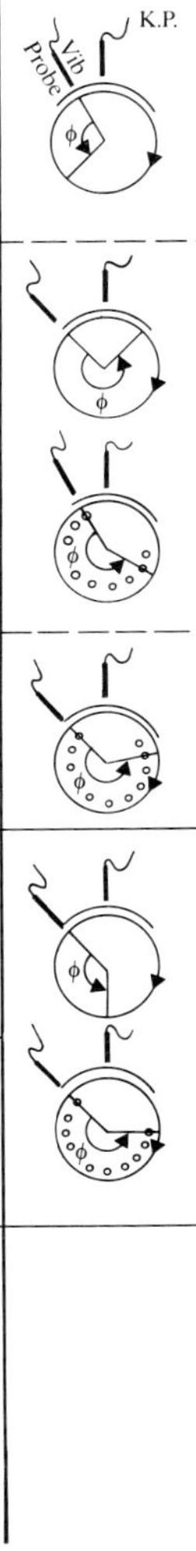

Machine identification 99DM-1 DATE 7/78

Run-out recorded @ 300 R.P.M.; "at-speed" data @ 31,000 RMP

Key phasor probe is @ 0° ; Probes @ 90 N& 90° F

- H.P. 67 in "run" press [f][Program]. . . read mag. card sides 1 & 2
- Press [f][a]

Enter run-out data

	Near plane Angle		Near plane Amount		Far plane Angle		Far plane Amount
1	272° R/S	2	0.5 R/S	3	123° R/S	4	0.4 R/S

- Press [f] [b] and enter "at speed" data

Orig. data	5	148° R/S	6	1.8 R/S	11	115° R/S	12	3.6 R/S
T.W. @ near plane	7	178° R/S	8	1.1 R/S	13	98° R/S	14	2.0 R/S
T.W. @ far plane	9	98° R/S	10	2.1 R/S	15	102° R/S	16	3.7 R/S

T.W. locations	17	120°	18	4.9 gms Units	19	220°	20	4.9 gms Units

- Press [A,B,C,D] to read out wt. add locations

Corr. weight + location	A	B	C	D
	7.49	84.96°	5.32	179.73°
	Amt	Loc	Amt	Loc

X →

- Press [E] Enter trim data

	Near plane Angle		Near plane Amount		Far plane Angle		Far plane Amount
1	268° R/S	2	0.6 R/S	3	0.85° R/S	4	0.5 R/S

- Press [A,B,C,D] to read out trim corrections

A	B	C	D
2.79	-47.11 312.89°	1.55	135.54°
Amt	Loc	Amt	Loc

X →

Note: At pts.(3) marked (X →) in left margin. System parameters may be recorded in run by [r] [w/data] and mag. card.

Notes

- Phase angles (all) taken against rotation from vib. data probe (sensor).
- Trial wts. (T.W.) are also placed against rotation from vib. data probe.
- Correction wts. & angles hold same logic.

Coefficients $\mathbf{R}_{11}$ and $\mathbf{R}_{22}$ are the direct or principal balancing coefficients, while coefficients $\mathbf{R}_{12}$ and $\mathbf{R}_{21}$ are the cross-coupling balancing coefficients. Note that both the cross-coupling coefficients $\mathbf{R}_{12}$ and $\mathbf{R}_{21}$ are greater than the principal balancing coefficients $\mathbf{R}_{11}$ and $\mathbf{R}_{22}$. When this occurs, it is difficult to balance the rotor by single-plane balancing using each plane in turn. This approach is only valid if the principal balancing coefficients (or influence coefficients) are larger than the cross-coupling coefficients.

The balance values are given by

$$\begin{aligned}\mathbf{U}_{b1} &= 1.724\angle 137° \times 2.12\angle 137° + 3.458\angle 334° \times 3.2\angle 114° \\ &= 3.65\angle 274° + 11.06\angle 88° = 7.49\angle 85° \text{ (against rotation)}\end{aligned}$$

$$\begin{aligned}\mathbf{U}_{b2} &= 3.69\angle 346° \times 2.12\angle 137° + 2.09\angle 147° \times 3.2\angle 114° \\ &= 7.822\angle 123° + 6.68\angle 261° = 5.3\angle 180° \text{ (against rotation)}\end{aligned}$$

Figure 3.45 represents the locations of the trial and final balance weights on the near and far balancing planes. On the near balancing plane, the direction of rotation, looking at the end of the rotor, is counterclockwise. The near balancing plane is shown labeled with 36 balancing holes. The trial weight of 4.9 g is placed in hole 13. The final balance weight is shown as being situated between holes 9 and 10. A vector split of the weight between holes 9 and 10 results in the computation of 3.79 g for hole 9 and 3.726 g for hole 10.

When the far balancing plane is viewed from the other end of the rotor, the direction of rotation is shown to be clockwise. In both balance planes, however, the holes are always labeled against rotation. The initial trial weight is shown located at hole 23, and the final balance should be 5.32 g located at hole 18. If the trial weight is left in hole 23, then a trim balance of 3.54 g is computed. This trim balance may be split between holes 12 and 13 with 1.32 g in hole 12 and 2.23 g in hole 13.

The combination of the trim plus the trial is equivalent to the predicted total required balance weight of 5.3 g in hole 18. After the balance weights were placed on the rotor, the resulting amplitudes $\mathbf{Z}_{nb} = 0.6\angle 268°$ and $\mathbf{Z}_{b} = 0.5\angle 0.85°$ are recorded.

A refined trim balance based on the $[R]$ balance matrix and the current-compensated vibration readings may be computed as follows:

$$\mathbf{U}_{n,\text{trim}} = \mathbf{R}_{11}\mathbf{Z}_{nbc} + \mathbf{R}_{12}\mathbf{Z}_{fbc} = 2.79\angle 313°$$

$$\mathbf{U}_{f,\text{trim}} = \mathbf{R}_{21}\mathbf{Z}_{nbc} + \mathbf{R}_{22}\mathbf{Z}_{fbc} = 1.55\angle 135.5°$$

This additional balance correction would reduce the final uncompensated rotor response to

$$\mathbf{U}_{n,\text{final}} = 0.5 \text{ mil } \angle 272°$$

$$\mathbf{U}_{f,\text{final}} = 0.40 \text{ mil } \angle 123°$$

The above values are identical to the initial slow-roll vectors.

3.4 MULTIPLANE FLEXIBLE ROTOR BALANCING

3.4.1 Introduction

All rotating shafts exhibit flexibility effects when spinning. A rotor is called *flexible* if the shaft does not appear to hold balance after being balanced as a rigid

Two plane balancing of 99DX-1 – (Jackson – *Practical Vibration Prixer*, pp. 103)
H = 3600 rpm, run-out recorded at 300 rpm, non-contact displacement probes
phase angles and trial weights measured against rotation from data probe

Balance plane 1: near balance plane

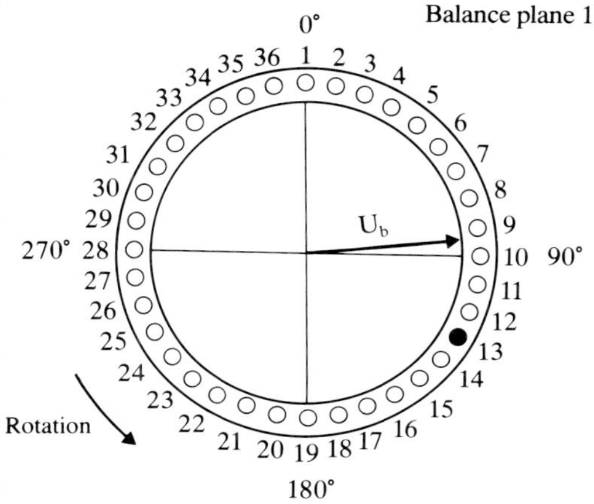

Trial weight (solid circle):
U_{trial} = 4.90 cms
Phase = 120.00°

Total balance (solid line):
U_{total} = 7.49 cms
Phase = 84.96°

Trim balance (dashed line):
U_{trim} = 7.49 cms
Phase = 84.96°

Hole split: Total
3.790 cms at hole 9
3.726 cms at hole 10

Two plane balancing of 99DX-1 – (Jackson – *Practical Vibration Prixer*, pp. 103)
H = 3600 rpm, run-out recorded at 300 rpm, non-contact displacement probes
phase angles and trial weights measured against rotation from data probe

Balance plane 2: far balance plane

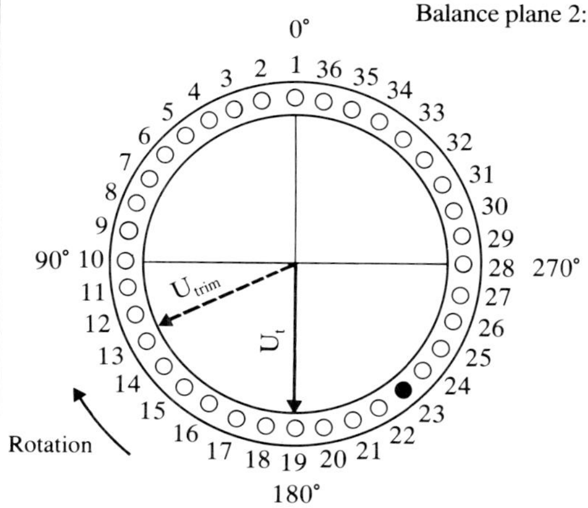

Trial weight (solid circle):
U_{trial} = 4.90 cms
Phase = 220.00°

Total balance (solid line):
U_{total} = 5.32 cms
Phase = 179.73°

Trim balance (dashed line):
U_{trim} = 3.54 cms
Phase = 116.28°

Hole split: trim
1.324 cms at hole 12
2.229 cms at hole 13

FIGURE 3.45 Location of trial, final, and trim balance weights on near and far balance planes.

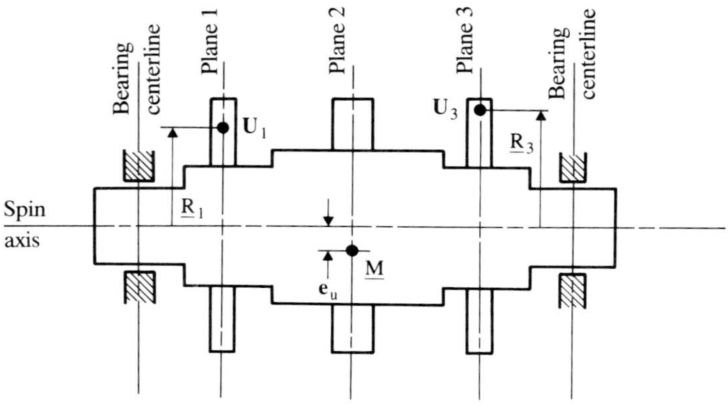

FIGURE 3.46 Multimass rotor with two-plane static balance correction.

body. Figure 3.46 represents a multistation rotor with the rotor mass center displaced from the axis of rotation by the distance $\mathbf{e}_u$ at rotor plane 2.

For this class of unbalance, the rotor rigid-body inertia axis will be shifted from the spin axis by the distance of $\mathbf{e}_u$. When this rotor is placed in a rigid-body balancing machine, either the hard or soft bearing type, the balance machine will indicate a two-plane correction in which the left and right planes will be in phase and of similar magnitude.

Correction weights of w_1 and w_3 are placed at planes 1 and 3, respectively, at radii R_1 and R_3. As a rigid body, the multistage rotor now appears to be perfectly balanced. The original rotor unbalance may be represented by

$$M\mathbf{e}_u = m_2\mathbf{R}_2 = \mathbf{U}_2 \tag{3.140}$$

The rotor is in static balance since it satisfies the rigid rotor balance conditions that

$$\sum \mathbf{U} = \mathbf{U}_1 + \mathbf{U}_2 + \mathbf{U}_3 = 0 \tag{3.141}$$

If the rotor is balanced as shown in Fig. 3.47 is operated at a sufficiently low speed, then the rotor will appear to be well behaved and balance will be maintained. As the rotor speed increases, the inertia loads caused by the unbalance distributions cause the shaft to deflect due to elasticity effects of the shaft. This will cause the shaft to bow while rotating, as shown in Fig. 3.48. These effects become more pronounced as the shaft speed exceeds 70 percent of the rotor critical speed. At speeds above this value, additional balancing must be performed at speed, called *trim balancing,* to maintain small rotor amplitudes of motion.

3.4.2 Types of Unbalance

The multimass flexible rotor may have a variety of unbalance distributions. Figure 3.49 represents four basic types of unbalance distributions encountered with a multimass rotor. There are many instances in which all four types of unbalance are encountered.

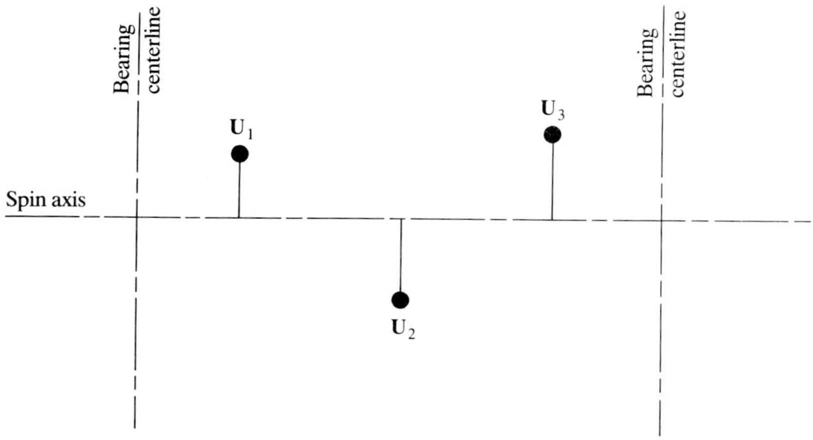

FIGURE 3.47 Equivalent multiplane unbalance distribution.

Figure 3.49*a* represents a continuous unbalance distribution along the shaft due to machining a concentric shaft with radial eccentricity. The second distribution, shown in Fig. 3.49*b*, includes radial unbalances such as encountered by assembled compressor turbine stages on a shaft. It is often desirable to individually balance each wheel before assembly on a shaft to miminize radial wheel unbalance.

Figure 3.49*c* represents a shaft with bow. In a bowed rotor, the mass-elastic centerline does not lie along the shaft axis of rotation. Bows may be introduced by nonuniform shrink fits, thermal effects, permanent sag due to gravitational effects, or excessive vibration. The initial bow may change with operating conditions, as is the case with all generator rotors. Excessive shaft bow is difficult to correct by simple balancing because it may require large correction weights.

The last case shown is the disk skew. The influence of disk skew is the most pronounced with overhung impellers and wheels. Disk skew may be induced by

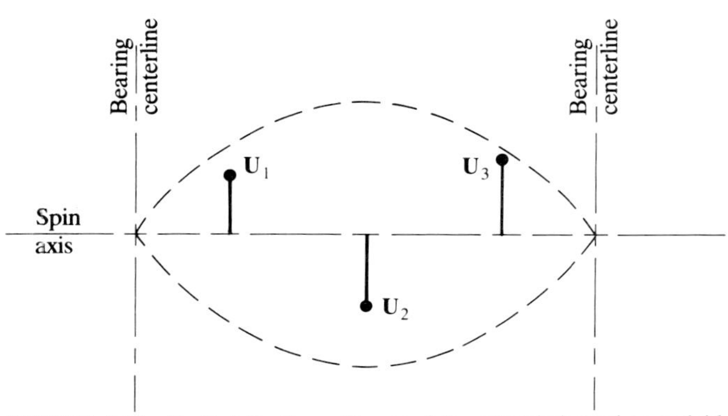

FIGURE 3.48 Shaft deflection with speed for rotor balanced as a rigid body.

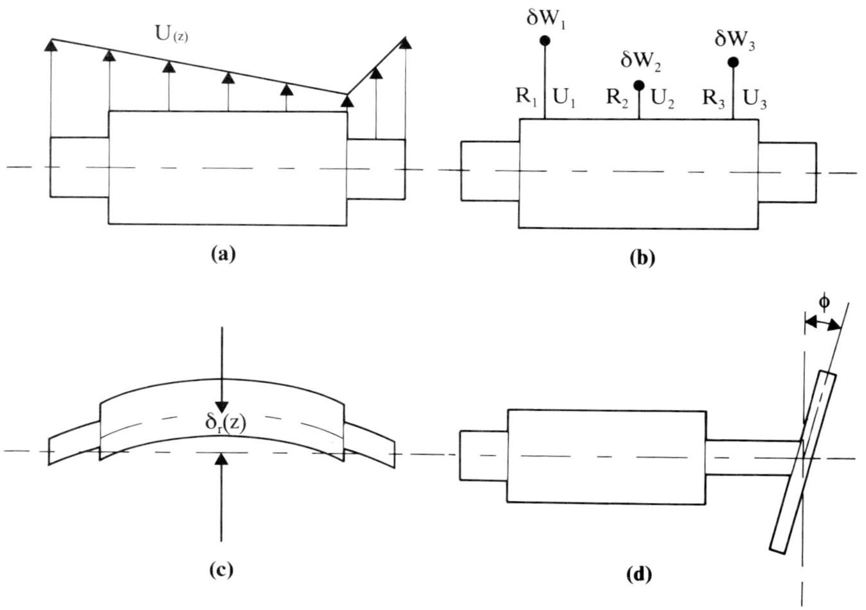

FIGURE 3.49 Types of unbalance distribution in a multimass flexible rotor: (*a*) continuous unbalance; (*b*) distributed radial point mass unbalance; (*c*) rotor row unbalance; (*d*) moment unbalance.

improper filtering of the impeller to the shaft. A shaft bow will also induce a disk skew effect in an overhung rotor.

The general rotor equations of motion may be expressed in matrix form as

$$[M]\{\ddot{\mathbf{Z}}\} + [G]\{\dot{\mathbf{Z}}\} + [C]\{\dot{\mathbf{Z}}\} + ([K][\{\mathbf{Z}\} - \{\mathbf{Z}_0\}]) = \{\mathbf{F}(t)\} \tag{3.142}$$

where $\{\mathbf{Z}\}^t = [\{X\}, \{Y\}, \{\Theta_x\}, \{\Theta_y\}]^t$
= horizontal and vertical displacements and rotations of shaft
$[G]$ = skew symmetric gyroscopic damping matrix
$\{\mathbf{Z}_0\}$ = initial shaft bow

The forcing function due to radial unbalance and disk skew is of the form:

$$\{\mathbf{F}(t)\}^t = [\{Ue^{-i\phi_m t}\}, \{\tau(I_p - I_t)e^{-i\phi_d t}\}]\omega^2 e^{i\omega t} \tag{3.143}$$

3.4.3 Critical Speeds of a Uniform Shaft and Reduction to a Point-Mass Model

To more fully understand some of the concepts of balancing a multistation flexible rotor with a discrete number of balance planes, it is necessary to introduce some concepts of continuum theory of shafts and the reduction of a continuum to a discrete point or lumped-mass stiffness rotor model.

Figure 3.50*a* represents a uniform beam of length L and diameter D. The

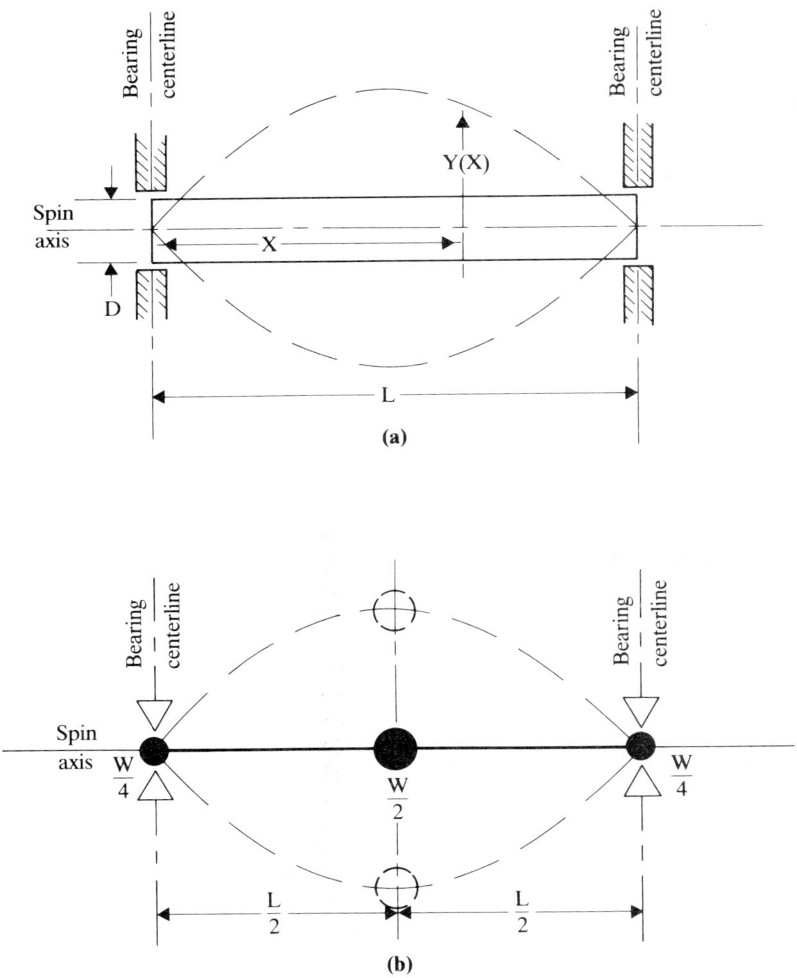

FIGURE 3.50 (*a*) First mode of a uniform beam; (*b*) uniform beam represented as a three-station lumped-mass model.

beam natural frequencies (considering ideal rigid supports) are given by

$$\omega_n = (n\pi)^2 \sqrt{\frac{EI}{ML^3}} \qquad \text{rad/s} \tag{3.144}$$

where $\omega_n = n$th natural frequency. Corresponding to each natural frequency (or eigenvalue) is a mode shape or eigenvector. The eigenvectors for a uniform beam are given by

$$\phi(x) = \sin \frac{n\pi x}{L} \tag{3.145}$$

The displacement mode shape does not represent the actual shaft deflection but represents a normalized shape that the shaft assumes at that frequency. If the uniform shaft is rotated about its axis, then any residual unbalance or shaft bow will cause the shaft to bow outward as the rotor speed approaches the shaft critical speed.

The actual whirl radius that the shaft will assume will be dependent on the magnitude of unbalance, the relationship of the rotor speed to the critical speed, and the amount of damping present in the system.

Modern rotor dynamic theory is based on representing a system as a set of discrete mass stations, as reviewed in detail in Chap. 2, rather than using continuum theory. With continuum theory it is difficult to accurately simulate a multimass rotor with an arbitrary unbalance distribution. Modal theory does, however, provide us with valuable insights on multiplane and modal balancing of flexible rotors.

Figure 3.50*b* represents the uniform rotor represented as a three-station model. A proper lumped-mass model to represent the rotor for the first critical speed requires three stations with weights of $W/4$ at the ends of the rotor and $W/2$ at the center of the shaft (assuming the weight distribution of $W/3$ at each station leads to an improper lumped-mass model). For the assumption of rigid bearings at the ends of the shaft, we arrive at a Jeffcott rotor simulating a uniform shaft by placing half of the total shaft weight at the middle of the bearing span. Note, however, that this lumped-mass model is valid only for motion through the rotor first critical speed.

Since the shaft stiffness of a uniform beam at the center span is given by

$$K_{\text{shaft}} = \frac{48EI}{L^3} \tag{3.146}$$

the natural frequency or critical speed of the lumped-mass representation of Fig. 3.50*b* is given by

$$\omega_1 = \sqrt{\frac{98EI}{ML^3}} \tag{3.147}$$

Example 3.12. Compute the critical speed of a uniform steel shaft with 4.00-in diameter and 60 in long, using continuum and lumped-mass models.

$$E = 30 \times 10^6 \text{ lb/in}^2$$

$$I = \frac{\pi D^4}{64} = \frac{\pi 4^2}{64} = 12.566 \text{ in}^2$$

$$M = \rho A L = \frac{0.283 \text{ lb/in}^3}{386 \text{ in/s}^2} \times \frac{\pi 4^2}{2} \times 60$$

$$= 1.106 \frac{\text{lb} \cdot \text{s}}{\text{in}^2} \text{ total shaft mass (426.75 lb)}$$

$$K = \frac{48EI}{L^3} = 83{,}773 \text{ lb/in}$$

$$\omega_{1,\text{uniform shaft}} = (3.142)^2 \sqrt{\frac{30.0 \times 10^6 \times 12.566}{1.106 \times 60^3}}$$

$$= 392.06 \text{ rad/s} = 3744 \text{ rpm}$$

$$\omega_{1,\text{lumped mass}} = \sqrt{\frac{83,773}{\left(\frac{1.106}{2}\right)}}$$

$$= 389.2 \text{ rad/s} = 3717 \text{ rpm}$$

From the above example, we see that the three-station lumped-mass rotor represents the uniform beam first critical speed with a deviation of less than 0.7 percent. It has been found that any continuous unbalance distribution along a rotor may be represented by a set of three discrete balance stations located along the shaft to simulate rotor bowing over the first critical speed.

To approximate the characteristics of the uniform beam for the second mode, five mass stations are required, as shown in Fig. 3.51. The ends have lumped weights of $W/8$, and the center sections have lumped weights of $W/4$. This results in a lumped-mass model which approximates the second rotor critical speed, also to less than 0.7 percent error. Note that a lumped-mass model in which each

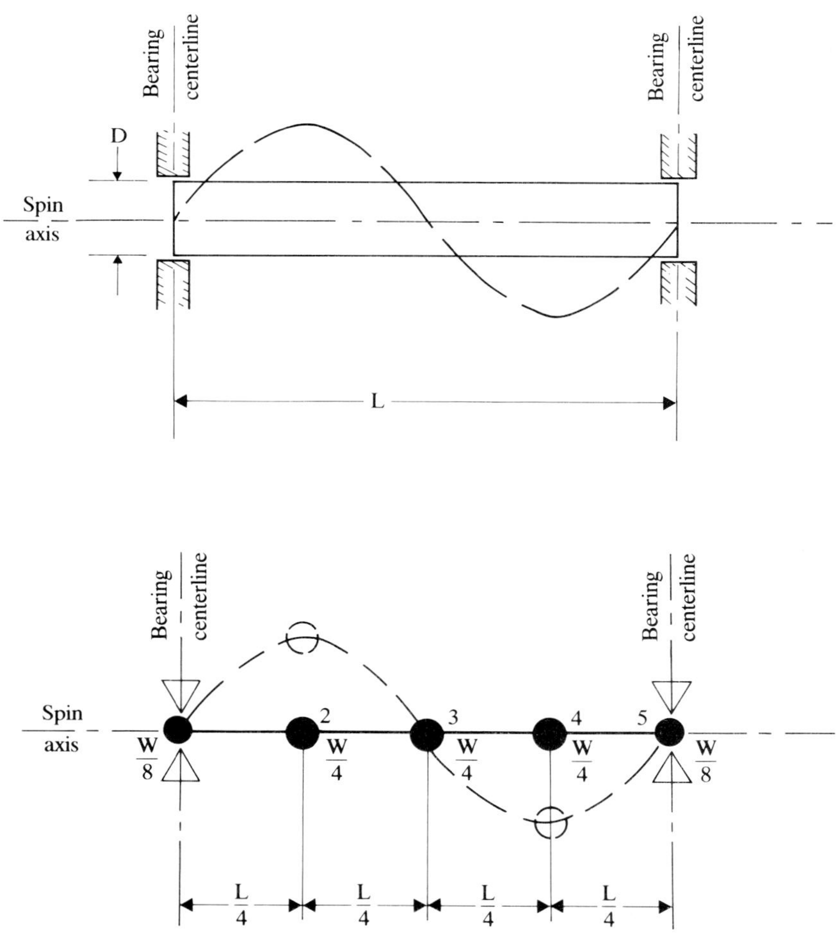

FIGURE 3.51 Lumped-mass model for second mode representation of a uniform beam.

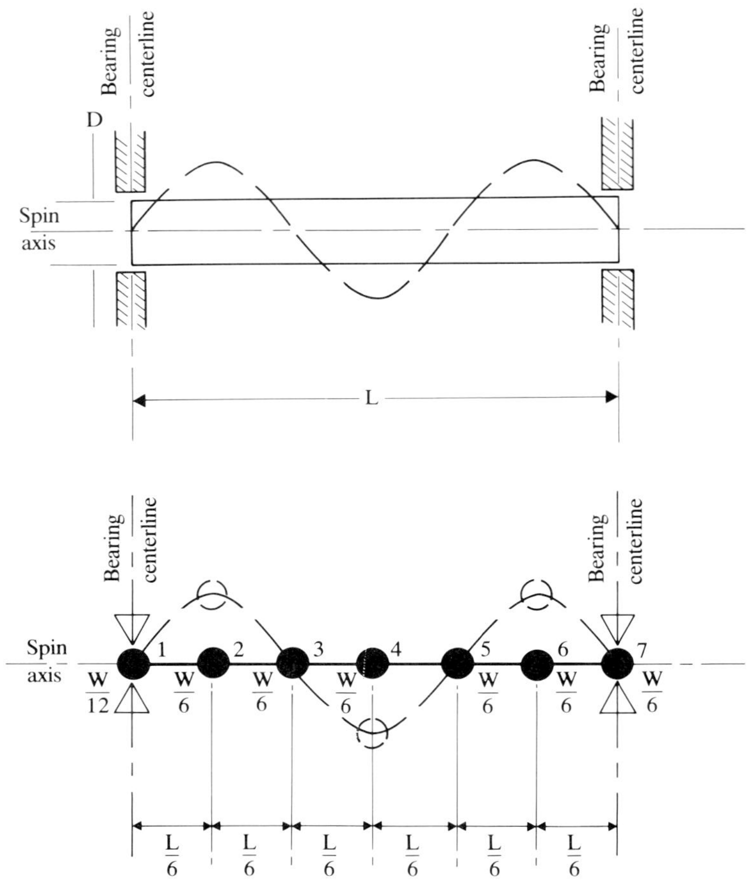

FIGURE 3.52 Lumped-mass model for third mode representation of a uniform beam.

station has a weight of $W/5$ would produce an incorrect rotor model in which the critical speed was 12 percent too high.

Figure 3.52 shows a seven-station lumped-mass model to represent the third mode of a uniform beam to less than 1 percent error. In this model, the lumped weights are $W/12$ at the end stations and $W/6$ at the interior stations. The analogy may be repeated for higher modes. Only $2N+1$ mass stations are required to represent the Nth critical speed of a uniform shaft to less than 1 percent error for all modes.

3.4.4 N or $N+2$ Planes of Balancing

For a uniform rotor with ideal rigid bearings, there will be $N+1$ node points at which zero motion occurs. If a rotor has a point along the shaft at which zero displacement occurs for a particular critical speed, it is called a *nodal point.* The placement of a balance weight at this nodal point will have no effect on the rotor balancing for speeds within 20 percent of this particular critical speed.

Hence, the number of active balance planes required for the idealized uniform rotor to balance the Nth critical speed is given by the number of required mass stations minus the zero-displacement nodal points as follows:

$$N_{\text{balance planes}} = (2N+1)_{\text{total nodes}} - (N+1)_{\text{zero disp. nodes}} = N \tag{3.148}$$

Thus for the first mode or critical speed ($N = 1$), only one balance plane is required. Two balance planes are required to balance the second mode, and three planes are required to balance the third mode. This is referred to as the *N-plane method* of balancing. Long turbogenerators usually follow this rule. There has been considerable discussion in the literature as to whether N or $N + 2$ balance planes are required for balancing a flexible rotor (e.g., Kellenberger, 1972; Bishop and Parkinson, 1965, 1972).

For a highly flexible rotor with little displacement at the bearing locations, only N balance planes are required. If the rotor is to operate near or above the third rotor critical speed, rigid-body balancing may actually result in higher vibrations near the third critical speed. Under these circumstances three-plane modal balancing must be employed.

A great deal of information, as to the required number of balancing planes and their most effective location, may be obtained by examining the critical speed map and corresponding mode shapes for a particular rotor.

Figure 2.36 represents a typical critical speed map of a uniform shaft for various bearing stiffness values, as shown by Rieger (1986). The first three critical speeds are plotted in log-log format as a function of bearing stiffness. From the chart we see that for bearing stiffness stress below 2×10^5 lb/in, the rotor first and second critical speeds are effectively straight lines on the log-log plot.

If the rotor speed is kept below the rotor third critical speed and the bearing stiffness is below 2×10^5 lb/in, then the rotor will operate essentially as a rigid body. The modes shown on the chart for this region represent rigid-body conical and cylindrical modes of motion. The rotor may be balanced as a rigid body in any two arbitrary planes.

As the rotor speed is increased, assuming that the bearing stiffness values do not change, the rotor third critical speed is encountered. The bearing stiffness has little influence on the third critical speed until stiffness values of over 0.5×10^6 are encountered. This third mode is referred to as the *free-free mode*. Gas bearing supported rotors have the characteristics of rigid-body behavior at the first two modes due to their low bearing stiffness values. The rotor will behave as if it is badly out of balance as it approaches the third or free-free critical speed. This will be particularly true if the original balance weights have been placed at the node points of the third mode. Three additional balance planes will be required to balance the third mode so as not to upset the first two modes. A modal balancing procedure will be required to balance this mode that uses three simultaneous balance weights.

The other extreme range of operation is the situation of high bearing stiffness relative to the shaft stiffness. For bearing stiffness values of above 5×10^6 lb/in, the bearings begin to approach nodal points of zero vibration. Under these circumstances, rigid-body balancing may not be beneficial. If, e.g., the rigid-body correction weights are placed in the rotor end planes, there is no improvement in the balance condition. A single balancing plane at the rotor center is required to balance the rotor for the first mode. To balance the second mode, two balance planes are required for the case of high bearing stiffness. These balance planes should not be placed at node points of the second mode, but rather at the locations corresponding to maximum amplitude.

3.4.5 Rotor Modal Equations of Motion and Modal Unbalance Distribution

The generalized equations of motion of a rotor bearing system (neglecting shaft bow) may be written in terms of the rotor mass, damping, and stiffness matrices as follows:

$$[M]\{\ddot{\mathbf{Z}}\} + [C]\{\dot{\mathbf{Z}}\} + [K]\{\mathbf{Z}\} = \omega^2\{\mathbf{U}\}\mathbf{e}^{i\omega t} \tag{3.149}$$

where $\{\mathbf{Z}\}$ represents the complex rotor displacements at the various n stations along the shaft.

The undamped system critical speeds are determined by setting the damping and external forcing functions to zero. This results in the standard eigenvalue problem for the determination of the rotor critical speeds:

$$[-\omega^2[M] + [K]]\{\mathbf{Z}\} = 0 \tag{3.150}$$

where

$$\{\mathbf{Z}\} = \sum_{i=1}^{m} q_i\{\phi\}_i = [\Phi]\{q\} \tag{3.151}$$

$$q_i = \text{modal coefficient for } i\text{th mode}$$

Since $[M]$ and $[K]$ are symmetric $n \times n$ matrices, there are only n natural frequencies associated with the system as compared to an infinity of solutions for a continuum such as our uniform beam example. From the standpoint of flexible rotor balancing, only the first several critical speeds are of interest.

By using only the first m modes (where $m < n$ stations), the rotor equations of motion are reduced further:

$$\Big[\underbrace{\Phi}_{n\times m}\Big] = \underbrace{\Big[\underbrace{\{\phi\}_1}_{n\times 1} \quad \underbrace{\{\phi\}_2}_{n\times 1} \quad \underbrace{\{\phi\}_m}_{n\times 1}\Big]}_{m \text{ modes}} \Big\} \; n \text{ stations} \tag{3.152}$$

In a typical multiplane balancing problem, $m = 2$ or 3 modes should be sufficient. Large turbine-generators, however, may require 4 to 5 modes for accurate multiplane balancing.

Assuming that three modes of motion will be sufficient to represent the rotor balancing problem, the modal properties are given by

$$\Big[\underbrace{\phi_1 \quad \phi_2 \quad \phi_3}_{m\times n}\Big]^T \Big[\underbrace{M}_{n\times n}\Big]\Big[\underbrace{\phi_1 \quad \phi_2 \quad \phi_3}_{n\times m}\Big] = \begin{bmatrix} M_1 & & \\ & M_2 & \\ & & M_3 \end{bmatrix}_{m\times m}$$

$$\Big[\underbrace{\Phi}_{m\times n}\Big]^T \Big[\underbrace{C}_{n\times n}\Big]\Big[\underbrace{\Phi}_{n\times m}\Big] = \begin{bmatrix} C_{11} & C_{12} & C_{13} \\ C_{21} & C_{22} & C_{23} \\ C_{31} & C_{32} & C_{33} \end{bmatrix}_{m\times m} \tag{3.153}$$

$$\Big[\underbrace{\Phi}_{m\times n}\Big]^T \Big[\underbrace{K}_{n\times n}\Big]\Big[\underbrace{\Phi}_{n\times m}\Big] = \begin{bmatrix} K_1 & & \\ & K_2 & \\ & & K_3 \end{bmatrix}_{m\times m}$$

Note that in the reduced modal $[M]$, $[C]$, and $[K]$ matrices, the mass and stiffness matrices are diagonal matrices because of the properties of orthogonality. In general, with discrete bearings providing the damping, the damping modal matrix is never diagonal. The damping matrix is diagonal only if it is proportional to the mass and stiffness matrix as follows:

$$[C]_{\text{diag}} = \alpha[K] + \beta[M] \tag{3.154}$$

This idealized assumption, called *proportional damping,* does not occur in real-life rotor systems. The concept of proportional damping is often used in structural damping but is not exact when applied to turborotors with substantial damping at the bearings. With symmetric rotors such as the uniform beam, the model cross-coupling between the first and the second modes C_{12} and C_{21} may be ignored, along with the cross-coupling between the second and third modes C_{23} and C_{32}. However, the modal cross-coupling coefficients between the first and third modes C_{13} and C_{31} are never zero as is true for the modal cross-coupling coefficients C_{24} and C_{42} between the second and fourth modes.

The oversimplification and neglect of these modal cross-coupling coefficients have caused much confusion and led to numerous misconceptions concerning modal balancing as compared to the influence coefficient method.

The reduced modal equations of motion using three modes only for a turborotor are of the form

$$\begin{aligned} M_1\ddot{\mathbf{q}}_1 + C_{11}\dot{\mathbf{q}}_1 + C_{13}\dot{\mathbf{q}}_3 + K_1\mathbf{q}_1 &= \omega^2\phi_1^T\mathbf{U}e^{i\omega t} \\ M_2\ddot{\mathbf{q}}_2 + C_{22}\dot{\mathbf{q}}_2 + K_2\mathbf{q}_2 &= \omega^2\phi_2^T\mathbf{U}e^{i\omega t} \\ M_3\ddot{\mathbf{q}}_3 + C_{31}\dot{\mathbf{q}}_1 + C_{33}\dot{\mathbf{q}}_3 + K_3\mathbf{q}_3 &= \omega^2\phi_3^T\mathbf{U}e^{i\omega t} \end{aligned} \tag{3.155}$$

By dividing by modal mass, the generalized equations become

$$\begin{aligned} \ddot{\mathbf{q}}_1 + 2\xi_{11}\omega_1\dot{\mathbf{q}}_1 + 2\xi_{13}\omega_3\dot{\mathbf{q}}_3 + \omega_1^2\mathbf{q}_1 &= \omega^2\mathbf{e}_1e^{i\omega t} \\ \ddot{\mathbf{q}} + 2\xi_{22}\omega_2\dot{\mathbf{q}}_2 + \omega_2^2\mathbf{q}_2 &= \omega^2\mathbf{e}_2e^{i\omega t} \\ \ddot{\mathbf{q}}_3 + 2\xi_{31}\omega_1\dot{\mathbf{q}}_1 + 2\xi_{33}\dot{\mathbf{q}}_3 + \omega_3^2\mathbf{q}_3 &= \omega^2\mathbf{e}_3e^{i\omega t} \end{aligned} \tag{3.156}$$

where $\mathbf{e}_i$ are the complex modal unbalance eccentricity vectors

$$\mathbf{e}_i = \frac{\phi_i^T}{M_i}\mathbf{U} \tag{3.157}$$

It is of interest that the second modal equation is essentially uncoupled from the first and third modal equations for most symmetric 2-bearing rotor systems. The second mode balancing follows identically the procedure used for balancing the single-mass Jeffcott rotor except that a second mode distribution of weights is applied. These two weights are usually taken as equal and are placed on the rotor out of phase to each other.

The first and third equations are modally coupled through the modal cross-coupling coefficients C_{13} and C_{31}. Only for the case where the bearing amplitudes of motion approach nodal points of zero motion do these coefficients disappear. This situation is approximated by long generators in which modal uncoupling may be approximated. The assumption of modal uncoupling leads to the general theory of modal balancing. The modal cross-coupling coefficients cause the rotor to form a nonplanar mode shape or corkscrew in space.

Example 3.13. For the uniform rotor of Example 3.12, assume unbalance components of $\mathbf{U}_1 = 100\ \text{g} \cdot \text{in}$, $\mathbf{U}_2 = -200\ \text{g} \cdot \text{in}$, and $\mathbf{U}_3 = 100\ \text{g} \cdot \text{in}$. Compute the modal unbalance eccentricity vectors $\mathbf{e}_i$ for the system.

Assuming that the balance weights are placed at $x_1 = L/4$, $x_2 = L/2$, and $x_3 = 3L/4$, the mode shapes for these positions for the three modes are given by

$$\{\phi\}_1^T = [0.707 \;\; 1.00 \;\; 0.707]$$

$$\{\phi\}_2^T = [1.00 \;\; 0.0 \;\; -1.00]$$

$$\{\phi\}_3^T = [0.707 \;\; -1.0 \;\; 0.707]$$

The modal weight for the uniform beam for the three modes is $W_1 = W_2 = W_3 = W_{\text{tot/2}} = 213.4\ \text{lb} = 97{,}000\ \text{g}$.

The modal unbalance eccentricities $\mathbf{e}_i$ in mils are given by

$$\mathbf{e}_1 = \frac{1}{97{,}000\ \text{g}} [0.707 \;\; 1 \;\; 0.707] \begin{Bmatrix} 100 \\ -200 \\ 100 \end{Bmatrix} \text{g} \cdot \text{in}$$

$$= \frac{70.7 - 200 + 70.7}{97{,}000} = -0.0006\ \text{in} = -0.6\ \text{mil} < 0°$$

$$\mathbf{e}_2 = \frac{1}{97{,}000\ \text{g}} [1.0 \;\; 0 \;\; -1.0] \begin{Bmatrix} 100 \\ -200 \\ 100 \end{Bmatrix} \text{g} \cdot \text{in} = 0$$

$$\mathbf{e}_3 = \frac{1}{97{,}000\ \text{g}} [0.707 \;\; -1.0 \;\; 0.707] \begin{Bmatrix} 100 \\ -200 \\ 100 \end{Bmatrix} \text{g} \cdot \text{in}$$

$$= \frac{70.7 + 200.0 + 70.7}{97{,}000} = 0.00352\ \text{in} = 3.52\ \text{mils} < 0°$$

In Example 3.13 we see that the modal unbalance eccentricity for the first mode $\mathbf{e}_1 = -0.6$ mil while the modal unbalance eccentricity for the third mode is $\mathbf{e}_3 = 3.52$ mils. The rotor response at the third critical speed may be considerable. Since its modal unbalance eccentricity is 3.52/0.6, or 5.8, times that at the first critical speed.

If it is assumed that the modes are uncoupled (all the modal cross-coupling coefficients are assumed negligible), then the rotor response $\mathbf{Z}$ is a function of the rotor modal amplification factors and the modal unbalance eccentricities:

$$\{\mathbf{Z}\} = \sum_{i=1}^{m \text{ modes}} Q_i(\omega)\mathbf{e}_i\{\phi_i\}e^{i(\omega t - \phi_m)} \tag{3.158}$$

where

$$Q_i(\omega) = \frac{(\omega/\omega_i)^2}{\sqrt{[1 - (\omega/\omega_i)^2]^2 + (2\xi_i(\omega/\omega_i))^2}}$$

At the critical speed where $\omega = \omega_i$, the rotor amplification factor is given by

$$Q_i(\omega)\big|_{\omega=\omega_i} = A_{ci} = \frac{1}{2\xi_i} \tag{3.159}$$

Neglecting the small influences of the other modes, the rotor response at the ith critical speed is given approximately by

$$\{\mathbf{Z}\}\big|_{\omega=\omega_i} = \frac{\mathbf{e}_i}{2\xi_i}\{\phi_i(x)\}e^{i(\omega t-90)} \tag{3.160}$$

The rotor motion at the critical speed (for modal uncoupling only) is equal to the products of the rotor modal unbalance eccentricity vector $\mathbf{e}_i$ times the modal amplification factor multiplied by the mode shape.

From Eq. 3.160 the displacements of the shaft are lagging the unbalance eccentricity vector **e** by 90° at the critical speed.

Example 3.14. Compute the rotor motions at the center of the shaft at the first and third critical speeds, assuming rotor amplification factors of $A_1 = 10$ and $A_3 = 8$ for the two modes.

At the first critical speed of $N_1 = 3717$ rpm, the rotor amplitude at the center (peak-to-peak motion) is given by

$$\begin{aligned}\{\mathbf{Z}\}\big|_{x=L/2} &= 2A_1\mathbf{e}_1\{\phi\}_1\angle -90^\circ \\ &= 20(-0.6\text{ mil})\left\{\phi\left(x=\frac{L}{2}\right)\right\}_1\angle -90^\circ \\ &= 12\text{ mils }\angle -270^\circ\text{ peak to peak}\end{aligned}$$

At the third critical speed of $N_3 = 33{,}453$ rpm the rotor amplitude at the center ($x = L/2$) is given by

$$\begin{aligned}\{\mathbf{Z}\} &= 2A_3\mathbf{e}_3\phi\angle -90^\circ \\ &= 16(3.52\text{ mils})\phi_3\angle -90^\circ \\ &= 56.3\text{ mils }\angle -90^\circ\text{ peak to peak}\end{aligned}$$

It is thus obvious that the rigid-body balancing procedure of Example 3.13 has created a set of unbalance weights which will have a substantial effect on the rotor at the first critical speed (12 mils peak to peak) and a catastrophic effect on the third critical speed (56 mils peak to peak). Without the rigid-body balancing, the rotor would have an amplitude of 32 mils peak to peak at the third critical speed but would have an amplitude of 41 mils at the first critical speed.

The rotor unbalance distribution $\{\mathbf{U}\}$ may be expressed in modal components as

$$\{\mathbf{U}\} = \mathbf{U}_{m1}\{\phi\}_1 + \mathbf{U}_{m2}\{\phi\}_2 + \mathbf{U}_{m3}\{\phi\}_3 \tag{3.161}$$

Example 3.15. For the uniform rotor of Example 3.12, determine the modal unbalance components $\mathbf{U}_{m1}$ for an unbalance of 100 g · in at $L/4$ and $L/2$ along the shaft, as shown in Fig. 3.53.

Case 1:

$$\{\mathbf{U}\} = 100\text{ g}\cdot\text{in} < 0^\circ \qquad \text{at } x = \frac{L}{4}$$

$$\{\mathbf{U}\} = \begin{Bmatrix}100\\0\\0\end{Bmatrix} = \mathbf{U}_{m1}\begin{Bmatrix}0.707\\1\\0.707\end{Bmatrix} + \mathbf{U}_{m2}\begin{Bmatrix}-1\\0\\1\end{Bmatrix} + \mathbf{U}_{m3}\begin{Bmatrix}-0.707\\1\\-0.707\end{Bmatrix}$$

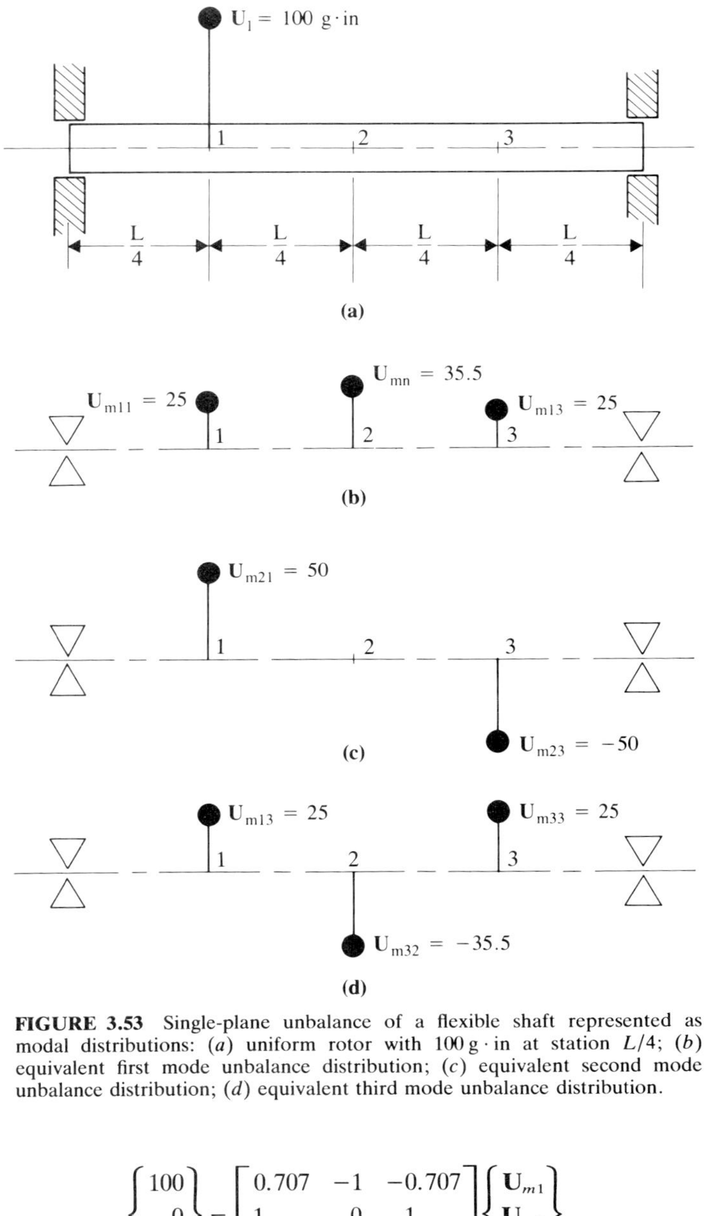

FIGURE 3.53 Single-plane unbalance of a flexible shaft represented as modal distributions: (*a*) uniform rotor with 100 g · in at station $L/4$; (*b*) equivalent first mode unbalance distribution; (*c*) equivalent second mode unbalance distribution; (*d*) equivalent third mode unbalance distribution.

$$\begin{Bmatrix} 100 \\ 0 \\ 0 \end{Bmatrix} = \begin{bmatrix} 0.707 & -1 & -0.707 \\ 1 & 0 & 1 \\ 0.707 & 1 & -0.707 \end{bmatrix} \begin{Bmatrix} \mathbf{U}_{m1} \\ \mathbf{U}_{m2} \\ \mathbf{U}_{m3} \end{Bmatrix}$$

Solving for the components $\mathbf{U}_{mi}$ gives

$$\mathbf{U}_{m1} = 35.5 \text{ g} \cdot \text{in} \qquad \mathbf{U}_{m2} = -50 \text{ g} \cdot \text{in} \qquad \mathbf{U}_{m3} = 35.5 \text{ g} \cdot \text{in}$$

The corresponding modal unbalance eccentricities are given by

$$\mathbf{e}_1 = 0.73 \text{ mils} \qquad \mathbf{e}_2 = -1.03 \text{ mils} \qquad \mathbf{e}_3 = -0.707 \text{ mil}$$

The relationship between the modal unbalance is

$$\mathbf{U}_{mi} = 4\mathbf{e}_i\mathbf{M}_i \tag{3.162}$$

For the unbalance of 100 g · in, placed at the quarter span along the shaft, all three critical speeds of the rotor will be excited.

Case 2:

$$\mathbf{U} = 100 \text{ g in} \qquad \text{at } x = \frac{L}{2}$$

$$\mathbf{U} = \begin{Bmatrix} 0 \\ 100 \\ 0 \end{Bmatrix} = \mathbf{U}_{m1}\{\phi\}_1 + \mathbf{U}_{m2}\{\phi\}_2 + \mathbf{U}_{m3}\{\phi\}_3$$

Solving for the modal unbalances $\mathbf{U}_{mi}$ gives

$$\mathbf{U}_{m1} = 50 \text{ g} \cdot \text{in} \qquad \mathbf{U}_{m2} = 0 \text{ g} \cdot \text{in} \qquad \mathbf{U}_{m3} = 50 \text{ g} \cdot \text{in} < 0°$$

From case 2 an unbalance located at the shaft center will cause an excitation of both the first and the third modes.

Modal Balancing of the Uniform Rotor

Figure 3.53 represents a uniform shaft with 100 g · in of unbalance at $x = L/4$. The single plane of unbalance may be represented as modal distributions located at three planes. Each modal unbalance distribution has the property that it has little or no effect on the excitation of the other modes. For example, with the balance distributions shown in Fig. 3.53*a* for the first mode, no rotor excitation will be experienced at the second or third modes. If the unbalances are doubled but the ratio of balance values is maintained, then only the response at the first critical speed will be doubled. No influence will be experienced at the higher modes.

To modally balance a flexible rotor such as given in Example 3.13, the balance weights should be selected as a modal distribution. The requirement for balancing a particular mode is that the modal unbalance eccentricity $\mathbf{e}_i$ be zero. Thus a rotor may be modally balanced for other modes. Modal balancing usually upsets the rigid rotor balancing.

The condition for modally balancing a rotor at the first mode is given by

$$\{\phi\}_1^T\left[\{\mathbf{U}\} + \sum_{bj}^{m} \mathbf{U}_{bj}\{\phi\}_j\right] = \mathbf{M}_1\mathbf{e}_1 = 0 \tag{3.163}$$

Expanding the above gives

$$\mathbf{U}_{m1}\phi_1^T\phi_1 + \mathbf{U}_{m2}\phi_1^T\phi_2 + \mathbf{U}_{m3}\phi_1^T\phi_3 + \mathbf{U}_{b1}\phi_1^T\phi_1 + \mathbf{U}_{b2}\phi_1^T\phi_2 + \mathbf{U}_{b3}\phi_1^T\phi_3 = 0$$

For the uniform beam, the displacement modes are orthogonal. (For rotors with unsymmetric axial mass distributions, this is not the case.) The first mode balance condition reduces to

$$\phi_1^T\phi_1[\{\mathbf{U}\}_{m1} + \{\mathbf{U}\}_{b1}] + \epsilon = 0$$

or

$$\{\mathbf{U}_b\}_1 = \mathbf{U}_{b1}\{\phi\}_1 = -\mathbf{U}_{m1}\{\phi_1\}_1 \tag{3.164}$$

Example 3.16. For the flexible rotor of Example 3.13 which was balanced as a rigid rotor, determine the first and third mode balancing components.

The first mode balancing components are given by

$$\phi_1^T[\{\mathbf{U}\} + U_{b1}\phi_1] = 0$$

$$[0.707 \;\; 1 \;\; 0.707]\left[\begin{Bmatrix} 100 \\ -200 \\ 100 \end{Bmatrix} + U_{b1}\begin{Bmatrix} 0.707 \\ 1.0 \\ 0.707 \end{Bmatrix}\right] = 0$$

$$-58.6 + 2\mathbf{U}_{b1} = 0 \qquad \mathbf{U}_{b1} = 29.3 \text{ g} \cdot \text{in} < 0°$$

The first mode balancing distribution to place on the shaft to balance out the rotor through the first critical speed is

$$\{\mathbf{U}_b\}_1 = 29.3\{\phi\}_1 = \{20.7 \;\; 29.3 \;\; 20.7\}^T \qquad \text{g} \cdot \text{in}$$

After first mode balancing, the unbalance distribution is now

$$\{\mathbf{U}\}_{\text{current}} = \begin{Bmatrix} 100 \\ -200 \\ 100 \end{Bmatrix} + \begin{Bmatrix} 20.7 \\ 29.3 \\ 20.7 \end{Bmatrix} = \begin{Bmatrix} 120.9 \\ -170.7 \\ 120.90 \end{Bmatrix}$$

The current rotor is no longer balanced as a rigid rotor, but is perfectly balanced to operate through the first two flexible critical speeds. At this point, further rigid-body balancing would destroy the modal balancing of the rotor.

To operate the rotor through the third critical speed, the rotor must be now modally balanced for the third mode. The third mode balancing component U_{b3} is given by

$$\phi_3^T[\{\mathbf{U}\} + \mathbf{U}_{b3}\phi_3] = 0$$

$$[-0.707 \;\; 1 \;\; -0.707]\left[\begin{Bmatrix} 120.9 \\ -170.9 \\ 120.90 \end{Bmatrix} + U_{b3}\begin{Bmatrix} -0.707 \\ 1.00 \\ -0.707 \end{Bmatrix}\right] = 0$$

$$-341.90 + 2\mathbf{U}_{b3} = 0 \qquad \mathbf{U}_{b3} = 170.9 \text{ g} \cdot \text{in} < 0°$$

The third mode balancing component is given by

$$\{\mathbf{U}_b\}_3 = 170.9\begin{Bmatrix} -0.707 \\ 1.00 \\ -0.707 \end{Bmatrix} = \begin{Bmatrix} -120.9 \\ 170.9 \\ -120.9 \end{Bmatrix}$$

The final balance is now

$$\{\mathbf{U}_b\}_{\text{final}} = \left[\{\mathbf{U}\} + \begin{Bmatrix} -120.9 \\ 170.9 \\ -120.9 \end{Bmatrix}\right] = \{0\}$$

Hence after the third mode balancing has been performed, the rotor is perfectly balanced. For very flexible rotors, the rotor is best balanced by using a modal procedure.

In actual balancing of a flexible rotor using the modal method, one must determine the modal balance components by using a modified form of the influence coefficient method in a procedure similar to that used with the single-mass Jeffcott rotor. Instead of applying only a single trial weight, a modal distribution of weights is applied to the shaft. The modal influence coefficients should be obtained by taking vibration measurements near the critical speed to be balanced. It is not necessary or desirable to take data at higher speeds near another critical speed if one is using a modal distribution of weights.

Figure 3.54 represents the experimental data obtained on a flexible three-mass rotor after modal balancing through two critical speeds. Initially the rotor had a high response while passing through the first critical speed. A set of modal weights was placed in phase at the three rotor stations. These modal weights were treated as if they were a single balance component. The first mode was balanced by using the single-plane influence coefficient method.

3.4.6 Multiplane Balancing by the Influence Coefficient Method

General Influence Coefficient Method of Balancing—Exact-Point Method

The two-plane method of balancing derived from Thearle's earlier work (1934) may be readily expanded to N planes of balancing. This method is referred to as

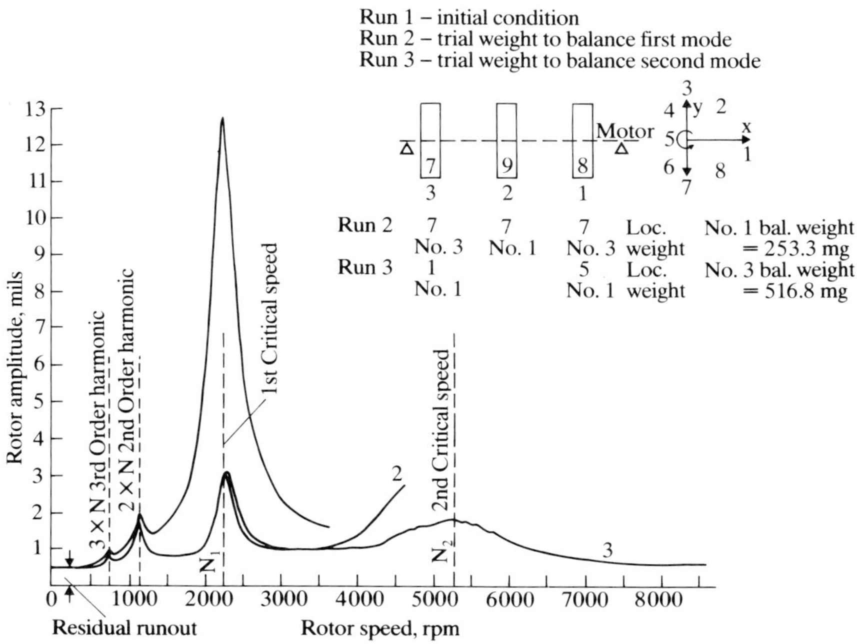

FIGURE 3.54 Modal balancing of a three-mass rotor (Gunter et al., 1976).

the *exact-point speed method* of balancing and was presented in detail by Tessarzik (1970).

In the exact-point speed method, the number of probes corresponds to the number of balance planes. The measurements are made at only one speed.

The vibration is assumed to be of the form

$$\begin{aligned} \mathbf{Z}_1 &= \mathbf{a}_{11}\mathbf{U}_1 + \mathbf{a}_{12}\mathbf{U}_2 + \mathbf{a}_{13}\mathbf{U}_3 \\ \mathbf{Z}_2 &= \mathbf{a}_{21}\mathbf{U}_1 + \mathbf{a}_{22}\mathbf{U}_2 + \mathbf{a}_{23}\mathbf{U}_3 \\ \mathbf{Z}_3 &= \mathbf{a}_{31}\mathbf{U}_1 + \mathbf{a}_{32}\mathbf{U}_2 + \mathbf{a}_{33}\mathbf{U}_3 \end{aligned} \tag{3.165}$$

In general matrix form, the equations of vibration are

$$\{\mathbf{Z}\}_n = [\mathbf{a}]_{n\times n}\{\mathbf{U}\}_n \tag{3.166}$$

where $[\mathbf{a}]$ is a square $n \times n$ matrix of influence coefficients. Since $[\mathbf{a}]$ is a square complex matrix, its complex inverse may be computed. The balance solution is given by

$$\{\mathbf{U}\}_b = -\{\mathbf{U}\}_n = -[\mathbf{a}]^{-1}_{n\times n}\{\mathbf{Z}\}_n \tag{3.167}$$

A total of n^2 influence coefficients must be determined to use this procedure.

The influence coefficients $\mathbf{a}_{ij}$ must be determined for a particular speed by placing a trial balance weight at the ith balance plane and recording the resulting amplitudes at the jth probe. For N planes of balancing, a total of $N + 1$ runs must be made to establish the coefficients for a given speed.

The exact-point and the least-squared-error procedures for flexible rotor balancing were investigated experimentally by Tessarzik and Badgley (1973). Figure 3.55 represents the test rig showing balance planes and probe locations. (Figure 3.56 represents the rotor mode shapes for the first three critical speeds. The first two critical speeds are essentially rigid-body cylindrical and conical modes. The third critical speed is equivalent to a free-free flexible mode with zero amplitude of motion at the bearings. The first two rigid-body modes may be

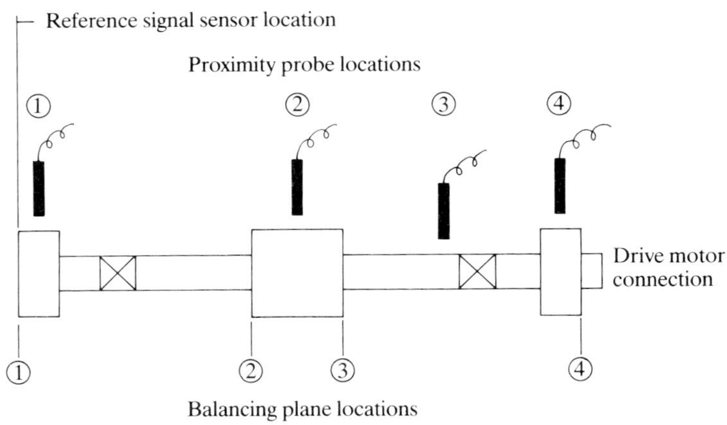

FIGURE 3.55 Tessarzik–Badgley (1973) test rig showing balance planes and probe locations.

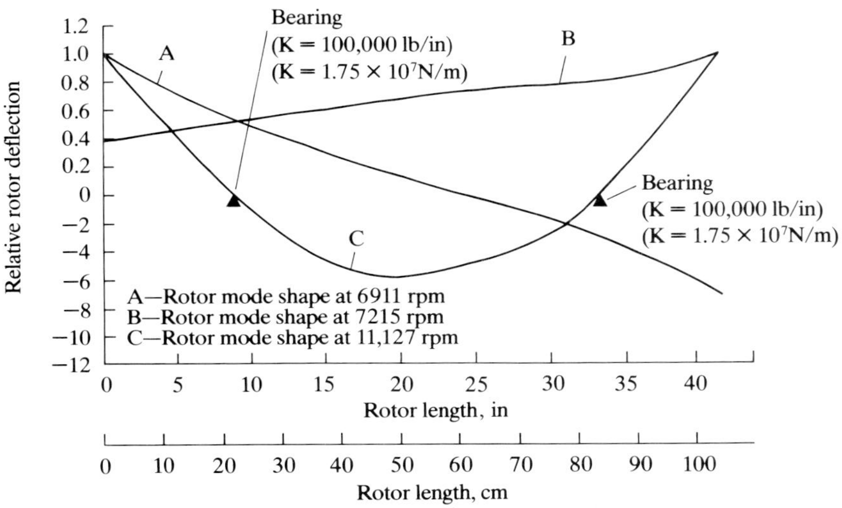

FIGURE 3.56 Critical-speed mode shapes for Tessarzik–Badgley (1973) test rig.

readily balanced, but the third free-free critical speed is extremely difficult to balance because of the high rotor sensitivity for this mode.

Figure 3.57 represents the rotor response of the Tessarzik and Badgley (1973) test rotor before and after three consecutive balancing runs by the exact-point speed influence coefficient method. Curve *A* represents the initial response of the rotor with a spiral or corkscrew unbalance distribution. The rotor rigid-body critical speeds at 6911 and 7215 rpm in the initial run *A* are not apparent due to the unbalance distribution used and the bearing damping.

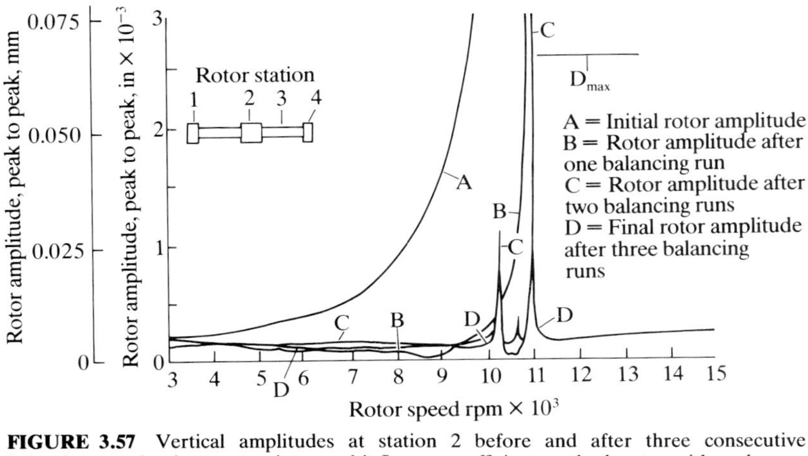

FIGURE 3.57 Vertical amplitudes at station 2 before and after three consecutive balancing runs by the exact-point speed influence coefficient method; rotor with corkscrew unbalance (Tessarzik and Badgley, 1973).

The rotor initial amplitude appears to increase with speed. Without a prior critical speed analysis of this rotor, from the appearance of curve *A* one would think that the rotor is approaching the first critical speed. One would then be tempted to balance the rotor by a single balancing plane near the rotor center. This would be incorrect.

The rotor third system critical speed encountered at 11,000 rpm is actually the first system flexible critical speed or free-free flexible mode. An analysis of the rotor strain energy distribution for the three modes indicates that the strain energy in the bearings is approximately 90 percent for the first mode, 96 percent for the second mode, but only 1 percent for the third mode. Hence, the first two modes are almost pure rigid-body modes, and the third mode is a pure bending mode requiring three planes of balance.

In Fig. 3.57 Tessarzik and Badgley (1973) have succeeded in reducing the rotor amplitude of motion over a wide speed range except at the sharp critical speed response at 11,000 rpm, as shown in curve *D*. This unusual mode, because of its extremely low damping, is extremely difficult to balance by either the exact-point or least-squared error balancing method.

To balance this mode, a specialized modal balancing procedure must be used employing the Nyquist or polar response diagram in conjunction the three-trial-weight method, using modal trial weight distributions with measurements taken at the critical speed.

Least-Squared-Error Multiplane Balancing Procedure

In the exact-point method of balancing using influence coefficients, one is limited to the computation of n planes of balance corresponding to one speed only. In a very flexible rotor, the rotor response is dictated by the modal distribution of the unbalances. If one measures the vibrations at speeds far removed from the critical speed, then an accurate estimation of the modal unbalance component cannot be predicted.

By taking measurements at a number of speeds, a best fit of the balance weights may be computed. The use of least-squared-error method with the data taken at multiple speeds will ensure that you do not overbalance at one speed, causing excessive vibrations at a second speed. The exact-point speed method is a subset of the generalized least-squared-error procedure with n probes and n balance planes.

The theory for the least-squared error method was first presented by Goodman in 1964 and expanded upon by Lund and Jonnessen in 1972 and Palazzolo and Gunter in 1977. The early work by Goodman represented a major advance in multiplane balancing.

In the least-squared-error method, n planes of balancing are assumed and m values of vibrations are taken in which $m > n$. In the least-squared-error method, one may use more or less probes than the number of balance planes n. In this method, the vibration may be taken on the same probes but for a number of speeds.

The vibration is assumed to be of the form

$$
\begin{aligned}
\mathbf{Z}_1 &= \mathbf{a}_{11}\mathbf{U}_1 + \mathbf{a}_{12}\mathbf{U}_2 + \cdots + \mathbf{a}_{1n}\mathbf{U}_n + \epsilon_1 \\
\mathbf{Z}_2 &= \mathbf{a}_{21}\mathbf{U}_1 + \mathbf{a}_{22}\mathbf{U}_2 + \cdots + \mathbf{a}_{2n}\mathbf{U}_n + \epsilon_2 \\
&\vdots \\
\mathbf{Z}_m &= \mathbf{a}_{n1}\mathbf{U}_1 + \mathbf{a}_{m2}\mathbf{U}_2 + \cdots + \mathbf{a}_{mn}\mathbf{U}_n + \epsilon_m
\end{aligned}
\tag{3.168}
$$

The measurements $\mathbf{Z}_1 - \mathbf{Z}_m$ may be a combination of only several probes but measured over a number of speeds. If measurements are not taken near a particular critical speed, then calculations of that particular nodal distribution may not necessarily be accurately performed.

The vibration in matrix form is given by

$$\{\mathbf{Z}\}_m = [a]_{m\times n}\{\mathbf{U}\}_n + \{\epsilon\}_m \tag{3.169}$$

where $m \geq n$ ($m = n$ is exact-point method).

The problem of least-squared-error analysis is identical to the problem of computing a pseudoinverse of unknowns with m equations. The procedure here is somewhat different since we are dealing with complex vectors and matrices.

The influence coefficient matrix must first be generated by the application of trial weights on the rotor. The resulting rotor responses due to the application of the trial weights must be computed at the same speeds as the initial readings. Also the slow-roll vector $\{\mathbf{Z}\}_0$ should be recorded before and after application of all weights to ensure that the shaft runout or bow is not substantially changing.

The transpose of the complex conjugate of the set of influence coefficients is computed next:

$$[\mathbf{T}]_{n\times m} = [\mathbf{a}]^{*t}_{m\times n}$$

Multiplying Eq. 3.168, we obtain

$$[\mathbf{T}]_{n\times m}\{\mathbf{Z}\}_m = [\mathbf{T}]_{n\times m}[\mathbf{a}]_{m\times n}\{\mathbf{U}\}_n + [\mathbf{T}]_{n\times m}\{\epsilon\}_m$$

The product of $[\mathbf{T}]_{n\times m} \times [\mathbf{a}_1]_{m\times n}$ is a square matrix of order n which may be inverted. Let

$$[\mathbf{D}]^{-1}_{n\times n} = [[\mathbf{T}]_{n\times m} \times [\mathbf{a}]_{m\times n}]^{-1}$$

The least-squared-error balancing matrix is given by

$$\{\mathbf{U}_b\}_n = [\mathbf{B}]_{n\times m}\{\mathbf{Z}\}_m \tag{3.170}$$

where

$$[\mathbf{B}]_{n\times m} = -[\mathbf{D}]^{-1}_{n\times n} \cdot [\mathbf{T}]_{n\times m}$$

For balancing by using runout compensation, the least-squared-error equation is given by

$$\{\mathbf{U}_b\}_n = [\mathbf{B}]_{n\times m}[\{\mathbf{Z}\}_m - \{\mathbf{Z}_0\}_m] \tag{3.171}$$

The procedure as outlined above may also be modified to include a weighting function to emphasize certain vibrations. The least-squared-error method is preferred to the exact-point method. Previously, this procedure required a mainframe computer to perform the complex calculations and matrix inversion. However, with current handheld calculators and portable computers, this procedure may be quickly accomplished in the field.

The application of the least-squared-error method for multiplane balancing of a bowed flexible rotor was extensively investigated by Palazzolo and Gunter in 1977. The synchronous amplitudes of motion for three stations along the rotor are shown in Fig. 3.58. The rotor has an initial shaft bow of 4 mils near the rotor center. With the original rotor unbalance distribution, the rotor amplitude at the first mode was almost 20 mils, and the rotor could not operate through the second critical speed because of excessive vibrations.

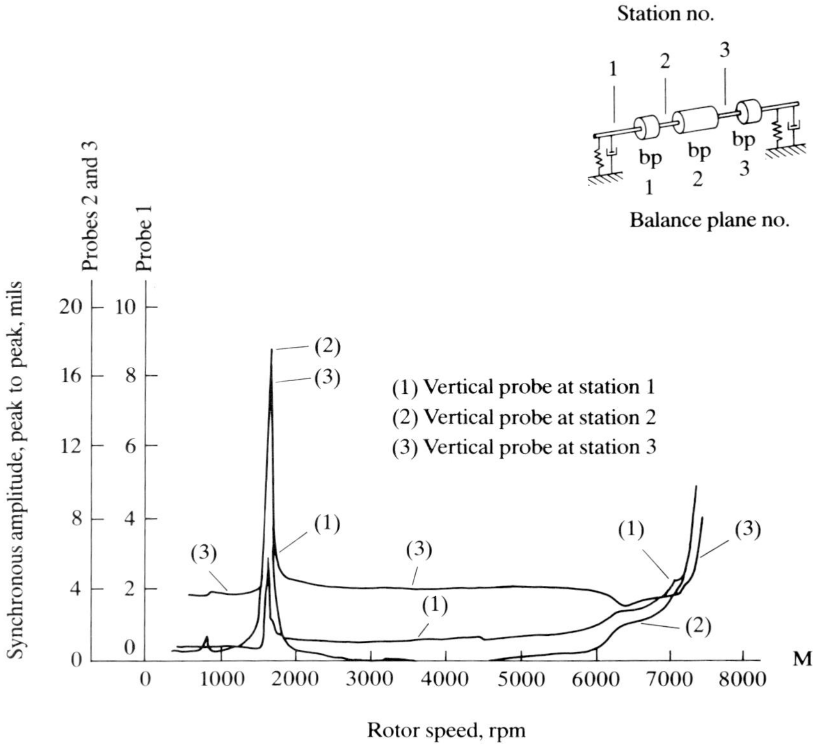

FIGURE 3.58 Synchronous unbalance response of a Bently three-mass rotor before least-squared-error balancing (Palazzolo and Gunter, 1977).

The least-squared-error method was used to compute the balance distribution in the rotor. Table 3.9 represents the calculated rotor three-plane balance values by the least-squared-error method for various combinations of speed data. Case 1, e.g., represents the balance distribution at 1200 rpm. This distribution of balance correction weights would give excellent balance at 1200 rpm, but the rotor would still be badly out of balance at the first critical speed and would not be able to traverse the second critical-speed region.

Case 5 represents the rotor balance computed in with five speed cases from 1200 to 1900 rpm. This represents speed cases taken below and above the first critical speed. This distribution represents an excellent balance for the first mode. If one were to apply this balance distribution to the rotor for first mode balancing, then the second mode could be balanced by the influence coefficient method using second-mode distributions of trial weights. This would represent a combined modal–influence coefficient method. Case 8 represents the best compromise balance set determined with eight speed cases from 200 to 7500 rpm. Figure 3.58 represents the motion of a three-mass rotor with shaft bow. The rotor has critical speeds at 1700 and 7500 rpm. It is impossible to operate the rotor above 7200 rpm because of the unbalance level.

TABLE 3.9 Calculated Rotor Three Plane Balance Values by the Least Squares Method for Various Combinations of Speed Data

	Speeds (rpm)	Correction unbalances					
		BP1		BP2		BP3	
Case	N	g · in,	deg	g · in,	deg	g · in,	deg
1	1200	0.683	207.045	0.110	121.33	0.811	60.75
2	1200, 1400	0.673	224.5	0.094	128.32	0.925	70.78
3	1200, 1400, 1500	0.855	185.9	0.161	119.9	0.705	35.73
4	1200, 1400, 1500, 1800	0.56	134.7	0.193	96.46	0.113	63.80
5	1200, 1400, 1500, 1800, 1900	0.513	117.45	0.172	91.15	0.134	94.27
6	1200, 1400, 1500, 1800, 1900, 6500	0.263	79.52	0.114	84.72	0.393	116.86
7	1200, 1400, 1500, 1800, 1900, 6500, 7000	0.178	87.73	0.106	97.7	0.419	106.6
8	1200, 1400, 1500, 1800, 1900, 6500, 7000, 7500	0.092	122.57	0.112	106.4	0.374	100.38

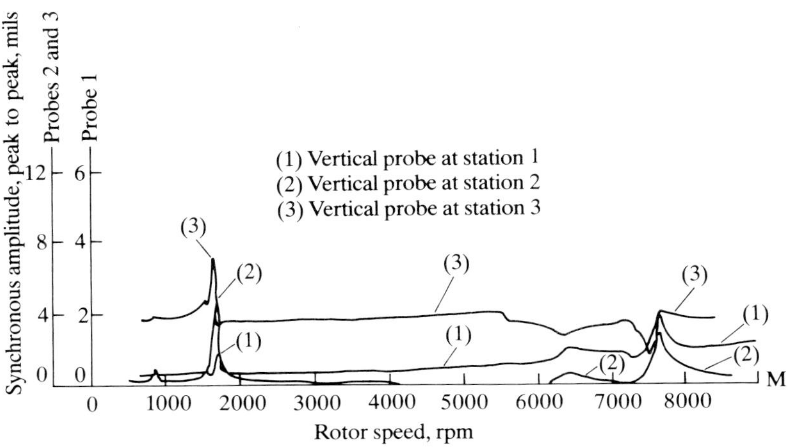

FIGURE 3.59 Synchronous unbalance response of a Bently three-mass rotor after least-squared-error balancing (Palazzolo and Gunter, 1977).

Figure 3.59 represents the motion of the bowed rotor after balancing with the weight distribution given in case 8 using eight speed cases. The rotor second mode has been completely balanced with some residual vibration remaining at the first mode. Note that Table 3.9 shows that the use of additional vibration data will, in general, constrain the values of the balance weights computed.

Thus, one will normally be on the conservative side with the use of the least-squared-error balancing procedure. The exact-point method should not be used to calculate balancing for flexible rotors operating through several critical speeds. The use of the exact-point method will lead to the calculation of excessive balance weights. The least-squared-error balancing method has been demonstrated by Gunter and Humphris (1986) to be highly effective for field balancing of gas turbine-generator systems with load-dependent thermal bows.

3.5 REFERENCES

Badgley, R. H., 1974, "Recent Developments in MultiPlane-Multispeed Balancing of Flexible Rotors in the United States," International Union of Theoretical and Applied Mechanics, Lyngby, Denmark.

Benson, R. C., 1974, "Dynamic Response of an Overhung, Unbalanced Skewed Rotor in Fluid Film Bearings," M.S. thesis, University of Virginia, Charlottesville.

Bently, D. E. 1982, *Polar Plotting Applications for Rotating Machinery,* Vibrations Institute, Clarendon Hills, Ill.

Bently-Nevada Co., 1984a, *ADRE Multi-Plane Balance Package,* Minden, Nev.

———, 1984b, *Automated Diagnostics for Rotating Machinery—ADRE Operators' Guide,* Minden, Nev.

Bishop, R. E. D., and A. G. Parkinson, 1963, "On the Isolation of Modes in the Balancing of Flexible Shafts," *Proceedings of the Institute of Mechanical Engineers,* 177: 407–423.

——— and ———, 1966, "Second-Order Vibration of Flexible Shafts," *Proceedings Royal Society of London,* 259: 1–31.

——— and ———, 1972, "On the Use of Balancing Machines for Flexible Rotors," *Journal of Engineering for Industry,* 94(2): 561–576.

Blake, M. P., 1967, "Use Phase Measuring to Balance Rotors in Place," *Hydrocarbon Processing,* August.

Darlow, M., 1989, "Flexible Rotor Balancing by the Unified Balancing Approach," Rensselaer Polytechnic Institute, Troy, N.Y.

Foiles, W., and E. J. Gunter, 1982, "Balancing a Three Mass Rotor with Shaft Bow," University of Virginia, Charlottesville.

Goodman, T. P., 1964, "A Least-Squares Method for Computing Balance Corrections," *Journal of Engineering for Industry,* 86(3): 273–279.

Gunter, E. J., 1966, "Dynamic Stability of Rotor-Bearing Systems," NASA Report SP-113, Washington, D.C.

———, L. E. Barrett, and P. E. Allaire, 1976, "Balancing of Multimass Flexible Rotors Part I Theory and Part II Experimental Results," Proceedings of the Fifth Turbomachinery Symposium, A&M University, College Station, Texas, October.

———, H. Springer, and H. H. Humphries, 1982, "Balancing of a Multimass Flexible Rotor-Bearing System Without Phase Measurements," Proceedings of Conference on Rotordynamics Problems in Power Plants, Rome, Italy, September.

——— and R. Humphris, 1986, "Field Balancing of 70 MW Gas Turbine Generators," Proceedings International Conference on Rotordynamics, JSME, (Tokyo), pp. 135–143.

International Standards Organization, 1973, *Balance Quality of Rotating Rigid Bodies,* ISO Document 1940, Geneva, Switzerland.

Jackson, C., 1971, "Using the Orbit to Balance," *Mechanical Engineering,* pp. 28–32, February.

———, 1979, *Practical Vibration Primer,* Guef Publishing Co., Houston.

Jeffcott, H. H., 1919, "The Lateral Vibration of Loaded Shafts in the Neighborhood of a Whirling Speed—The Effect of Want of Balance," *Phil. Mag.,* 37: 304.

Kellenberger, W., 1972, "Should a Flexible Rotor Be Balanced in N or $(N+2)$ Planes?" *Journal of Engineering for Industry,* 94: 548–560.

Kirk, R. G., and E. J. Gunter, 1973, "Nonlinear Transient Analysis of Multimass Flexible Rotors—Theory and Applications," NASA CR-2300, Washington, D.C.

Little, R. M., 1971, "Current State of the Art of Flexible Rotor Balancing Technology," Ph.D. thesis, University of Virginia, Charlottesville.

Lund, J. W., and J. Tonnesen, 1972, "Analysis and Experiments on Multi-Plane Balancing of Flexible Rotors," *Journal of Engineering for Industry,* 94(1): 233.

Mechanalysis Inc., "Vector Calculations for Two Plane Balancing," Applications Report 327, IRD, Columbus, Ohio.

Nicholas, J. C., E. J. Gunter, and P. E. Allaire, 1976, "Effect of Residual Shaft Bow on Unbalance Response and Balancing of a Single Mass Flexible Rotor: Part I—Unbalance Response; Part II—Balancing," *Journal of Engineering for Power,* 98(2): 171–189.

Palazzolo, A. B., and E. J. Gunter, 1977, "Multimass Flexible Rotor Balancing by the Least Squares Error Method," Vibration Institute, Clarendon Hills, Ill.

———, 1982, "Modal Balancing of a Multimass Flexible Rotor Without Trial Weights," *A.S.M.E.,* Gas Turbine Paper 82-GT-267.

Rieger, N., 1986, *Balancing of Rigid and Flexible Rotors,* SVN 12, The Shock and Vibration Center.

Salamone, D. J., and E. J. Gunter, 1978, "Effect of Shaft Warp and Disc Skew on Synchronous Unbalance Response of a Multi-Mass Rotor in Fluid Film Bearings," *Topics in Fluid Film Bearings and Rotor Bearing Systems Design and Optimization,* ASME, New York, pp. 79–107.

Schenek Trebel Corp., 1980, "Fundamentals of Balancing," Deer Park, Long Island, New York.

Tessarzik, J. M., 1970, "Flexible Rotor Balancing by the Exact Point-Speed Influence Coefficient Method," NASA, Cleveland, Ohio, CR-72774, October.

——— and R. H. Badgley, 1973, "Experimental Evaluation of the Exact Point-Speed and Least Squares Procedures for Flexible Rotor Balancing by the Influence Coefficient Method," ASME 73-DET-115, New York.

Thearle, E. L., 1934, "Dynamic Balancing of Rotating Machinery in the Field," *Transactions of ASME,* 56: 745-753.

Thomson, W. T., 1965, *Vibration Theory and Applications,* Prentice-Hall, Englewood Cliffs, N.J.

Winkler. A. F., 1983, "High Speed Rotating Unbalance, Coupling or Rotor," *Proceedings Vibration Institute,* Nassau Bay Conference, Vibration Institute, Clarendon Hills, Ill.

CHAPTER 4

PERFORMANCE VERIFICATION, DIAGNOSTICS, PARAMETER IDENTIFICATION, AND CONDITION MONITORING OF ROTATING MACHINERY

Dr. Ronald L. Eshleman
President and Director Vibration Institute

Charles Jackson, P.E.
Turbomachinery Consultant

4.1 INTRODUCTION

The availability of sophisticated electronics has greatly increased the vibration-testing capabilities of engineers. The application of microprocessors to vibration analysis has simplified evaluation of the characteristics and condition of rotating machinery. Improvements in sensors, signal conditioning, and data storage devices have made vibration work relatively trouble-free.

The increased availability of electronic hardware has also motivated the development of experimental techniques. Today the *fast Fourier transform* (FFT) (Cooley and Tukey, 1965) is the foundation for most vibration data processing. Diagnostic and modal analysis techniques are presently being refined to take advantage of FFT capabilities. Because microprocessors are now so widely used, digital processing has become the most widely used method for data reduction.

This chapter deals with vibration-testing techniques of rotating machinery. Included are applications of instrumentation and testing techniques to verification of performance, diagnosis of faults, parameter identification, and condition monitoring. Instrumentation including transducers as well as systems used to acquire, process, and store data is covered in Sec. 4.2. Use of the instrumentation is outlined in later sections. Data acquisition procedures and display formats are reviewed in Sec. 4.3.

The parameter identification techniques used to determine vibratory characteristics of a rotating machine are described in Sec. 4.4. These techniques provide data for developing and validating models that can be used by those concerned with such characteristics as critical speeds, mode shapes, and response levels. Modal testing techniques provide detailed modeling data; critical speed and resonance techniques yield global properties of the machine.

New or reconditioned rotating machines are subjected to performance testing to ensure that they are free from design and manufacturing faults. Vibration-oriented performance-testing procedures, test codes, and standards are the topics of Sec. 4.5.

The vibration tests now used in fault diagnosis of rotating machines have become routine. Component faults from wear, installation, and design can be isolated by nonintrusive methods when the properties of the machine are known and modern testing techniques are applied. Section 4.6 describes testing techniques and methods for identifying faults in various types of machine.

The sophistication of vibration instrumentation and the increased data storage capacity have changed the philosophy of machine maintenance. Preventive maintenance, which meant little more than routinely changing machine components, has evolved into predictive maintenance, the techniques of which are used to determine in advance when components should be replaced. Condition monitoring programs, hardware, techniques, and criteria are outlined in Sec. 4.7. Both permanent monitors, which protect a machine from sudden changes in process and sudden development of faults, and analytic capabilities are described.

4.2 VIBRATION INSTRUMENTATION

Vibration information is acquired by transducers positioned at optimal locations within a mechanical system. Transducers (Norton, 1969) convert mechanical responses to electric signals that are conditioned and processed by a wide variety of electronic instruments. The instruments provide the information necessary to monitor machine condition, verify performance, diagnose faults, and identify parameters. The same instruments may also be used for in situ balancing of the rotor, as described in Chap. 3. Transducer selection (Eshleman, 1987a) is based on the sensitivity, size, measurement parameter, and frequency response. Because each transducer type has inherent limitations, it is important to choose the proper transducer for a given task. Filters and amplifiers are used, when necessary, to eliminate unwanted noise before an analysis is carried out. A straightforward analysis of a raw or processed signal can be performed by observing it on an oscilloscope, which shows the time-varying signal. The raw or filtered signals from two transducers located 90° apart and displayed as the x and y drivers on an oscilloscope reflect the motion (orbit) of a shaft. More complex signals require additional processing or filtering. This section deals with the description, selection, and use of vibration instrumentation.

4.2.1 Vibration transducers

Vibration transducers convert mechanical motion to an electric signal. Their selection is based on the assigned task—monitoring, analysis, performance testing, parameter identification, or verification. If shaft motion is of interest, proximity probes (Jackson, 1979) that measure relative displacement should be used. If a bearing housing or structural motion is the best indicator of vibratory information, a seismic transducer (Norton, 1969) sensitive to velocity or acceleration is used.

Important considerations in transducer selection are the frequency response and sensitivity of the transducer and the strength of the signal being measured. Frequency response is a measure of how the transducer or other electronic signal processing equipment represents the amplitude of the signal at various frequencies used in the vibration analysis. An instrument may not provide the exact amplitude at the output as seen in the input because of its response characteristics. This must be noted and accounted for in the analysis. A frequency-response curve that gives the ratio of output amplitude to input amplitude versus frequency provides this information. Figure 4.1 shows the frequency response of typical seismic (acceleration and velocity) transducers. Accelerometers provide flat frequency response at low frequencies; however, Fig. 4.2 shows that the signal strength of acceleration is low at low frequencies. On the other hand, signal strength for the proximity probe is low at high frequencies, but its frequency response is flat at frequencies where the signal strength is good. At high frequencies, therefore, noise could dominate the signal and cause a low, undesirable signal-to-noise ratio. The accelerometer has the same problem at low frequencies. A velocity signal is maintained throughout the range of frequencies of interest, but the frequency response of the transducer rolls off at low and high frequencies; i.e., the transducer sends a signal of lower amplitude than that present at the measurement point. Thus, the transducer type and sensitivity as well as the signal-to-noise ratio must be considered in selecting a transducer that will obtain the best signal for processing. It is especially important that the frequency response of the transducer correctly report amplitude when vibration severity is a major concern. The three basic transducers are proximity probes, velocity transducers, and accelerometers.

Proximity Probes

The proximity probe (Jackson, 1979) measures static and dynamic displacement of a shaft relative to the bearing housing or other static part in close proximity. It is permanently mounted on many machines for monitoring (protection) and analysis. Applications of the probe to vibration measurement in radial and axial positions are covered in detail in API 670 (1986). The probe is a coil of wire surrounded by a nonconductive plastic or ceramic material contained in a threaded body. An oscillator-demodulator, often referred to as a *driver* or *proximitor,* is required to excite the probe at about 1.5 MHz. The resulting magnetic field radiates from the tip of the probe. When a shaft is brought close to the probe, eddy currents are induced that extract energy from the field and decrease its amplitude. This decrease in amplitude provides an alternating current (ac) signal directly proportional to vibration. The direct current (dc) voltage from the oscillator-demodulator varies in proportion to the distance between the probe tip and the conducting material. The sensitivity of the probe is generally 200 mV/mil (8 mV/μm) with a gap range (distance from the tip of the probe to

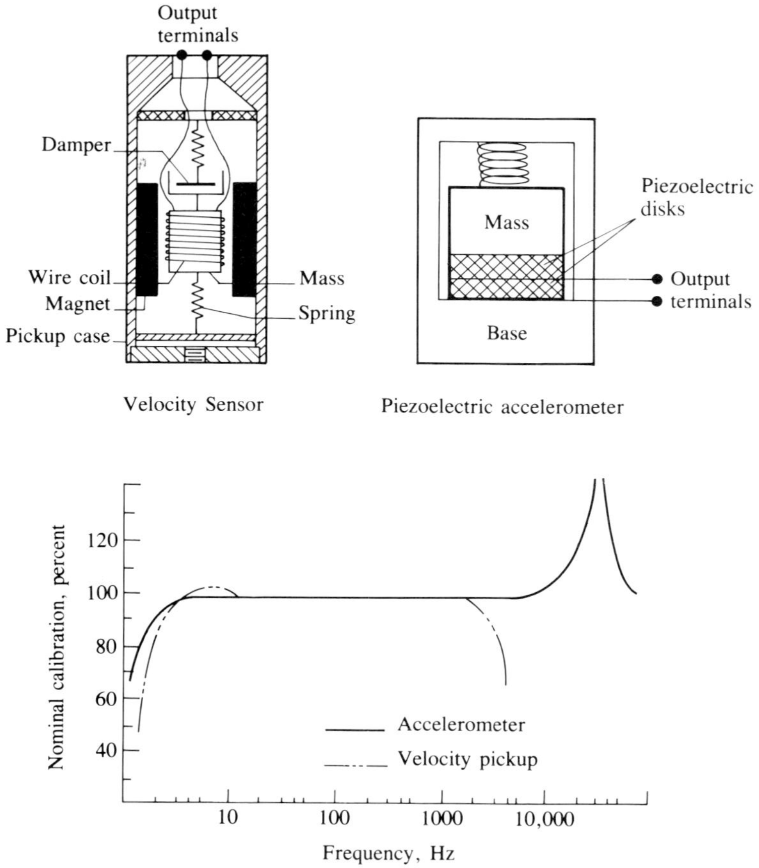

FIGURE 4.1 Frequency response of accelerometer and velocity pickup.

the shaft) of 0 to 80 mils. The oscillator-demodulator requires a supply of −18 or −24 V dc. Probe sensitivity varies with shaft conductance (API 670, 1986) and is sensitive to changes in shaft properties—called *electrical runout*—as well as to surface scratches and mechanical runout. As is true of all transducers, shielding and grounding are important.

Velocity Transducers

The velocity transducer (Norton, 1969) is a seismic transducer used to measure vibration levels on casings or bearing housings in the range from 10 to 2000 Hz. It is a self-excited active transducer—i.e., it requires no power supply—and consists of a permanent magnet mounted in springs encased in a cylindrical coil. Motion of the coil relative to the magnet generates a voltage proportional to the vibration velocity. This self-generated signal can be directly passed through an oscilloscope, meter, or analyzer for evaluation. A typical velocity transducer generates 500 mV/(in/s) except at frequencies below 10 Hz, where the natural frequency of

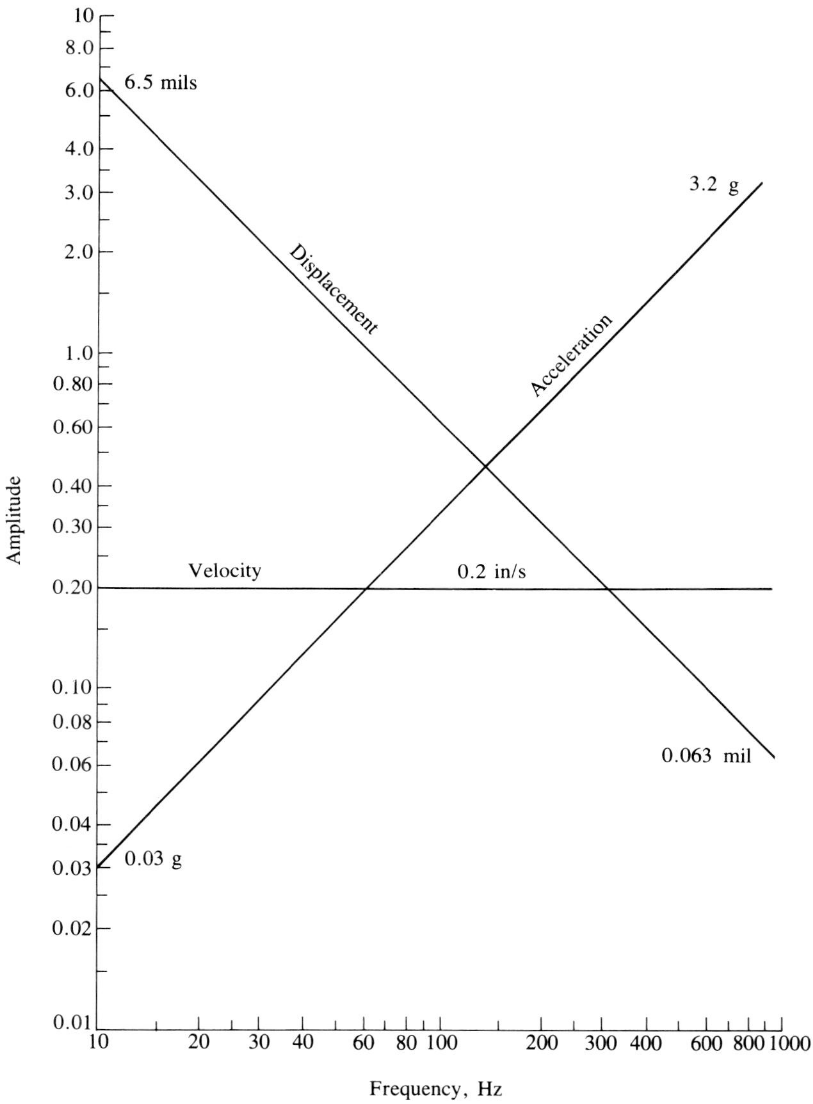

FIGURE 4.2 Sinusoidal displacement and acceleration magnitude versus frequency for constant velocity of 0.2 in/s (Eshleman and Jackson, 1989).

the active element occurs. The fact that a reduction in output of the transducer occurs below 10 Hz requires that a frequency-dependent compensation factor (Fig. 4.1) be applied to the amplitude of the signal. The measured phase also changes with frequency at low frequencies because of the natural frequency of the transducer. Vibration velocity is used to measure vibration severity because the overall vibration level for a given machine condition is independent of the frequency in the range applicable to the velocity transducer. These transducers are often larger than accelerometers. Care must be taken in their location and shielding because they can generate false electric signals (e.g., at power supply frequency of 60 Hz) around motors and generators. Velocity transducers can be used to measure shaft vibration with a "fish tail," a simple wooden device that attaches to the transducer; a vee notch permits the fish tail to ride on the rotating shaft. Keys and other variations of the shaft surface pose safety hazards.

Accelerometers and Force Gauges

Accelerometers (Mitchell, 1981) are used to measure vibration levels on casings and bearing housings. An accelerometer consists of a small mass mounted on a piezoelectric crystal that produces an electrical output proportional to the strain in the crystal induced by the force induced by the force acceleration of a small attached mass. Force transducers such as modal hammers also contain a piezoelectric crystal. In this case, however, the electrical output of the crystal is proportional to the force applied to the crystal by the hammer stroke.

The piezoelectric crystal actively generates a high-impedance signal that must be modified by charge or voltage conversion to low impedance. The size of an accelerometer is proportional to its sensitivity. Small accelerometers—the size of a pencil eraser—have a sensitivity of 5 mV/g ($1\,g = 386\,\text{in/s}^2$) and a flat frequency response up to 25 kHz. A 1-V/g accelerometer, which is used for low-frequency measurement, may be as large as a velocity transducer; however, the limit of its usable frequency range is 3000 Hz. If vibration velocity is desired, the signal is usually integrated before it is recorded or analyzed. Accelerometers are recommended for permanent seismic monitoring because of their extended life (API 678, 1981). In addition, transverse sensitivity is low; however, cable noise, transmission distance, and temperature sensitivity must be carefully evaluated. Excellent guidelines are available for accelerometer use (Bruel and Kjaer, undated).

Optical Pickups

The optical pickup is most often used to obtain the once-per-revolution reference signal required to measure phase angle between a once-per-revolution vibration peak and a piece of reflecting tape on a shaft. The pickup sends a voltage pulse to the oscilloscope or analyzer when energized by light pulses from the reflecting tape. The analyzer compares the timing of the shaft reference pulse (tape) to other marks on the shaft or to vibration peaks to give phase relations or to itself on successive cycles of the shaft to effect a speed measurement.

The optical pickup has also been used to observe time elapsed between equally spaced marks on a rotating shaft in torsional vibration measurements (Eshleman, 1987b). The optical system includes a pickup mounted adjacent to the shaft, reflective tape on the shaft, and a power supply or amplifier.

Magnetic Pickups

The magnetic pickup, which is self-excited, is often used as a shaft timing mark because a voltage pulse occurs when it encounters a discontinuity such as a keyway. The pickup is typically placed 20 to 80 mils from the shaft. Magnetic pickups are used in torsional vibration measurements (Eshleman, 1987b) to produce a series of pulses proportional to the shaft speed. If torsional vibrations are present, the time between pulses varies, producing frequency modulation.

One disadvantage of the magnetic pickup is that the magnitude of the voltage is dependent on the speed; signal conditioning is sometimes difficult under these circumstances. The proximity probe, which is powered, does not have this disadvantage.

4.2.2 Data Acquisition System

Signals acquired from a vibration transducer can often be viewed directly on an oscilloscope, spectrum analyzer, or meter. In some cases it is advantageous to record or condition signals prior to analysis. However, preanalysis conditioning should be minimized because it can affect frequency response, shift phase, and introduce noise. The analyst must be aware of these pitfalls and take them into consideration when processing data.

Data Conditioners

Conditioners, which include filters, analog integrators, and amplifiers, are used to enhance data, but all have frequency-response characteristics that must be considered during analysis. Filters (Mitchell, 1981) are used when dynamic range is a consideration. Dynamic range is the measure of the ability to distinguish low-amplitude signals from noise in the presence of high-amplitude signals. Filters can eliminate noise from a signal or increase the efficiency of processing. The frequency-response characteristics of a filter change in magnitude and phase for some components of a signal; the changes are similar to those shown in Fig. 4.1 for accelerometers. Analog integration converts a measured acceleration signal to velocity or a velocity signal to displacement. Integration of acceleration to velocity decreases the dynamic range of a signal; the decrease in dynamic range is advantageous in tape recording. If low-amplitude low-frequency components of vibration are present along with important high-amplitude high-frequency components in an acceleration signal, integration performed prior to recording prevents loss of the low-frequency information in the noise. Amplifiers increase the gain (amplification) of the measured signal and are often used with tape recorders. Because noise introduced during the recording process is sometimes of the same order of magnitude as the signal, processing for good resolution is difficult. It is therefore important to increase the gain of the signal before recording. It is necessary in the case of torsional vibration measurements to use a frequency demodulator to process a signal from a toothed wheel-proximity probe sensor arrangement (Eshleman, 1987b).

Data Recording

Recording data is often convenient and sometimes mandatory in the data acquisition process, e.g., when transient vibration phenomena are measured. Digital computers, digital recorders, and tape recorders are used for this function. Digital recording devices require analog-to-digital conversion prior to recording signals from analog transducers (Baxter and Eshleman, 1985). Analog recorders operate in the direct and frequency modulated (FM) mode; the advantages of each depend on the frequency range of the data being acquired. Recording data saves acquisition time as well as machine time and money because many points can be monitored simultaneously for later analysis. Another advantage is that recorded data can be analyzed in many different ways without time or environmental pressures. In addition, tapes provide a permanent record.

Data Collectors

In recent years, electronic data collectors (Henson, 1987) have increased the productivity of predictive maintenance programs in which machine condition is monitored and analyzed. Depending on its design, the electronic data collector, which is programmed to accept data from a large number of points on a defined route in a plant, can provide instant analysis of data being acquired and stored. All data collectors have the capability of storing root-mean-square (rms) or peak data that can be transferred to a digital computer; trends can then be established and reports generated. The levels of filtering and analysis provided by data collectors are variable. Some display 400- to 3200-line spectra at the measurement site and can also store them. Storage capacity is limited, however, because the collector is usually small; tradeoffs are thus necessary in the data storage process. Spectra are often stored in response to a cue from a high-amplitude overall measurement. Time-domain data are rarely stored because of the considerable space they require; thus, it is impossible to reanalyze stored data from an electronic data collector.

4.2.3 Data Processing

Vibration data can be processed in various devices to obtain the frequency and phase information necessary to analyze complex machinery (Eshleman and Jackson, 1989). Although data can be processed in both analog and digital instruments, the capabilities, versatility, size, and relatively low cost of digital instruments have made them the instrument of choice for most users.

Tunable and Swept-Filter Analyzers

Analog filters indicate the rms or peak vibration level in a band limited by the frequency width of the filter. The bandpass filter can be either mechanically tuned by hand to the center frequency of the filter or set to sequentially sweep across a selected frequency range. The output is displayed on a meter or plotted if a permanent record is desired. The sweep rate is consistent with the time necessary to acquire and filter the data (Mitchell, 1981).

Tracking Filters

Tracking filters are used to obtain a single component of vibration that is referred to a fixed or variable frequency—usually the shaft speed or one of its multiples. The tracking filter is used in balancing and modal testing.

FFT Spectrum Analyzers

The microprocessor-based FFT spectrum analyzer is the preferred tool for analyzing vibration data because of its superior resolution, dynamic range, and display capabilities. FFT analyzers (Brigham, 1974) acquire a block of data over a designated frequency range during a period of time, digitize the data, and perform a frequency analysis using the microprocessor-based algorithm. FFT analyzers are available in single- and multichannel configurations; however, no additional processing modes are available beyond the dual-channel analyzer. Thus, more than two channels merely permit faster simultaneous signal processing and display. Single-channel analyzers provide displays that can be analyzed in detail with cursors that use absolute, relative, differential, and harmonic identification of time, amplitude, frequency, and phase (with respect to an input trigger signal). Many analyzers contain buffers capable of storing large quantities of data and can also produce waterfall, or cascade, diagrams, i.e., amplitude versus frequency for various times or speeds (Eshleman and Jackson, 1989). These analyzers can perform integration (digital on the frequency domain only) after the FFT analysis, carry out mathematical manipulation and rms band analysis, and compute power spectral density. The magnitudes of rms-based bands can be displayed in a linear or logarithmic format. Peripheral capabilities vary among FFT analyzers, but all have a frequency range of 20 kHz or more, at least 10 frequency spans, and a dynamic range of 70 dB or more. The FFT algorithm used in modern analyzers provides 400 or more lines (number of digitized points) of data for a selected frequency range in the spectrum. A zoom feature is usually present that allows resolution to tenths of a hertz. Resolution is the ability to distinctly resolve components of vibration that have frequencies very close to each other. Resolution to tenths of a hertz is essential for accurate frequency analysis of such equipment as electric motors, in which vibration components are very closely spaced.

Modern FFT analyzers have a digitized time-domain display. This feature allows accurate measurement of the interval between repetitive events. Most analyzers display two windows simultaneously. Dual-channel analyzers used for advanced diagnosis and modal analysis can compare data processed in channel A and channel B according to amplitude (transfer function) and phase as a function of frequency; they can also display the dual time domain. The power relationship between the two channels, known as *coherence,* is also shown. Any displayed plot can be digitally transferred to a plotter through an IEEE 488 interface.

4.2.4 Data Display

Several methods are available for displaying and permanently recording processed and unprocessed data. Most instruments used for analysis, including oscilloscopes, have a cathode-ray tube (CRT) display; a few simpler instruments display rms, peak, or average values of measured vibration. Strip charts (with good response from dc to 120 Hz) and oscillographic recorders (with response from 0 to 5 kHz) record a raw or filtered signal directly from a transducer and trace it on a chart. Analog and digital plotters provide hard copies of CRT-displayed data. The analog *xy* plotter can be used in the time-based mode (0.5 to 100 s/in) to record the time waveform of low-frequency signals.

4.3 DATA ACQUISITION PROCEDURES

This section outlines data acquisition procedures used to generate displays of vibration data for analysis. Brief descriptions of the instrumentation required to acquire and analyze data are also given (see Fig. 4.3). Vibration data can be either analyzed directly after they have been acquired from a transducer or recorded for later analysis. When data are recorded, proper recording procedures (Baxter and Eshleman, 1985) must be observed. In fact, pitfalls in tape recording can severely limit the usefulness of data. An analog oscilloscope is usually used to display instantaneous time waveforms or the orbits of measured signals. Additional information from FFT spectrum analysis, transfer functions, and phase analysis can be derived from the digitized signal of a *digital signal analyzer* (DSA). The digitized time waveform and orbit can also be displayed on the CRT of an

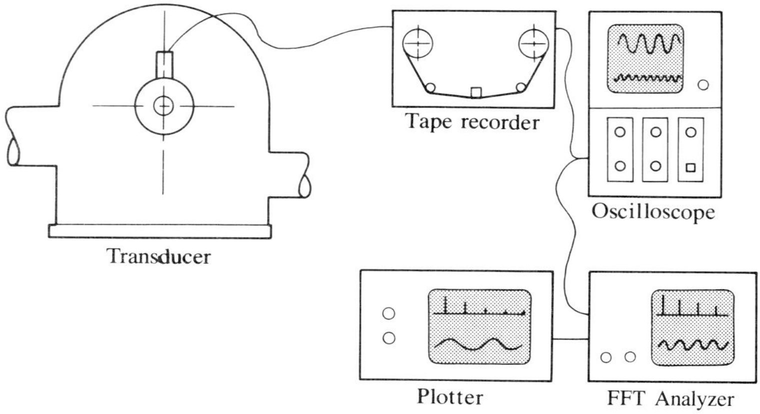

FIGURE 4.3 Instrument setup for vibration data acquisition and analysis (Eshleman and Jackson, 1989).

FFT analyzer. Hard copy of a CRT display is obtained from a digital plotter, usually by passing the data through an IEEE 488 interface. Although spectra can be obtained directly from packaged swept-filter analyzers, the resolution, dynamic range, and display capabilities are severely limited.

The analyst must record the sensitivities of the transducers used and consider their frequency-response characteristics. The units of transducer sensitivity are millivolts (mV) per engineering unit, e.g., mils, inches per second, *g*'s, pounds, degrees, pound-inches. Any amplification or attenuation used to condition a signal must be reflected in the sensitivity (Eshleman and Jackson, 1989). The frequency range of a signal, which is usually determined by the operating speed of the machine, must be compatible with the frequency-response characteristics of the transducer, tape recorder, and any signal conditioning instruments used. When the frequency response curve of the instrument rolls off (output amplitude is less than input amplitude), the amplitudes of the analyzed data must be adjusted at distinct frequencies to reflect the lower output of the conditioning equipment or transducer. For example, if a velocity transducer were used on a machine operating at 300 rpm, the component of vibration at operating speed would have to be raised by a factor of 3.0 (Fig. 4.1). Similar corrections are necessary for other frequency components in the rolloff region and for other instruments.

The amplification or attenuation of a conditioning instrument is often given in decibels (dB). The formula for decibels in terms of voltage is

$$\mathrm{dB} = 20 \log \frac{V}{V_{\mathrm{ref}}} \tag{4.1a}$$

$$\frac{V}{V_{\mathrm{ref}}} = 10^{\mathrm{dB}/20} \tag{4.1b}$$

For example, if an amplifier shows a gain of 9 dB, the transducer sensitivity used in the spectrum analysis must be increased by an amplification factor of $10^{9/20}$ or 2.82. Typical values of the relationship between the dB (log) and (V/V_{ref}) (linear) scales are as follows:

dB (log)	V/V_{ref} (linear ratio)
6	2:1
10	3:1
12	4:1
20	10:1
40	100:1

Signal conditioning is usually carried out before data are tape-recorded (Fig. 4.3) with the exception of torsional vibration measurements; in this case the raw signal, which may be of the order of 500 to 1000 mV, should be recorded because of the low sensitivity (~5 mV) of some torsional signal conditioners.

4.3.1 Constant Operating Speed

Vibration data from a machine running at constant operating speed are generally repetitive; small variations occur as a result of the influences of load, temperature, and process. Environmental and load conditions should be noted when the data are taken.

Time Domain

The time domain reflects the physics of the machine in the vibration signal—particularly if it is displayed on an analog oscilloscope. A time-domain signal of a blower (Fig. 4.4*a*) is shown as it appears on an FFT spectrum analyzer. The time domain is useful in identifying unique events in a machine and the rate at which they are repeated. The length (in seconds) of the display of data from the time domain given in Table 4.1 depends on the information sought; this length is normally related to the operating period T of the machine, where T in seconds is equal to 60/rpm. The best phase-angle resolution for basic balancing is obtained by using the fundamental period T for display. In Fig. 4.4*a* the shape of the time waveform is obtained from a $12T$ (400-ms display/33.3-ms fundamental period) display. Details of very high-frequency signals in the presence of a low-frequency rpm are best seen in the range from $T/50$ to $T/100$, where one to five events might be displayed. If very detailed examination of a waveform is necessary, one or two repetitions of the event should be displayed; a very high-frequency oscilloscope may be required. Higher multiples of T are used for long-term and phase trends.

The differential time feature of the FFT spectrum analyzer is shown in Fig. 4.4*b*, a signal obtained from an impact test on a turbine casing. A marker set at an original time serves as a reference; the cursor can be moved between events to accurately measure the time between them.

Trend Analysis

Plots of trends of such important parameters as peak casing vibration, peak-to-peak shaft vibration (Fig. 4.5), or bearing temperature, which provide records at

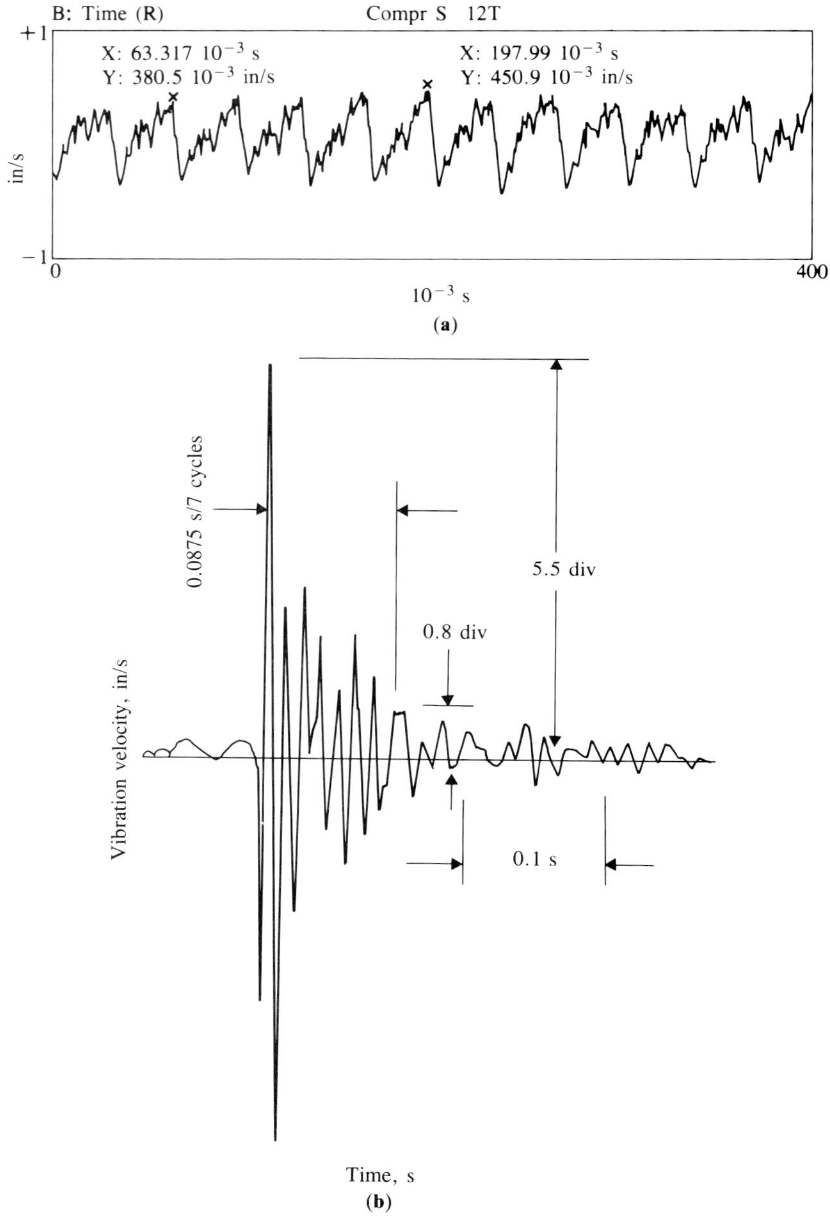

FIGURE 4.4 (*a*) Time-domain display of a compressor; (*b*) time-domain data from impact test on turbine casing.

TABLE 4.1 An Approach to the Presentation of Standardized Waveforms (After Catlin, 1987)

Time, s	Display	Purpose
$T/100$ $T/50$ $T/20$	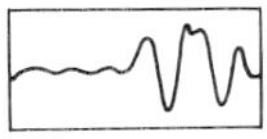	Details of higher frequency
$T/10$ $T/5$ $T/2$	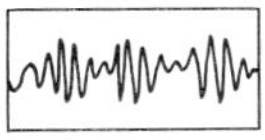	Trends of higher frequency
T $2T$ $5T$	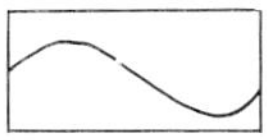	Balancing/phase
$10T$ $20T$ $50T$	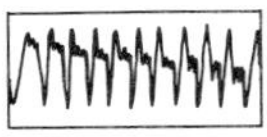	Phase trends
$100T$		Long term time trends

$T \equiv$ period of fundamental frequency.

specific intervals, are often called *extended time domain*—hours, days, years. The data are obtained by monitoring and recording, either continuously or periodically. Continuous records require permanent machine monitors; discrete data can be unloaded from electronic data collectors for periodic monitoring. These records are used in predictive maintenance programs when decisions on machine condition must be made—e.g., when to provide surveillance or recommend shutdown for repair. Figure 4.5 shows a record of turbine shaft vibration rising in days due to bearing wear.

Frequency Domain

A display of vibration magnitude versus frequency by a *swept-filter analyzer* (SFA) or an FFT analyzer is called a *frequency domain,* or *spectrum.* Analyses by an SFA and an FFT analyzer are shown in Fig. 4.6. No information about sidebands is available from the SFA, but it was this information that allowed identification of the vibration source.

An SFA requires about 1 min to perform an analysis; the sweep time required depends on the inverse of the bandwidth of the filter used. The magnitude of vibration is assessed by an SFA for all frequencies in a selected range as the filter passes. If no vibration is occurring when the filter passes a specific frequency, no vibration will be seen. The FFT analyzer, on the other hand, processes a block of data. The time span is dependent on the frequency span chosen for the spectrum and the number of lines used by the analyzer to process the data:

$$f_{\max} = \frac{N}{2}\frac{1}{T} \tag{4.2}$$

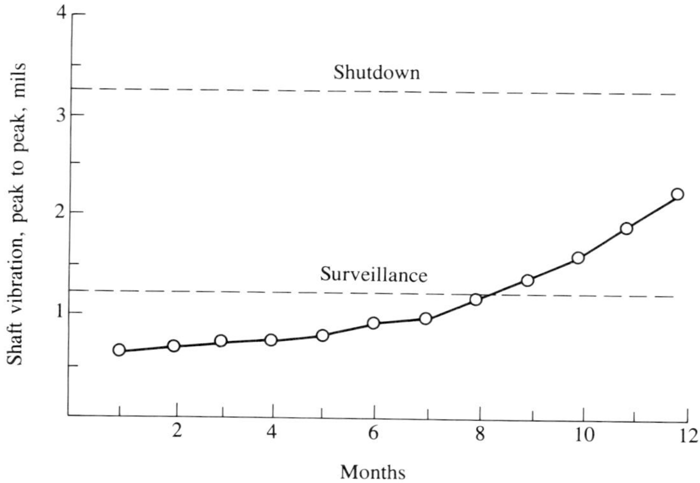

FIGURE 4.5 Trend plot for turbine shaft vibrations.

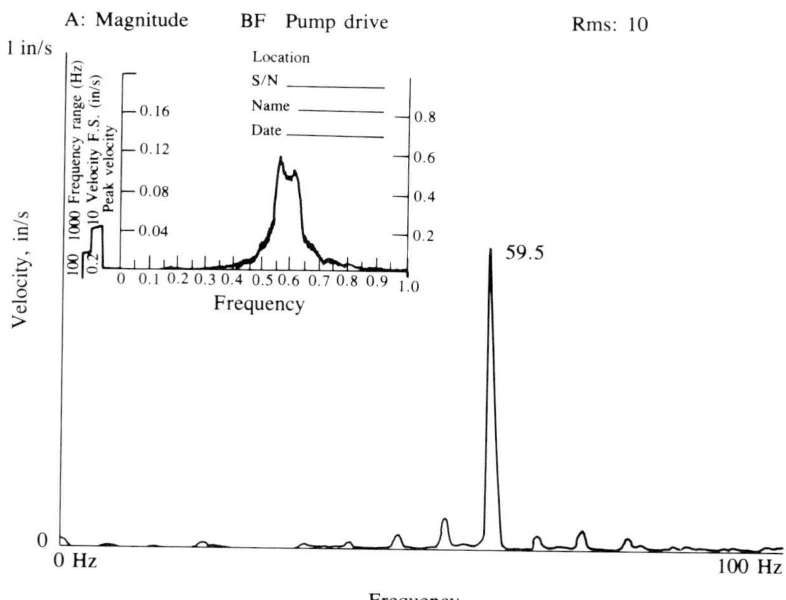

FIGURE 4.6 FFT and swept-filter analysis of shaft vibration of a boiler feed pump drive.

where T is the period of the time record, f_{max} is the highest frequency measured, and N is the number of samples (usually 1028). This sampling rate is subject to the constraints of aliasing (Hewlett-Packard, 1982); i.e., if samples are taken at too low a rate, erroneous frequency components are introduced into the spectrum. Because the FFT analyzer averages a block of data, it is certain to detect any event in those data. From Eq. 4.2 we can see that the time record is long at low frequencies, regardless of the processing time required by the computer. This feature is a serious disadvantage when variable-speed data are being processed.

Phase Analysis

A phase angle quantifies the timing of two signals. They can be vibration signals, a vibration signal and a force signal, or a signal from a shaft-speed reference mark and a vibration signal (see Fig. 4.7). The last is used to obtain the phase relationship between force and vibration required in balancing.

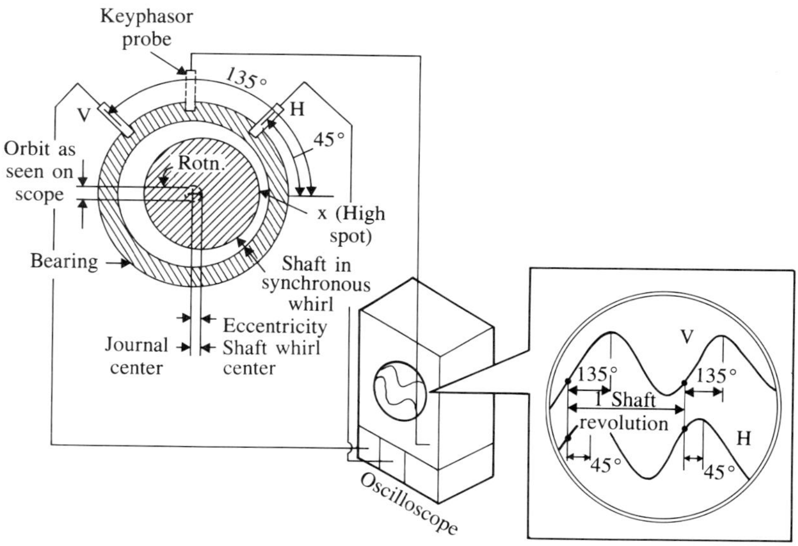

FIGURE 4.7 Proximity probe and Keyphasor® configuration for rotor vibration measurement (Jackson, 1979). Keyphasor® is a registered trademark of the Bently Nevada Corp.

A phase analysis can be conducted on an oscilloscope, tracking filter, or dual-channel analyzer. The phase angle is derived from an oscilloscope by setting the time scale (horizontal) so that about 1 cycle of vibration takes place over the measurement span of the oscilloscope—a 1 T-period time domain is used. The phase angle is the ratio of the number of divisions between the trigger and the next vibration peak to the number of divisions in T, multiplied by 360°.

The phase angle between other signals at different locations on the machine provides information on misalignment and balance. The phase angle on an oscilloscope can be seen more clearly by blanking the screen of the oscilloscope (Fig. 4.7) when the shaft reference mark passes the fixed sensor. (The signal on the screen is blanked by placing an ac-coupled—no direct current is present—signal on the Z axis of the oscilloscope.) The phase angle can also be seen with a strobe light: The firing strobe light is synchronized to a filtered vibration signal to display an angle on a fixed or rotating protractor.

Orbital Analysis

The data from proximity probes necessary to carry out an orbital analysis are shown in Fig. 4.7. Because the bearing split precludes the mounting of a horizontal probe, the probes are oriented at an angle of 45° (see Fig. 4.7) as prescribed by API 670 (1986). The horizontal probe is mounted 90° to the right (clockwise) of the vertical probe regardless of the orientation around the bearing or the direction of shaft rotation. Probes should be placed in or adjacent to the bearings and where rotor motion is greatest, i.e., couplings, seals, splines, governor drive gears, and rotor flanges. It is obvious that nodal points on flexible rotors should be avoided. The orbital domain is examined for shape, which provides evidence of preload, and loops, which indicate the direction of precession of the shaft and whirl frequency. An orbit provides a fundamental picture of the motion of the journal in its bearing. Orbit plus dc gap shows where the journal is moving with respect to the bearing.

Modal Analysis

Modal analysis (Ewins, 1984) involves the examination of the modal properties of mass, stiffness, and damping to obtain the dynamic characteristics of machines and structures. Analysis of the data, which are comprised of an applied force and the resulting vibration, requires a dual-channel analyzer. The force is applied at one position (see Fig. 4.8), and vibration responses are measured at locations on the machine that describe the modal properties of the system. The applied force can be a known mass unbalance applied to the balance plane of the machine to excite the system from the rotor. An instrumented hammer containing a force transducer can also be used to apply force to the bearing pedestals or the frame of

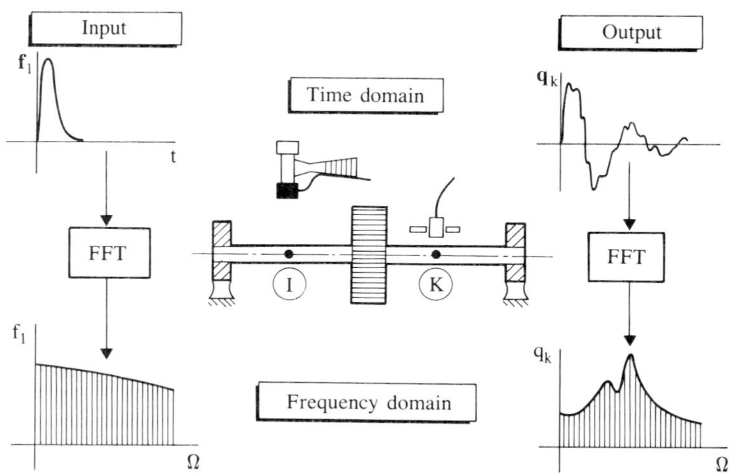

FIGURE 4.8 Test procedure for modal analysis of rotating machinery (Nordmann, 1984).

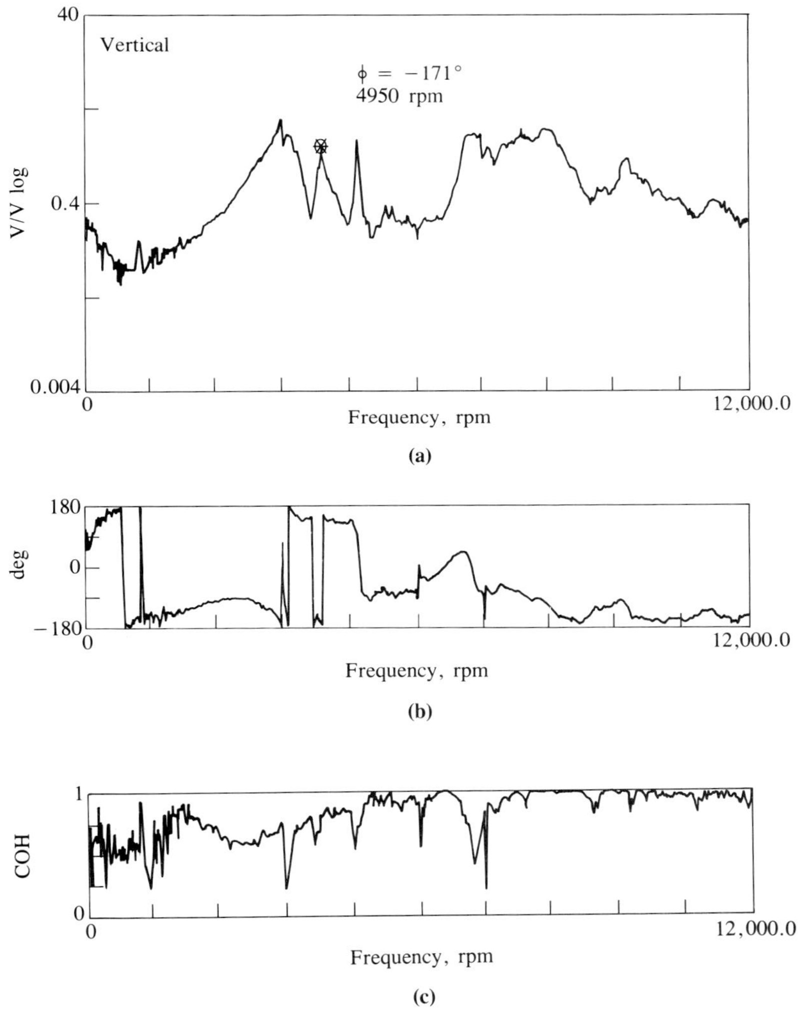

FIGURE 4.9 Dual-channel processing of force and vibration signals (Guy et al., 1988). (*a*) Transfer function versus frequency; (*b*) phase versus frequency; (*c*) coherence versus frequency.

a machine. Accelerometers and proximity probes on the machine provide signals that describe the vibration response to the applied force. Shakers that apply a controllable variable frequency are also used in modal testing. The transfer function, phase, and coherence (Fig. 4.9) are found from processing the force and vibration signals in a dual-channel analyzer. The transfer function—a plot of the ratio of vibration response to excitation (or the inverse) versus frequency—represents the cause-and-effect relationship between force and vibration for the frequencies that are excited. A phase plot shows the phase between force and

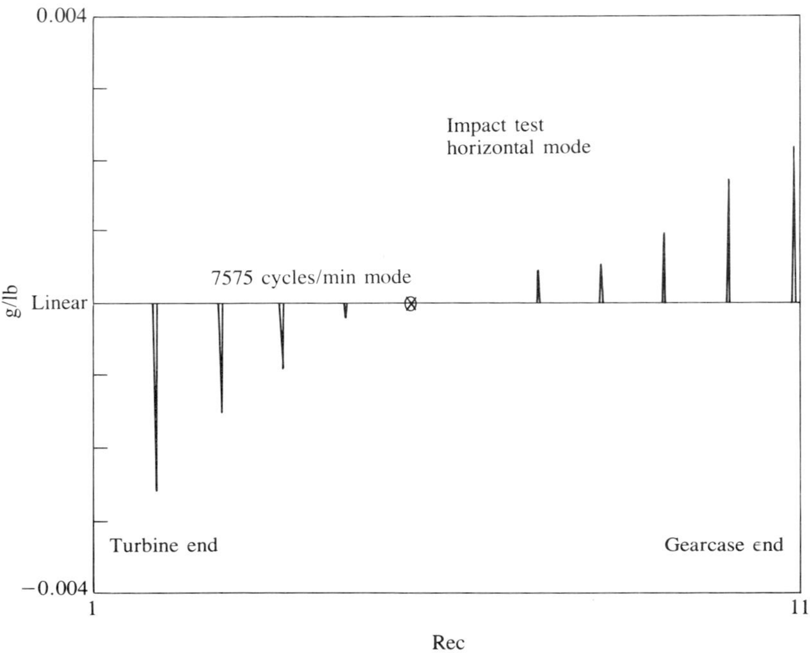

FIGURE 4.10 Imaginery response of a turbine rotor (Guy et al., 1988).

vibration; coherence represents the portion of measured vibration attributable to the force. Coherence varies from 0 to 1. Zero indicates that none of the vibration results from the applied force; a coherence of 1 indicates that all the vibration is due to the applied force. The real and imaginary plots (Fig. 4.10) shown are obtained from the transfer function and the phase angle. Damping can be obtained from the real plot; normalized imaginary values plotted at selected positions provide the mode shape of the rotor or structure.

4.3.2 Variable Speed and Frequency

Considerable information about a machine or structure can be derived from variable-speed and frequency tests, e.g., natural frequencies, critical speeds, damping, and diagnostic information about rotor bow and cracks. Modal testing techniques provide mode shapes and information about modal damping and stiffness and mass properties. Forces are induced either by the machine (e.g., mass unbalance, misalignment) or by a shaker. Shakers are more commonly used on structures for frequency sweeps, however. Force is not measured during typical critical speed testing, but in modal testing, in which vibration response is compared to known forces, the force must be measured and quantified.

Bode Plots

The Bode plot (Fig. 4.11*a*) is a graph of amplitude (usually relative displacement) versus speed and phase (between the shaft reference mark and peak vibration) versus speed. The peak value on the turbine in Fig. 4.11*a* is 5800 rpm; thus the critical speed is discerned directly. If a proximity probe is used, shaft runout must be subtracted to obtain an accurate plot. A tracking filter is used to process data from the transducer and the reference mark on the shaft.

Polar Plots

The test and instrument setup for a polar plot is similar to that for a Bode plot (Jackson, 1982), as illustrated in Fig. 4.11*b*. A polar plot is a graph of amplitude versus phase for various machine speeds. The plot is obtained by simultaneously plotting real and imaginary responses of a system on orthogonal axes. Natural frequencies appear as loops. Modal properties can be extracted from polar plots. Polar plots used in modal testing are called *Nyquist diagrams*. They show amplitude versus phase at various excitation frequencies.

Peak-Hold Plots

A graph similar to a Bode plot can be obtained on an FFT analyzer without an external trigger if the start-up and coast-down are not too rapid; see Fig. 4.12. Tracking filters will track and record events occurring in almost any rotating machine during start-up and coast-down—0 to 10,000 rpm in 10 s is not uncommon—within the narrow frequency range being tracked.

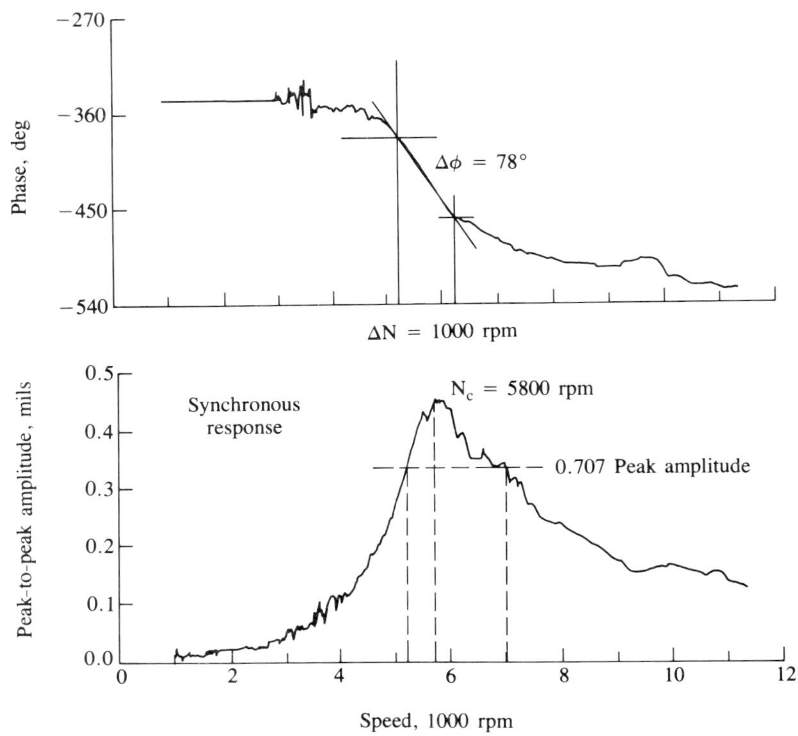

(a)

FIGURE 4.11 Shaft vibration of a turbine during coast-down: (*a*) Bode plot, (*b*) polar plot.

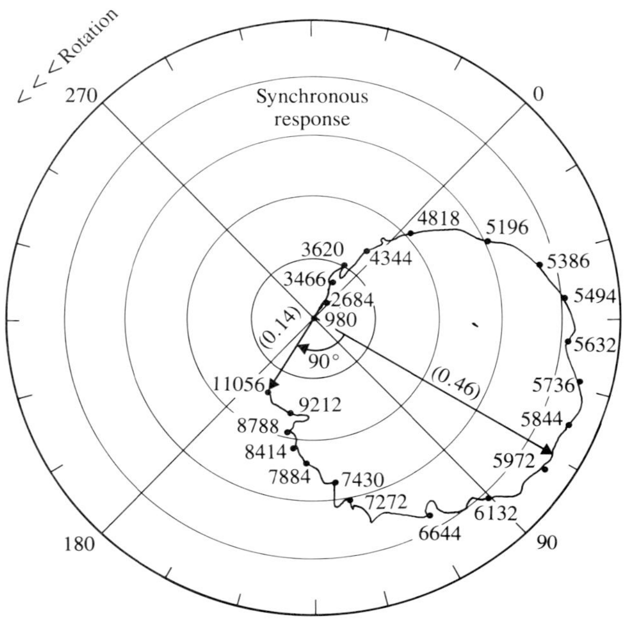

Full scale amp = 0.5 mils, peak-to-peak amp per div = 0.02 mils, peak to peak

(b)

FIGURE 4.11 (*Continued*)

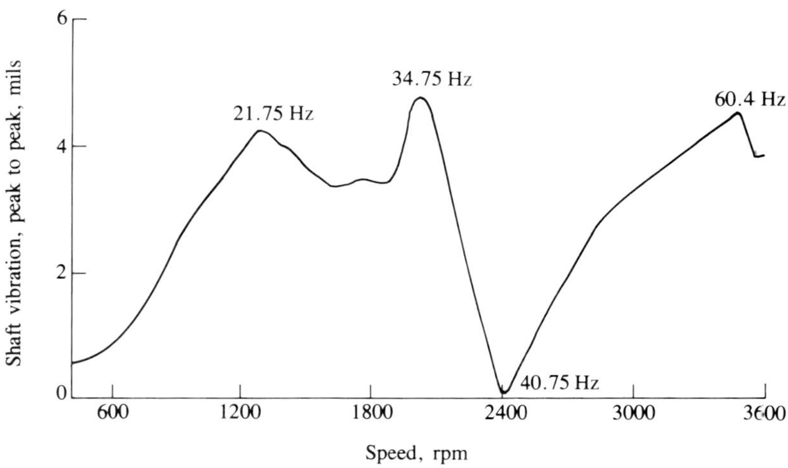

FIGURE 4.12 Use of peak hold feature of an FFT spectrum analyzer to measure vibration at varied frequencies.

However, an FFT analyzer is not able to track rapid start-ups because it samples a block of data, the time of which depends on the frequency span setting and the number of samples per block of data. If a 400-line FFT analyzer is set on a frequency span of 100 Hz (6000 rpm), the time required for a data sample t_{max} or number of lines per frequency span will be 400 lines/100 Hz or 4 s. Many samples are necessary to obtain a smooth curve on an FFT analyzer during start-up or coast-down; otherwise data will drop out, and the record will be unreliable. For example, a 10-s start-up is not uncommon in a two-pole (3600-rpm) electric motor. The FFT analyzer set on 100 Hz (6000 cycles/min) would provide 2.5 data points, or 1 point for each 1440 rpm. Obviously, the critical speed obtained from such data would not be accurate. The problem can be partially or entirely overcome by overlap processing and the use of fewer lines of resolution. In the overlap processing mode, an FFT analyzer processes a sample containing only a selected portion of the new data. The rate of data display thus increases because new samples are available in a short time. From the formula it is apparent that sampling time decreases with resolution. If 100 lines are used, only 1 s of data is needed per sample, and samples are taken every 360 rpm.

Waterfall Diagrams

The waterfall diagram (Fig. 4.13) is a three-dimensional plot of spectra at various machine speeds. The sequential spectra can be separated by uniform increments

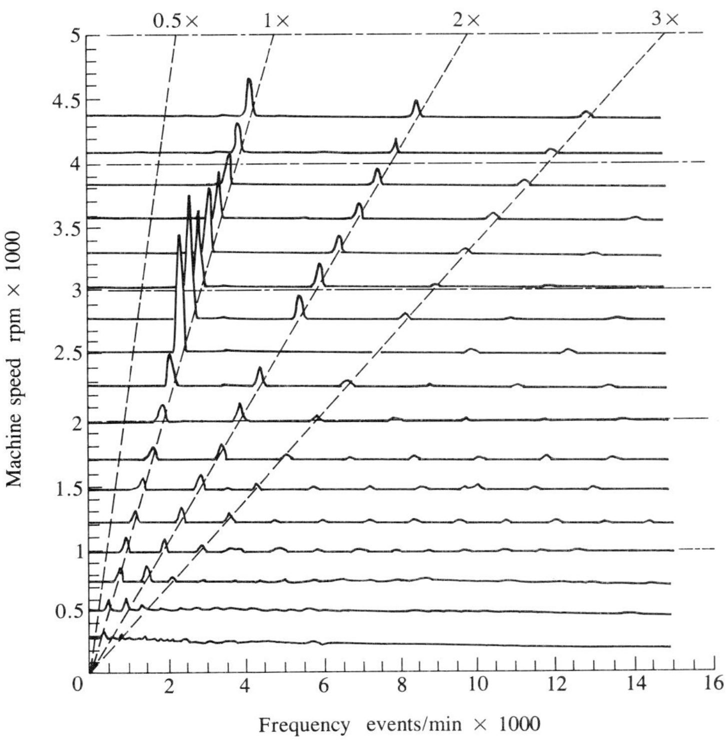

FIGURE 4.13 Waterfall diagram of a machine start-up.

of time or, when the data are acquired by relatively slow acceleration, by uniform increments in speed. In the latter case, one dimension (the ordinate) is equivalent to the speed of rotation, so that the plot suggests the frequency-versus-speed-of-rotation interference diagram (or, more precisely, a 90° inversion of it). The plot allows the analyst to evaluate the various frequency components of vibration as the machine runs up to operating speed. Such diagrams can be plotted from a tape-recorded signal and an FFT analyzer. Modern FFT analyzers store the data

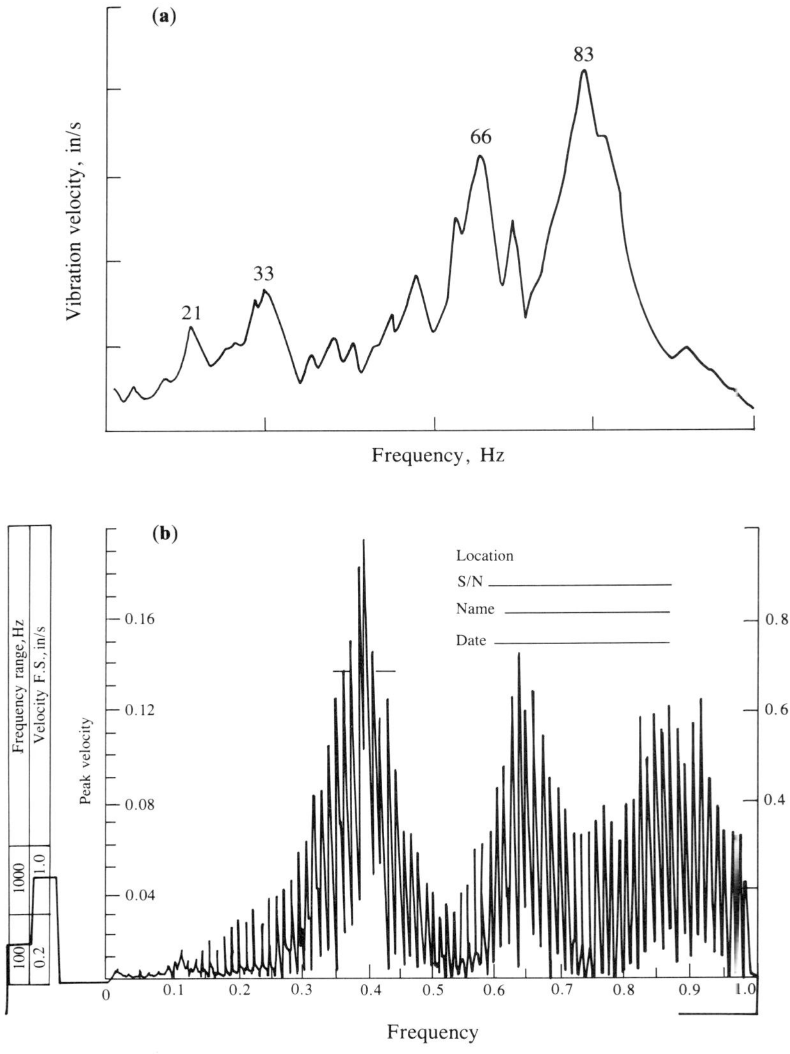

FIGURE 4.14 Frequency analysis of an impact: (*a*) Steam-turbine pedestal using FFT processing, (*b*) sheet-metal structure using a swept-filter analyzer.

in a buffer and reconstruct the diagram after data acquisition; individual spectra can be examined on the FFT because it can remove a selected record from storage.

Transient Data

Transient data, which are obtained by single or repeated impacts, can be examined in the time domain (as in Fig. 4.4*b*) or the frequency domain, as in Fig. 4.14*a*, where steam-turbine pedestal response is shown. Transient data provide records that can be used to determine natural frequencies and damping. The impact testing method is currently the most popular way to obtain structural modal properties. The reason is that, theoretically, all frequency responses occur simultaneously. Data acquisition is similar to other tests, but tape recorders and buffers are valuable aids because timing is essential. The bump test can be used with an SFA (Fig. 4.14*b*) on a sheet-metal structure. The structure is continuously struck as the analyzer sweeps through the frequency range. The structure responds at its natural frequencies; the result is an envelope of the response that defines those frequencies.

4.4 PARAMETER IDENTIFICATION

Some dynamic characteristics of a machine or structure must be known before a vibration problem can be diagnosed or corrected. These dynamic characteristics, or parameters, include natural frequencies, mode shapes, damping, and such modal properties as stiffness and mass. Parameters are quantified by exciting the machine with a hammer, shaker, or motion of the machine itself. The magnitude of this stimulus may be known or unknown. Response is measured with accelerometers, velocity transducers, or proximity probes. Oscilloscopes, tracking filters, swept-filter analyzers, and FFT spectrum analyzers are then used to analyze the data.

4.4.1 Natural Frequencies

Natural frequencies, which are unique to a machine or structural design, are important because, when they are excited by impact mechanisms and vibratory forces, large amplitudes result. The phenomenon is known as *resonance*. Natural frequencies are obtained experimentally from impact tests with hammers, variable-frequency shakers, or tests in which the speed of the machine is varied.

The natural frequencies of a machine can change with the machine speed because they are influenced by the stiffness of the supporting system, which, in turn, is influenced by the speed of rotation. In machines with self-acting fluid-film bearings, bearing stiffness (Eshleman and Jackson, 1989) is a function of the viscosity of the fluid and the load; these properties are also influenced by the machine speed. Because the stiffness of fluid-film bearings is load-dependent, the natural frequency of a machine operating at constant speed can vary if the load changes because of excessive mass unbalance, misalignment, steam-induced forces from uneven admission, thermal distortion, or foundation decay. The natural frequencies of machines with large overhung disks and wheels are speed-dependent because the gyroscopic moments (a mass effect) vary with speed. In additon, nonlinearities in the support stiffness (many support stiffnesses are harder for the large amplitudes of actual operation than for the small amplitudes of modal testing) will also cause deviations between statically measured and dynamically measured natural frequencies.

Natural frequencies of machines are typically determined by measuring vibration levels during start-up and coast-down. However, these natural frequencies must be associated with the speeds at which they occur. Impact and shaker tests are less commonly used. The hammer or shaker is applied to the housing of a bearing or to the machine to vibrate the rotor at various frequencies. The results of a surge test on an axial compressor are shown in Fig. 4.15 (Jackson

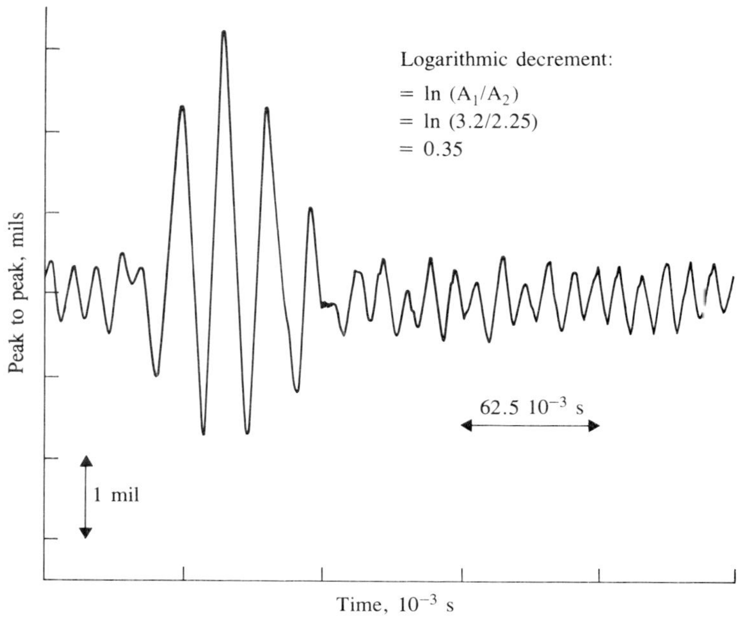

FIGURE 4.15 Natural frequency of an axial compressor obtained by a surge test.

and Leader, 1983). In this case the natural frequency is excited at operating speed. If a nonoperating rotor is bumped, the natural frequency is that of the rotor resting in its bearings. The natural frequency of a rotor mounted in rolling-element bearings is close to that obtained at the operating speed even though the stiffness of such bearings increases with load. But data obtained when the rotor is resting in its bearings as well as information obtained from rotors bumped in a sling in the free-free mode (horizontally)—although not indicative of any parameter at operating speed—are valuable for modeling the rotor and its support bearings.

Couplings can also influence the natural frequencies of machines. The overhung mass can significantly decrease the natural frequency from values obtained for the machine without the coupling. In addition, couplings that are not flexurally compliant, such as gear couplings, can stiffen a machine system and raise its natural frequencies (Eshleman, 1970).

Natural frequencies of machine foundations, casings, and piping are usually obtained from impact tests. The size of the impactor must, of course, be adjusted to the size of the apparatus to be excited. Either a single impact or repeated impacts can be used. If the force must be controlled and measured, instrumented hammers and shakers are required. The force should not exceed that necessary for a desired effect. Excessive force can drive the system into nonlinear behavior and change its dynamic characteristics, i.e., natural frequencies, mode shape, and damping.

The record from a swept-filter analyzer when a structure is repeatedly bumped during a frequency scan is given in Fig. 4.14*b* for a typical system. As the analyzer moves through the natural frequencies of 40, 64, and 88 Hz, the vibration amplitude increases. The result is an envelope that identifies the frequencies. Time and frequency domains of a signal from a steam-turbine pedestal during an impact test are shown in Figs. 4.4*b* and 4.14*a*, respectively. The frequency domain (Fig. 4.14*a*) shows the natural frequencies directly; they are at 21, 33, 66, and 83 Hz. The damped period of the first natural frequency is simple to calculate from the time domain; it is 0.085 s/7 cycles, or 0.012 s/cycle. However, the damped periods of the less predominant natural frequencies are more difficult to calculate; a method is given in Sec. 4.4.4 for calculating damping from this plot.

4.4.2 Resonance Testing

Variable-frequency shakers are used to conduct resonance tests on machines, structures, and piping. Mechanical, electrohydraulic, and electrodynamic shakers are available. The frequency is varied through a specific range until the amplitude peaks, the phase changes sharply, or both occur. The peak-hold plot obtained on a spectrum analyzer or a tracking filter related to the force signal can be used to identify the natural frequencies.

4.4.3 Critical-Speed Testing

The various natural forces within a machine can be used to excite natural frequencies for critical-speed tests. An interference diagram for a motor-driven overhung fan with natural frequencies that vary with speed is given in Fig. 4.16. Natural and forcing frequencies are plotted as functions of speed; the points at which the curves intersect (interference points) are called critical speeds. In Fig. 4.16 the natural frequency of the fan at rest is 15 Hz. During operation, however, it is close to 20 Hz; the fan thus operates at a critical speed. The increase in natural frequency is caused by the stiffening of gyroscopic forces. The seriousness of critical speeds depends on the vibratory force level and the damping. If a system is well damped at a natural frequency, that frequency may not show up as a critical speed. If no vibratory force at that frequency is present, the system will not respond at its natural frequency. Controlled tests are difficult, however, because both the natural frequencies and the forces of the exciters vary with speed. The magnitude of the mass unbalance force increases as the square of speed; such unbalances are often used to excite critical speeds in rotating machines.

Note that components of a forcing function at frequencies 2X, 3X, etc., of

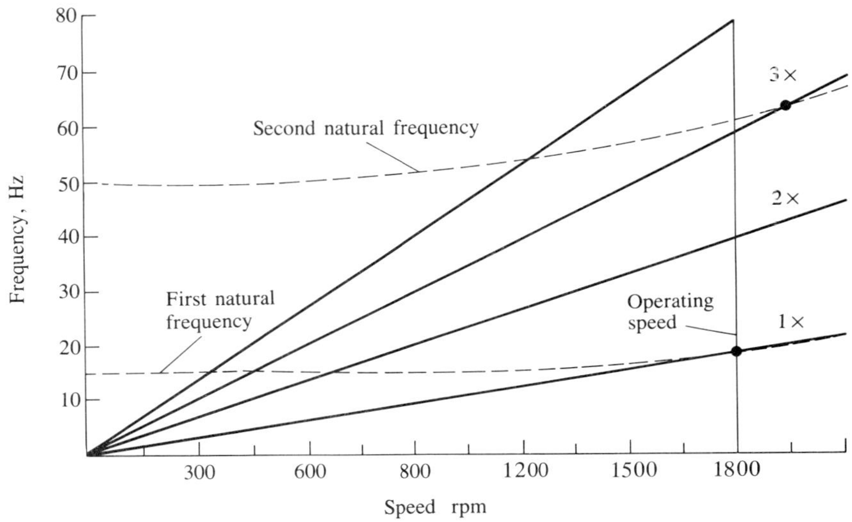

FIGURE 4.16 Interference diagram for a motor-driven overhung fan.

running speed can also excite critical response. The gravity critical speed (Den Hartog, 1956) is a 2X excitation arising as a result of raising and lowering the mass unbalance force in the gravity field; it occurs at exactly one-half the normal first critical speed of a rotor. Asymmetric shaft stiffness can also cause a 2X excitation. Many mechanisms in rotating machinery produce a synchronous critical speed. (See Sec. 1.6 and 1.7 for a description of self-excited and nonlinear responses.)

As described in Sec. 1.8.1, the multifrequency, impactlike forcing functions of reciprocating engines and compressors are related to running speed and thus induce many critical speeds. Indeed, the two rotations of the crankshaft in the basic force cycle of a four-stroke cycle engine yield both integer orders and one-half orders of engine speed. The force necessary to excite specific critical speeds depends on several factors related to the design of an engine—its firing order, arrangement and number of cylinders, and crankshaft design.

A star diagram (Den Hartog, 1956) can be used to estimate the severity of a force that usually excites torsional (twisting) natural frequencies of a system. These frequencies are determined by gradually passing the engine through its operating speed range from low to high idle speeds while measuring torsional response.

During synchronous motor start-ups, a torque that varies from 120 Hz (at rest) to 0 Hz (at operating speed) at a frequency equal to twice the slip frequency tends to severely excite the first torsional natural frequency of a system. This response is short but has been known to spin couplings, twist off shafts, and cause gear sets to fail. The designer can often impose a limited number of starts on a unit by considering low-cycle fatigue. Good design practices limit the magnitude of torsional excitation (Pollard, 1980).

4.4.4 Damping

Damping is important in assessing the sensitivity of the system response to such forcing functions as mass unbalance, misalignment, and thermal distortion. Information about damping will reveal potential problems of a machine passing through or dwelling on a critical speed. Operating a machine with significant damping at a critical speed is not necessarily harmful; on the other hand, low damping can sometimes mean disaster. Evaluation of sensitive machines—i.e., those subject to mass unbalance from the process and those operating close to a critical speed—requires data on damping.

Damping for a single mode can be measured in the domains of time, frequency, or phase. Modal testing is used to evaluate multimode damping. The time domain of vibration of a steam-turbine pedestal is shown in Fig. 4.4*b*. The damped natural period T is obtained directly from the plot and is equal to 0.012 s/cycle, giving a frequency f or $1/T$ equal to 83.3 Hz. The damping ratio—a measure of modal viscous damping—can be calculated from Fig. 4.4*b*. The damping ratio is the ratio of system damping to critical damping, i.e., the minimum magnitude of damping that will not allow vibration. The *logarithmic decrement* is defined as

$$\delta = \frac{1}{n} \ln \frac{x_0}{x_n} \tag{4.3}$$

where δ = logarithmic decrement
n = number of cycles of vibration measured
x_0 = amplitude of vibration on initial cycle
x_n = amplitude of vibration on cycle n
ln = natural logarithm to base e

The logarithmic decrement is related to the damping ratio by

$$\delta = \frac{2\pi(c/c_c)}{\sqrt{1-(c/c_c)^2}} \tag{4.4a}$$

$$\delta \approx 2\pi \frac{c}{c_c} \qquad \text{for } \frac{c}{c_c} < 0.5 \tag{4.4b}$$

or

$$\frac{c}{c_c} \approx \frac{\delta}{2\pi} \tag{4.4c}$$

where δ = logarithmic decrement
c/c_c = damping ratio
c = actual damping, lb · s/in
$c_c = 2m\omega_n$ = critical damping
m = modal mass, lb · s/in
$\omega_n = 2\pi f_n$ = natural frequency, rad/s
f_n = natural frequency, Hz

The damping ratio varies from 0 (no damping) to 1 (no vibration). Typical damping ratios are as follows:

Material or system	Damping ratio
Steel	0.001
Rubber	0.05
Rolling-element bearing machines	0.025
Fluid-film bearing machines	0.03–1.0
API specification	0.0625

The logarithmic decrement δ for the dominant pedestal mode at 83 Hz (Fig. 4.4*b*) is $\frac{1}{7}\ln(5.5/0.8)$, or 0.275, so that, from Eq. 4.4*c*, The damping ratio c/c_c is $0.275/(2\pi)$, or 0.043.

Another method for evaluating damping for a single mode of vibration utilizes the frequency domain (see Fig. 4.11*a*). Half-power points, that is, 0.707 peak amplitude, of vibration response around a critical speed or resonance are used:

$$Q = \mathrm{AF} = \frac{f_n}{f_2 - f_1} = \frac{1}{2(c/c_c)} \tag{4.5a}$$

or

$$\frac{c}{c_c} = \frac{1}{2\mathrm{AF}} \tag{4.5b}$$

where Q = quality factor
AF = amplification factor
f_n = natural frequency, cycles/min, rpm, or Hz
f_1 = first half-power point; i.e., lower frequency at which a vibration response of 0.707 peak vibration occurs. Peak vibration occurs at a natural frequency or critical speed consistent with f_n.
f_2 = second half-power point, i.e., higher frequency at which a vibration response of 0.707 peak vibration occurs, consistent with f_n

From Fig. 4.11*a* we see that, for $f_n = 5800$ rpm, $f_1 = 5200$ rpm, and $f_2 = 7100$ rpm, AF = 3.05 which gives, from Eq. 4.5*b*, $c/c_c = 0.16$.

With the third method for evaluating damping, the phase display associated with a frequency-response plot or start-up coast-down test is used. The Bode plot shown in Fig. 4.11*a* contains a phase-angle change with respect to speed as the rotor passes through a critical speed. The formula for the amplification factor is

$$\mathrm{AF} = \frac{\pi N_c}{360} \frac{d\phi}{dN} \tag{4.6}$$

where AF = amplification factor
N_c = critical speed, rpm
$\frac{d\phi}{dN}$ = rate of change of phase angle with respect to speed around critical speed, deg/rpm

The rate of change $d\phi/dN$ is approximated by finite differences $\Delta\phi/\Delta N$.

From Fig. 4.11*a* then, with N_c equal to 5800 rpm and $\Delta\phi/\Delta N$ equal to $(458-380)/(6250-5250)$, or 0.078, we find the amplification factor is $(5800\pi/360)(0.078)$, or 3.95.

4.4.5 Modal Testing Techniques

Modal testing techniques (Ewins, 1984) are widely used for structural studies, but various problems must be overcome before they can be applied to rotating machinery. First, the rotor is not easily accessible for taking the many measurement points required to describe its mode shapes. In addition, the applied vibratory force is difficult to quantify. Measurements must be taken at speed because the dynamic properties of the rotor depend on speed (see Sec. 4.4.1); however, in addition to the imposed controlled vibratory excitation forces, vibratory forces intrinsic to the operational system are acting. The rotor response stimulated by the known excitation must thus be separated from the residual response. Extended testing techniques have been developed to overcome these problems (Nordmann, 1984; Bently et al., 1985).

Modal testing of rotating machines and the foundations and piping associated with them has been motivated by several factors. Such test data as natural frequencies and mode shapes allow validation of computer models. Measured damping data are important because theoretical values are uncertain. In addition, modal test data are used experimentally to formulate models of components that can be incorporated into the machine assembly by substructuring processes. The data can also be used with a modal model to determine dynamic forces that cannot be measured directly. Also results of modal testing can be used to infer critical and limiting operating parameters throughout a machine—e.g., closure of close operating clearances, limiting bearing loads—from the often modest amount of data typically available on an operating machine that is being monitored. Finally, modal data are valuable for in-depth diagnosis of rotating machinery. The computer model, in combination with modal test data, allows the analyst to evaluate such components of design as critical speeds, resonances, damping, and modes as well as the machine condition (e.g., excessive forces caused by mass unbalance, misalignment, bearing wear, and looseness).

The analyst who performs a modal test on a rotating machine must understand mechanical vibration theory—including rotor and structural dynamics—and the importance of accurate vibration measurements and realistic and detailed data analysis. Accurate modeling of a machine requires background knowledge of the physics of bearings, rotors, structures, and piping as well as modeling and computational techniques. Preparing for modal testing is not a simple process, but both the computational and experimental tasks involved can be performed on relatively inexpensive microcomputers with commercially available software.

Frequency-Response Function Measurement

From the general derivatives of rotor behavior developed in analytic models such as are discussed in Chap. 2, we see that the modal parameters of rotating machinery can also be determined from a measured set of *frequency-response functions* (FRFs) for which force, amplitude, and phase are available. The coherence function is used to establish the validity of the data.

Several experimental methods (Nordmann, 1984) have been developed for acquiring the data necessary to formulate FRFs. A force from a hammer or shaker can be applied, or mass unbalance can be calibrated. In the latter case, response on the rotor is measured with a proximity probe to obtain displacement; an accelerometer is mounted on the structure. When mass unbalance is used, a set of selective deflection measurements is made over the entire system for a range of frequencies of interest. Both the component of deflection in phase with excitation and that 90° out of phase are defined. The background response of the rotating machine must be eliminated. Artificial excitation using swept sine forces from a shaker is time-consuming but is potentially more accurate because only one frequency is excited at a time. However, impact and random forces that simultaneously excite many modes are more often used today in structural modal testing. FRFs vary with rotor speed and are computed directly from the ratio of Fourier transforms of the measured output (response) and input (force) signals. In impact tests the frequency content and amplitude of the force signal are influenced by the hammer mass, flexibility of the impact cap, and impact velocity. A long impulse yields low energy and a wide frequency range. A long impulse is required to excite low frequencies.

The standard FRF is typically response divided by force (receptance, mobility, inertance). Arguments have been made (Bently et al., 1985) for using the inverse FRFs. Dynamic stiffness, mechanical impedance, or apparent mass is used for such discrete-parameter identification in rotating machines as bearing stiffness and damping.

Graphical displays of FRF data are complex; they are represented by three quantities that include frequency and two parts of a complex function. The three common displays are the Bode plot (FRF magnitude, phase, and coherence versus frequency; see Fig. 4.9) real and imaginary parts (Fig. 4.10) of the FRF, and the Nyquist plot (real versus imaginary plot), which does not explicitly contain frequency information. The Bode plot of Fig. 4.11*a* is a special plot generated by a force applied as a function of the rotating speed. The polar plot (Fig. 4.11*b*) is a special case of the Nyquist plot (FRF versus phase is equivalent to the real and imaginary parts versus frequency). Once-per-revolution rotating speed is the variable frequency. The coherence plot (Fig. 4.9*c*) is used to establish the portion of the measured response attributable to the applied force and thus checks the validity of the generated FRF. The natural frequencies are taken directly from any of the three frequency-response plots. The mode shape of the rotor can be obtained by stacking the imaginary components (Fig. 4.10) of the FRF along the rotor in a waterfall. The envelope of the component is the mode shape. The real display of the FRF provides damping data; the half-power method is used.

Data Analysis

The procedures of experimental modal analysis are applied to modal test data to identify the parameters of a theoretical model that will simulate machine behavior. The data must be matched or curve-fitted to a modal equation such as the modal response of a machine

$$Y(\omega) = i\omega \sum_{j=1}^{N} \frac{A_j}{\lambda_j^2 - \omega^2} \tag{4.7}$$

where $Y(\omega)$ = measured modal values at various frequencies
A_j, λ_j = modal parameters related to properties of machine

The nature of the data dictates which of several procedures will be used to extract the modal parameters. The most widely used approach is the single-degree-of-freedom curve fit. In systems with well-spaced modes, acceptable values for the parameters can be found by using test data around the natural frequency. The single-degree-of-freedom constants plus an offset term, which accounts for the contributions at other modes, can be used effectively.

The Nyquist plot is generated from modal data for a single mode (Ewins, 1984). The process is repeated for each mode until the complete frequency range is covered. A theoretical regeneration can then be performed to establish the entire mobility curve; circle-fitted data are used. Modes that are very close to other modes—identifiable on the Nyquist plot by lack of a circular configuration—indicate that more complex processes are occurring; a so-called multiple-degree-of-freedom curve fit must then be used that simultaneously includes the effects of many modes (Ewins, 1984). The type of damping and slightly nonlinear behavior must be considered in the curve-fitting process.

Nordmann (1984) has published a procedure for identifying modal parameters of rotors that utilizes analytical functions fitted to measured data. FRFs are measured at various speeds between excitation points l and measurement points k. Analytical FRFs, which depend on eigenvalues and right- and left-hand eigenvectors, are fitted to the measured data by varying modal parameters. The parameters that allow a fit of the measured data to the analytical functions represent the dynamic characteristics of the rotor-bearing system.

The modal parameter estimation procedure (Nordmann, 1984) selected depends on the degree of modal coupling in the rotor system. According to the theory of modal analysis, the total frequency response at a measurement point consists of the weighted sum of the contributions of all modes of the system. The contribution of a particular mode is usually greatest close to a natural frequency. When modal coupling is light and modal cross-coupling is minimal, the single-degree-of-freedom model provides a good approximation of modal parameters of the system for that natural mode. Thus, if the modes of a system are well separated, data analysis—Bode, Nyquist, real, and imaginary plots—can be used to determine modal parameters directly from natural frequencies and amplitudes at resonance. When heavy modal damping is present (as is sometimes the case when journal bearings support a rotor), a multiple degree-of-freedom curve-fitting technique must be used based on the representation of the error function:

$$E = \sum_{p}^{P} \{\mathrm{Re}\,[H'_{kl}(\Omega_p)] - \mathrm{Re}\,[H_{kl}(\Omega_p, \lambda_j, \Phi_j)]\}^2 + \sum_{p}^{P} \{\mathrm{Im}\,[H'_{kl}(\Omega_p)] - \mathrm{Im}\,[H_{kl}(\Omega_p, \lambda_j, \Phi_j)]\}^2 \tag{4.8}$$

where H_{kl} = analytical frequency-response equation
H'_{kl} = measured frequency response
P = number of lines in measured FRFs
Re = real part
Im = imaginary part
p = line number
j = mode number
k = measured points
l = excitation points
Ω = shaft speed
λ = eigenvalue
Φ = eigenvector

Test data are curve-fitted to the analytical expression to obtain modal parameters. A least-squares procedure for minimizing the error function E is commonly used to extract modal parameters. The error function is differentiated with respect to each unknown and set equal to zero; the unknowns of the resulting set of equations are the modal parameters. The equations are nonlinear in the eigenvalue parts (Nordmann, 1984) and must be solved by an iterative procedure.

The methods for extracting modal parameters from experimental data described above deal exclusively with compliance functions—receptance, mobility, and inertance. It has been suggested (Bently et al., 1985) that the dynamic stiffness functions (the inverse of compliance) are more useful in that direct numerical values of such specific physical properties of the systems as discrete mass, stiffness, or damping can be calculated. The method would be advantageous for direct experimental characterization of mechanical components, e.g., bearings, dampers, pedestals, rotors. This method can also be used to develop modal models.

In a test set up for determining the dynamic stiffness of a rotor using mass unbalance as the rotating excitation force, flexural vibrations of the shaft are measured in two lateral directions with two noncontacting shaft displacement probes mounted orthogonally. A third displacement probe used as the Keyphasor® relates the phase angle between the rotating force (mass unbalance) and the vibration response. A known mass unbalance is used in this instance to identify parameters of the system. (The same setup is typically used to balance a rotor with unknown mass distribution.)

The parameters of a rotor that is straight and well balanced could be characterized by introducing a known mass unbalance at a known location on the rotor and then varying the rotor speed from zero to the desired value. Unfortunately, the rotor also responds to forces other than the known mass unbalance. Therefore, calibration tests and subtraction of the residual response are necessary. It has been suggested (Bently et al., 1985) that vectorial subtraction of start-up and coast-down test data over the speed range of the test will eliminate unwanted residual response. The known mass unbalance in the experimental setup is 180° out of phase for the two tests.

The data from the once-per-revolution excitation in the test setup are restricted to that frequency. For example, the properties of fluid-film bearings are frequency-dependent. But data of the excitation from mass unbalance are within the frequency range of the highest test speed of the rotor. At a minimum, the speed range of the rotor should include at least one critical speed.

The digital vector filter provides information on the amplitude and phase of the vertical and horizontal rotor responses to a varied synchronous speed excitation. Figure 4.17 shows plots of direct (k_D) and quadrature (k_Q) dynamic stiffness versus speed. These results were obtained from test data and the following relationships:

$$k_D = \frac{\text{force amplitude} \times \cos\text{(phase)}}{\text{response amplitude}} \tag{4.9a}$$

$$k_Q = \frac{\text{force amplitude} \times \text{sine (phase)}}{\text{response amplitude}} \tag{4.9b}$$

The mass in each direction is calculated from the slope of the k_D-versus-speed-squared curve; the stiffness k_v is derived from the intercept of the graph with the vertical axis. Damping is derived from the slope of the curve of k_Q versus speed. Results using this approach are excellent for near-symmetric rotors.

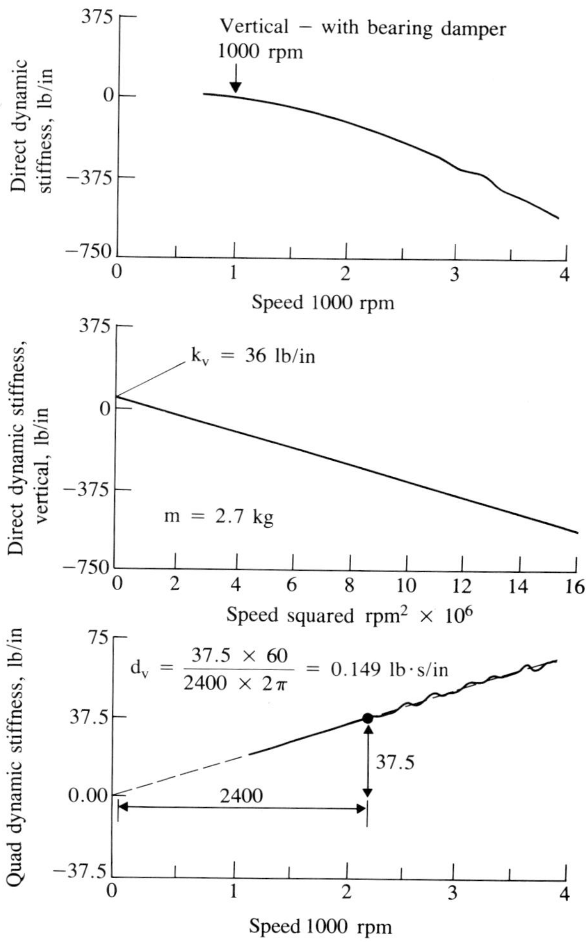

FIGURE 4.17 Direct and quadrature dynamic stiffness versus speed (after Bently et al., 1985).

4.5 PERFORMANCE TESTING

Performance testing is conducted on new equipment in the shop and during commissioning. Vibration levels, bearing temperatures, and seal leakage are measured during a mechanical performance test. The shop test provides the initial data for evaluating a design. These data are used to establish the locations of critical speeds and resonances and whether unwanted vibration response occurs at operating speed. The evaluation process is not complete, however, until the equipment is commissioned and full loading is possible.

Data taken during commissioning are used to evaluate the overall performance of a complete machine system. The components, e.g., foundations and piping, are now in their actual environment and can be loaded and cycled. Performance tests are a check on the entire design and installation processes and also provide an opportunity to identify any deficiencies in the system.

4.5.1 Objectives

Shop testing, which is carried out after equipment is constructed but before it is commissioned, is a time for handling problems prior to start-up and avoiding start-up delays. The process is similar to commissioning but allows the owner and the owner's contractor or agents to perform additional tests and collect good data without production interference. However, it is difficult to assess everything in a builder's shop. A previously agreed-to written test plan is important. Good test data can be obtained on a test stand and in the field. Field data are presented in this section and should be considered equivalent to shop data.

During a performance test such preanalysis work as that done on rotordynamics can be evaluated (Jackson and Leader, 1979a, b). Figure 4.18 shows the

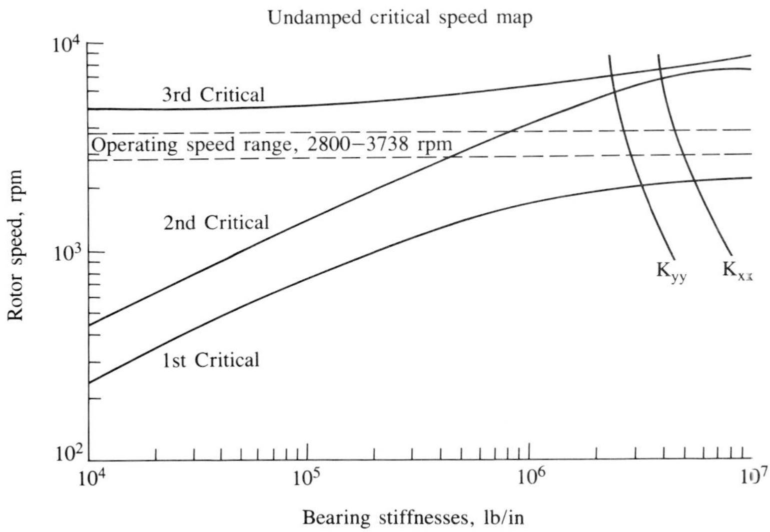

FIGURE 4.18 Undamped critical-speed map for a 20,000-hp propylene refrigeration compressor.

undamped critical-speed map of a 20,000-hp propylene refrigeration compressor mounted in five-lobe tilting-pad bearings. The high bearing stiffness from the 0.4 preload places the critical speed in a rotor-dependent position. Such a situation is not advisable because a bearing change can only lower the critical speed. The Bode plot (Fig. 4.19) confirms the critical-speed analysis and the fact that the critical speed is not within the operating speed range. Separation margins of the critical speeds from the intended operating speed range have been defined by the American Petroleum Institute (API); the resonances must be 20 percent above

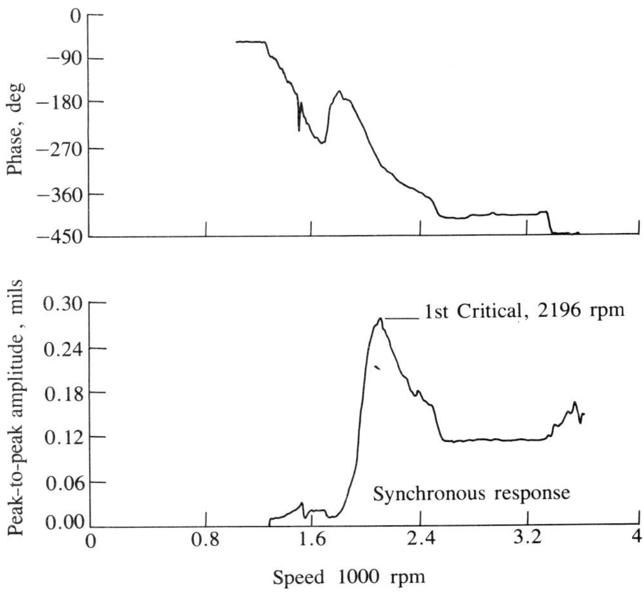

FIGURE 4.19 Compensated Bode plot of a 20,000-hp propylene refrigeration compressor run-up.

the maximum continuous speed and/or 15 percent below the operating speeds. Torsional resonances must be 10 percent from operating speed.

According to past API standards, the rotor, because it is thermally stable after 4 h of operation, must have an amplification factor (AF) of less than 8 on deceleration and with runouts vectorially compensated—less than 5 is preferred. An AF of 8 implies a damping ratio of 0.0625 or about 6 percent of critical damping for a rotor system. The AF of a 20,000-hp refrigeration centrifugal compressor, under surge, at the running speed of 3531 rpm is about 7.0 (Fig. 4.19), or 7 percent of critical damping. The implied logarithmic decrement is 0.44. A resonance response from 1945 to 2048 cycles/min (Fig. 4.19) confirms the critical speed as calculated and measured; the logarithmic decrement is 0.28, or 4.5 percent of critical damping.

Present API standards require complex data from the manufacturer; acceptability will depend on calculated deflections at each seal along the rotor as a percentage of the total clearance.

Good baseline data can be recorded for an acceptable new machine during a shop test. Two plots that should be generated from such data are shown in Fig. 4.20. The data are from a third-body cracked gas compressor with a critical speed of 2600 rpm when operating at 4400 rpm. The first plot (Fig. 4.20*a*) is a Bode plot of amplitude versus speed. Superimposed on that plot are a synchronous plot and a compensated once-per-revolution (1X) plot of amplitude and phase versus speed. Any residuals at the speed of compensation are removed. In the compensated Bode plot, the slow-roll runouts have been vectorially compensated at 700 to 800 rpm. This phase plot, which seems more realistic, is an example of improper data processing when significant runout exists. The plot should be a

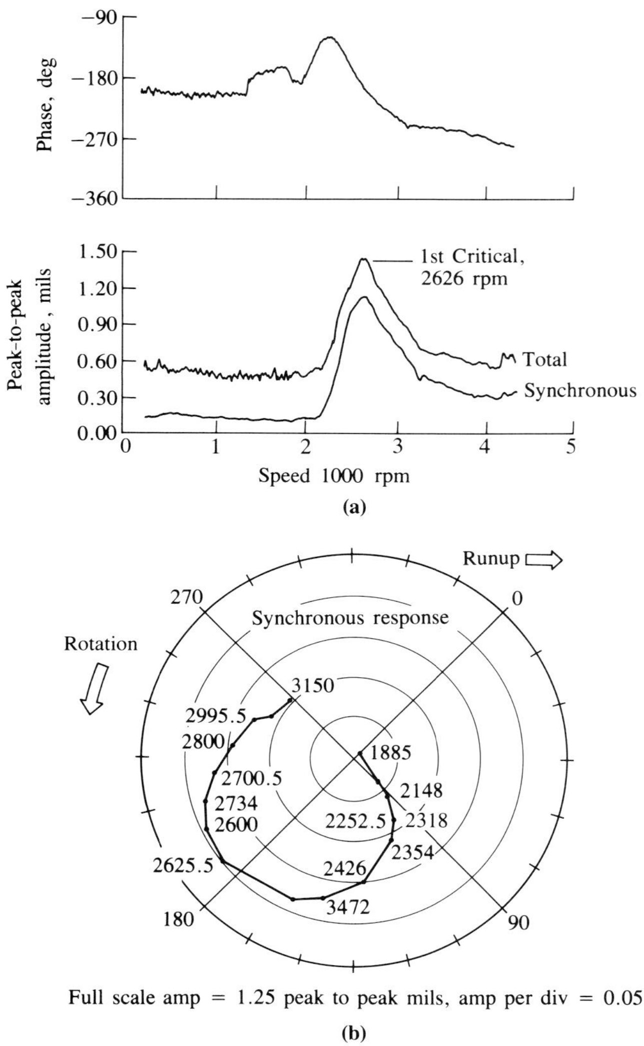

FIGURE 4.20 Bode and polar plots of a third body cracked gas compressor—raw and synchronous.

compensated polar plot (Fig. 4.20*b*). Unique phenomena can be easily seen in a polar plot; e.g., the effects of a temporary or permanent bow in a rotor are obvious. The final plot that should be used, a waterfall plot (Fig. 4.13), is the best single hard-copy plot of the overall condition of a machine during start-up or coast-down. The primary frequency that excites any amplitude is the synchronous (1X) frequency. A small amount of 2X is seen. No subsynchronous oil whirl or oil whip is observed.

A rotor rub may not be detected if only a tracking filter is being used. An oscilloscope should be used because it shows the output of the vibration probes in the unprocessed time domain. The return of slow-roll data to the prestart-up condition can be seen on the oscilloscope during coast-down. If the slow-roll information gradually disappears under no load, a rub may have occurred.

Bearing temperatures are recorded, and leakage rates of mechanical seals are measured during a shop test. The leakage for a 4-h running test should be monitored. Bearings and journals as well as the immediate seal area are inspected after a test.

4.5.2 Facilities and Equipment

Performance testing of such rotating machines as steam turbines and centrifugal compressors is usually organized at the manufacturer's facility. Although the tests can be done in the field, aerodynamic performance cannot be assessed unless the equipment has been installed according to such codes as ASME PTC-6 for steam turbines and ASME PTC-10 for compressors; these installations require too much space in an operating plant.

Various instruments are needed to carry out a mechanical test. Using a multichannel FM tape recorder to record data from all sensors saves test time and permits data to be coordinated. Certain digital recorders provide equivalent results. Data are extracted from the tapes without repeating the tests. When problems exist, in-depth analysis can be carried out easily. Oscilloscopes are needed to display direct and filtered real-time data. Direct data can be used to detect instabilities, which do not show up at synchronous, or 1X, frequencies. Orbits traces are necessary to detect the maximum vibration levels (see Sec. 4.5.3).

A synchronous tracking filter, or vector filter, uses a timing mark for tracking. The mark is a once-per-turn phasor pulse from the actual machine during a test. The vector filter can process a minimum of two orthogonal signals from eddy-current proximity probes in the direct or filtered mode and in the runout mode, either compensated or uncompensated.

An FFT spectrum analyzer is used to evaluate the frequency domain within a very narrow range and to expand certain regions to improve resolution. The analyzer can detect orders, i.e., multiples of frequencies of running speed. Instruments such as strip chart recorders provide continuous data on bearing temperatures, operating pressures, and axial position of the rotor. These data are related to both speed and load and can be related to each other by time, speed, and horsepower.

4.5.3 Procedures

Many purchasers require that equipment meet API standards. According to API, the minimum time for a mechanical test of a steam turbine or centrifugal compressor is 4 h. The machine must be operated at either 10 percent above the maximum continuous speed (105 percent of rated speed) or 115.5 percent of rated speed. Three trips at ± 1 percent of rated speed are usually required and should be carried out at the beginning of the test for safety reasons.

The machine is operated within about 10 percent of rated speed but not at resonant speed to allow sufficient time—about 10 or 15 min—for stabilization and recording of data. The 4-h run is an endurance test. Malfunctions, bad design, poor assembly, or lack of cooling or lubrication often becomes evident during the test and voids it. The test can also be run at partial or full load depending on the purchaser's option.

Transient data can also be evaluated during testing. The amplification factor can be measured in several ways (see Sec. 4.4.4). Correctly plotted Bode plots (Fig. 4.11*a*) and polar plots (Fig. 4.11*b*) should provide similar values of damping. A complete envelope will not be obtained if the amplification is lowered. However, because most rotors are not single-mass, single-degree-of-freedom systems, response is often more complex; resonance envelopes are not as clean as shown in Fig. 4.11 because contributions to the response are obtained from several modes. The Q, or bandwidth, definition of the amplification factor is shown in Fig. 4.11.

A fast rate of phase change (Fig. 4.11*a*) indicates a high response at the critical speed, i.e., a high AF. Conversely, little damping is available. The expression for AF based on rate of phase change per rate of speed change is given in Sec. 4.4.4.

All the amplification factors shown are close to each other and were measured correctly; i.e., the start-up was not too fast, and the full envelope was plotted. Recent changes in the API standard require that measurements during shop testing use from 60 to 100 percent of the plotter's display. The shape factor of the complete resonance response envelope is important because the AF is equal to 0.5ξ, where ξ is the ratio of actual damping to critical damping.

It is wise to measure the natural frequency and the rate of decay (logarithmic decrement) while a compressor is being subjected to surge at running speed. Because the rotor is operating above the critical speed, damping at the natural frequency of the rotor is a fair assessment of resistance to oil whip. The response of a compressor rotor to surge while operating at full speed is shown in Fig. 4.15; the time signal was captured at the instant of surge.

4.5.4 Standards

Vibration standards for acceptability include those developed by the API, International Standards Organization (ISO), and American National Standards Institute (ANSI); others include manufacturers' standards and company operating standards. Requirements for shop testing, the mechanical running test, are best described in the API standards for refineries. Confusion can arise with the various standards. Some are shaft-relative values; others pertain to bearing cap and seismic measurements. Utilities sometimes combine measurements from two probes—one on the shaft that measures relative motion and one on the bearing that measures absolute motion—to obtain absolute shaft motion.

Limits for mechanical or electrical runouts must be accepted prior to an actual test. Mechanical runouts are often limited to 0.5 mil peak to peak, e.g., thrust collar faces. Radial runouts of a rotor shaft may be about one-half that amount. The combined electrical and mechanical runout of proximity probe measurements on the surface of a rotor shaft is limited to 25 percent of the acceptable vibration limit for the rotor at speed. The current vibration limit for compressors (API 617, 1979) is vibration in mils peak to peak equal to $(12{,}000/\Omega)^{1/2}$, where Ω is rotor speed in revolutions per minute. For a speed of 5500 rpm the vibration limit throughout the operating speed range would be 1.476 mils peak to peak. The combined electrical and mechanical runout would be 25 percent of 1.476, or 0.37 mil peak to peak. The runout would apply only to proximity probes because seismic probes mounted on bearings have no runout.

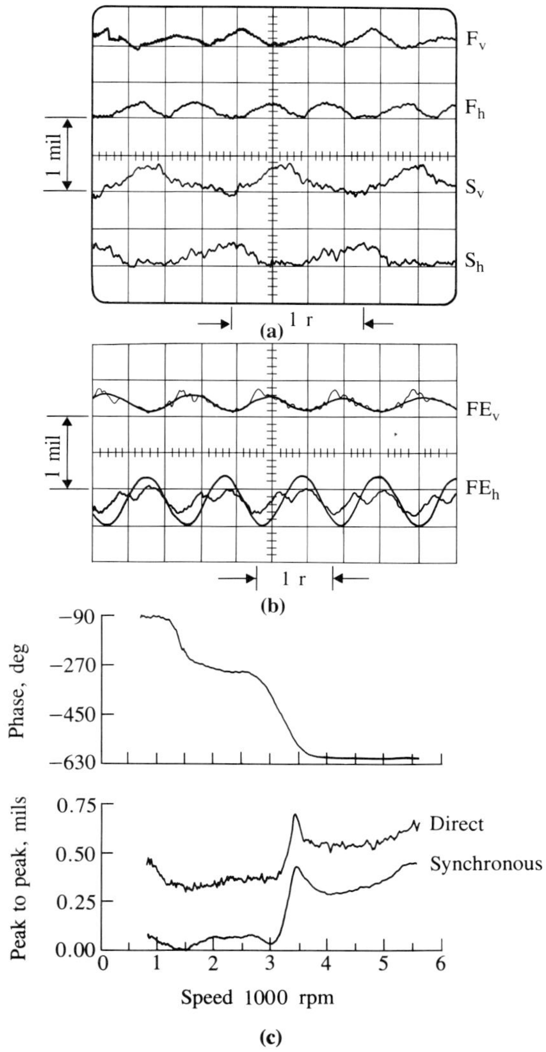

FIGURE 4.21 Slow-roll, transient, and running data with a 2X runout: (*a*) Slow roll—500-rpm rotor 105; (*b*) running data—5614-rpm rotor 105; (*c*) transient data.

Figure 4.21*a* and *b* is a photograph of a 2X unprocessed runout from a compressor that differs from the synchronous filtered 1X plot but is within test specifications. Figure 4.21*c* shows that the runout can cause the Bode plot in a coast-down test to be wrong. The first critical speed is at not 3400 rpm but 2600 rpm. The second critical speed is at 3400 rpm. A bow is incorrectly identified at 1500 rpm. The amplification factor is also wrong, as is the phase change through resonance. This plot was taken during the first commissioning.

The residual unbalance in a rotor affects its acceptability. However, the value for residual unbalance in API 612 (1979) for steam turbines is the old Navy standard without a speed-squared term in the denominator. The unbalance in ounce-inches is equal to $4W/N$; W is the rotor or journal weight in pounds, and N is the speed in rpm.

An upper speed–minimum unbalance limit has been established for high-speed compressors, typically the four-poster designs in API 672 R/2 (1979). These compressors are arranged so that the driver is connected to a gear that can drive two pinion shafts; an overhung three-dimensional impeller can be located at each end of the pinions to complete the four posts. The minimal limit of residual unbalance is 25 μin multiplied by the weight of the rotor in ounces, which gives a figure in units of ounce-inches at the balancer manufacturer's guaranteed capability.

Values for residual unbalance can be taken from various standards. The original API value was about grade 2 to 2.5 mm/s in ISO terms. The new Navy value of $4W/N$ is equivalent to grade 0.67 and is much tighter. A rotor tested in a full-speed vacuum chamber is often tested to ISO grade 1.0, or 1 mm/s. The eccentricity of unbalance in millimeters is obtained by dividing the grade in millimeters per second by the operating speed in radians per second. The balance tolerance is in kilogram-millimeters per kilogram or millimeters of eccentricity. If eccentricity is converted to inches, the balance tolerance can be expressed in pound-inches of rotor, or inches of eccentricity.

A 1000-lb (455-kg) rotor operating at 10,000 rpm (1047 rad/s) and balanced to ISO, grade 1.0 (1 mm/s) will have an eccentricity of 0.95 μm, or 37.6 μin. The calculation is (1 mm/s)/(1047 rad/s). The balancing tolerance of 37.6×10^{-6} lb · in/lb equals 0.6 oz · in/1000 lb, or 0.6 oz · in. The expected vibration from such unbalance would be twice the eccentricity, or 2 times 37.6 μin which is 75.2 μin or 0.075 mil peak to peak. The vibration level is thus very low, as is the residual unbalance. Most balancers guarantee 25 μin.

API Standard 670 (1986) is a document on the correct installation of probes and monitors. It also includes axial position monitoring and guidelines as well as temperature sensors for both radial and thrust bearings. API Standard 678 (1981) covers seismic requirements for mounting, wiring, and specifications.

4.6 DIAGNOSIS OF ROTATING MACHINERY MALFUNCTIONS

Only since the late 1960s have experimental and analytical techniques for assessing vibration levels been used routinely for diagnosing faults in operational rotating machinery. Until that time these techniques were used almost exclusively in the course of development by designers and developers to detect and correct design features causing excessive vibration. The diagnosis of vibration problems in machines depends on good measurements to obtain data from transducers during machine operation as well as diagnostic tests, parameter identification tests, and knowledge of the machine itself. Because the foundation of diagnostics is frequency information, the analyzers used to process data must have adequate resolution. Successful implementation of most diagnostic techniques requires familiarity with the frequencies of the components of the machine as well as its dynamic properties. Parameter identification techniques are used to quantify natural frequencies, mode shapes, damping, and stability and frequency-response characteristics. Time-domain analysis—the study of signatures directly from a transducer—and orbital analysis are used to diagnose faults when motions of a machine or a rotor provide direct information about dynamic behavior. Modern diagnostic techniques and the identification of specific malfunctions of rotating machines and their components are described in this section.

4.6.1 Measurement Procedures and Locations

Methods used to obtain data are described in Sec. 4.3. The procedure used to gather data for fault diagnosis depends on the equipment type and, to some extent, the nature of the problem. Data from the machine during operation are essential. If the surroundings might cause difficulties in data interpretation, environmental data should be collected when the machine is not operating. If all the parameters of the machine have not been quantified, parameter identification tests (Sec. 4.4) might be necessary. Complete familiarity with the working mechanisms of the machine enables the diagnostician to establish cause (machine fault) and effect (machine response) relationships.

Transducer locations and positions are selected to obtain information from the machine by the most direct route. The distance traveled by a vibration signal from its source to a transducer should be as short as possible to avoid signal amplification or attenuation, introduction of excessive noise, and waveform distortion; unfortunately, transducer placement is usually limited by availability of access and the experience of the diagnostician. Displacement sensors should be used, if possible, when the relative motion between rotor and bearing support structures is large. They provide a direct signal of rotor motion and an orbit if two sensors are mounted orthogonally (Fig. 4.7). Seismic sensors (velocity or acceleration) are usually used on machines with stiff bearings. Measurements should be taken on the bearing caps in the x, y, z directions. Otherwise, the sensors should be placed as close to the bearings as possible. Diagnosis of vibration in a structure requires information about panels, structural supports, pedestals, and foundations during operation. Identification of faults that require phase data, e.g., misalignment, is dependent on the simultaneous recording or analysis of data from two sensors.

4.6.2 Diagnostic Techniques

The techniques used to diagnose machinery listed in Table 4.2 depend on processed and unprocessed vibration signals. These signals are examined principally for frequency information that can be related to a machine fault.

Time-Domain or Waveform Analysis

The time domain (Fig. 4.4) is an unprocessed record of the vibration signal from a transducer. (Both digital and analog displays of time domains are considered unprocessed data.) The time domain contains information about the physical behavior of a machine but is limited to viewing a single plane of motion and by the fact that it can be too complex for analysis if excessive noise, signal modification, or several frequencies are present; in such cases processing is necessary. But processing should be minimized whenever possible so that errors are not introduced into the data.

TABLE 4.2 Diagnostic Techniques for Rotating Machinery

Technique	Use	Description	Instrument type
Time-domain analysis	Modulation, pulses, phase, truncation, glitch	Amplitude versus time	Analog and digital oscilloscope, FFT spectrum analyzer
Orbital analysis	Shaft motion, subsynchronous whirl	Relative displacement of rotor bearing in *XY* direction	Digital vector filter, oscilloscope
Spectrum analysis	Direct frequencies, natural frequencies, sidebands, beats, subharmonics, sum and difference frequencies	Amplitude versus frequency	FFT spectrum analyzer
Cepstrum analysis	Sideband and harmonic frequency measurement; accurate quantification of sideband and harmonic severity	Inverse Fourier transform of logarithmic power spectrum	FFT spectrum analyzer

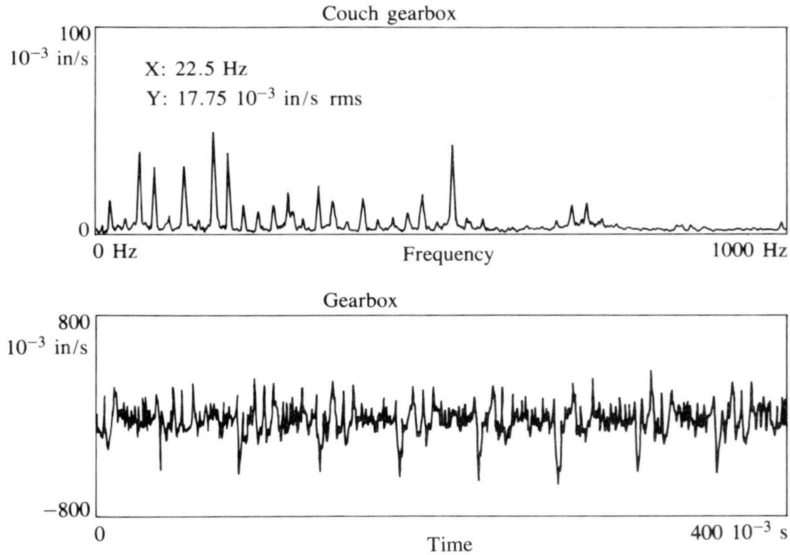

FIGURE 4.22 Pulses from two chipped teeth in a gearbox (after Szrom, 1984).

For fault diagnosis the two important factors in evaluating the time domain are the frequency and phasing of events. Valuable information is provided by signal shape, i.e., truncation (Fig. 4.4*a*), pulses (Fig. 4.22), modulation (Fig. 4.23), and "glitch", or shaft-induced signals obtained from a proximity probe that are caused by scratches on the shaft and/or electrical variations (Fig. 4.24). The time between events (Fig. 4.4*a*) represents the frequency of vibration and can be related to information about the frequencies of components within the machine. The 134.7 ms figure represents 4 cycles of the vibration of the compressor at operating speed (29.7 Hz). The phase (Fig. 4.25) between two signals provides information about vibratory behavior that can be used to diagnose a fault such as misalignment.

The time domain in Fig. 4.4, which is truncated, or clipped, shows orders of the fundamental, or once-per-revolution, frequency (33.66 ms, or 29.7 Hz) that are lower in amplitude than the fundamental frequency. The differential time feature of the analyzer provides better frequency resolution with the time domain than does the frequency domain (spectrum) which shows 30 Hz for the once-per-revolution vibration. The time domain in Fig. 4.26 is not truncated but contains naturally occurring orders of the fundamental frequency. The higher orders have higher amplitudes than the fundamental frequency. However, the amplitude depends on the strength of the excitation and its amplification; thus, higher-order frequencies can have higher or lower amplitudes than the fundamental frequency. The vibration plotted in Fig. 4.26 was measured on a rocking motor that was not properly bolted to its foundation.

The signal in Fig. 4.26 is synchronous because the phase of the higher-order components does not change with respect to the fundamental frequency. The second-order component in Fig. 4.27 is asynchronous with the fundamental frequency because the two do not move in phase; a two-pole electric motor with a

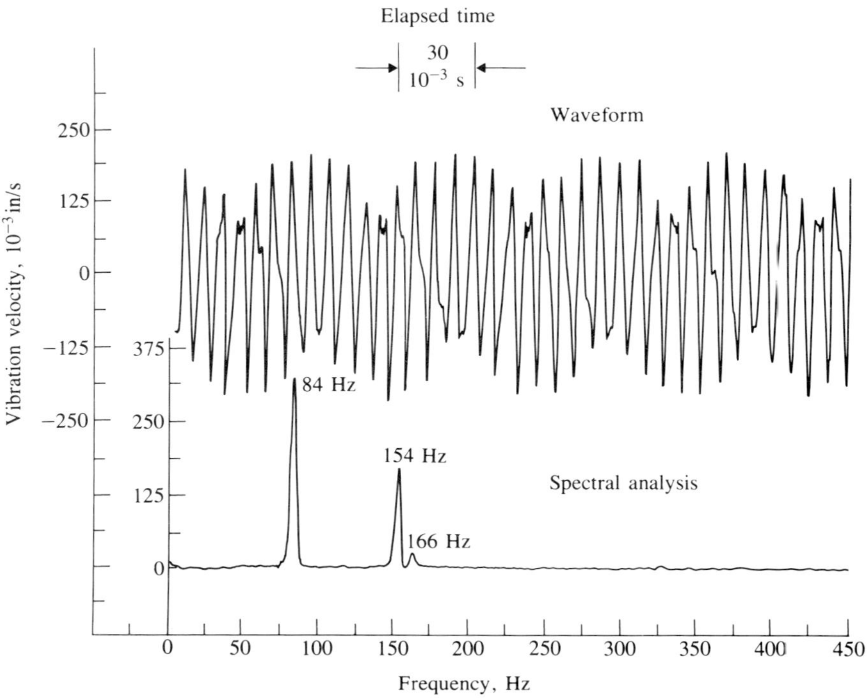

FIGURE 4.23 Amplitude modulation of a steam-turbine governor.

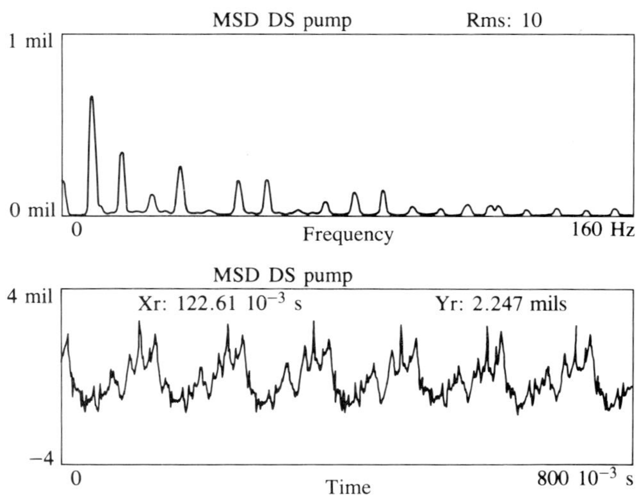

FIGURE 4.24 Pump shaft scratch adjacent to a proximity probe.

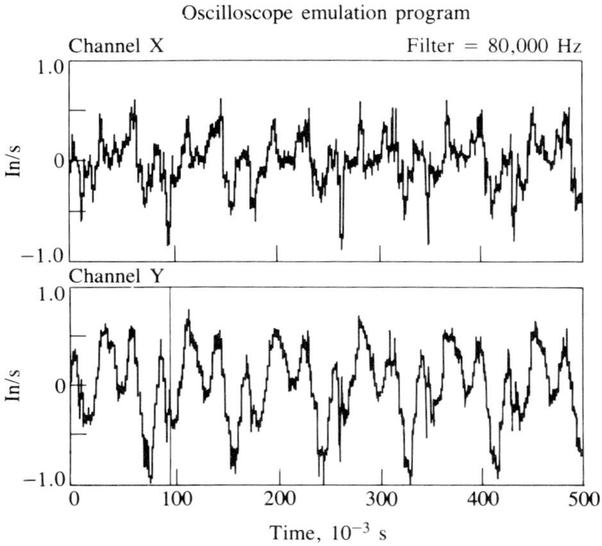

FIGURE 4.25 Phase measurement using a dual time domain.

stator or air-gap variation problem is typically asynchronous because the magnetic vibration generated at 120 Hz is greater in magnitude than that at twice the operating speed.

The signal in Fig. 4.23 is an example of amplitude modulation, i.e., change in amplitude with time. The second order of the operating speed of a steam

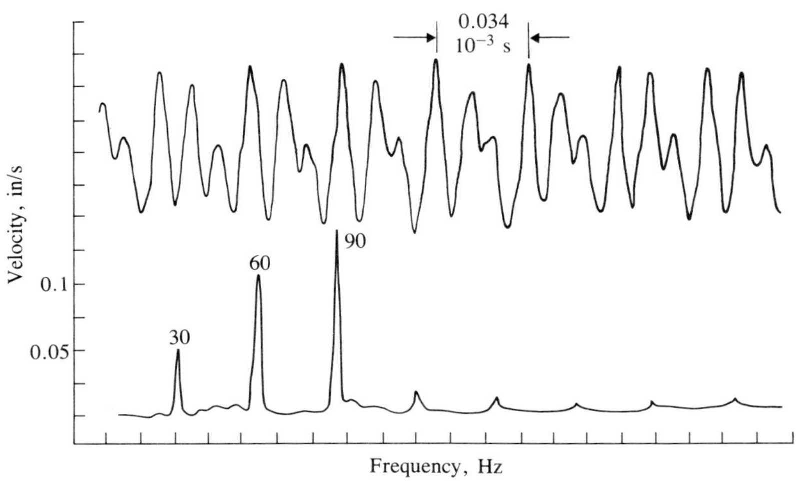

FIGURE 4.26 Vibrations measured on a rocking motor—synchronous orders.

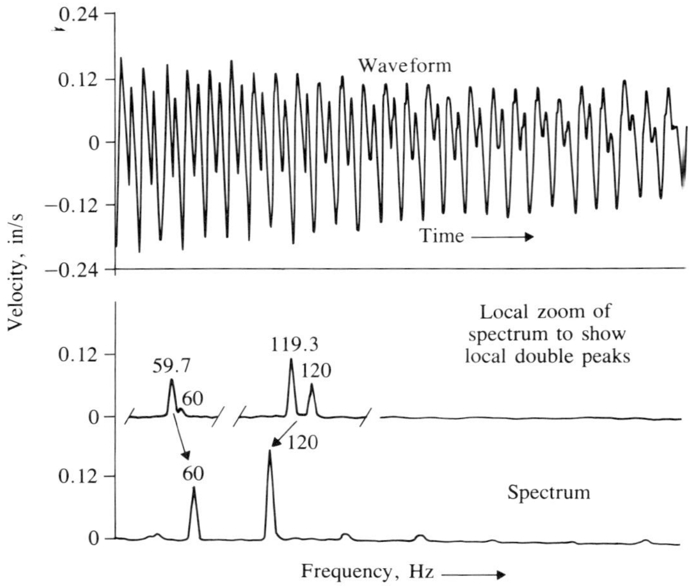

FIGURE 4.27 Vibrations measured on a 4000-hp induction motor—nonsynchronous orders.

turbine is beating with a structural natural frequency of the governor. The time domain shows the first and second orders as well as the period of the beat (0.0965 s). The period is related to the difference between the natural frequency of 156 Hz and 2 times the operating speed of 166 Hz.

Frequency modulation cannot usually be observed in the time domain because the frequency changes are small. Figure 4.28 is an example of torsional vibration of a printing press. In this case frequency modulation is not apparent in the time domain but can be seen in the sidebands of the spectrum. Sidebands occur as a result of both amplitude modulation and frequency modulation; the time domain can thus be used to characterize the modulation.

The shape of pulses in the time domain provides physical insight into the machine condition and dynamic behavior. The abrupt entry into a defective bearing race by a roller can manifest pulses. Pulses excite natural frequencies that are usually seen in the frequency domain (spectrum). Pulselike activity in the same time domain from chipped gear teeth is shown in Fig. 4.22; this activity excited natural frequencies with sidebands in the spectrum.

Severe runout (once-per-revolution component in the frequency spectrum), or glitch (Fig. 4.24), arises from scratches on the shaft or electrical runout and is sensed by proximity probes; it can be easily detected in the time domain. Severe runout behavior is repetitive; i.e., it occurs once per revolution. The long string of multiple orders evident in Fig. 4.24 is produced by the FFT process but could be mistaken for impact or looseness.

Phase angles between two signals or a Keyphasor® (the signal referenced to a mark on the shaft) and a vibration signal can be seen in the time domain of Fig. 4.7. The phase angle can be calculated from the divisions on the time-domain

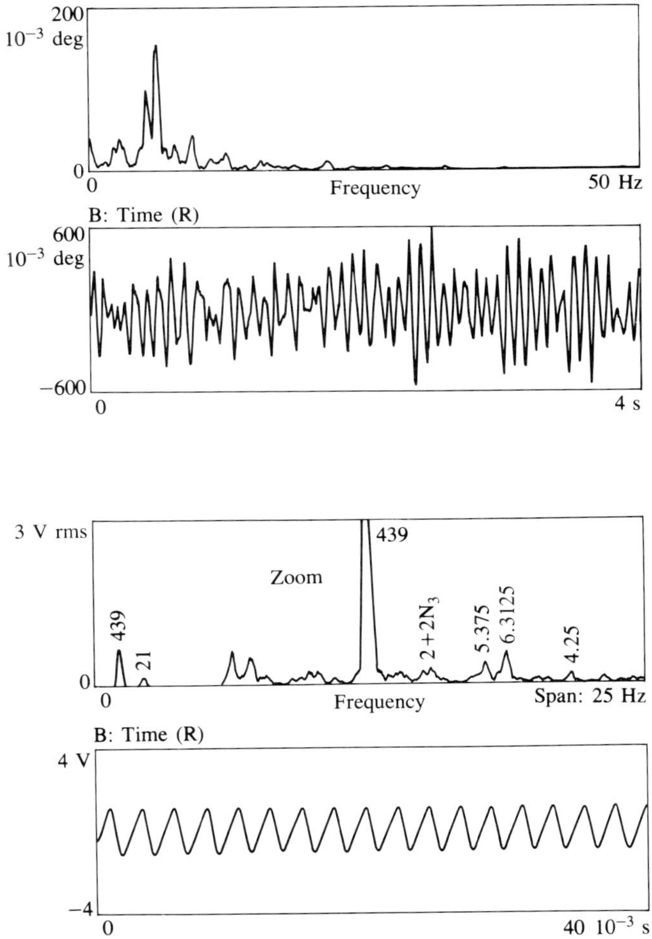

FIGURE 4.28 Torsional vibration, a form of frequency modulation.

display. The interval between the Keyphasor® mark and the next positive peak represents the phase angle used in balancing. The determination of phase angle is shown in Fig. 4.7.

Orbital Analysis

The orbit shown in Fig. 4.29 is called a *Lissajous pattern* in the context of its electronic display. The x and y motions of the rotor with respect to a sensor mounted in the journal bearing are simultaneously displayed on the horizontal

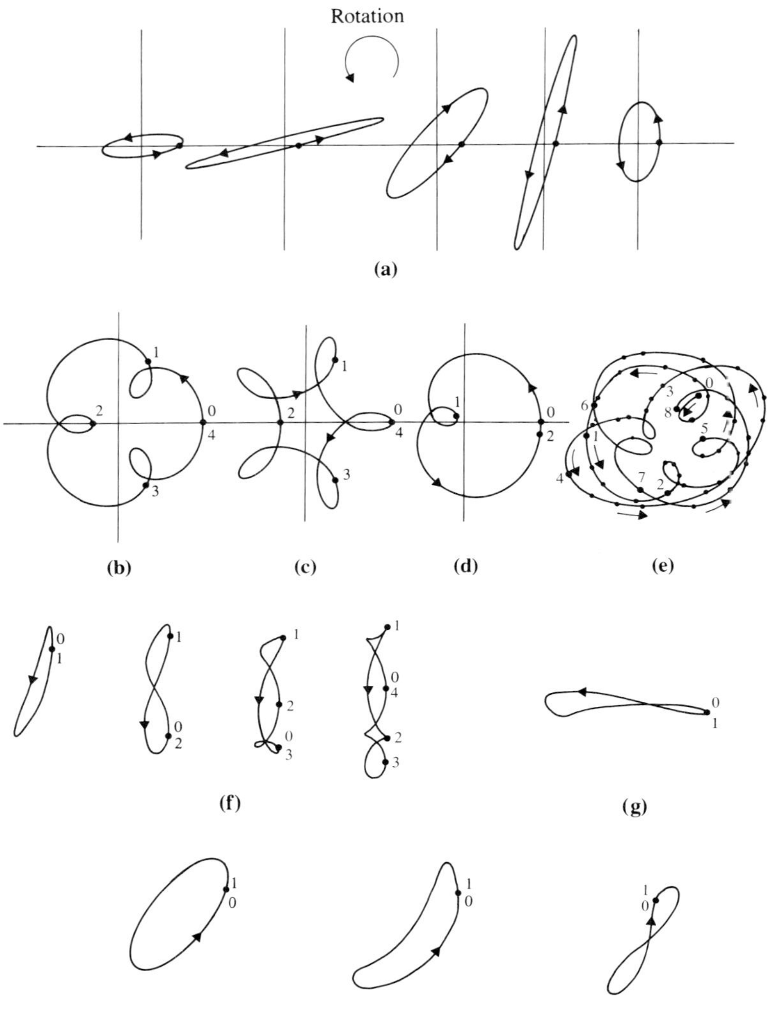

FIGURE 4.29 Shaft orbits of rotor vibration measured with a proximity probe.

and vertical axes of an oscilloscope. The orbit represents an instantaneous position of the rotor. Timing of orbital motion is established by adding a once-per-revolution blanking mark (Fig. 4.7) to the orbit. When the mark, or keyway, passes the fixed sensor, a once-per-revolution voltage pulse is produced that passes to the z axis of the oscilloscope. The energy from the voltage pulse shows up as a blanking mark on the oscilloscope. This mark is helpful in calculating the whirl ratio, which is the ratio of the number of whirl cycles to shaft rotations.

The information about rotor motion provided by an orbit is similar to that obtained from a time domain. Both indicate rotor performance directly. The orbit can be used to assess oil whirl and other asynchronous motions as well as synchronous phenomena such as mass unbalance and misalignment. Orbital analysis is possibly the best way to distinguish between mass unbalance and misalignment, both of which cause once-per-revolution forces (Fig. 4.29). Note the elliptical shape of the orbit from mass unbalance as opposed to the flat and double-looped shape of the orbit from misalignment (Fig. 4.29*h*).

For synchronous rotor whirling in a symmetric bearing or casing or foundation machine that is generated by unbalance or misalignment, the orbit will be a circle with a single once-per-revolution time marker. The radius of the circle increases with speed to a maximum value at the first critical speed and then declines and increases again as successive critical speeds are passed. For an asymmetric machine (i.e., in which the natural frequency of the system on one principal axis is dissimilar to the natural frequency of the system on the orthogonal principal axis) the orbit will be an elongated ellipse as it passes through each natural frequency (Fig. 4.29*a*). The major axis of the ellipse will be along the principal axis associated with the critical natural frequency being encountered. It is possible to encounter backward synchronous whirl in the speed region between the pair of critical speeds.

Most asynchronous phenomena occur at a speed above the critical speed and result in whirling at the critical frequency (see Sec. 1.6). The whirl speed is then a fraction (less than 1) of the rotor speed. The whirling motion may be in the same direction as rotation, i.e., forward whirl, as is encountered in oil whip, hysteretic whirl, and whirl caused by fluid trapped in the rotor. Backward whirl caused by a dry friction rub between rotating and static parts occurs less frequently. In most real situations the asynchronous whirling motion is accompanied by synchronous whirling associated with residual unbalance or misalignment. The orbit of such motion is a more complex pattern, such as shown in Fig. 4.29*b* and *c*. In Fig. 4.29*b* a pattern resulting from the superposition of once-per-revolution whirling with asynchronous forward whirl at one-fourth the rotation speed and at comparable amplitude is obtained. Note the four once-per-revolution timer marks in a complete pattern and three internal loops. Figure 4.29*c* shows a pattern resulting from the superposition of a once-per-revolution whirling with asynchronous backward whirl at one-fourth the rotation speed and at comparable amplitude. Four once-per-revolution timer marks are in a complete cycle, and five external loops are obtained. The number of internal loops (oil whirl) plus 1 equals the frequency ratio for forward procession. The number of external loops minus 1 equals the frequency ratio of backward procession with the exception of two and one.

Some instances of asynchronous whirl occur at a nonrational fraction of rotation speed, so that the orbit is not a simple closed regular pattern. Figure 4.29*d* shows a typical orbit for oil whip, in which the whirl direction is forward and the whirl speed is less than one-half the speed (0.47 in this example). Figure 4.29*e* (from Yamamoto, 1959) shows the orbit caused by forward whirl at 0.38 times the speed associated with slight differences in ball diameters of a ball-bearing-supported rotor.

Another class of orbits associated with subharmonic vibration (as described in Sec. 1.7.4) is caused by the rotor bouncing at its natural frequency when operating at some higher speed—a whole-number multiple of the natural frequency—when there is looseness in the foundation or bearing support system, when a bearing or seal that normally runs out of contact with the stator comes into local contact with it, or when pedestal flexibility governs support stiffness. The orbit is unique because the bouncing motion is planar rather than circular. When superimposed on the synchronous once-per-revolution response that is actually driving the subharmonic response, the Nth-order orbit—at a speed N times the natural bounce frequency—will be an elongated figure (aligned in the plane of the bounce), with N once-per-revolution timer marks and N convolutions. This is shown in Fig. 4.29*f* for $N = 1, 2, 3, 4$ and has been computed (Ehrich, 1988) and observed (Muszynska, 1984). Figure 4.29*g* shows another aspect of the response—that at a speed one-half of the natural frequency, or twice per revolution—when a small asynchronous response occurs at the natural frequency. The orbit has two convolutions as in the case of the one-half-per-revolution orbits, but there is only a single once-per-revolution timer mark.

As stated above, orbital analysis in combination with dc gap measurement is perhaps the best way to distinguish between mass unbalance and misalignment. Note the elliptical shape of the orbit for mass unbalance and misalignment in its early stages in Fig. 4.29*a*. After more severe misalignment occurs, the bearing restrains the rotor so that it cannot make a complete elliptical orbit, as shown in Fig. 4.29*h*. This figure shows stages of progressively increasing misalignment. Note that only a single once-per-revolution timer mark occurs with one and two loops. In the latter case a 2X vibration component is shown in the spectrum. The orientation of the orbit within the bearing shows the direction of the preload (perpendicular to flattened shape) caused by the misalignment.

Spectrum Analysis

A spectrum analysis is conducted with an FFT algorithm or filters; the processed signal is an amplitude-versus-frequency display. Frequencies of vibration response can be related to direct excitation frequencies or their orders, natural frequencies, sidebands, subharmonics, and sum and difference frequencies.

Frequencies can be identified directly in a linear system because the frequencies of the measured vibration response are equal to those of the forces causing the vibration. When a minimal number of frequencies are present, e.g., at operating speed, little resolution of the signal is required, and electronic filters can be used to analyze the spectrum. The operating speed of the machine is usually the fundamental frequency used in the analysis. The frequency at operating speed is identified and then linked to other frequencies in the spectrum such as direct orders (multiples) of operating speed. A spectrum with three major frequencies is shown in Fig. 4.26; they occur at operating speed (30 Hz), second order (60 Hz), and third order (90 Hz). The frequencies and relative magnitudes of the orders identify the cause of the vibration; the amplitudes and any sideband activity establish its severity. In the time domain (Fig. 4.30), the variation in amplitude of the gear mesh caused by the fault can be seen directly. It is manifest in the spectrum by sidebands (often called *difference frequencies*), as described in Sec. 1.7.5.

Vibration at frequencies that are orders of the operating speed is an example typically resulting from truncation of the signal (Fig. 4.4*a*). Truncation can be attributed either to nonlinear behavior of the machine or to modification of the signal at interfaces (bolted joints) within the machine during transmission from the source to the measurement point. Amplification can occur if a natural frequency happens to be located at an order frequency. Orders can also result from shaft asymmetry and at universal joints and couplings when forces are generated that are exact multiples of the shaft speed.

Relative amplitudes of orders are important in the diagnostic process. The

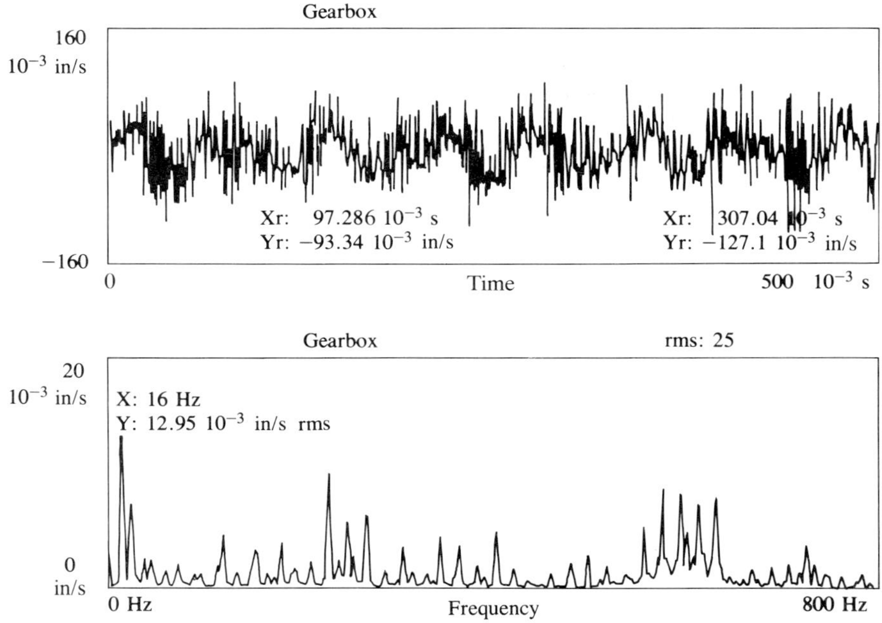

FIGURE 4.30 Sideband generation from a misaligned gear set.

shape of the spectrum is different for displacement, velocity, and acceleration. Low frequencies are emphasized when displacement is measured. High frequencies are emphasized in an acceleration spectrum.

Pulselike forces in mechanical systems cause vibration responses at natural frequencies (Fig. 4.22). The responses are not changed by varying the forcing frequency (speed of the machine). However, such variations modify the frequencies of any sidebands. Sidebands (Fig. 4.30) are thus a very effective diagnostic tool. Sidebands result from either amplitude or frequency modulation of the center frequency. For example, the variation of forces in the meshing of gears will cause amplitude-modulated gear mesh frequencies (Fig. 4.30). The amplitudes of the sidebands with respect to the amplitude of the center frequency are usually indicative of the severity of the condition.

Subharmonic excitations are vibration components that occur at exactly one-fourth, one-third, and one-half the operating speed. They result from forced vibration of a nonlinear system, e.g., fans mounted on nonlinear springs; rotors on nonlinear pedestals, rolling-element bearings, loose supports, or baseplates. If the natural frequency of a system coincides with a subharmonic excitation, subharmonic resonance occurs.

Beats are caused by a periodic pulsation of vibration amplitude resulting from the addition and subtraction of two signals with excitation frequencies close to each other. If the amplitudes of the two signals are equal, pure amplitude modulation occurs. If the components of vibration are not equal, both frequency modulation and amplitude modulation occur simultaneously. In Fig. 4.31 two vacuum pumps are mounted close to each other; the vane-pass frequencies of the

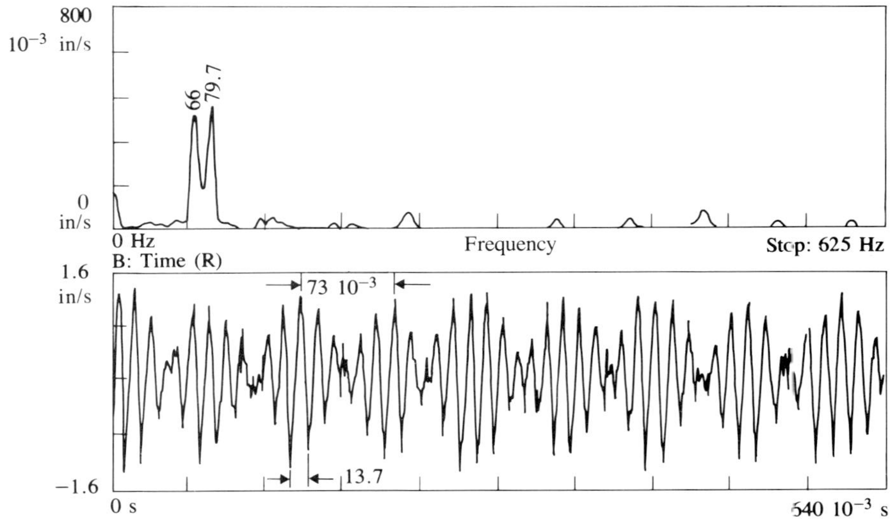

FIGURE 4.31 Beats generated by two vacuum pumps.

pumps are shown.

$$v = a_1 \sin \omega_1 t + a_2 \sin \omega_2 t \tag{4.10}$$

where v = total vibration signal, V
a_1 = amplitude of signal 1, V
a_2 = amplitude of signal 2, V
ω_1 = frequency of component 1, rad/s
ω_2 = frequency of component 2, rad/s
t = time

The equation can be rearranged to give an apparent vibration signal at a mean frequency $(\omega_1 + \omega_2)/2$—only if $a_1 = a_2$—as well as the pulsation in amplitude (amplitude modulation) at the difference frequency $(\omega_2 - \omega_1)$. Beat problems can be identified in either the time domain or the frequency domain (Fig. 4.31). The length of the period of the beat is inversely related to the closeness of the frequencies of the components; i.e., the closer the frequencies, the longer the period.

The vibration response shown in Fig. 4.32 is a truncated beat-generated waveform. The pure beat phenomenon is a periodic pulsation in vibration amplitude caused by the simultaneous response of a machine to two excitation frequencies ω_1 and ω_2. A number of sum and difference frequencies can be calculated from truncated beat signals (Wirt, 1962; Ehrich, 1972b). The cause of the truncation can be misalignment, looseness, stiffness nonlinearities, or signal modification at joints and interfaces of the structure.

A truncated waveform is a periodic function of the beat frequency $\omega_2 - \omega_1$ that can be expressed as a series of harmonic functions by using a Fourier series.

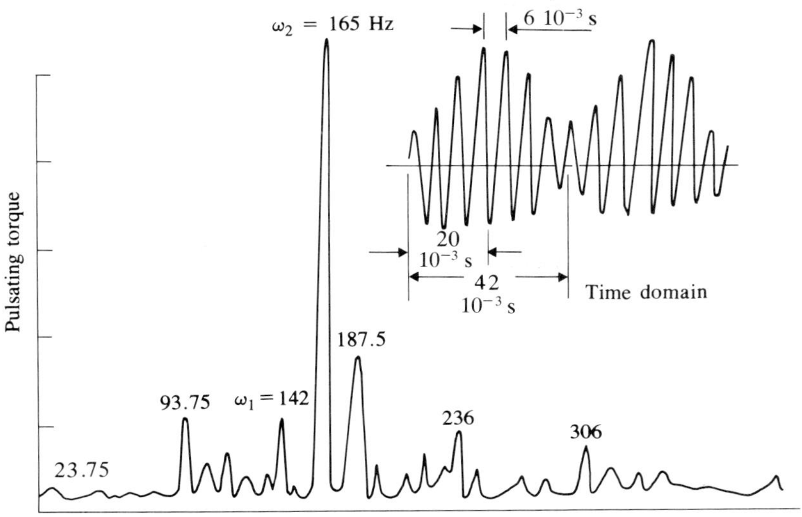

FIGURE 4.32 Sum and difference frequencies in an engine pump unit.

Spectral analysis results in a number of discrete sum and difference frequencies that depend on the exact shape of the vibration response. A mathematical analysis of this signal in the time domain has been published by Ehrich (1972b). The truncation of the beat frequency induces strong harmonic excitations at the sum and difference frequencies $\omega_1 + \omega_2$ and $\omega_2 - \omega_1$. Third harmonics $2\omega_1 + \omega_2$ and $2\omega_2 + \omega_1$ are generated. The components of the sum and difference frequencies are accompanied by sidebands separated by $\omega_2 - \omega_1$ from the center bands. Table 4.3 shows the pattern of frequency components that can evolve from this analysis.

Figure 4.33 is a schematic of the process involved in generating sum and difference frequencies in rotating machines. Vibration measured on the stator is represented by the truncated beat. The degree and shape of the truncation are

TABLE 4.3 Pattern of Frequency Components in a Truncated Beat Frequency Waveform (after Ehrich, 1972b)

	Sideband frequencies				Center frequencies			Sideband frequencies			Harmonic zone no.
			—		0		$\omega_2 - \omega_1$				0
		$2\omega_1 - \omega_2$		ω_1		ω_2		$2\omega_2 - \omega_1$			1
	$3\omega_1 - \omega_2$		$2\omega_1$		$\omega_1 + \omega_2$		$2\omega_2$		$3\omega_2 - \omega_1$		2
$4\omega_1 - \omega_2$		$3\omega_1$		$2\omega_1 + \omega_2$		$\omega_1 + 2\omega_2$		$3\omega_2$		$4\omega_2 - \omega_1$	3
					etc.						

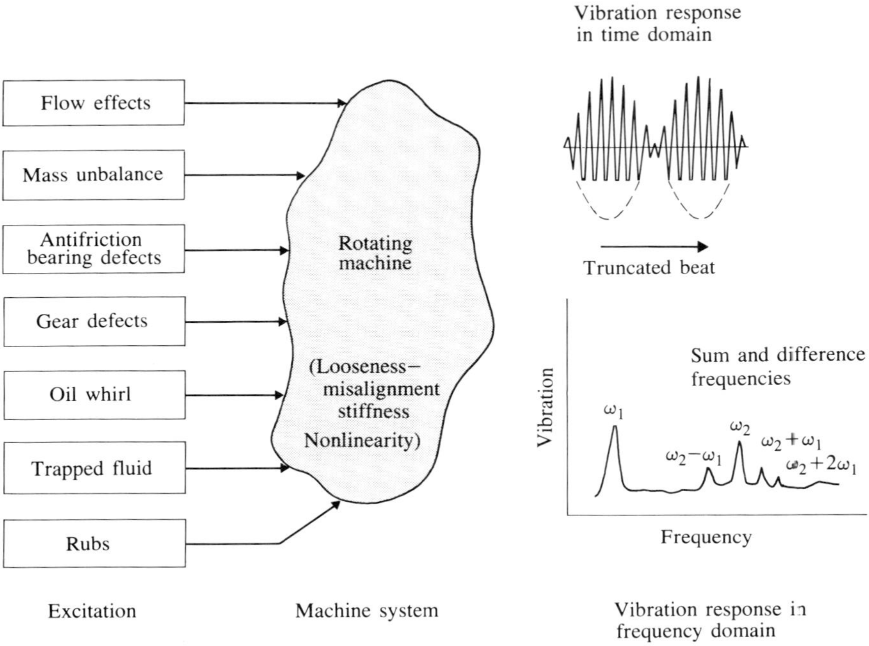

FIGURE 4.33 Generation of sum and difference frequencies (after Eshleman and Jackson, 1989).

shown in the time domain; they depend on such physical characteristics as stiffness, damping, and mass as well as the transmission path between the rotor and stator. The degree and the shape of the truncation in turn determine the amplitudes of the sum and difference frequencies in the frequency domain. The existence of sum and difference frequencies and their amplitudes allows the analyst to pinpoint the source of a vibration problem.

Figure 4.32 is an example of torsional sum and difference frequencies from an internal combustion engine driving a piston pump through a gearbox with a 1.517 reduction in speed. The two-cycle eight-cylinder engine, which operates at a speed of 2136 rpm, vibrates at 4 times (142.5 Hz) and 8 times (285 Hz) the running speed. The torsional excitation of the piston pump is at 7 times (165 Hz) the running speed. Figure 4.32 represents the shear strain measured on a load cell inserted between the gearbox and the piston pump; strain gauges were used. The signal and bridge excitations were provided by telemetry. The excitation of the piston pump at 165 Hz is about 5.3 times that of the engine signal. The result is extensive sum and difference frequencies from the combined 4-times-running-speed signal of the engine (142.5 Hz) and 7-times-running-speed signal of the piston pump (165 Hz). The large vibration in the pump causes truncation in the time-domain signal and probably induces stiffness nonlinearities in the gearbox. The lack of high-amplitude harmonic excitations from the piston pump—low-level 330- and 493-Hz signals are present—indicates lack of impact excitation, i.e., unloading of gear teeth. The first sum and difference frequency is the basic difference frequency at 23.75 Hz.

Cepstrum Analysis

The *cepstrum* (Randall, 1987) is defined as the inverse Fourier transform of the logarithmic power spectrum commonly used in vibration analysis. The complex cepstrum is reversible to a time signal. The terminology used in cepstrum analysis is analogous to that used in spectrum analysis—*cepstrum* from spectrum, *quefrequency* from frequency, and *rahmonic* from harmonic. A cepstrum domain consists of data processed one step further than the frequency domain. A cepstrum is effective for accurately measuring frequency spacing—harmonic and sideband patterns—in the power spectrum. One component of a cepstrum represents the global power content of an entire family of harmonics or sidebands. The severity of a defect such as that in a rolling-element bearing is provided by one component of the cepstrum display. (In the power spectrum, the severity is represented by a number of harmonics or sidebands for which the total power is not easily obtained.)

4.6.3 Identification of Malfunctions

The ease with which a fault can be identified from good test data is directly proportional to the information available on the design of a machine and its working mechanisms. This is especially true when similar frequencies are obtained for different faults, e.g., mass unbalance and misalignment. The operating speed is usually the reference frequency for diagnostic techniques. Other frequencies are either related to the operating speed or shown to be unrelated. Table 4.4 lists malfunctions that can be associated with the machine speed.

TABLE 4.4 Identification and Correction of Malfunctions of Rotating Machinery—Operating Speed Effects

Fault	Frequency*	Spectrum, time domain, orbit shape	Correction
Mass unbalance	1X	Distinct 1X with much lower values of 2X, 3X, etc., elliptical and circular orbits	Field or shop balancing
Misalignment	1X, 2X, etc.	Distinct 1X with equal or higher values of 2X, 3X, etc., figure-eight orbit	Perform hot and/or cold alignment
Shaft bow	1X	Dropout of vibration around critical speed in Bode plot	Heating or peening to straighten rotor

TABLE 4.4 (*Continued*)

Fault	Frequency*	Spectrum, time domain, orbit shape	Correction
Steam loading	1X	Load-sensitive 1X	Modify admission sequence; repair diaphragms; install nozzle blocks properly
Bearing wear	1X, subharmonics, orders	High 1X, high $\frac{1}{2}$X, sometimes $1\frac{1}{2}$ or orders; cannot be balanced	Replace bearing
Gravity excitation	2X	$\frac{1}{2}$ critical speed appears on Bode plot (unfiltered)	Reduce eccentricity by balancing
Asymmetric rotor	2X	$\frac{1}{2}$ critical speed appears on Bode plot (unfiltered)	Eliminate asymmetry
Cracked rotor	1X, 2X	High 1X, $\frac{1}{2}$ critical speeds may show up on coast-down	Remove rotor
Looseness	1X plus large number of orders, $\frac{1}{2}$X may show up	High 1X with lower-level orders, large $\frac{1}{2}$ order	Shim and tighten bolts to obtain rigidity
Coupling lockup	1X, 2X, 3X, etc.	1X with high 2X similar to misalignment; start and stops may yield different vibration patterns	Replace coupling or remove sludge
Thermal instability	1X	1X has varying phase angle and amplitude	Compromise balance or remove problem

* Given terms of operating speed; 1X = one times operating speed.

Critical Speeds

An excitation with a frequency close to or equal to a natural frequency under conditions of low damping (less than 15 percent of critical damping) is defined as a *resonance*. The interference diagram shown in Fig. 4.16 illustrates the concept of an excitation equal to a natural frequency at various rotor speeds. The term *critical speed* is used to describe the condition in a rotating machine. A shaft with a cross section which is not axisymmetric (e.g., one that is cracked) operating with a horizontal rotational axis and in the presence of gravity can excite critical speeds at multiples *N* of the operating speed (Table 4.4) because excitation occurs at *N* times the rotor speed. The critical speeds appear at lower speeds than normal, that is, $\frac{1}{4}$, $\frac{1}{3}$, and $\frac{1}{2}$ for 4X, 3X, and 2X, respectively.

Critical speeds can affect rotor performance. If the operating speed is higher than one or more critical speeds, the rotor must be able to pass through them with acceptable vibration levels; i.e., there must be excellent balance, and some damping must be present in the bearing. In the past the API recommended a peak amplification factor of no more than 8 (see Sec. 4.5) for this parameter. It is true that rotors can be driven through their critical speeds by applying sufficient power and that vibration response is attenuated by the fast start-up; however, the rotor coasts down at a rate dependent on its inertia and any frictional or aerodynamic forces present. Because no control is possible, a rotor can remain at a critical speed for a dangerously long time during coast-down, especially if it is operating in a low-density fluid; such is the case for a steam turbine during shutdown.

Mass Unbalance

Mass unbalance occurs when the geometric center and the mass center of a rotor do not coincide. It is a once-per-revolution (frequency of rotor speed) fault that is difficult to distinguish from misalignment and steam whirl (Table 4.4). Mass unbalance has a fixed phase angle with respect to a shaft reference mark and a spectrum with low-amplitude higher-order frequency components. Under normal conditions, sinusoidal signals are obtained. When excessive mass unbalance is present, however, nonlinear behavior of a bearing or pedestal can cause truncated signals that introduce higher-order vibrations (for example, 2X, 3X) with amplitudes less than 1X.

Misalignment

Misalignment in a redundantly supported rotor (i.e., a rotor with three or more radially loaded bearings) causes a rotating preload in the bearings, shaft, and external couplings that varies at the frequency of shaft speed; the magnitude of the resulting vibration is dependent on the stiffness of the system. Severe misalignment can cause nonlinear bearing behavior in one or both directions, depending on asymmetry in the bearing and stiffness of the pedestal and foundation. The nonlinear behavior causes clipped waveforms and second- and higher-order components of shaft vibration. The second-order component of vibration in cases of severe misalignment can exceed the first-order component; the result is an orbit with a figure-eight loop (Fig. 4.29*h*). The orbit for minor misalignment is composed largely of once-per-revolution vibration. High first-order axial vibration (out of phase) is also a symptom of misalignment. A proximity probe measures rotor vibration with respect to the bearing. The gap voltage reflects the position of the journal in its bearing and is thus a good way to diagnose a journal moved out of position by forces due to misalignment. The misalignment can be aggravated or relieved by thermal changes in the machine or its foundation because the position of one machine is changed with respect to the next machine. Thermal changes and twisted casings cause internal distortion, or internal misalignment, of gearboxes and other machine components. Casings can be twisted as a result of improper bolting, shims, and inadequate bolting.

Rotor Bow or Bent Shaft

A rotor bow or bent shaft usually causes a preload on the bearings of both units, depending on the flexibility of the coupling. The mass center of a bent shaft can be sufficiently removed from the geometric center to cause some mass unbalance. If the machine passes through a critical speed during start-up or coast-down, a diagnostic test can be performed to determine the nature of the bow. A dip in vibration level (Fig. 4.12) is a sign of rotor bow (Nicholas et al., 1976). The response on a polar plot is a trace close to the origin. The severity of the bow can be assessed below or above a critical speed. The bow is a constant force, but mass unbalance varies with the square of speed; if the dip in vibration response occurs above the critical speed, the effect of the bow is greater than that of mass unbalance. Of course, the phasing of the mass unbalance and the bow affects the vibration response.

The causes of rotor bow include unequal thermal conditions, rotor sag, and alteration of metallurgical properties due to rubbing. Unequal temperatures resulting from short-circuited generator coils are common. Compromise balancing can be performed to allow continued operation for a finite period. Unequal temperatures at the top and bottom of a rotor that is undergoing convective cooling during the period following shutdown in a steam or gas turbine can distort the rotor and lead to rotor bow. Rotors must be slow-rolled for a period after shutdown to avoid severe vibration when the rotor is started again. The rotor can be damaged during the process of rolling out the bow. Rubs can cause bows and unstable conditions below the critical speed because the motion of the shaft is in phase with the forces causing vibration. Heavy rubs and mishandling can cause permanent rotor bow; such bows can sometimes be removed by thermal soaking and peening.

Steam Loading

Steam loading affects bearing performance and can cause excitation through diaphragms, nozzle blocks, and blading. Unloading of the rotor from unequal conditions of steam admission can change bearing stiffness and damping—thus changing critical speeds—and induce oil whirl or oil whip instabilities. Tilted diaphragms and skewed nozzle blocks cause a once- or twice-per-revolution excitation because the steam loading of the rotor varies with rotation. Pulses in steam forces may induce subharmonic resonance. Unevenly loaded blades can cause high-frequency excitation.

Bearing Wear

Journal bearing wear results in excessive clearance that causes a high once-per-revolution vibration with an unstable phase angle. Although the data may seem to indicate mass unbalance, attempts to balance the rotor will fail because the influence coefficient process will not work. Trial weights will cause vibration changes in magnitude and phase angle atypical of a valid balance. Excessive clearance will occasionally change the stiffness of the bearing which in turn changes the critical speeds of the machine. Excessive clearance can also induce subharmonic vibrations; if a natural frequency is present at one-fourth, one-third, or one-half of the operating speed, subharmonic resonance occurs.

Cracked Rotor

A crack in a rotor causes asymmetric stiffness. Such characteristics of the crack as its depth, shape (whether it is open or breathing), and location govern the response measured on the rotor or bearing housing. A displacement probe usually provides the greatest sensitivity when relative motion of the rotor is measured. Stiffness changes cause a twice-per-revolution vibration from a balanced asymmetric shaft.

The success of tests conducted to detect a cracked rotor depends on the nature of the crack. Unexplained, excessively high once-per-revolution vibration is a primary indicator of a cracked rotor. Twice-per-revolution vibration may excite half critical speeds during start-up and coast-down.

Coupling Lockup

The preload imposed on bearings by coupling lockup is similar to rotor misalignment. In coupling lockup the twice-per-revolution vibration is higher than that at once per revolution. If the machine is stopped and restarted, the coupling may lock up in a different position, resulting in a different orbit and vibration signature than the original. This difference in signature is one way to distinguish between coupling lockup and misalignment.

Resonance

Natural frequencies excited by such forces as mass unbalance and its orders amplify vibration. The degree of amplification depends on the magnitude of the force, the damping, and the proximity of the forcing frequency to the natural frequency. A critical speed is a special type of resonance excited by a rotor.

Looseness and Impact

Loose mechanical equipment causes impacts that can be identified in the spectrum as once-per-revolution vibration plus orders. Depending on the nature of the machine's support, subharmonics of one-fourth, one-third, and one-half order occur. Orders close to or on natural frequencies have the highest magnitude regardless of whether they are orders or subharmonics because the forces are amplified by resonance.

Subsynchronous Faults

The malfunctions that appear as fractional frequencies of operating speed are listed in Table 4.5. They include oil whirl, oil whip (Sec. 1.6.2), rubbing of journal to bearing (rubs), rubbing between built-up rotors (hysteresis), trapped fluid, and subharmonic resonance (Sec. 1.7.4).

Oil Whirl and Oil Whip

Oil whirl is a malfunction associated with fluid-film bearings. It appears at a frequency below 0.5 times the operating speed and tracks operating speed at 0.35X to 0.47X to a frequency equal to the first natural frequency of the system. At this point oil whirl becomes an instability known as *oil,* or *resonant, whip.* The frequency of oil whip does not increase with shaft speed beyond the first natural frequency, which is thus its onset, or threshold, speed.

An instability such as oil whip differs from a forced vibration such as mass unbalance. A forced vibration is a response to a vibratory force imposed on a system. Oil whip is a form of self-excited vibration that draws energy from the spinning rotor.

Oil whirl typically occurs in response to a change in the condition of a sensitive rotor-bearing system; loss of load on the bearing and wear (increased clearance) are the two most common causes of oil whirl when bearing design is not a problem. The whirling motion results from the inability of the bearing to produce

TABLE 4.5 Identification and Correction of Malfunctions of Rotating Machinery—Fractional Frequency Effects

Fault	Frequency	Spectrum, time domain, orbit shape	Correlation
Oil whirl	0.35X to 0.47X	Subsynchronous component less than $\frac{1}{2}$ order informal loop in orbit	Temporary: load bearing heavier, correct misalignment; long term: change bearing type
Oil whip	fn1*	Subsynchronous component does not change with speed	Change bearing type
Subharmonic resonance	$\frac{1}{2}$X, $\frac{1}{3}$X, $\frac{1}{4}$X, and higher	Subsynchronous vibration depending on natural frequency	Remove looseness, excessive flexibility; change natural frequency so it does not match fractional frequencies
Hysteresis	0.65X to 0.85X	High-magnitude fractional frequency (greater than $\frac{1}{2}$X)	Eliminate or secure built-up parts
Rubs	$\frac{1}{4}$X, $\frac{1}{3}$X, $\frac{1}{2}$X, or orders	External loops in orbit	Eliminate condition such as thermal bow and mass unbalance that causes rub
Trapped fluid	0.8X to 0.9X	Beating—sum and difference frequencies	Remove fluid in rotor if possible; otherwise eliminate 1X component

* Fn1 stands for first natural frequency.

a damping force sufficient to suppress the pressure field induced as the journal bearing rotates.

The fluid-film bearing is a self-acting bearing; i.e., it generates its own pressure field. This pressure field supports the weight of the rotor and equalizes other forces, e.g., gear separation, misalignment, mass unbalance. However, in the process of generating the pressure field, the bearing may also generate a potentially destabilizing tangential force (Ehrich, 1972a) which must be controlled by damping in the bearing. If the bearing fails to develop sufficient damping, oil whirl or oil whip occurs, depending on the rotor speed and location of the natural frequency.

Oil whirl and oil whip can be eliminated by loading the bearing properly or returning the oil temperature, oil viscosity, and clearance to specification. If design is a problem, a pressure-dam or tilting-pad bearing can be used.

Subharmonic Resonance

Subharmonic resonance, which is often confused with oil whirl, is a subsynchronous phenomenon that occurs at a frequency exactly equal to one-third, one-fourth, or one-half of the operating speed. One of the components of the nonlinear system usually vibrates at a natural frequency of the machine system; the resulting amplified vibration is termed a *subharmonic resonance*. The condition can be corrected in several ways: by moving the natural frequency of the machine system, by removing the source of the nonlinear behavior (e.g., support pedestal, bearing), or by reducing the vibratory force that usually induces the nonlinear behavior.

Hysteretic Whirl

Hysteretic whirl is a subsynchronous vibration. The forces originate in friction either between rotor components or in the material of the rotor and create a phase lag of the strain (motion) in the stressed (force) rotor (Ehrich, 1972a). Hysteretic whirl is generally observed at frequencies approximately and often greater than one-half of the operating speed.

Rubs

Depending on the nature of the contact between a rotor and its bearings or casing, rubs produce unsteady subharmonics or orders of operating speed. A rub below the first critical speed produces heat that destabilizes the rotor. The rub causes the rotor to bow and increases the once-per-revolution forces. The bow causes more rubbing, which leads to an unstable situation.

Trapped Fluid

Fluid trapped in a rotor generates forces at a frequency lower than operating speed (Ehrich, 1972*a*) because the rotating fluid cannot match the operating speed of the rotor. The slower speed of the fluid causes a subsynchronous fractional excitation sufficiently close to operating speed to cause beats and/or sum and difference frequencies.

Rolling-Element Bearings

Defects in rolling-element bearings are readily analyzed (Berggren, 1988) today with modern FFT spectrum analyzers. When a rolling element passes over a bearing defect in the races, balls, or cages (retainer), pulselike excitation forces are generated that result in one or a combination of the following bearing frequencies (Blake and Mitchell, 1972).

$$\text{ftf} = \frac{\Omega}{2}\left(1 - \frac{B}{P}\cos\theta\right) \tag{4.11a}$$

$$\text{bpfi} = \left(\frac{N}{2}\right)\Omega\left(1 + \frac{B}{P}\cos\theta\right) \tag{4.11b}$$

$$\text{bpfo} = \left(\frac{N}{2}\right)\Omega\left(1 - \frac{B}{P}\cos\theta\right) \tag{4.11c}$$

$$\text{bsf} = \left(\frac{P}{2B}\right)\Omega\left[1 - \left(\frac{B}{P}\right)^2 \cos^2\theta\right] \tag{4.11d}$$

where bpfi, bpfo = frequencies generated by balls of rollers passing over defective inner and outer races

bsf = frequencies generated by ball or roller defects

ftf = frequencies generated by cage defects or improper movements

N = number of balls or rollers

θ = contact angle, or angle between lines perpendicular to shaft and from center of ball to point where arc of ball and race contact

P = pitch diameter of rollers or balls (use average values if exact value is not known)

B = diameter of ball or roller or average value for tapered bearings

Ω = speed of rotating unit in revolutions per second or hertz, which is equivalent to rpm/60

Modulation of the four frequencies by the speed of the rotating unit causes sidebands. The ball pass frequencies can be modulated by the fundamental train or ball spin frequencies, which can in turn modulate natural frequencies. The frequencies generated by defective bearings are thus combinations of bearing frequencies, natural frequencies, and frequencies of the rotating unit. Diagnosis of a bearing defect or bearing condition therefore depends on the combination of frequencies present. It is often best to measure vibration in the axial direction because that is where the transmissibility of the signal to the bearing cap or structure is strongest.

A small defect on the inner or outer race of a bearing produces discrete spectral lines at the appropriate bearing frequency and its orders. The ball pass frequency of the inner race (42 Hz) and its orders (84, 124, and 166 Hz), shown in Fig. 4.34, were measured on a bearing that had shallow flaking. The vibration from the roller passing through the defect can be seen on the time domain. Sidebands appear as the condition of the bearing deteriorates. Figure 4.35 is a spectrum of a bearing that failed 2 weeks after the analysis was made. Note the center bearing frequencies and their orders surrounded by sidebands whose difference frequencies are the speed of the shaft. Bearing analysis involves the calculation of bearing frequencies, measurement and analysis of vibration signals, identification of sidebands in the spectrum, and evaluations of the spectrum, shape of the time domain, energy, and amplitude. Table 4.6 summarizes the faults of rolling-element bearings and lists vibratory symptoms in the time and frequency domains.

Gearboxes

A list of possible gearbox faults and symptoms is given in Table 4.7. Both the time domain and the frequency domain of the analyzer must be used. The time domain provides the best information for identifying broken, cracked, or chipped teeth. Pulses will appear at a frequency equal to the number of defective teeth multiplied by the shaft speed (Fig. 4.22), unless more than one faulty tooth is in mesh simultaneously. Problems with misalignment and distortion are generally identified in the time domain as modulation of the gear-mesh frequency (Fig.

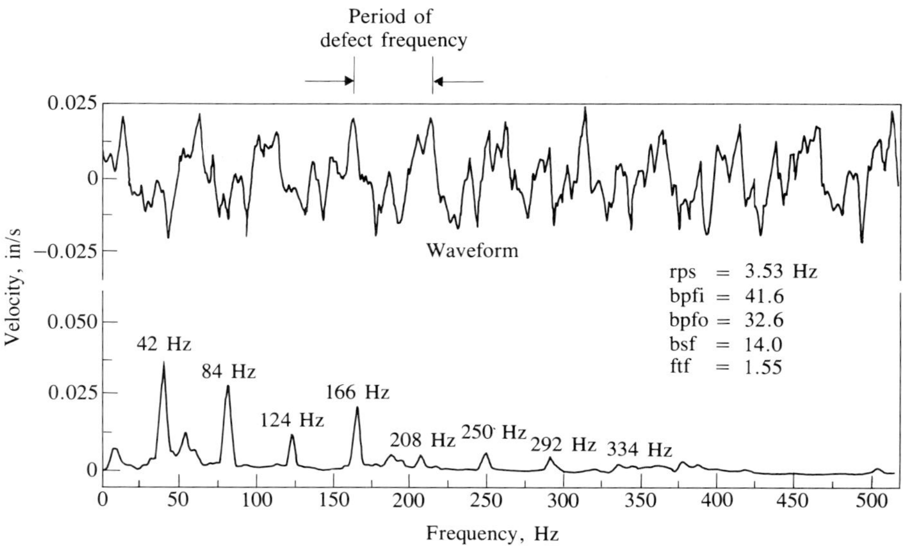

FIGURE 4.34 Shallow flaking on the inner race of a rolling-element bearing.

4.30). Gear mesh problems attributable to uneven wear, improper backlash, scoring, and eccentricity generally appear in the spectrum as gear mesh with sidebands at the frequency of the speed of the faulty shaft. The best signal strength for herringbone and helical gears is usually obtained with an axial measurement; gearboxes with spur gears should be measured in the radial direction.

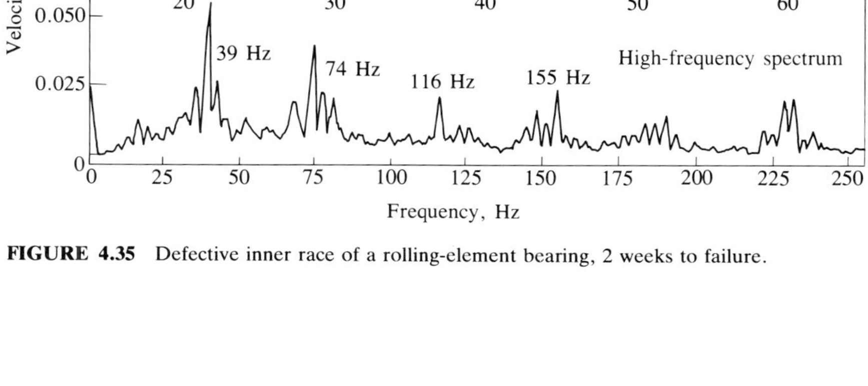

FIGURE 4.35 Defective inner race of a rolling-element bearing, 2 weeks to failure.

TABLE 4.6 Identification of Malfunctions of Rolling-Element Bearings

Fault	Frequency	Spectrum, time domain, shape	Comments
Defects on inner/outer race	BP frequencies and multiples	Shape of spectrum depends on size and severity of defect(s)	Sidebands of operation speed increase with severity
Defects on or broken retainer	ftf	FTF frequencies and orders or ball pass frequencies modulated by fundamental train	Serious—remove bearing
Ball or roller defects	Ball spin frequencies or ftf	Natural frequencies modulated by ftf, ball pass frequencies modulated by ball spin frequencies	Severe defects—remove bearing
Excessive clearance in bearing	Natural frequencies	Natural frequencies modulated by operating speed—no bearing frequencies	Pulses caused by impacting in bearing
Excessive clearance in housing	Operating speed plus orders	Operating speed plus decreasing amplitude orders	Pulses caused by impacting of bearing in housing
Corrosion	BP frequencies, BS frequencies	Many frequency components and orders	Nature of corrosion in bearing determines frequencies
Races slipping on shaft or housing	Variable bearing frequencies	"Smeared" frequencies and very low-frequency sidebands	Frequency modulation

BP = ball pass; ftf = fundamental train frequency; BS = ball spin.

TABLE 4.7 Identification of Malfunctions in Gears and Gearboxes

Fault	Frequency	Spectrum, time domain, style
Eccentric gears	Gear mesh	Gear mesh with sidebands at frequency of eccentric gear
Gear mesh wear	Gear mesh	Gear mesh with sidebands at frequency of worn, scored, or pitted gear(s)
Improper backlash or end float	Gear mesh	Gear mesh with orders and sidebands at frequency of pinion or gear
Broken, cracked, or chipped gear teeth	Natural frequencies	Pulses in time domain; natural frequencies in spectrum
Gearbox distortion	Gear mesh	Gear mesh and orders in spectrum; varying gear-mesh amplitude in time domain—shaft frequency plus low-amplitude orders
Resonance	Natural frequencies	Natural frequencies and sidebands

TABLE 4.8 Identification and Correction of Motor Malfunctions—Electrical Effects

Fault	Frequency	Spectrum, time domain, orbit shape	Correction
Air-gap variation	120 Hz	120 Hz plus sidebands, beating 2X with 120 Hz	Center armature relieving distortion on frame; eliminate excessive bearing clearance and/or any other condition that causes armature to be off center with stator
Broken rotor bars	1X	1X plus sidebands equal to numbers of poles X slip frequency	Replace loose or broken rotor bars
Eccentric rotor	1X	1X, 2X/120-Hz beats possible	May cause air-gap variation
Stator flexibility	120 Hz	2X/120-Hz beats	Stiffen stator structure
Short-circuited laminations	1X	1X (load-sensitive)	Turn rotor to remove short circuits
Off magnetic center	1X, 2X, 3X	Impacting in axial direction	Remove source of axial-constraint bearing thrust, coupling

Electric Motors

The mechanical malfunctions that affect rotating machines also cause problems with electric motors, e.g., mass unbalance, looseness, resonance, misalignment, eccentricity, bearing defects, and oil whirl. In addition, electric motors are sensitive to common mechanically induced electrical faults that generate mechanical vibrations (Table 4.8). They include variation in air gap, failure to stay on magnetic center, stator flexibility, broken or loose rotor bars, and short-circuited laminations. Campbell (1985) has published a complete chart on causes, checks, and cures for mechanical and electrical problems in ac electric motors. Mechanical malfunctions induce vibrations at twice the line frequency (120 Hz) and sidebands at the number of poles times the slip frequency. Vibrations at frequencies equal to the number of slots of rotor or stator times motor speed occur with an eccentric motor or if the number of slots is similar in the stator and in the rotor.

The most typical abnormal vibration in two-pole induction motors is associated with the variation in air gap. A motor with an unbalanced or eccentric armature or some other mechanical condition, e.g., flexible stator, that causes a variation in the air gap with rotation generates vibration at exactly twice the line frequency (Fig. 4.27). If a mechanical component of vibration is present at twice the operating speed, beats occur because the frequencies of the two components are very close. As might be expected, electrical problems disappear upon shutdown. It is wise, therefore, to shut down the motor and then observe the vibration spectrum.

Such faults as a broken or loose rotor bar or a short-circuiting ring connection to the rotor bar are evident only when the motor is under load. Broken rotor bars cause sidebands equal to the number of poles times and slip frequency on the operating speed. Short-circuited laminations cause local hot spots in the rotor that in turn bow the rotor; the result is a once-per-revolution vibration.

4.7 CONDITION MONITORING

Continuous monitoring for machine protection and condition analysis has been available for more than 20 years. But only in the past 3 years—with the increasing use of digital electronic data collectors and processors—have predictive maintenance programs based on periodic monitoring of machine vibrations become cost-effective and practical in most operational installations. Condition monitoring of critical equipment requires hard-wired permanently mounted sensors. They protect the equipment from unexpected sudden failure and provide the capability for condition analysis and diagnosis. The expense of continuous monitors is often more than justified by avoidance of repair costs and loss of production after failures. It is a fact that unexpected failures are much more serious and costly than parts replacement on a scheduled basis as a result of condition monitoring.

The primary goal of a good predictive maintenance program is efficient establishment of the mechanical condition of a machine so that decisions about repairs can be made. The electronic data collector has become the cornerstone of efficient data acquisition; an evaluation of data collector systems has been published (Henson, 1987).

This section is concerned with the development of predictive maintenance programs based on continuous and periodic condition monitoring of machines. Justification and goals, monitoring techniques and equipment, type and frequency of measurement, transducer location, personnel needed, and standards and criteria are discussed.

4.7.1 Objective

The objective of any condition monitoring program is to maintain or increase the availability of production equipment and avoid costs of failures. The cost of instrumentation and personnel involved in such a program must be justified by savings in machinery repair costs and decreases in lost production time. Savings must be documented so that management is aware that the investment has been worthwhile. The return on investment in protective monitoring is significant. The cost of $500 per point of hardware with an equal installation cost is a total of $1000 per point. The potential loss could start at $50,000 per day for a production outage that usually exceeds the cost for repairing the equipment. Such figures are typical of outages in the petrochemical and petroleum industry. Costs to utilities could easily be higher. Indeed, the return on investment in predictive maintenance programs is high—savings of as much as $25 per horsepower-year have been documented by Eshleman and Jackson (1989).

Management must establish an operational philosophy and be willing to make a commitment in test equipment and personnel that allows for development and operation of a program. Moreover, after a program has proved successful, commitment to it must continue or the program will falter. Those responsible for condition monitoring programs must thus continuously document savings and returns on money invested.

4.7.2 Development of Monitoring Programs

The development of a condition monitoring program requires careful planning. Criteria and goals must be established for the program. Criticality of equipment must be established, and the equipment ranked accordingly. The ranking will determine the resources, e.g., work force and instrumentation, that can be devoted to any one machine. Critical equipment for which no backup exists, such as large turbines, should be monitored continuously; two levels of alarms—alert and shutdown—should be used. Small pumps, motors, and fans that are not critical and can be spared can be monitored on a periodic basis by data collection devices. The evaluation of the criticality of equipment in order to allocate resources for protection is not unlike the purchase of insurance for a specific reason.

Decisions must be made concerning monitoring techniques and instrumentation that will be used as well as transducer locations and the frequency of measurements. Careful planning in the initial stages can save thousands of dollars as a program is developed.

Personnel are an important consideration. Guidelines for selecting personnel have been developed that are based on the program's organization and goals (Maxwell, 1984). The greater the emphasis on analysis, the more thoroughly trained the personnel must be. The number of persons required for a program depends on several factors: the number of machines being monitored, the number of data points per machine, and the frequency of measurements.

The degree of sophistication of monitoring techniques and the instrumentation and personnel needed to carry them out depend on the type and reliability of the machinery being monitored in addition to the program goals. Sophisticated monitoring programs increase the time available for planning repairs before failures occur. At the same time, the likelihood of a sudden failure that interrupts production will decrease. Unreliable machinery may require frequent monitoring and sophisticated techniques; installation of reliable machinery will thus probably save money in a predictive maintenance program.

Eshleman (1986) has written an overview of the methods used in periodical predictive maintenance. The techniques of screening, data acquisition, and analysis, as applied to a program, are examined.

4.7.3 Permanent Monitoring Equipment

Critical equipment and those items for which no spares are available such as turbine generators, turbine compressors, and mechanical drive turbines and motors must be permanently monitored. Certain equipment for which spares may be available should also be permanently monitored such as boiler feed pumps, injection pumps, reactor circulating pumps, plant and instrument air compressors, breathing air compressors, fire water pumps, oxygen compressors, forced- and induced-draft fans, emergency generators, and protective power supply systems.

A protective monitoring system has two set levels. The first is the warning (alert), and the second is danger (shutdown). The first level should give operating personnel sufficient time to solve the problem—high temperature, extreme thrust, decrease in speed—before the shutdown level occurs. The instrumentation used to monitor a machine must be reliable. An analog network is often considered essential for protection; monitored data are processed for analysis and reduced for presentation. There is no disagreement that digital processing is best used in data processing for analysis. However, any protection system should be hard-wired. Computer systems do not have the degree of reliability (99.9 percent) that is required and has been claimed. Components should be heat-tested, heat-cycled, treated with fungicide, and free from such interference as electrical transmission. Unnecessary shutdowns are costly and decrease the credibility of such programs and the instrumentation associated with them.

A typical control room in a large ethylene plant might include protection monitoring for such parameters as vibration levels, bearing temperatures, process temperatures and pressures, and speed. These parameters are typical of processes controlled from the control room rather than the equipment platform. The rule should always be to give higher priority to monitoring equipment condition than to the process being controlled. Placement of monitors in an air-conditioned, humidity-controlled, air-filtered environment greatly improves the reliability of the monitoring.

The Process

Data on vibration level and axial position of the rotor are typically taken from eddy-current proximity probes. The warning level might be 2.5 mils peak to peak and the shutdown level 4.5 mils peak to peak. The full-scale vibration meter would be 125 μm (5 mils). Major divisions can be readable in either English or metric units; 1 in on the scale is equal to 1-mil or 25-μm peak-to-peak vibration amplitude.

Good monitoring design involves good sensor locations and access for predictive monitoring or troubleshooting. Patch panels should be mounted at the rear of the console panels; BNC connectors should be shielded and grounded and arranged in a process flow scheme for simplicity and accuracy. Data taken in the control room are then free from climatic conditions and interference by the operator. Instrumentation for analysis should be available for connection to such permanent patch panels.

Accelerometers are usually installed according to API 678 (1981). Most seismic monitors are built in a dual-path configuration; i.e., a manually controlled switch allows two readouts per meter. A dual-path monitor is configured to take a signal from one probe, an accelerometer in this case, and display the signal in terms of acceleration, or g's, and, after integration, in peak velocity (inches per second). The monitor thus has two scales. The acceleration signal from the gearbox of a motor (4-pole, 1780 rpm) gear-driven compressor (11,000 rpm) would be displayed in a frequency band of 1 to 10 kHz. The integrated velocity is displayed in a frequency band from 5 Hz to 2 kHz, an overlap of 1 kHz. The frequency band can be narrowed to the gear-mesh frequency with a select switch.

The motor has the same dual path as the gearbox but not the narrowband select feature. For an accelerometer that has a resonant frequency of 24 kHz, displaying 10 kHz will provide good results. Although not within 20 percent of the resonant frequency band, 10 kHz is well within the 50 percent resonant frequency and does not overlap any resonant-frequency-excited sensor responses, i.e., errors in amplified response. The accelerometer might be mounted in accordance with API 678 (1981). The electrical impedance of the accelerometer is matched to that in the accelerometer case to avoid problems from noise and cable whip. The accelerometer is typically accurate below 250°F, has a voltage sensitivity of 25 mV/g, and has an interface device (voltage amplifier ratio of 4:1). The signal that reaches the monitor in the control room 600 ft away would be at 10 mV/g.

For proximity probes used on the governor end of a steam turbine, armored coaxial cable, rather than the more conventional flexible conduit connections, is used from the 1-m-long sensor to the 4-m-long extension cable. This cable is connected to the oscillator-detector-driver and transmitted to the monitors in the control room at 200 mV/mil. Two redundant probes in the axial position might be mounted to and through the endplate of the blind end of the turbine endplate. The same plate would provide access to up to five speed sensors mounted to a base ring that monitor a notched wheel driven by the rotor of the steam turbine. The redundant speed sensors are electromagnetic proximity probes; they monitor overspeed and the electrohydraulic control of the inlet valve servos of the governor.

API 670 (1986) and API 678 (1981) specify the many requirements necessary for a monitoring system. These standards have been put together by experienced users and with the cooperation of vendors and contractors. Among the factors included in the standards are displays, probes, field modifications, buffered design features, physical sizes of monitors, error detection for component failure, isolation of the system from noise and other interference, power supplies, connectors, and cable specifications.

Motors

The probes used on a motor depend on its size and criticality. A keyphasor, once-per-revolution probe should be provided. One accelerometer should be mounted in a horizontal position at each bearing. Pressure-fed hydrodynamic bearings should be filtered (10 μm) with ISO 68, 46, or 32 oil depending on the speed, the components of the train, and the lubrication system. Temperature probes should be placed in the babbitt backing metal at the eccentricity position of the rotor—typically 10° to 20° attitude angle. Two *resistance temperature detectors* (RTDs) should be embedded in each of the three phases of the stator windings.

If the motor is larger than 1000 hp, two proximity probes should be mounted 90° apart on each bearing. This is especially important for two-pole motors.

Steam Turbines

Steam turbines should have dual thrust probes at the axial position of the rotor mounted close to the thrust bearing. The probe should monitor an integral part of the rotor—usually the shaft end or a fixed collar close to the shaft end. The probes should be in a dual voting logic; 15-mil movement in either the active thrust direction (toward the exhaust) or the inactive direction (toward nozzles) should activate the warning alarm. An increase in movement to 25 mils should activate the shutdown alarm. The limits should be recorded on a data-managing system. Movement is measured from a reference such as the rotor reference mark, active-to-inactive thrust float at the bearings—usually 12 to 14 mils—or the setup positions or the first commissioning load position.

A position reference is necessary for the steam inlet and exhaust valves. It can be the cam lift, bar lift, or a controlled stroke. A local indicator can be used, or the position can be transmitted to the control room or be part of the electrohydraulic feedback signal in the speed-control scheme.

Dual-tip RTD or thermocouple sensors should be placed in the backing metal of the bearings but not in contact with the babbitt, in accordance with API 670 (1986). Fifty percent of the active self-leveling articulating thrust pads should be monitored. A minimum of two probes (top and bottom) should be used; three probes 120° apart are preferred. The maximum number is four probes 90° apart.

Two proximity probes should be placed in each bearing region to sense shaft-related vibration. The probes should be 90° apart, straddle the vertical centerline of the rotor or bearing, and be held in retractable nonresonant probe holders.

Many utilities used the dual-probe assembly for large steam turbine-generator trains and gas turbine-generator drives. Two probes are involved, and three measurements are taken. The assembly is usually mounted to the structure—typically the bearing housing. One probe senses shaft motion relative to the bearing. The second probe is seismically mounted to detect the absolute bearing vibration relative to space. The absolute vibration can be determined by vector addition of the measurements.

A probe can be installed in the axial position at the exhaust end of a rotor to measure growth with temperature. The transient increase in heat in the rotor or casing can thus be monitored. Two additional sensors can be installed to measure growth away from the fixed support on the exhaust end of the high-pressure bearing pedestal. The movement can exceed 0.5 in and requires a long-range sensor, vernier scales, or a linear variable differential transformer (LVDT) or its equivalent mounted on each side of the high-pressure pedestal.

It is also wise to record the first-stage pressure of the steam turbine relative to total steam flow or load on a condensing turbine. The pressure is a good indicator of blade fouling due to deposits of salt or silica. These data in combination with the positions of the steam inlet valves, thrust movement of the rotor, and temperatures of the thrust bearings in the active thrust bearing pads can be used to confirm that a turbine is dirty and that thrust overload exists.

Gas Turbines

The monitoring of a gas turbine is similar to that of a steam turbine. Several additional accelerometers are needed at the gas generator and the power turbine.

Their locations are determined according to the sensitivity points stated by the manufacturer, e.g., inlet scroll, transition section, crossovers, and power turbine exhaust housing. High temperatures prohibit the use of impedance matching accelerometers with piezoelectric amplifier circuits. The installation requires insulation and cooling of the mountings and charge amplifiers close to the accelerometers.

Gears

One accelerometer is installed on each low-speed bearing at the most sensitive position (API 678, 1981)—often horizontal for a horizontal box. Pairs of eddy-current proximity probes are placed 90° apart on the pinion in the area of each bearing. The mounts probably cannot straddle the vertical centerline but can be 90° apart. Thrust position probes are installed at the thrust bearing end, which is generally the low-speed shaft. One thrust probe is placed at the blind end of the pinion to provide information about axial motion, i.e., pinion shuttling.

Proximity probes are mounted in the gear case. The probe holder is mounted so that the gear and pinion teeth can be monitored. The mounting can be installed in a 1-in National Pipe Thread (NPT) coupling or boss tap at the time all drillings are made in the gearbox. API 613 (1988) for measuring torsional vibration in gears specified that the sensor output must be processed through an FM demodulating integrating instrument to obtain angular displacement. Such a sensor is not permanent but is used only at start-up, during commissioning, or for troubleshooting.

Temperature sensors should be placed at the bearings. It may be necessary to analyze the eccentricity position to determine the proper angular location. For example, if the pinion is up mesh in design—i.e., the pinion is in torque—the sensor is lifted vertically and separated horizontally from whatever pressure angle is used for tooth engagement. The attitude angle, which is always measured from the load vector, might be at 2:00 or 3:00 o'clock or 9:00 or 10:00, looking down the train. The temperature sensors for the bearing should be placed according to these calculations.

Once-per-revolution phase, triggering, or speed referencing is needed if a gear reducer or increaser is incorporated in the train. It is best to indicate with a timing mark the first piece of equipment upstream in a change of speed. It is a good idea to be consistent because various parts of a train are commissioned with certain couplings not in the system.

Centrifugal- and Axial-Flow Compressors

The procedures used to measure the shaft vibration of a centrifugal compressor are relatively similar to those used with a steam turbine. Active thrust is toward the active thrust bearings; some compressor manufacturers counterbalance the aerodynamic thrust to place the rotor initially on the inactive side of the float zone. As labyrinths and other seals wear, the rotor eventually wanders across to the active position.

The balance piston pressure should be measured either statically in the chamber or actively in the balance line. Additions to API 617 (1979) allow a flow measurement in the balance line to permit better assessment of deterioration of the thrust balance. Such deterioration occurs when the balance piston labyrinths wear and more flow is required in the balancing line.

Surge detection devices determine reverse flow and surge and are used to control aerodynamic performance. They affect mechanical performance and can cause large increases in vibration levels. The response to surge on an axial compressor causes much more damage than that which occurs on a centrifugal compressor. The rotor resonance and the effective damping at the resonance frequency will be excited. The logarithmic decrement is easily measured during surge. (See Sec. 4.5.)

Centrifugal Pumps
Accelerometers are often used to monitor pumps; sometimes pumps critical to a process are monitored with proximity probes. The accelerometer can be used as a permanent system because it has a long life, low output, and high sensitivity. Velocity sensors, which usually fail within the first 2 years of continuous duty (depending on the vibration frequency), put out a strong signal and are self-generating but lack low-frequency response accuracy. The amplitude can fall off at 600 cycles/min (10 Hz) and below; phase angle informaton may also begin to shift in the range from 900 to 2000 cycles/min (15 to 33 Hz).

4.7.4 Screening

Screening or examining the condition of machines with portable, uninstalled monitors is a way to determine at relatively low cost when a problem is developing in a specific machine so that there is adequate time for analysis and to prepare for repairs. Screening techniques vary in sophistication and consequently in effectiveness. The effectiveness depends on the screening device used and the type of machine monitored. Allowances must be made for changing operating conditions that affect overall vibration levels. Such changes may be caused by changes in process or environmental conditions. It is thus good monitoring policy to carry out a thorough vibration analysis before initiating a maintenance action.

Simple Devices
The earliest screening devices—subjective judgments from listening devices such as stethoscopes—were used to detect faults in rolling-element bearings. These tools are still used in conjunction with meters in many plants and mills. Relatively simple quantitative screening devices consist of a true rms meter and a velocity pickup or an accelerometer or accelerometer response filtered to read spike energy (Berggren, 1988). Vibration level increases by factors of 2 or 3 usually indicate that some action is necessary, e.g., more detailed vibration analysis or initiation of repair.

Other simple instruments such as the spike energy meter provide an accelerometer response that measures pulses resulting from a distinct fault in a specific machine component. Simple instruments can be adequate screening devices if nondestructive pulses and noise are not present at the measurement point. For example, the changes in vibration that occur when faults arise in a well-balanced motor-driven fan with a low vane-pass frequency will be detected because bearing-associated pulses can be sensed.

However, a pulse-indicating instrument may not detect the changes necessary to distinguish a new fault if the pulse level of the fault is low with respect to the gearing in a machine. In addition, a failure mechanism that is causing low signals may be masked by the normal vibration of another component. A spectrum from a gearbox in which gear mesh masked a bearing failure is shown in Fig. 4.36. The overall levels of pulses and vibration did not change when the bearing failed; yet spectrum analysis showed that a bearing failure was imminent. When random noise and vibration are present, simple screening methods may be ineffective—especially those that depend on pulses.

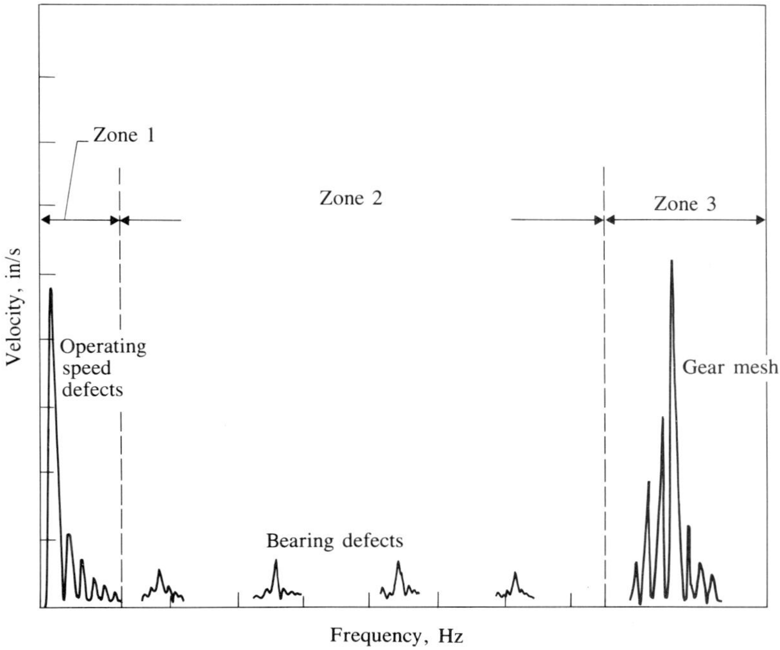

FIGURE 4.36 Spectrum of a bearing failure masked by gear mesh (after Eshleman, 1987a).

More Elaborate Devices

A more sophisticated level of screening involves filtering i.e., display changes in vibration in distinct frequency bands. The vibration spectrum shown in Fig. 4.36 has been divided into frequency bands that separate faults due to mass unbalance or misalignment (1X, 2X, 3X, and 4X) and bearing and gear meshing frequencies. Without analog or digital filtering a defect on a rolling-element bearing that was initiated and progressively worsened (Fig. 4.36) was masked by the gear mesh frequency or mass unbalance. As a result, even though the overall changes indicated by the meter did not indicate an impending disaster, filtered results did. Distinct frequency ranges can be screened with an electronic data collector; however, such screening is possible with less automation when bandpass filters that show only selected frequency bands are used.

Electronic Data Collectors

The new electronic data collectors provide fast, efficient screening and have some trending and analytical capabilities (Henson, 1987; Berggren, 1989). The electronic data collector used for data acquisition is paired with a host computer. It stores routinely collected data and provides current status reports, trends, and in-depth information in the form of spectra. Most electronic data collectors provide spectrum analysis—and a few have the capability to provide limited time waveforms—after the collected data have been deposited in the host computer. Some data collectors have a high-resolution spectrum and can even provide high-resolution spectra in the field. Electronic data collectors will also indicate and store vibration levels in selected frequency ranges for trending.

The nature of the fault and the criticality of the machine may require more information for a management decision about shutdown and repair or determination of when repair action should be initiated than the collector is capable of providing. But one advantage of collectors is that they eliminate the need for personnel to prepare tables and generate charts of data; in other words, they have automated the labor-intensive process of monitoring. One result is large savings in personnel costs that more than offset the costs of the data collector and the associated computer.

To date, however, only one electronic data collector is available that can store the large quantities of data necessary for complete analysis and balancing (e.g., spectrum, time, orbital, cascade, polar, and Bode plots) as well as review the frequency domain in different frequency spans. Because data collectors in general have not replaced tape recorders and FFT spectrum analyzers for detailed vibration analysis, these instruments must still be considered in the selection of test equipment.

4.7.5 Measurement

Measurement of machine vibration with a transducer placed at a sensitive location is the key to a successful predictive maintenance program. The frequency of measurement determines the efficiency of the program—the return on investment is low if data are collected too frequently. On the other hand, infrequent and disorganized data acquisition can result in costly failures and lost operational time.

Where to Measure

Measurement locations on any machine are limited by casings, guards, and piping. One goal of good monitoring is to measure a parameter that will provide the best indication of mechanical condition. The location should be as close as possible to the phenomenon being monitored; a remote location risks the possibility of signal attenuation and distortion in the transmission path from the source to the transducer. In some cases experimentation early in the program may be necessary, particularly if the machine is not a generic class—motor, pump, or fan—the operation of which has been measured for many years. If the design of the apparatus does not already incorporate embedded transducers or a provision for mounting transducers, a drawing should be examined to determine the best transmission path of the signal (Fig. 4.37). Such information will prevent interference by local resonances, nonlinear response, or complete lack of any signal indicating a mechanical fault.

Radial measurements in the horizontal direction are usually best for detecting radial problems, such as mass unbalance, because of the lack of stiffness in that direction.

In cases in which radial-to-axial coupling occurs, such as in angular contact bearings, an axial measurement in the load zone will provide the best indication of faults. Herringbone and helical gears are also best monitored in the axial direction if thrust is taken by the casing.

Transducers

Transducer selection and parameter selection depend on the physics of the vibration signal. The transducer selected should be the one capable of providing

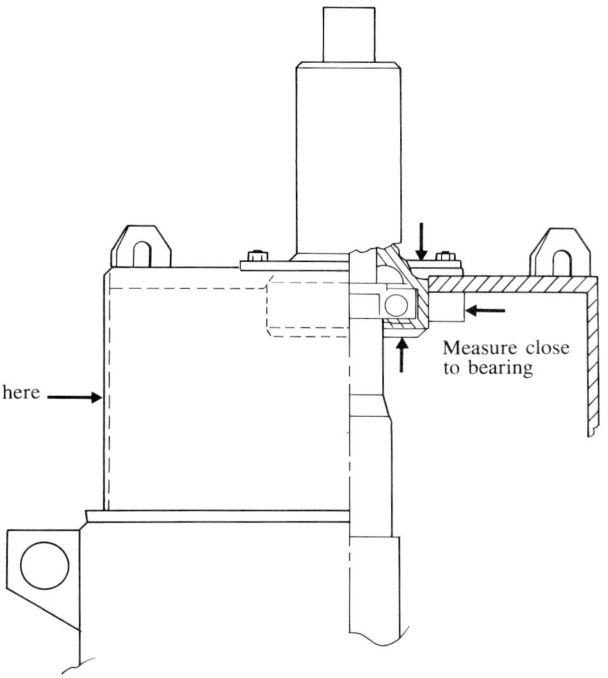

FIGURE 4.37 Measurement locations (after Eshleman, 1987a).

the most information per monitoring channel consistent with transducer performance, cost, and reliability. The transducer gathers the data; if, therefore, it does not have the proper frequency response or sufficient sensitivity, valuable information will be lost.

No single transducer is best for all applications; however, each type of transducer has a best application. It is thus important to choose the proper transducer. One consideration is the type of vibration to be measured, i.e., proximity probe or accelerometer. Selection of one type over another depends on the ratio of the weight of the machine casing to the rotor as well as the bearing stiffness.

If the rotor is light compared to the casing, shaft vibration should be measured. The proximity probe performs well and is reliable and cost-effective for shaft vibration measurements. It indicates relative shaft vibration and the position of the shaft relative to the bearing. The probe must be permanently mounted under controlled conditions.

In systems with high-frequency vibrations, the proximity probe is not adequate, not because the probe does not respond, but because the displacements are small. Nor is the probe adequate for machines with stiff bearings, again because little relative motion occurs. In these cases accelerometer measurements must be made on the casing because they provide the most active signal.

The pitfalls in casing measurements are many. The transducer selected should, of course, be effective, reliable, and relatively inexpensive. Velocity is usually measured (acceleration is integrated to get velocity) for frequencies up to 2 kHz because the monitoring levels that indicate machine condition are independent of frequency. The fact that the response of the velocity transducer rolls off at low and high frequencies (even to the vanishing point) is undesirable. In addition, the type of transducer is often sensitive to transverse vibration that may bind up moving parts. Phase shifts and amplitude rolloff can occur at low frequencies. However, the velocity transducer is rugged, self-excited—i.e., it requires no power supply—and easy to use.

The accelerometer is more costly initially and more difficult to use than the velocity transducer—it is not self-excited and thus requires a power supply. In addition, amplification may be required because of the low accelerations encountered at low frequencies. (The acceleration signal at low frequencies can be integrated to obtain the velocity.) Phase shifts and integration noise can be problems. For frequencies above 25 Hz, the signal from an accelerometer is acceptable and need not be processed.

A single transducer—usually an accelerometer because of its small size and frequency characteristics—is provided with some electronic data collectors. The frequency-response characteristics of the unit must be assessed, however, so that the user will not try to detect vibrations to which the collector does not respond. For instance, if a typical collector with an accelerometer is set up to respond to frequencies up to 8 kHz and a gearbox has gear mesh at 10 kHz, the signal will not reflect changes in the condition of the gear mesh.

Despite the fact that acceleration is measured, most collectors provide readout in acceleration or velocity. The parameter selected depends on the criteria established (Eshleman, 1988).

Frequency of Measurement

The mean time to failure of machine components, the cost of failures, and the cost of monitoring should be considered in selecting the frequency of measurement—rather than arbitrarily choosing some interval such as 1 week, 1 month, or 3 months.

Monitoring records for a given machine should be reviewed to determine how often it has failed in the past. Questions that must be answered include the following: How much does a failure cost in lost production, equipment, and repair personnel? Such potential costs must be weighted against the cost of increasing the frequency of monitoring. Does a spare exist? If not, monitor more often—perhaps every month; monitor less frequently if the machine is performing well.

Any program that is undertaken should be carried out properly. The number of machines monitored might be decreased if the schedule is such that the work cannot be done consistently. Chronic problems should be solved—periodic monitoring is an expensive way to compensate for unreliable machinery. High-speed machines go through many cycles in a short time and may have to be monitored more frequently.

Extrapolated trends are not dependable (Eshleman, 1986). The mechanisms that cause failures either are not well known or are complicated; there is thus no scientific basis for extrapolated trends.

4.7.6 Criteria and Limits—Guidelines

The transducers and electronics now used in monitoring systems provide data that must be evaluated on the basis of criteria and limits to ascertain machine condition. Computerized monitoring systems can evaluate data on the basis of overall parameter levels and the spectral shape.

Overall parameter levels are typically judged in terms of limits, e.g., normal, surveillance, and shutdown. These levels are followed for a time to establish trends. Overall parameter levels can be expressed in terms of the following:

- Peak or peak-to-peak overall vibration
- Rotor position
- Peak frequency component of vibration
- Root-mean-square vibration

The parameter used should be based on the machine sensitivity; i.e., the greatest change in parameter magnitude is obtained for a known change in machine condition.

Spectra are used when detail is required, principally in surveillance; each line of a spectrum is compared—either by computer or manually—to a standard or to baseline data. Automatic monitoring systems are capable of comparing point by point the shapes of spectra and overall parameter levels to baseline or standard data. This procedure requires a large capacity for data storage unless the machine is monitored only during surveillance and abnormal situations that are corrected without delay (Myrick, 1982).

Depending on the design of a machine and the speed at which it is operated, one or more of the following vibration parameters is monitored:

- Relative displacement
- Velocity
- Acceleration

Design factors and operational parameters that influence the condition of a machine include the speed and fatigue strength of the rotor; such bearing characteristics as clearance, size, geometry ($L \times D$), stiffness, damping, fatigue strength, and eccentricity ratio; Sommerfeld number for hydrodynamic bearings; capacity number; *DN* (diameter times rotation speed) for rolling-element bearings; and static and dynamic loads on the machine.

The diversity of machine designs, installations, and operational conditions has made impossible the development of absolute standards, levels, and guidelines that can be used in conjunction with monitoring systems to protect machines. Therefore, even though systems that monitor machine condition can accurately gather data very quickly, these data are of value for comparison and interpretation only if criteria and limits have been developed for a class of machines or on an individual machine basis during operation. Fortunately, general guidelines are available that can be used to develop scientific criteria and limits (Eshleman, 1986). This section deals with existing guidelines and techniques by which vibration criteria and limits are developed for specific machines.

Guidelines for acceptable vibration levels used in condition monitoring are based on shaft or casing vibration measurements. Shaft vibration is used to assess the condition of machines with large relative motions in the bearings and a high ratio of casing weight to rotor weight. Included are many machines with fluid-film bearings; an exception is the centrifugal pump. Casing and bearing cap vibrations are used in a condition-monitoring program to evaluated machines with stiff bearings. Both rolling-element and fluid-film bearings can be stiff or rigid, however, and their flexibility with respect to the rest of the system is important.

Shaft Vibration

Shaft vibration is measured with proximity probes at or close to the bearings. Such measurements are useful if relative motion can adequately provide sensitivity; machines with stiff bearings, e.g., are not sufficiently sensitive.

If two probes are used at the bearing, an orbit of its motion as well as the position of the journal inside the bearing can be obtained during operation. Proximity probe measurements (dc gap) establish the equilibrium position of the journal; the dynamic signal gives the instantaneous position. From this information an accurate assessment of bearing condition can be made directly from measurements. A guideline for evaluating shaft vibration based on bearing measurements is shown in the severity chart (Fig. 4.38) for normal, surveillance, and shutdown limits.

The ISO (1986) has developed a standard for evaluating peak-to-peak shaft vibration. If rotor displacement is measured in two directions that are perpendicular and in the same axial plane, the rotor orbit can be obtained by adding the output of each direction.

Measurement of relative and absolute shaft vibration should be broadband and cover a frequency range from at least 30 percent to at least 6 times the frequency of shaft rotation. The vibration level of a machine can be evaluated by using the

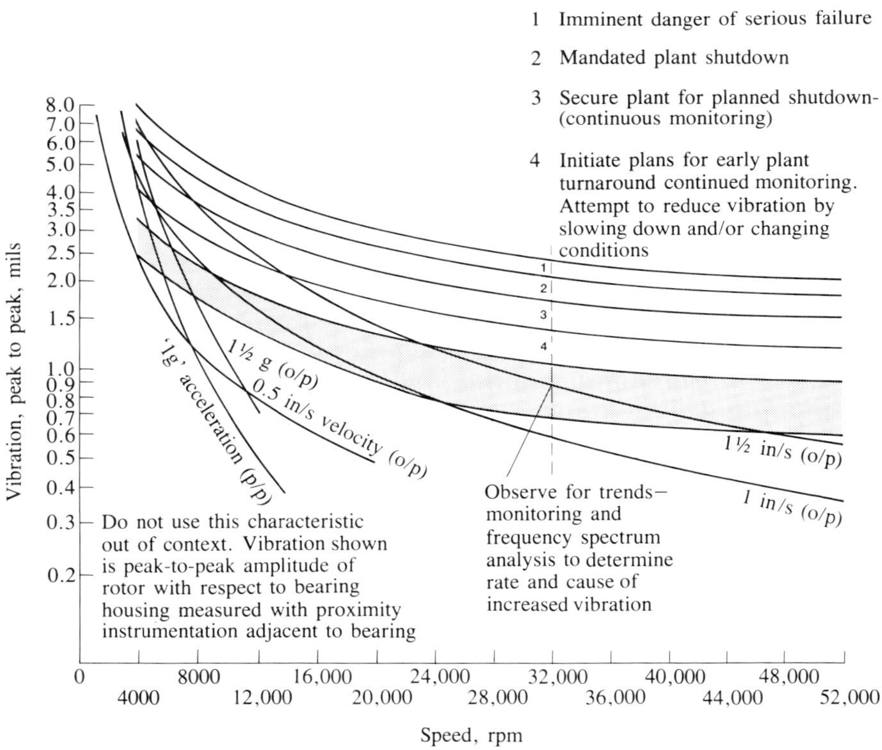

FIGURE 4.38 Dresser-Clark chart for proximity shaft vibration measurement on turbomachinery (Jackson, 1979).

absolute motion of the rotor shaft, kinetic load on the bearing, and clearances of rotor shaft relative to stationary elements.

Absolute shaft motion is an indicator of the level of bending stress on the rotor. Depending on the properties of the machine, changes in absolute shaft vibration can be used as criteria. The type of measurement used depends on the vibration level of the structure supporting the transducer used to measure relative motion. The following guidelines have been established by Maedel (1987).

1. If the vibration level of the support structure is less than 20 percent of the relative vibration of the shaft, either relative or absolute shaft vibration can be measured.
2. If the vibration level of the supporting structure is greater than 20 percent of the relative shaft vibration, absolute vibration must be measured. If larger than relative vibration, absolute vibration will be the more valid measurement.
3. If the kinetic load of the bearing is necessary to protect the bearing, vibration of the shaft relative to the bearing structure should be monitored as the most important criterion.
4. If clearances of the rotor shaft relative to the stationary elements are the criteria, the type of measurement used depends on the vibration level of the structure supporting the transducer used to measure relative motion.

   If the vibration level of the support structure is less than 20 percent of the relative vibration of the shaft, relative vibration is a measure of the clearance absorption.

   If the vibration level of the support structure exceeds 20 percent of the relative vibration level of the shaft, the latter may still be used as a criterion unless the vibration of the supporting structure is not representative of the vibration of the total stator housing. In such a case special measurement of vibration of the stator housing must be made at critical clearances.

Acceptable vibration levels that meet all these requirements depend on the size and mass of the vibrating body, characteristics of the mounting system, and the output and use of the machine. All the requirements must therefore be considered when ranges of vibration levels for a specific class of machinery are specified.

Casing Vibration

In some machines large vibratory forces are transmitted through the bearings to the casing. Vibration measurements should be made at the bearing cap or casing with velocity pickups or accelerometers. The type of measurement depends on the design and operating conditions of the machine. Maxwell (1982) has examined the theoretical basis of vibration severity limits and the relationship between shaft and casing measurements.

Most of the many tables and charts available for assessing vibration limits on bearing caps (Blake and Mitchell, 1972) apply to small general-purpose machines. The limits are based on overall peak or rms measurements. It is important when these limits are used as guidelines to remember that they were developed for the once-per-revolution component of vibration.

The severity chart shown in Fig. 4.39 was developed by Erskine (1981) for a wide variety of general industrial equipment. The severity level is obtained by dividing the vibration reading by a service factor before the chart is used. Possible alarm and shutdown levels are indicated on the chart. The compressor used above

To use chart

a. Decide machine group
b. Divide readings by service factor
c. Determine from chart the region in which corrected readings fall
d. For barrel casings divide by 1.5

Group K:	Small machines up to 20 HP
Group M:	Machines between 20–100 HP or machines on unusual (special) foundations up to 400 HP
Group G:	Large machines with rigid and heavy foundations whose natural frequency exceeds machine speed
Group T:	Large machines on foundations whose natural frequency is below machine speed

Machine group	Service Factor
K	1.0
M	1.6
G	2.5
T	4.0

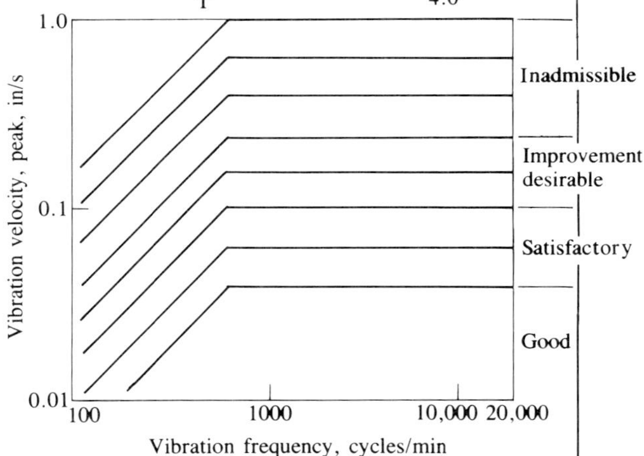

FIGURE 4.39 Machinery vibration limits (Erskine, 1981).

as an example would fit into group G, with a service factor of 2.5. A peak velocity of 0.4 and a service factor of 2.5 are equivalent to an effective vibration of 0.16. This value is in the surveillance region of the Erskine chart.

The ANSI S2.44 (1979) standard shown in Table 4.9 covers a wide variety of equipment. The data were developed from experience with vibration on bearing caps measured as the rms vibration velocity. The limits are based on whether the machine is soft-mounted or hard-mounted.

TABLE 4.9 Quality Judgment of Vibration Severity (ANSI S2.44, Part 1, 1986)

RMS velocity ranges of vibration severity		Vibration severity* for separate classes of machines			
mm/s	in/s	Class I	Class II	Class III	Class IV
0.28	0.01				
0.45	0.02	A			
0.71	0.03		A		
1.12	0.04	B		A	
1.8	0.07		B		A
2.8	0.11	C		B	
4.5	0.18		C		B
7.1	0.28	D		C	
11.2	0.44		D		C
18	0.71			D	
28	1.10				D
45	1.77				

* The letters A, B, C, and D represent machine vibration quality grades, ranging from good (A) to unacceptable (D).

4.7.7 Criteria and Limits—Techniques

The condition of a machine can be more accurately evaluated if data available from past operation can be compared with signals received from a monitoring system. In addition, the characteristics of the machine being monitored must be known. Such information can be obtained by evaluating parametric measurements and from rotordynamic studies.

Sensors and diagnostic equipment can be used to obtain the following machine characteristics: critical speeds, natural frequencies, mode shapes, bearing behavior, and sensitivity to mass unbalance and misalignment. Operating mode shapes can be plotted during typical operation; critical speeds are obtained during start-up and coast-down; balancing provides sensitivity data.

Limits at the bearing can be measured accurately from sensors located there. Dynamic studies are necessary to characterize rotor behavior at other locations, e.g., rotor and seal or rotor and casing. Such studies involve computer simulations based on a mathematical model derived from drawings and dimensions of various components—rotors, bearings, couplings, and pedestals. The model, which provides information about motions and forces for operational, fault, and normal conditions, must be validated, using test data from the machine. Software now available can be used to conduct such studies on microprocessors. The feasibility of using microprocessors to merge condition-monitoring data and rotordynamic studies is an exciting possibility for predictive maintenance in the near future.

Shaft Measurements

A method for estimating severity limits for specific machines has been developed by McHugh (1985) for synchronous and subsynchronous shaft vibrations within fluid-film journal bearings. His severity limits are based on the interference between journal and bearing and on babbitt fatigue. He uses the basic bearing parameters of stiffness and damping as well as measurements of shaft vibration to calculate static and dynamics forces (pressures) on the bearing. These pressures are compared to allowable babbitt-surface fatigue strengths to evaluate the effect of measured vibration on bearing deterioration. It is assumed that linearized coefficients are valid to 40 percent of the bearing clearance and that the orbital response of the shaft to a dynamic load is a tilted ellipse.

Because specific bearing parameters are involved in the analysis, calculations are necessary for each bearing and operating condition. In the chart shown in Fig. 4.40, the amplitude parameter—the ratio of static pressure to dynamic pressure multiplied by the ratio of vibration to radial clearance—is related to the eccentricity ratio.

The operating eccentricity ratio is established by either proximity probe measurements (dc gap) or calculations. Measurement is preferred. If calculations are used, data must be taken from the relationship between the capacity number and the eccentricity ratio of a specific bearing. The capacity number depends on the dimensions of the bearing, oil viscosity, load, and speed.

An example is given below for proximity probe measurements. The Sommerfeld number S for the bearings of a compressor is calculated from

$$S = \frac{\mu NLD(R/c)^2}{W} \tag{4.12}$$

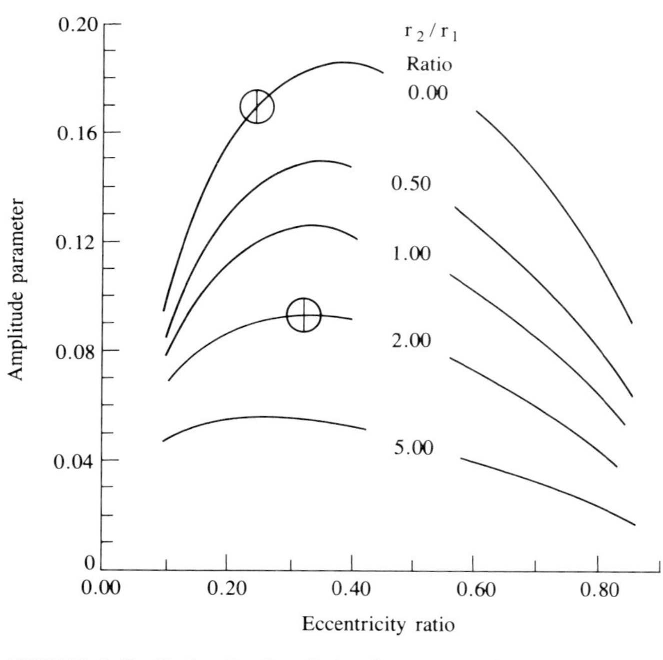

FIGURE 4.40 Shaft vibration limits for a two-axial-groove bearing (McHugh, 1985).

The values

μ = viscosity = 1.4×10^{-6} lb · s/in (SUS320 at 100°F)

N = speed of rotation, rps or rpm/60

L = bearing length = 7.0 in

D = bearing diameter = 10 in

R = bearing radius = $D/2$ = 5.0 in

C = bearing clearance = 13 mils = 0.013 in

c = radial clearance = $C/2$ = 6.5 mils = 0.0065 in

W = rotor weight on bearing = 9000 lb plus mass unbalance

yield a Sommerfeld number of 0.39. The capacity number is $(L/D)S$, or $\frac{7}{10} \times 0.39$, or 0.23. The eccentricity ratio is 0.25. An amplitude parameter (AP) is defined as

$$\text{AP} = \frac{P_s}{P_d}\frac{r}{c} \tag{4.13a}$$

$$\frac{r}{c} = \frac{\text{AP}}{P_s/P_d} \tag{4.13b}$$

where P_s = static pressure = $W/(LD)$
P_d = dynamic pressure in units consistent with P_s
r = peak shaft vibration
c = radial clearance in units consistent with r

and it can be read from Fig. 4.40 as a function of the ratio r_2/r_1, eccentricity ratio. For the example (AP) is found to be 0.17 and P_s is found to be $9000/(7 \times 10)$ or 129 lb/in^2. The allowable bearing fatigue strength suggested by McHugh (1985) is equivalent to a P_d value of 450 lb/in^2. The ratio of peak shaft vibration to radial clearance is calculated from the equation for the amplitude parameter so that, from Eq. 4.13b, we find r/c equal to $0.17/(\frac{129}{450})$, 0.59. However, McHugh (1985) suggests an r/c value of no more than 0.3 for long-term operation. Erskine (1981) suggests r/c ratios up to 0.35 for machines that operate at 3000 rpm (3600 rpm in the United States); his values are based on experience (see Table 4.10).

For a compressor with a radial clearance of 0.0065, an r/c limit of 0.35, or 0.0023 in (4.5 mils peak to peak), would be acceptable according to Erskine. The r/c ratio of 0.59 indicates that 7.6 mils peak to peak would be tolerable; this value could be beyond the linear assumptions of the analysis. If minor harmonics of the once-per-revolution component are present in the data, however, linearity is preserved. A vibration level of 7.6 mils could be tolerated for short periods provided the rotor did not rub the casing or seals.

Charts similar to that shown in Fig. 4.40 can be developed and used to establish vibration criteria for rotors when parameters other than bearings govern operating vibration limits, e.g., seals or internal rotor-to-stator rubs. Simulations would be used to obtain admissible vibration levels on the bearings.

TABLE 4.10 Shaft Vibration Limits (Erskine, 1981)

Severity	Ratio of vibration displacement to diametral clearance	
	50 Hz	200 Hz
Good	0.2	0.1
Satisfactory	0.35	0.25
Inadmissible	0.5	0.5

McHugh (1985) has also developed a method for estimating the dynamic load imposed on a bearing by combinations of synchronous and subsynchronous vibrations. Curves for several bearing geometries can be used to estimate the maximum dynamic load; they can also be used to estimate the combined subsynchronous and once-per-revolution components of shaft vibration permissible for a limiting dynamic load on a bearing.

A new amplitude parameter given by McHugh (1985) involves peak-to-peak synchronous whirl (see Fig. 4.40 for data for a two-axial-groove bearing with a length-to-diameter ratio $L/D = 1$):

$$A = \frac{P_s}{P_d}\frac{r_1}{c} \tag{4.14}$$

where $2r_1$ = synchronous whirl, peak-to-peak vibration amplitude
c = radial clearance
P_s = static pressure (steady load)
P_d = dynamic pressure

McHugh (1985) has published curves of various ratios of subsynchronous to synchronous vibration r_2/r_1. These curves reflect the fact that allowable synchronous vibration decreases as the ratio of subsynchronous vibration to synchronous vibration increases. For example, a compressor has an eccentricity ratio of 0.32 and 4 mils of synchronous vibration. The amplitude parameter is (129/450)(2/6.5), or 0.088. From Fig. 4.40 the value of the ratio r_2/r_1 is 2. Eight mils of subsynchronous vibration peak to peak are thus allowable.

Based on experience Sohre (1980) has established the following limits for subsynchronous whirl: a peak shaft vibration r/c of 0.05 is normal; $r/c = 0.2$ indicates shutdown for inspection; $r/c = 0.3$ means shut down immediately. This guideline allows only 4 mils of subsynchronous vibration.

After the alarm level has been reached, instrumentation used to monitor vibration should examine spectrum shape and fractional frequency as well as the once-per-revolution component and its orders. Absolute synchronous vibration is compared to the ratio of subsynchronous to synchronous vibration to determine if shutdown is necessary. The techniques developed by McHugh (1985) to obtain the sensitivity of bearings to severity limits are applicable to all fluid-film bearing-supported rotors. The techniques can be used in conjunction with Fig. 4.40 to develop severity charts for individual machines.

Casing Measurements

Criteria and limits for vibration levels measured on the casing have not been formalized because of the many different levels of measurement and screening

(Sec. 4.7.4). Overall guidelines have been established for evaluating once-per-revolution faults by Blake and Mitchell (1972) and Erskine (1981), who use a service factor approach. Both sets of charts are based on peak vibration velocity measured on the casing. The charts have been customized by comparing measured data from specific machines with known problems to the levels given in the charts. Service factors are calculated by dividing the allowable vibration levels on the charts by the measured values. In subsequent measurements, therefore, the effective vibration evaluated on the chart is obtained by multiplying the measured vibration by the service factor. For example, if 0.3 in/s were found to be satisfactory for a rotary blower, a service factor of $0.1/0.3 = 0.33$ would be established for the Erskine chart. Obviously service factors are not established on the basis of one measurement. A statistical sample of the relationship between machine condition and measured vibration is used.

4.7.8 Conclusions

The tools for performing periodic and continuous condition monitoring, which is the foundation of a predictive maintenance program, are available. Efficient, cost-effective protection, data collecting, processing, and display are possible. However, satisfactory results will be obtained only if the program is well planned and executed.

4.8 REFERENCES

ANSI S2.44, 1979, Part 1-1986, Measurement and Evaluation of Mechanical Vibration of Non-Reciprocating Machines, as Measured on Rotating Shafts, American National Standards Institute, New York.

API 612, 1979, Special-Purpose Steam Turbines for Refinery Services, American Petroleum Institute, Washington.

API 613, 1988, Special Purpose Gear Units for Refinery Service, American Petroleum Institute, Washington.

API 617, 1979, Centrifugal Compressors for General Refinery Services, American Petroleum Institute, Washington.

API 670, 1986, Vibration, Axial-Position, and Bearing Temperature Monitoring System, 2d ed., American Petroleum Institute, Washington.

API 672 R/2, 1979, Packaged Integral Geared Centrifugal Plant and Industrial Air for General Refrigeration Services, American Petroleum Institute, Washington.

API 678, 1981, Accelerometer-Based Vibration Monitoring System, American Petroleum Institute, Washington.

Baxter, N. L., and Ronald L. Eshleman, 1985, "Portable Magnetic Tape Recorders for Vibration Analysis and Monitoring," *Vibrations,* 1(3): 4–11.

Bently, D. E., 1983, "Studies Reveal Physical Phenomena of Rotor Rubs," *Orbit,* Bently-Nevada Corporation, Minden, Nev.

———, A. Muszynska, and D. I. G. Jones, 1985, "Some Aspects of the Application of Mechanical Impedance for Turbomachinery and Structural System Parameter Identification," *Third International Modal Analysis Conference,* International Society for Modal Testing and Analysis, Union College, Orlando, Fla.

Berggren, J. C., 1988, "Diagnosing Faults in Rolling Element Bearings, Part I: Assessing Bearing Condition," *Vibrations,* 4(1): 5–15.

———, 1989, "Diagnosing Faults in Rolling Element Bearings, Part III: Electronic Data Collector Applications," *Vibrations,* 5(2): 8–21.

Blake, M., and W. Mitchell, 1972, *Vibration and Acoustic Measurement Handbook,* Sparten Books, New York.

Brigham, E. O., 1974, *The Fast Fourier Transform,* Prentice-Hall, Englewood Cliffs, N.J.

Bruel and Kjaer, undated, *An Introduction to Vibration Measurement,* DK 2850, Naerum, Denmark.

Campbell, W. R., 1985, "Diagnosing Alternating Current Electric Motor Problems, Part II: Electromagnetic Problems," *Vibrations,* 1(3): 12–15.

Catlin, J. B., 1987, "The Use of Time Waveform and Phase in Vibration Analysis, Part I: Theoretical Considerations," *Vibrations,* 2(4): 7–11.

Cooley, J. W., and J. W. Tukey, 1965, "An Algorithm for the Machine Calculation of Complex Fourier Series," *Mathematics of Computation,* 19(90): 297–301.

Den Hartog, J. P., 1956, *Mechanical Vibrations,* 4th ed. McGraw-Hill, New York.

Ehrich, F. F., 1972a, "Identification and Avoidance of Instabilities and Self-Excited Vibrations in Rotating Machinery," ASME Paper 72-DE-21, American Society of Mechanical Engineers, New York.

———, 1972b, "Sum and Difference Frequencies in Vibration of High Speed Rotating Machinery," *ASME Journal of Engineering for Industry,* 91(1): 181–184.

———, 1988, "High Order Sub-harmonic Response of High Speed Rotors in Bearing Clearance," *ASME Journal of Vibration, Acoustics, Stress and Reliability in Design,* 110(1): 9–16.

Erskine, J. B., 1981, "Condition Monitoring in the Heavy Chemical Industry Using Noise and Vibration Measurements," *Proceedings, Machinery Vibration Monitoring and Analysis Meeting,* Vibration Institute, Clarendon Hills, Ill., pp. 17–33.

Eshleman, R. L., 1970, "Effects of Axial Torque on Rotor Response: An Experimental Investigation," ASME Paper 70-WA/DE-14, American Society of Mechanical Engineers, New York.

———, 1986, "Techniques for the Development of Criteria and Limits for Monitoring Rotating Machinery," *Vibrations,* 2(2): 5–12.

———, 1987a, "Periodic Monitoring for Predictive Maintenance Programs," *Vibrations,* 3(1): 3–8.

———, 1987b, "Torsional Vibrations in Machine Systems," *Vibrations,* 3(2): 3–10.

———, 1988, "Machinery Condition Analysis," *Vibrations,* 4(2): 3–11.

———, and C. Jackson, 1989, *Machinery Vibration Analysis I Notes,* Vibration Institute, Clarendon Hills, Ill.

Ewins, D. J., 1984, *Modal Testing: Theory and Practice,* Research Studies Press, Wiley, New York.

Guy, K. R., R. L. Eshleman, and Charles Jackson, 1988, "Auxiliary Turbine Subsynchronous Vibration," *Proceedings, 12th Annual Meeting,* Vibration Institute, Clarendon Hills, Ill.

Henson, G., 1987, "An Evaluation of Electronic Data Collector Systems," *Vibrations,* 2(4): 3–6.

Hewlett-Packard, 1982, *The Fundamentals of Signal Analysis,* Application Note 243, Palo Alto, Calif.

ISO 7919, 1986, "Mechanical Vibrations of Non-Reciprocating Machines," International Standards Organization, Geneva, Switzerland.

Jackson, C., 1979, *The Practical Vibration Primer,* Gulf Publishing, Houston.

———, 1982, "The Practical Vibration Primer No. 13, Mechanical Shop Testing," *Machinery Vibration Monitoring and Analysis, Mini Course Notes,* Vibration Institute, Clarendon Hills, Il.

———, and M. E. Leader, 1979a, "Rotor Critical Speed and Response Studies in Equipment Selection," *Proceedings, Machinery Vibration Monitoring and Analysis Meeting,* Vibration Institute, Clarendon Hills, Ill. pp. 43–50.

———, and ———, 1979b, "Turbomachines: How to Avoid Operating Problems," *Hydrocarbon Processing,* Gulf Publishing, Houston.

———, and ———, 1983, "Design, Testing, and Commissioning of a Synchronous Motor-Gear-Axial Compressor," *Twelfth Turbomachinery Symposium,* Texas A&M University, College Station, pp. 97–111.

Maedel, P. H. Jr., and R. L. Eshleman, 1987, "Vibration Standards," *Shock and Vibration Handbook,* 3rd ed., C. M. Harris, ed., McGraw-Hill, New York, pp. 19-1–19-14.

Maxwell, A. S., 1982, "Some Considerations in Adopting Machinery Vibration Standards," *Proceedings, Machinery Vibration Monitoring and Analysis Meeting,* Vibration Institute, Clarendon Hills, Ill., pp. 97–107.

Maxwell, J. H., 1984, *Machinery Vibration Analysis Notes,* Vibration Instiute, Clarendon Hills, Ill.

McHugh, J. D., 1985, "Estimating the Severity of Synchronous and Subsynchronous Shaft Vibration within Fluid-Film Journal Bearings," *Vibrations,* 1(1): 4–8.

Mitchell, J. S., 1981, *An Introduction to Machinery Analysis and Monitoring,* PennWell Publishing Co., Tulsa, Okla.

Muszynska, A., 1984, "Partial Lateral Rotor to Stator Rubs," IMechE Paper No. C281/84, Institution of Mechanical Engineers, London.

Myrick, S. T., 1982, "Survey Results on Condition Monitoring of Turbomachinery in the Petrochemical Industry, I: Protection and Diagnostic Monitoring of 'Critical' Machinery," *Proceedings, Machinery Vibration Monitoring and Analysis Meeting,* Vibration Institute, Clarendon Hills, Ill., pp. 59–86.

Nicholas, J. C., E. J. Gunter, Jr., and P. E. Allaire, 1976, "Effect of Residual Shaft Bow on Unbalance Response and Balancing of a Single Mass Flexible Rotor, Part 1: Unbalance Response," *ASME Journal of Engineering for Power,* 98(2): 171–181.

Nordmann, R., 1984, "Identification of Modal Parameters of an Elastic Rotor with Oil Film Bearings," *ASME Journal of Vibration, Acoustics, Stress and Reliability in Design,* 106(1): 107–112.

Norton, H. N., 1969, *Handbook of Transducers for Electronic Measuring Systems,* Prentice-Hall, Englewood Cliffs, N.J.

Pollard, E. I., 1980, "Synchronous Motors—Avoid Torsional Vibration Problems," *Hydrocarbon Processing,* Gulf Publishing Co., Houston, 59(2): 97.

Randall, R. B., 1987, "Vibration Measurement Equipment and Signal Analyzers," *Shock and Vibration Handbook,* 3rd ed., C. M. Harris and C. E. Crede, ed., McGraw-Hill, Chap. 13, pp. 13–43.

Sohre, J. S., 1980, "Turbomachinery Problems and Their Correction," *Sawyers Turbomachinery Maintenance Handbook,* Turbomachinery International Publications, Norwalk, Conn., part. 2, chap. 7.

Szrom, D. B., 1984, "Analysis and Correction of Gearbox Defects," *Proceedings, Machinery Vibration Monitoring and Analysis Meeting,* Vibration Institute, Clarendon Hills, Ill., pp. 147–152.

Wirt, L. S., 1962, "An Amplitude Modulation Theory for Gear Induced Vibrations: Case Study," *Strain Gage Readings, Stein Engineering Newsletter,* 5(4): 3–9.

Yamamato, T., 1959, "On Critical Speeds of a Shaft Supported by a Ball Bearing," *Journal of Applied Mechanics,* June, 26(2): 199–204.

Index

About the Editor

Fredric F. Ehrich, a registered professional engineer and a senior staff engineer for General Electric Aircraft Engines, received his B.S. and Sc.D. degrees in mechanical engineering from M.I.T. Dr. Ehrich is the author of over 40 published technical papers on rotordynamics and related topics, holds nine issued patents on aircraft gas turbine apparatus, and has contributed to the *McGraw-Hill Encyclopedia of Science & Technology* and the *Shock and Vibration Handbook* (McGraw-Hill).